AF323832

An Introduction to the Method of Fundamental Solutions

An Introduction to the Method of Fundamental Solutions

Alexander H.-D. Cheng
University of Mississippi, USA

C.S. Chen
University of Southern Mississippi, USA

Andreas Karageorghis
University of Cyprus, Cyprus

World Scientific

NEW JERSEY · LONDON · SINGAPORE · BEIJING · SHANGHAI · HONG KONG · TAIPEI · CHENNAI · TOKYO

Published by

World Scientific Publishing Co. Pte. Ltd.

5 Toh Tuck Link, Singapore 596224

USA office: 27 Warren Street, Suite 401-402, Hackensack, NJ 07601

UK office: 57 Shelton Street, Covent Garden, London WC2H 9HE

Library of Congress Cataloging-in-Publication Data
Names: Cheng, A. H.-D., author. | Chen, C. S. (Ching-Shyang), author. |
  Karageorghis, Andreas, author.
Title: An Introduction to the method of fundamental solutions / Alexander H-D Cheng,
  University of Mississippi, USA, C. S. Chen, University of Southern Mississippi, USA,
  Andreas Karageorghis, University of Cyprus, Cyprus.
Description: New Jersey : World Scientific, [2025] | Includes bibliographical references and index.
Identifiers: LCCN 2024035238 | ISBN 9789811298479 (hardcover) |
  ISBN 9789811298486 (ebook for institutions) | ISBN 9789811298493 (ebook for individuals)
Subjects: LCSH: Boundary value problems. | Differential equations, Partial--Numerical solutions.
Classification: LCC QA379 .C44 2025 | DDC 515/.35--dc23/eng/20250121
LC record available at https://lccn.loc.gov/2024035238

British Library Cataloguing-in-Publication Data
A catalogue record for this book is available from the British Library.

For any available supplementary material, please visit
https://www.worldscientific.com/worldscibooks/10.1142/13993#t=suppl

Desk Editors: Kannan Krishnan/Lai Fun Kwong

Typeset by Stallion Press
Email: enquiries@stallionpress.com

To our families and our esteemed colleagues
who inspired it all,
Graeme Fairweather
and Michael Golberg

Preface

The method of fundamental solutions (MFS) is a heuristic numerical method
for the solution of partial differential equations (PDEs). Its origin may be
traced to Theodore von Kármán in 1927 at the Technical University of
Aachen. Commissioned by the Zeppelin Airship Company, he calculated
the pressure distribution around the axisymmetric body of the SR-III airship, based on the potential flow theory, and compared his results with the
experimental ones.[1] Similar to the heroic effort of Walther Ritz in the Ritz
method,[2] these computational efforts were ahead of their time; hence did
not propagate right away.

Since the advent of electronic computers, there have been a few efforts
in distributing singular solutions to solve boundary value problems (BVPs)
in the 1960s and 1970s. The modern day revival of the MFS, however, can
be attributed to Mathon and Johnston in 1977.[3] The MFS has since been
applied to a wide range of problems.

In this book, we define the MFS as a meshless method in which singular
solutions are distributed outside of the solution domain and then collocated
for the boundary conditions. The definition can be extended to include

---

[1]von Kármán, T. (1927). "Berechnung der druckverteilung an luftschiffkörpern (Calculation of pressure distribution on airship hulls)." *Abhandlungen aus dem Aerodynamischen
Institut an der Technician Hochschule Aachen* **6**: 3–13.

[2]Ritz, W. (1908). "Über eine neue methode zur Lösung gewissen variations—Problems der
mathematischen physik (Using a new method for solving certain variations of mathematical
physik problems)." *Journal für die reine und angewandte Mathematik* **135**: 1–61.

[3]Mathon, R. and R. L. Johnston (1977). "Approximate solution of elliptic boundary-value
problems by fundamental solutions." *SIAM Journal on Numerical Analysis* **14**(4): 638–650.

nonsingular general solutions, leading to the Trefftz collocation method. These solutions automatically satisfy the homogeneous governing equations and possess exponential convergence properties. The inhomogeneous part of the solution can be treated by the method of particular solutions (MPS).

Because the MFS is both a meshless and boundary method, the discretization is easily performed. Moreover, as it uses the strong rather than weak formulation, the solution matrix can be generated without integration. When partial differential operators contain spatially varying coefficients and closed form fundamental solutions are not available, the MFS can be formulated as a local cloud method that uses overlapping covers. Within each cover the coefficients are approximated, and fundamental solutions can be among the best local basis functions for approximation.

This book consists of two parts. *Part I: Fundamentals* is aimed at providing the theoretical support of the MFS for newcomers to the method. The presentation however is intended to be practical rather than theoretical. *Part II: Advanced Topics* gives a sample of applications together with MATLAB® codes,[4] aimed at assisting users to build their own advanced codes.

In Part I, Chapter 1 gives a historical background of the MFS and its shared theoretical foundation with integral equations. Chapter 2 summarizes a wide range of fundamental solutions in closed form. Chapter 3 discusses other basis functions that can be used for the approximation of a function, such as T-complete general solutions, polynomials, and radial basis functions. For the MPS, we present in Chapter 4 closed form particular solutions corresponding to a broad range of partial differential operators and basis functions. In Chapter 5, general concepts for solving PDEs are reviewed, starting from the approximation by basis functions, using global or local interpolations, and fulfilling constraints of BVPs by the strong or weak formulation. Pros and cons of each approach are discussed. These fundamental concepts pave the way for the presentation of the MFS in Chapter 6. In this chapter, several issues regarding the MFS, such as solvability, degenerate scales, spurious frequencies, and stability are investigated with remedies offered. The accuracy and efficiency of the MFS for solving harmonic and nonharmonic BVPs are demonstrated by examples. Finally, a localized MFS is introduced to solve PDEs with variable coefficients.

---

[4]In addition to the listings in the book, these codes are also posted on the Internet site https://aquila.usm.edu/datasets/12/ for free download. These codes are organized by Examples in the book as shown in the README.txt file.

Part II of the book begins with Chapter 7, which presents BVPs for elliptic PDEs. The purpose of this chapter is to provide elementary MATLAB®️ codes that can be used as building blocks for advanced applications. Chapter 8 gives a demonstration of the MPS-MFS for solving inhomogeneous BVPs. Chapter 9 introduces matrix decomposition algorithms for the efficient solution of axisymmetric problems. Chapter 10 presents the recent development of using fundamental solutions as local, rather than global, basis functions in the localized method of fundamental solutions (LMFS). Chapter 11 covers the important topic of inverse problems, and the advantages of the MFS in solving them. Finally, Chapter 12 explores the use of the MFS for computer geometric modeling.

A.K. was introduced to the MFS by Graeme Fairweather while at the University of Kentucky in 1985 and has been collaborating with him on the MFS and other numerical methods since. C.S.C. started working on the MFS with the late Michael Golberg at the University of Nevada, Las Vegas in 1993 and collaborated with him on the MFS and related meshless methods until his untimely death in 2006. For their contribution, inspiration, mentorship and friendship we wish to dedicate this book to them.

We hope readers will find this book useful in their research and applications.

Alexander H.-D. Cheng
C.S. Chen
Andreas Karageorghis
2025

About the Authors

Alexander H.-D. Cheng is Emeritus Dean of Engineering and Professor of Civil Engineering at the University of Mississippi. His research covers the boundary element method, method of fundamental solutions, radial basis function collocation method, groundwater flow, saltwater intrusion, poromechanics, and nanomechanics. He has authored or co-authored five books, including one on *Trefftz and Collocation Methods*, and co-edited four books and eighteen conference proceedings books. He has published more than 180 refereed journal articles. He currently serves as the Editor-in-Chief of *Engineering Analysis with Boundary Elements* (Elsevier). He was formerly an Associate Editor for several journals, including *Transport in Porous Media* (Springer) and *Journal of Engineering Mechanics* (ASCE). He was the recipient of the George Green Medal, the Maurice A. Biot Medal, and the Walter L. Huber Civil Engineering Research Prize of ASCE.

C.S. Chen is currently Professor of Mathematics at the University of Southern Mississippi where he served as the Chair of the Department of Mathematics during 2005–2010. His main research interests lie in meshless methods and, more specifically, in using radial basis functions and the method of fundamental solutions for solving partial differential equations. He is currently serving as an Associate Editor in *Engineering Analysis with Boundary*

Elements and *Advances in Applied Mathematics and Mechanics*. He has published 170 research papers and co-authored 3 research monographs. He was a recipient of the George Green Medal and has received a Fulbright Specialists Award provided by the U.S. Department of State and the J. William Fulbright Foreign Scholarship Board.

Andreas Karageorghis completed both his undergraduate and graduate studies at the University of Oxford. After holding positions at the University of Kentucky (USA), the University of Wales (UK), and Southern Methodist University (USA), he joined the Department of Mathematics and Statistics of the University of Cyprus where he is currently Professor. His research interests include numerical algorithms and scientific computing, and particularly the method of fundamental solutions. He was an early developer of the MFS starting in the 1980s and one of the most prolific authors on the subject. He has published over 170 research papers in international journals. He co-organized and hosted the Workshop on Method of Fundamental Solutions in Ayia Napa, Cyprus, 2007, the first such international gathering on the subject. The results were published in the edited book *The Method of Fundamental Solutions—A Meshless Method* (2008). He is currently serving on the editorial boards of *Engineering Analysis with Boundary Elements* and *Numerical Algorithms*.

Contents

Part I

Fundamentals

Chapter 1

Introduction

1.1 What is a Fundamental Solution

In mathematics, a fundamental solution is a singular solution of a linear partial differential equation (PDE), following the concept of Green's function, named after the British mathematician George Green (1793–1841). Green in 1828 self published *An Essay on the Application of Mathematical Analysis to the Theories of Electricity and Magnetism* [347]. In that essay, he investigated, among other problems, the force field generated by a single electrical charge situated inside a grounded hollow spherical metal shell. In present-day mathematical terms, this is equivalent to solving the Poisson equation over a closed domain, with a Dirac delta function as its right-hand side (RHS), subject to a zero Dirichlet boundary condition (BC). Green called the resultant solution a potential function, which was the first time the term "potential" was used in mathematics and physics. Bernhard Riemann studied such a problem in his inauguration dissertation [754], and called the solution Green's function [276, 491].

Green's purpose for seeking such a special solution was to solve boundary value problems (BVPs) with arbitrary Dirichlet BCs. To achieve it, he developed a set of integral equation tools, known as Green's first, second, and third identities. The solution of a BVP is given by the integration of the product of the BC and the normal derivative of Green's function, over the boundary of the domain.

The fundamental solution of a linear partial differential operator is based on the same idea as Green's function, except that the former does not need to satisfy any BC; hence the fundamental solution is also called a free-space Green's function. While Green's function is unique, the fundamental

3

solution is a particular solution, and any general solution can be added to it. For both Green's function and the fundamental solution, the concept extends beyond the Laplacian operator. They can be associated with any linear partial differential operator with a Dirac delta function in space and/or time variables, as its RHS. These concentrated or impulse inhomogeneous terms typically have physical meanings, such as a fluid mass source, an electrical charge, a magnetic dipole, and a point force. These solutions can be directly used to simulate physical phenomena, or be utilized as a tool for obtaining analytical or numerical solutions of mathematical problems. A major goal of this book is to explore the use of fundamental solutions for the efficient and accurate numerical solution of PDEs.

1.2 Its Heritage

Many developments in mathematics have their origin in physics. For the concept leading to Green's function, we may trace its origin to Isaac Newton [676] and his law of universal gravitation:

$$F = -\frac{Gm_1m_2r}{r^3},\tag{1.1}$$

where F is the gravitational force, G the gravitational constant, m_1 and m_2 are two concentrated masses, and r is the distance vector between the two masses, with r its magnitude, such that $-r/r$ gives the unit vector that defines the force direction. Equation (1.1) is the well-known inverse square law of gravitation.

Lagrange [512] was the first to recognize the existence of a mathematical function

$$\phi = \frac{1}{r},\tag{1.2}$$

whose gradient gave the gravitational force field, as

$$F = Gm_1m_2\nabla\phi.\tag{1.3}$$

Unlike (1.2), which gives the force at a point, the function $\phi(x)$, called the gravitational potential, is continuously defined over space. If we place mass m_1 at the origin, then the force field exists for the entire space, wherever mass m_2 is placed.

Subsequently, Laplace in 1782 in his study of celestial mechanics [518] demonstrated that the gravity potential (1.2) satisfied an equation, first presented in spherical coordinates as

$$\frac{1}{r^2}\frac{\partial}{\partial r}\left(r^2\frac{\partial\phi}{\partial r}\right) = 0, \tag{1.4}$$

and then in Cartesian form in 1787 as

$$\frac{\partial^2\phi}{\partial x^2} + \frac{\partial^2\phi}{\partial y^2} + \frac{\partial^2\phi}{\partial z^2} = 0. \tag{1.5}$$

The above equation is known as the Laplace equation. It is easy to show that (1.2) satisfies (1.5) everywhere, except at the origin, where it is singular; hence the function (1.2) is a fundamental solution of the Laplacian operator in 3D.

The Laplace equation applies not only to the gravitational potential, but also to other potentials, such as that from the electrostatic force field. When there exist distributed electrical charges within a conducting body, Poisson [724] modified the Laplace equation to

$$\nabla^2\phi = -\frac{\rho}{\epsilon}, \tag{1.6}$$

where ϕ represents the electrical potential, ρ is the charge density, that is, charge per unit volume, and ϵ the permittivity. This equation is known as the Poisson equation.

Following Green's modus operandi, we may consider the simulation of a concentrated charge, such as a single electron, by shrinking the volume of the distribution to a point, yet keeping the amount of charge to one unit. The charge density then goes to infinity. At Green's time, there was neither notation, nor rules, for mathematical operations for such an object. It took a century, until the physicist Paul Dirac introduced the symbol δ, later called the Dirac delta function, to have a "convenient notation" [260]. Its mathematical operations, however, were still ad hoc and unsupported. Two decades later, Schwartz [787] presented the theory of distributions (generalized functions) that established the mathematical formality. The lack of rigor, however, did not prevent Green from solving the electrical field problem induced by a concentrated charge located inside a conducting sphere, whose surface was grounded to the Earth.

In present-day notation, the above problem is equivalent to solving the Poisson equation with a Dirac delta function as its RHS:

$$\nabla^2 V(\boldsymbol{x},\boldsymbol{x}') = \delta(\boldsymbol{x},\boldsymbol{x}'), \quad \boldsymbol{x},\,\boldsymbol{x}' \in \Omega, \tag{1.7}$$

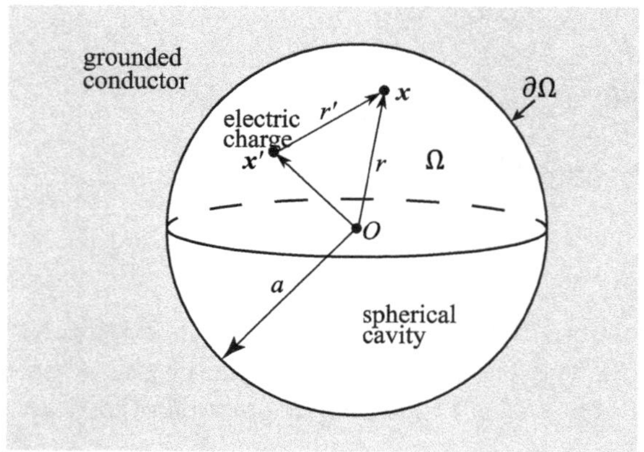

Figure 1.1. A single electrical charge in a spherical cavity inside a grounded conductor.

in which Ω is the domain of a sphere centered at the origin with radius a, and x' is the location of the singularity. The potential is subject to the null BC

$$V(x) = 0, \quad x \in \partial\Omega, \tag{1.8}$$

where $\partial\Omega$ is the surface of Ω (see Fig. 1.1). Green's function for this problem has been derived in physics and electrodynamics textbooks [409] using the method of images. Here we simply present the result:

$$V = -\frac{1}{4\pi}\left[\frac{1}{r'} - \frac{a}{\|x' - (a^2/r^2)x\|r}\right], \tag{1.9}$$

where $r' = \|x - x'\|$, and $r = \|x\|$, as illustrated in Fig. 1.1.

Green's purpose, however, was to solve a Dirichlet BVP, given as

$$\nabla^2\phi(x) = 0, \quad x \in \Omega, \tag{1.10}$$

subject to the BC

$$\phi(x) = f(x), \quad x \in \partial\Omega. \tag{1.11}$$

To solve such a problem, Green developed his integral identities. Green's second identity is

$$\int_\Omega \left(\psi\nabla^2\phi - \phi\nabla^2\psi\right) dx' = \int_{\partial\Omega}\left(\psi\frac{\partial\phi}{\partial n} - \phi\frac{\partial\psi}{\partial n}\right) dx', \tag{1.12}$$

in which Ω and $\partial\Omega$ below the integral sign denote domain and boundary integration, respectively, and n is the direction of the boundary outward normal. We note that the above equation is an identity for any functions ϕ and ψ that are twice differentiable. Substituting ϕ from (1.10) for ϕ, and V from (1.7) for ψ, and utilizing the property of the Dirac delta function (see Section 2.1), we obtain

$$\phi(x) = \int_{\partial\Omega} f(x') \frac{\partial V(x, x')}{\partial n(x')} \, dx', \quad x \in \Omega. \tag{1.13}$$

As the integrand contains all known quantities, the above equation formally gives the solution of the BVP defined by (1.10) and (1.11). If the functions and the geometry are amenable to analytical integration, we obtain the exact solution. If not, numerical integration can be performed to obtain an approximate solution.

Although in the above presentation we used a specific spherical geometry, Green's idea applies to any smooth geometry. The existence of Green's function for the Laplacian operator for an arbitrary geometry in 2D has been intensively investigated [346, 881]. The proof stems from Riemann's mapping theorem [754] stating that an arbitrary simply connected region of the plane can be conformally mapped one-to-one onto a circle. Osgood [696] established the existence of Green's function for any simply connected plane domain, whose boundary is more general than a Jordan curve.

The reciprocity relation (1.12) is not limited to the Laplacian operator and it can be derived for many other partial differential operators. Therefore, BVPs can be similarly solved. A key to the success of Green's methodology is the availability of Green's function in analytical form. There have been efforts to compile the various analytical or semi-analytical Green's functions [95, 788], or to numerically construct them [626, 637]. As each geometry has its own Green's function, this is a tedious task. In the application using (1.13), the value of Green's function needs to be evaluated at every quadrature point for the solution at a point x. If the evaluation of Green's function itself requires a significant amount of numerical effort, the methodology is not efficient, therefore undesirable. This leads to the idea of using fundamental solutions instead. As there is no geometry involved, fundamental solutions can be derived once and for all for each partial differential operator.

1.3 Use of Fundamental Solutions in Integral Equations

Similar to Green's functions, fundamental solutions can be utilized to find solutions of BVPs for PDEs. Our focus is on the numerical solution methodology. The advantage of using fundamental solutions is that the governing equation is automatically satisfied, and only BCs need to be enforced. In terms of numerical strategy, only the boundary variables and geometry are discretized, resulting in a much smaller solution system. This advantage has been exploited in solutions of integral equations. We shall discuss a few such cases in the sequel.

1.3.1 *Green's theorem*

In (1.12), if we replace ψ by the fundamental solution $1/r$, and ϕ by the solution of the Laplace equation (1.10), we obtain Green's third identity

$$c\phi(x) = \frac{1}{4\pi} \int_{\partial\Omega} \left[\frac{1}{r} \frac{\partial\phi(x')}{\partial n(x')} - \phi(x') \frac{\partial (1/r)}{\partial n(x')} \right] dx', \qquad (1.14)$$

where $r = \|x - x'\|$. In the above, the constant c is dependent on the location of x; it takes the value 1 for $x \in \Omega$, 1/2 for $x \in \partial\Omega$, and 0 otherwise. The above equation has been employed in the boundary element method (BEM) for the numerical solution of PDEs [12, 108, 923]. The strategy is to approximate ϕ and $\partial\phi/\partial n$ on the boundary $\partial\Omega$ by discrete values on a set of nodes, and interpolate between them. By placing x in turn on the N nodes where the discrete values are provided, we can numerically integrate (1.14) to obtain N linear equations in the ϕ and $\partial\phi/\partial n$ values at the nodes. In a well-posed BVP, either a Dirichlet type or a mixed Dirichlet–Neumann type BC is given on the boundary; hence half of the discrete values are known, and the other half (N of them) are unknown. The resulting $N \times N$ system can then be solved for the unknowns. Once the whole boundary data are available, (1.14) gives the solution at any point $x \in \Omega$. The details of such a procedure can be found in the books cited above and in the literature.

The use of fundamental solutions is not limited to solving the Laplace equation. It can be extended to other partial differential operators. Consider the generalized Green's second identity [655]

$$\int_{\Omega} \left[v\,\mathfrak{L}u - u\,\mathfrak{L}^*v \right] dx = \int_{\partial\Omega} \left[v\,\mathfrak{B}u - u\,\mathfrak{B}^*v \right] dx. \qquad (1.15)$$

In the above, u and v are two independent vectors, $\mathfrak{L}$ is a linear partial differential operator, $\mathfrak{L}^*$ is its adjoint operator, $\mathfrak{B}$ is the generalized boundary normal derivative, and $\mathfrak{B}^*$ is its adjoint operator. We note that the adjoint operator $\mathfrak{L}^*$ is obtained by integration by parts that transfers the differentiations from the variable u to v, and the boundary integrals on the RHS are consequences of that operation. If the resultant $\mathcal{L}^*$ is the same as $\mathcal{L}$, such as in the case of the Laplace equation, the operator is called self-adjoint [655].

Let us assume that $\boldsymbol{u}$ is the solution of the set of homogeneous equations

$$\mathcal{L}u_i = 0, \quad i = 1, 2, 3, \tag{1.16}$$

for 3D problems, subject to a set of well-posed BCs, and $\boldsymbol{v}$ is the solution of

$$\mathcal{L}^*u_{ik}^* = \delta_{ik}\delta, \quad i, k = 1, 2, 3, \tag{1.17}$$

where δ_{ik} is the Kronecker delta, and u_{ik}^* are the fundamental solutions. We note that as the variables are vectors, the RHS forcing function should also be a vector. We can introduce the forcing function one component at a time, which produces three sets of fundamental solutions for $k = 1, 2, 3$. Substituting the above into (1.15) then yields the boundary integral equations

$$cu_k = \int_{\partial\Omega} \left[u_i \, \mathcal{B}^* u_{ik}^* - u_{ik}^* \, \mathcal{B}u_i \right] dx, \tag{1.18}$$

which can be exploited for a numerical solution.

As an example, we may consider $\mathcal{L}$ as the Cauchy–Navier (linear elasticity) operator:

$$\mathcal{L}u \equiv \mu\, u_{i,jj} + \frac{\mu}{1 - 2v} u_{j,ji} = 0, \tag{1.19}$$

where u_i for $i = 1, 2, 3$, are components of the displacement vector $\boldsymbol{u}$, μ the shear modulus, and v the Poisson ratio. The point force solution in the k-direction is governed by

$$\mathcal{L}^*\boldsymbol{u}_k \equiv \mu\, u_{ik,jj}^* + \frac{\mu}{1 - 2v} u_{jk,ji}^* = -\delta_{ik}\delta, \tag{1.20}$$

which gives the well-known Kelvin solution

$$u_{ik}^* = \frac{1}{16\pi\,\mu(1 - v)} \frac{1}{r} \left[\frac{(x_i - x_i')(x_k - x_k')}{r^2} + (3 - 3v)\delta_{ik} \right]. \tag{1.21}$$

Substituting the solutions of (1.19) and (1.20) into (1.15), we obtain the Somigliana integral equations

$$cu_k(\boldsymbol{x}) = \int_{\partial\Omega} \left[t_i(\boldsymbol{x}')u_{ik}^*(\boldsymbol{x}, \boldsymbol{x}') - t_{ik}^*(\boldsymbol{x}, \boldsymbol{x}')u_i(\boldsymbol{x}') \right] d\boldsymbol{x}'. \tag{1.22}$$

In the above,

$$t_i = \sigma_{ij} n_j \quad \text{and} \quad t^*_{ik} = \sigma^*_{ijk} n_j \tag{1.23}$$

are the boundary tractions, and σ_{ij} and σ^*_{ijk} are obtained from u_i and u^*_{ik}, respectively, based on the constitutive equations [99]. The above equation can be exploited for the numerical solution of BVPs, similar to (1.14).

1.3.2 *Fredholm integral equations*

The Fredholm integral equations of the first and second kind are, respectively,

$$\int_a^b K(x, x')\, \mu(x')\, dx' = f(x), \quad a \le x \le b, \tag{1.24}$$

and

$$\mu(x) - \lambda \int_a^b K(x, x')\, \mu(x')\, dx' = f(x), \quad a \le x \le b. \tag{1.25}$$

In the above, the kernel function $K(x, x')$ and $f(x)$ are known continuous functions, λ is a constant, and $\mu(x')$ is an unknown function to be determined. We notice that when the RHS $f(x)$ is set to 0, the second kind equation (1.25) has the form of an eigensystem [442, 643], which has non-trivial solutions only for certain values of λ, known as characteristic values, or eigenvalues, associated with the kernel function K.

The difference between the first and second kind equations is that for the former, the unknown μ is contained only inside the integral, while for the latter, it is found both inside and outside the integral. Both integral equations can be used for the numerical solution of BVPs. The second kind is supported by an existence and uniqueness theorem [442, 814] and is guaranteed to have a solution (except for cases with a special kernel and geometry such that eigenvalues exist for the homogeneous equation [371, 412]). The first kind, however, does not possess existence and uniqueness properties [400] and its numerical solution can be ill-conditioned [350, 905].

For potential problems, we can utilize Green's theorem to formulate a first kind integral equation, as well as a second kind. Consider an interior domain Ω, bounded by $\partial\Omega$, with its exterior denoted as Ω'. For an interior domain BVP, we can express its potential ϕ at a point $x \in \Omega$ using the integral equation representation (1.14) as

$$\phi(x) = \frac{1}{4\pi} \int_{\partial\Omega} \left[\frac{1}{r} \frac{\partial \phi(x')}{\partial n(x')} - \phi(x') \frac{\partial (1/r)}{\partial n(x')} \right] dx', \qquad (1.26)$$

in which $r = \|x - x'\|$. With the same boundary $\partial\Omega$, we can also define an exterior domain BVP with potential ϕ'. Again utilizing Green's formula (1.14), this time for ϕ', while fixing $x \in \Omega$, it becomes

$$0 = \frac{1}{4\pi} \int_{\partial\Omega} \left[\frac{1}{r} \frac{\partial \phi'(x')}{\partial n'(x')} - \phi'(x') \frac{\partial (1/r)}{\partial n'(x')} \right] dx', \qquad (1.27)$$

where n' is the direction of the outward normal of Ω' on $\partial\Omega$, and thus $n' = -n$. We note that the left-hand side (LHS) of the above equation is set to zero because $x \notin \Omega'$. If we add these two equations, we obtain

$$\phi(x) = \frac{1}{4\pi} \int_{\partial\Omega} \left\{ \frac{1}{r} \left[\frac{\partial \phi(x')}{\partial n(x')} - \frac{\partial \phi'(x')}{\partial n(x')} \right] \right.$$
$$\left. - [\phi(x') - \phi'(x')] \frac{\partial (1/r)}{\partial n(x')} \right\} dx', \qquad (1.28)$$

where we have converted n' to n.

For a given interior problem, we can create an exterior problem with the Neumann BC, taking the same value as the normal derivative of the interior problem,

$$\frac{\partial \phi'(x)}{\partial n(x)} = \frac{\partial \phi(x)}{\partial n(x)}, \quad x \in \partial\Omega. \qquad (1.29)$$

The exterior problem has a unique solution ϕ' that decays at infinity. Equation (1.28) then becomes

$$\phi(x) = \int_{\partial\Omega} \mu(x') \frac{\partial (1/r)}{\partial n(x')} dx', \quad x \in \Omega, \qquad (1.30)$$

where

$$\mu(x') = -\frac{1}{4\pi} [\phi(x') - \phi'(x')], \quad x' \in \partial\Omega. \qquad (1.31)$$

We note that the kernel function $\partial(1/r)/\partial n$ is a "dipole", that is, the field resulting by bringing a positive and a negative charge infinitely close together. Equation (1.30) is called a "double layer" potential integral equation [400, 814]. The function μ defined in (1.31) is the distribution density. As the exterior problem is a created one, we may simply consider μ as an "artificial density", and ignore its definition [105, 412].

Equation (1.30) may be used to solve an interior domain Dirichlet BVP. If we place x on the boundary $\partial\Omega$, we obtain the following equation:

$$\phi(x) = -2\pi \mu(x) + \int_{\partial\Omega} \mu(x') \frac{\partial(1/r)}{\partial n(x')} dx', \quad x \in \partial\Omega. \tag{1.32}$$

We note that the leading term on the RHS is produced by the integration of the $1/r^2$ singularity over an infinitesimal hemisphere based on the Cauchy principal value. For a Dirichlet BVP, ϕ is prescribed on $\partial\Omega$ and is a known function. We recognize (1.32) to be a multi-dimensional Fredholm integral equation of the second kind. We can discretize the boundary into a set of nodes to form area elements. By placing x in turn on these nodes, we can form a system of linear equations for the unknown discrete values of μ on the nodes. Once it is solved, (1.30) gives the solution of ϕ in the domain. As mentioned above, the second kind integral has a strong existence and uniqueness theoretical foundation; hence it is popular in the mathematical community for solving Dirichlet BVPs [643].

In a mixed BVP, a Neumann BC is also involved. To utilize the normal derivative data, (1.32) needs to be differentiated in the direction n of the boundary normal. The resultant kernel function $\partial^2(1/r)/\partial n(x')\partial n(x)$ contains a non-integrable hypersingularity of order $1/r^3$, which needs to be treated by the Hadamard finite part integral [152, 698]. The treatment, however, is not easy on a curved element or a corner node. Therefore, in practice, the "simple layer" (or "single layer") potential method, involving the integral equation of the first kind, is generally used, despite its theoretical deficiency.

Referring to (1.28), if we create an exterior problem having the same potential as the interior one, $\phi'(x) = \phi(x)$, on $\partial\Omega$ as the BC, we obtain a simple layer potential integral equation [14, 105, 412]

$$\phi(x) = \int_{\partial\Omega} \sigma(x') \frac{1}{r} dx', \quad x \in \Omega, \ x' \in \partial\Omega, \tag{1.33}$$

where σ is associated with the jump of normal derivatives between the interior and exterior solutions, and is a density to be determined. For a Dirichlet BVP, (1.33) can be used as (1.32) in a numerical procedure. However, as

(1.33) is an integral equation of the first kind, the solution system can suffer from nonuniqueness and instability. This problem was studied by Fichera [301] and others [400, 401], and remedies have been offered [364, 866, 905].

For mixed BVPs, we can differentiate (1.33) to obtain

$$\frac{\partial \phi(x)}{\partial n(x)} = -2\pi \sigma(x) + \int_{\partial \Omega} \sigma(x') \frac{\partial(1/r)}{\partial n(x)} \, dx', \quad x \in \partial \Omega. \tag{1.34}$$

Again, the leading term on the RHS comes from the Cauchy principal value integration. The pair of integral equations, (1.33) and (1.34), can be used to solve mixed BVPs.

Of the three methods employing fundamental solutions to solve PDEs, the one using Green's third identity as presented in Section 1.3.1 is referred to as the "direct boundary element method", for its solution of the potentials and its normal derivatives from the integral equation. The two methods presented in the current section, the simple layer and the double layer potential method, are referred to as the "indirect boundary element methods", because intermediate variables, the fictitious densities, need to be determined first.

In the above we have used the Laplace equation as an example. For other operators, such as the elasticity operator, we may find the equivalent of the simple layer potential method as the stress discontinuity method [193, 229], and the double layer potential method as the displacement discontinuity method (DDM) [193, 228, 799].

To gain an understanding of the numerical implementation, we shall give a simple example below. For a 3D potential problem with a Dirichlet BC, we may subdivide the surface $\partial \Omega$ in (1.33) into N area elements $\partial \Omega_j$, each with a center node x_j, with $j = 1, \ldots, N$. In the lowest order approximation, we assign $\sigma(x)$ within each $\partial \Omega_j$ by a yet unknown constant σ_j. Equation (1.33) then becomes the following:

$$\hat{\phi}(x) = \sum_{j=1}^{N} c_j(x) \sigma_j, \tag{1.35}$$

where

$$c_j(x) = \int_{\partial \Omega_j} \frac{1}{\|x - x'\|} \, dx', \quad x \in \Omega, \ x' \in \partial \Omega_j. \tag{1.36}$$

To satisfy the BC, we can place x in turn on nodes $x_i, i = 1, \ldots, N$, where the boundary value is given as $\phi(x_i) = \phi_i$. We then obtain a set of linear equations

$$\phi_i = \sum_{j=1}^{N} a_{ij}\,\sigma_j, \quad i = 1, \ldots, N. \tag{1.37}$$

In the above, $a_{ij} = c_j(\boldsymbol{x}_i)$ can be evaluated analytically for some simple-shaped area elements or numerically, and the σ_j are unknown. The system can be solved for the σ_j. Once the σ_j are available, (1.35) and (1.36) allow the approximate solution to be evaluated at any $\boldsymbol{x} \in \Omega$.

If for a moment we ignore the integral equation background of (1.35) and (1.36), we can take a heuristic view of the above procedure. We observe that each c_j in (1.36) is a source distributed over an area. The RHS of equation (1.35) is a summation of such sources at different locations, which automatically satisfies the governing equation. Equation (1.37) is an effort to satisfy the BC. Now, let us move these area sources away from the boundary, to a set of nodes outside of Ω. These area elements can take any shape and do not need to cover a full surface, and can even be shrunk to points. The RHS of equation (1.35) still satisfies the governing equation. We can perform the same collocation procedure as (1.37) to satisfy the BC, and the BVP can be solved in a similar way. This is in essence the method of fundamental solutions (MFS).

1.3.3   *Null-field integral equations*

A number of studies on the MFS [17, 94, 334, 443, 613] attributed its origin to Kupradze and his coworkers [505–507, 509]. Kupradze called his method the method of functional equations, and based its theoretical foundation on the Fischer-Riesz representation theorem [508, 731]. In fact, Kupradze's method is a null-field integral equation method. It has more influence on the early development of the boundary integral equation method (BIEM) [231, 758, 904] than on the MFS [200, 203].

The method of functional equations has been applied to potential and elasticity problems. In the following we shall give a brief description of how to solve potential problems. In a null-field integral equation method, we utilize Green's third identity (1.14), and place the field point $\boldsymbol{x}$ outside Ω, so that the LHS vanishes

$$0 = \int_{\partial\Omega} \left[\phi(\boldsymbol{x}') \frac{\partial G(\boldsymbol{x}, \boldsymbol{x}')}{\partial n(\boldsymbol{x}')} - G(\boldsymbol{x}, \boldsymbol{x}') \frac{\partial \phi(\boldsymbol{x}')}{\partial n(\boldsymbol{x}')} \right] d\boldsymbol{x}', \quad \boldsymbol{x} \in \Omega', \tag{1.38}$$

where Ω' is the exterior of Ω. In the above, we have replaced $1/r$ by G, the fundamental solution of the partial differential operator. For a mixed BVP, a numerical implementation can be performed as follows [506]. For an interior domain problem with domain Ω and boundary $\partial\Omega$, we can create an auxiliary boundary $\partial\Omega'$ that encloses $\partial\Omega$. On $\partial\Omega'$, we can select a set of nodes x_i, with $i = 1, \ldots, N$, and sequentially place the center of the fundamental solution on these nodes. (We note that the creation of an auxiliary boundary is for convenience only, as the nodes x_i can be distributed in any pattern in Ω' and need not be associated with a surface.) Assuming that the boundary $\partial\Omega$ can be parameterized, and a certain quadrature rule can be applied using a set of quadrature points x'_j, $j = 1, \ldots, N$, and weights w_j, the integration of (1.38) yields

$$\sum_{j=1}^{N} w_j\, G(x_i, x'_j) \frac{\partial \phi(x'_j)}{\partial n(x'_j)} = \sum_{j=1}^{N} w_j\, \phi(x'_j) \frac{\partial G(x_i, x'_j)}{\partial n(x'_j)},$$

$$x_i \in \partial\Omega', \quad x'_j \in \partial\Omega, \quad \text{for } i = 1, \ldots, N. \tag{1.39}$$

We note that of the discrete values ϕ and $\partial\phi/\partial n$ on a node x'_j, one is the given BC, and the other is unknown. Equation (1.39) consists of N linear equations with N unknowns, and is a sufficient solution system.

In a general geometry, it will be difficult to parameterize the surface. The surface $\partial\Omega$ is then subdivided into subsurfaces $\partial\Omega_j$, with $j = 1, \ldots, N$. Using the simplest interpolation scheme, we assume that ϕ and $\partial\phi/\partial n$ take constant values on each element $\partial\Omega_j$. Then (1.38) can be integrated to give

$$\sum_{j=1}^{N} a_{ij} \phi_j = \sum_{j=1}^{N} b_{ij} \left(\frac{\partial\phi}{\partial n} \right)_j, \tag{1.40}$$

where

$$a_{ij} = \int_{\partial\Omega_j} \frac{\partial G(x_i, x')}{\partial n(x')}\, dx', \tag{1.41a}$$

$$b_{ij} = \int_{\partial\Omega_j} G(x_i, x')\, dx'. \tag{1.41b}$$

Again, either ϕ_j or $(\partial\phi/\partial n)_j$ is given as boundary data, and the system can be solved for the unknown parts. To increase the accuracy, more complicated interpolation schemes can be used, particularly those with inter-element continuity properties. In that case, element connectivity needs to be constructed, and assemblage of unknowns be performed. This procedure then yields a linear system similar to (1.40). We may notice a special advantage of the null-field method. As the singularity is outside of Ω and away from $\partial\Omega$, the integration is entirely regular, and no special treatment of singular integrals is needed.

For exterior domain wave problems, the null-field method was pioneered by Waterman [901–903] for solving electromagnetic, acoustic, as well as elastic wave scattering problems in the form of the T-matrix method. The method has been popular in many fields [83, 161, 432]. Martin [625] showed that for exterior domain wave scattering problems, the same null-field solution satisfies a Fredholm integral equation of the second kind; hence it is guaranteed to be solvable. In fact, it is not only solvable, but also free from the spurious frequency issues plaguing integral equations of both the first and the second kind in 2D. Doicu and Mishchenko [262, 263] further proved the existence and uniqueness of the null-field integral equation solution for both the interior and exterior domain wave scattering problems.

1.4 Origin of the MFS

Rankine in 1864 and 1871 [746, 747] demonstrated that by placing discrete sources and sinks of equal strengths along an axis, and superposing on them a uniform flow, one can generate the ideal flow field around 2D symmetric and 3D axisymmetric bodies, known as Rankine bodies. This methodology, however, cannot control the shape of the body generated. An inverse procedure is needed.

Inspired by Rankine's work [880], von Kármán in 1927 devised a numerical technique that can construct the velocity field of a far-field uniform flow passing a body of given shape [878, 879]. Using the axisymmetric body of a Zeppelin ZR-III airship as an example, the pressure distribution around the hull was calculated and compared with that of the experiment. Due to its historical significance, we shall give a brief description of the methodology below.

For an axisymmetric flow, we shall use the cylindrical coordinates (r, θ, z). In the r–z plane, the potential field due to a source located on the z-axis with the coordinate $(0, z')$ can be expressed as

$$\phi(r, z, z') = -\frac{1}{4\pi\rho},\tag{1.42}$$

where $\rho = \sqrt{r^2 + (z - z')^2}$, while the corresponding stream function is

$$\psi(r, z, z') = -\frac{1}{4\pi}\left(1 + \frac{z - z'}{\rho}\right). \qquad (1.43)$$

The above expression with the intensity $1/a$ can be integrated with respect to z' from $z' - a/2$ to $z' + a/2$ to create a line source of length a as

$$\psi^\ell(r, z, z') = -\frac{1}{4\pi}\left(1 + \frac{\rho_1 - \rho_2}{a}\right), \qquad (1.44)$$

where $\rho_1 = \sqrt{r^2 + (z - z' + a/2)^2}$ and $\rho_2 = \sqrt{r^2 + (z - z' - a/2)^2}$ (see Fig. 1.2). We may divide the axis of the body into N segments of length a, and place the line source on them. Together with a uniform flow parallel to the z-axis with velocity U,

$$\psi_o(r) = \frac{U r^2}{2}, \qquad (1.45)$$

the combined stream function is

$$\hat{\psi}(r, z) = \psi_o(r) + \sum_{j=1}^{N} Q_j \psi^\ell(r, z, z'_j), \qquad (1.46)$$

in which the z'_j are the centers of the line sources, and the Q_j are undetermined source strengths. To enforce the BC that the hull surface on the

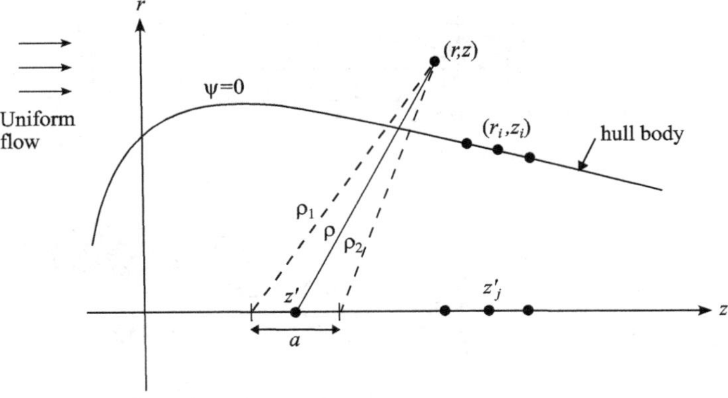

Figure 1.2. Source distribution on the r–z plane for an axisymmetric body.

r–z plane is a stream line, we select N points on the contour, (r_i, z_i), $i = 1, \ldots, N$, and require that the stream function takes the value zero there,

$$\psi_o(r_i) + \sum_{j=1}^{N} Q_j \, \psi^\ell(r_i, z_i, z_j') = 0, \quad i = 1, \ldots, N. \tag{1.47}$$

The above set of linear equations allows us to solve for the Q_j values. Once the stream function (1.46) is defined, we can find the velocity from

$$V_r = -\frac{1}{r}\frac{\partial \psi}{\partial z}, \tag{1.48a}$$

$$V_z = \frac{1}{r}\frac{\partial \psi}{\partial r}. \tag{1.48b}$$

The pressure can then be determined from the Bernoulli equation. With a heroic hand calculation effort, the pressure distribution on the hull surface was calculated and compared with that of the experiment (see Fig. 1.3).

The above procedure can be modified by using potentials instead of stream functions, and dipoles and vortices instead of sources. With the

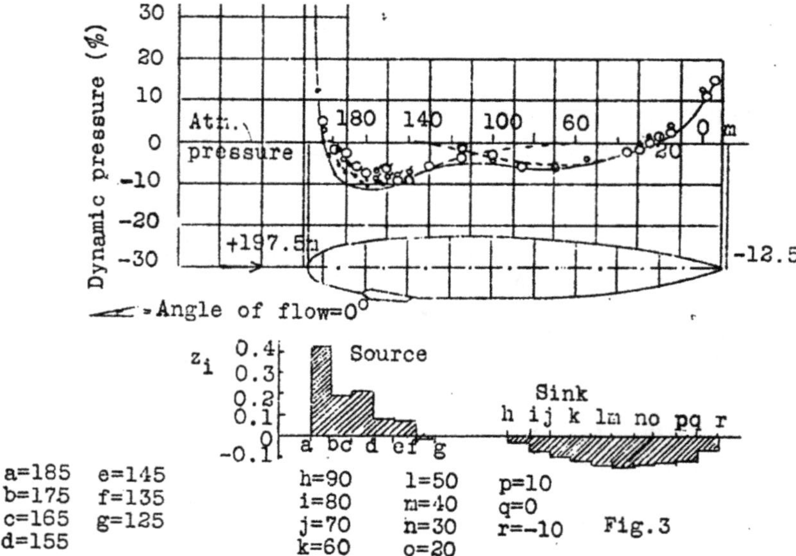

Figure 1.3. Distributing sources and sinks of undetermined intensity to calculate the pressure distribution of a uniform flow passing the body of Zeppelin ZR-III.

Source: Taken from von Kármán [879].

advent of electronic computers, the method has evolved to become the Panel Method in the aeronautic community, in which vortices are integrated on surface panels to cover the aircraft body. Subsonic as well as supersonic flows have been simulated this way [282, 382–384].

We recognize that the above procedure is exactly the modern day MFS, except that point singularities, rather than integrated ones, are typically used in present day implementations. The simplicity of the concept of adding up singular solutions to approximate the solution of a PDE has inspired many applications in different fields. For example, Oliveira [687] in 1968 distributed point forces on an auxiliary contour to solve elasticity problems. Singer *et al.* in 1974 [805] solved the external potential field created by electrical charges on the surface of a conductor, using a distribution of line and ring charges, respectively for 2D and 3D axisymmetric problems. Such ideas can be traced back to even earlier times, such as in 1950 by Loeb *et al.* [600] without a computer, and in 1968 using an electronic computer [1]. The method, known as the charge simulation method, has become popular in electrical engineering [610, 805, 813]. Moreover, the method of auxiliary sources used for electromagnetic wave scattering modeling is also just the MFS [710, 956, 957]. In fact, we may attribute the original idea of distributing electrical charges as a means of solving electrical potential problems to George Green [347], by his introduction of Green's function.

The modern day revival of the MFS, however, should be attributed to Mathon and Johnston [630], who in 1977 proposed the distribution of fundamental solutions exterior to the solution domain to approximate the potential

$$\hat{\phi}(x) = \sum_{j=1}^{N} c_j G(x, x'_j), \quad x \in \overline{\Omega}, \ x'_j \in \partial\Omega', \tag{1.49}$$

as a means of solving elliptic BVPs. In the above, G is the fundamental solution of the partial differential operator, the c_j are the undetermined coefficients, $\overline{\Omega} = \Omega \cup \partial\Omega$ with Ω the solution domain, which is bounded by $\partial\Omega$, and $\partial\Omega'$ is a surface enclosing Ω and $\partial\Omega$ (see Fig. 1.4). To give an illustration, we shall describe the procedure for solving a mixed BVP in the following.

Given ϕ satisfying the Laplace equation (1.10) and the BCs

$$\phi(x) = f(x), \quad x \in \partial\Omega_D, \tag{1.50a}$$

$$\frac{\partial\phi}{\partial n}(x) = g(x), \quad x \in \partial\Omega_N, \tag{1.50b}$$

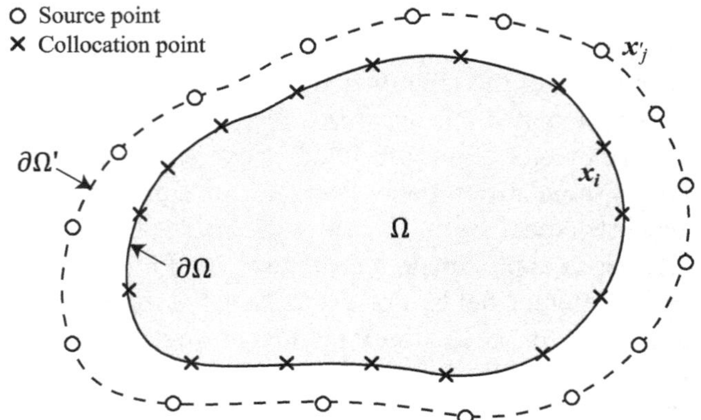

Figure 1.4. Distribution of source and collocation points in the MFS scheme.

where $\partial\Omega_D \cup \partial\Omega_N = \partial\Omega$ and $\partial\Omega_D \cap \partial\Omega_N = \emptyset$, the approximation (1.49) automatically satisfies the governing equation. To enforce the BCs in (1.50), we select M_1 points on $\partial\Omega_D$, and M_2 on $\partial\Omega_N$, with $M_1 + M_2 = M$. On the points x_i, $i = 1, \ldots, M$, we enforce the following conditions

$$\sum_{j=1}^{N} c_j\, G(x_i, x'_j) = f(x_i), \quad i = 1, \ldots, M_1, \ \ x'_j \in \partial\Omega', \ \ x_i \in \partial\Omega_D,$$

$$(1.51)$$

$$\sum_{j=1}^{N} c_j \frac{\partial G(x_i, x'_j)}{\partial n(x_i)} = g(x_i), \quad i = M_1 + 1, \ldots, M, \ \ x'_j \in \partial\Omega', \ \ x_i \in \partial\Omega_N.$$

$$(1.52)$$

The above equations can be put into matrix form as

$$A c = b, \tag{1.53}$$

where A is an $M \times N$ matrix assembled from G and $\partial G/\partial n$ in (1.51) and (1.52), b is an $M \times 1$ vector assembled from f and g, and c ($N \times 1$) contains the unknown source strengths. It is necessary that $M \geq N$, that is, there are at least as many collocation points as source points, or else the system is underdetermined. When $M = N$, the system should be uniquely solvable for the N coefficients c_j, barring the special case of degenerate scale in 2D (see Section 6.4). The approximate solution (1.49) is then continuously defined for all $x \in \overline{\Omega}$. The convergence of solution is achieved by letting the number of sources $N \to \infty$, and making the collocation nodes dense over

$\partial\Omega$. In the case that there are more collocation nodes than source nodes, $M > N$, the system becomes overdetermined. An optimal solution can be obtained by a linear least squares procedure, as described in Section 6.7.

1.5 Link Between the MFS and Integral Equations

To further explore the link between the MFS and integral equations, we shall follow Christiansen [208] and present two integral equation methods. First, let us consider two contours/surfaces, as schematically shown in Fig. 1.4, in which $\partial\Omega$ is the boundary of the BVP to be solved, and $\partial\Omega'$ is an auxiliary boundary enclosing $\partial\Omega$. In Method 1, the sources are distributed on $\partial\Omega'$, and collocation is conducted on $\partial\Omega$; and in Method 2, the roles are reversed.

Method 1. Consider solving a BVP with the Dirichlet BC

$$\phi(x) = f(x), \quad x \in \partial\Omega, \tag{1.54}$$

for which there exists a unique solution $\phi(x)$ for $x \in \Omega$. Following the potential theory [481], we can generate a potential field Φ by integrating the fundamental solution G over a closed surface with a continuous intensity σ, as

$$\Phi(x) = \int_{\partial\Omega'} \sigma(x') G(x, x') \, dx', \quad x \in \Omega', \tag{1.55}$$

where we have chosen the auxiliary boundary $\partial\Omega'$ for such a distribution, and Ω' is the domain enclosed by $\partial\Omega'$.

Christiansen [208] demonstrated that, barring some special situations, we can find a unique distribution of $\sigma(x')$, $x' \in \partial\Omega'$, such that $\Phi(x) = \phi(x)$ for all $x \in \Omega$. One such special situation to be excluded is the following. If the solution ϕ is generated by a singularity located exterior to $\partial\Omega$, but interior to $\partial\Omega'$, then such a σ cannot be found. However, as $\partial\Omega'$ is our selection, we can always choose it such that it excludes all such singularities (with the exception that a singularity is located on the boundary $\partial\Omega$).

Given the existence and uniqueness of σ, we can devise a numerical scheme for its approximate solution $\hat{\phi}(x)$. To do so, we subdivide the surface

$\partial\Omega'$ into N surface elements $\partial\Omega'_j$, and represent σ by an interpolation of discrete values on the elements, to obtain

$$\hat{\phi}(x) = \sum_{j=1}^{N} \int_{\partial\Omega'_j} \left[G(x, x') \sum_{k=1}^{n} N_k(x')\sigma_{jk} \right] dx'. \qquad (1.56)$$

In the above, the N_k are interpolation functions, which in the finite element method (FEM) are called shape functions, and σ_{jk} are the n discrete values on the element j. Connectivity data are needed to identify shared discrete values among elements to guarantee C^0 continuity.

In a simplification, we can assume that σ takes a constant value on each element, thus creating discontinuous elements; and we obtain

$$\hat{\phi}(x) = \sum_{j=1}^{N} \sigma_j \int_{\partial\Omega'_j} G(x, x')dx'. \qquad (1.57)$$

We may make another simplification by arguing that the distributed singularities behave like point singularities at a large distance. We then arrive at a summation formula without integration, as

$$\hat{\phi}(x) = \sum_{j=1}^{N} c_j\, G(x, x'_j), \qquad (1.58)$$

where the c_j are the products of constant densities σ_j with the areas of the elements $\partial\Omega'_j$, and are undetermined coefficients. Comparing (1.58) with (1.49), we observe that we have recovered the MFS approximation. Equation (1.58) can be used for the collocation of BCs on $\partial\Omega$ to create a linear system for the determination of the c_j.

Method 2. In the second method, we may distribute sources and dipoles on the boundary $\partial\Omega$ to create the following potentials:

$$\phi_s(x) = \int_{\partial\Omega} \mu(x')\, G(x, x')\, dx', \qquad (1.59)$$

$$\phi_d(x) = \int_{\partial\Omega} f(x') \frac{\partial G(x, x')}{\partial n(x')}\, dx', \qquad (1.60)$$

in which μ is the unknown density, and f is the Dirichlet BC. The above equations can be used to represent both the internal and external potential fields. Here we shall select a point $x \in \partial\Omega'$, external to the solution domain. From Green's third identity (1.38), we find that if we require $\mu = \partial\phi/\partial n$

on $\partial\Omega$, then the above potentials can be equated

$$\int_{\partial\Omega} \frac{\partial\phi(x')}{\partial n(x')} G(x, x') \, dx' = \int_{\partial\Omega} f(x') \frac{\partial G(x, x')}{\partial n(x')} \, dx', \quad x \in \partial\Omega'. \quad (1.61)$$

This is exactly the integral equation (1.38) used in Kupradze's method. If we place x at N points on $\partial\Omega'$, we can create a linear system to solve for N discrete $\partial\phi/\partial n$ values.

Method 3. Following Method 2, if we bring the contour $\partial\Omega'$ to coincide with $\partial\Omega$, we then obtain the boundary integral equation (1.14) used in the BEM, with the LHS generated by the Cauchy principal value integration of the singularity on $\partial\Omega$.

We observe in the above that there exist certain connections between the integral equation method and the MFS formulations; though they are not equivalent. Integral equation methods have a longer history of development than the MFS. This gives us the opportunity to "borrow" the theoretical results of existence and uniqueness, and remedies used in the integral equation method solutions, for use in the MFS, as we shall demonstrate in Chapter 6.

1.6 Simple MFS Codes

Before presenting a full discussion of theoretical background of the MFS in Chapters 2–5, leading to the computational aspects in Chapter 6, it is beneficial to give a quick demonstration of the simplicity of computer coding based on the methodology. In the following, we present two generic MATLAB® codes for solving BVPs for the Laplace equation in 2D and 3D.

■ **Example 1.1**

We solve the Laplace equation in the square $(0, 1) \times (0, 1)$, which is subject to the mixed BCs

$$\frac{\partial\phi}{\partial n} = 0, \quad \text{on } x = [0, 1], \ y = 0, \quad \text{and} \quad x = 0, \ y = [0, 1];$$

$$\phi = \cosh\frac{\pi}{2}\cos\frac{\pi y}{2}, \quad \text{on } x = 1, \ y = [0, 1];$$

$$\phi = 0, \quad \text{on } x = [0, 1], \ y = 1.$$

For checking the error, we note that the exact solution of the problem is

$$\phi = \cosh \frac{\pi x}{2} \cos \frac{\pi y}{2}.$$

To solve the problem, we use the MFS approximation (1.49) and (1.50) with $G = \ln r$. The detailed procedure is described in Sections 5.5.1 and (5.95)–(5.104). The MATLAB® code `MFSLap2D.m` is created for solving the problem and listed below. The program requires the input of collocation and source points, boundary condition type, boundary value, etc. To keep the code generic, data are separately prepared and stored in a text file `xy.csv`.

For the present solution, a coarse discretization using 16 points is shown in Fig. 1.5. The BC types are marked by different symbols. We notice that at the two corner nodes, $(1, 0)$ and $(0, 1)$, both the Dirichlet and Neumann BCs are present. The duality of BC can create ambiguity for some numerical methods, and "double nodes" were

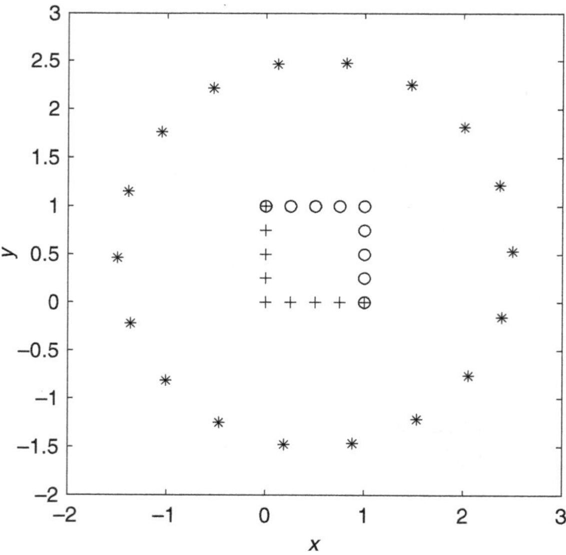

Figure 1.5. Example 1.1: Collocation and source points. ○: Collocation points for Dirichlet BC; +: collocation points for Neumann BC; and ∗: source points.

Table 1.1. Comparison of the MFS and exact solution.

(x, y)	$(0, 0)$	$(0, 0.5)$	$(0.5, 0)$	$(0.5, 0.5)$
MFS	0.9999984	0.7071065	1.324609	0.9366402
Exact	1.000000	0.7071065	1.324609	0.9366401

sometimes used [807], in which two nodes close to each other were assigned in the proximity of the corner node. One is assigned the Dirichlet condition and the other the Neumann condition. The duality of BC is not an issue for the MFS. Collocation of the two BCs can be performed at the same location to generate two independent equations. In the current case, the 16 nodes generate 18 collocation equations.

A circle of radius 2 is used for the distribution of 18 equally spaced sources. For the Neumann BC, the directional cosines of the outward normal are needed. These and other data are prepared in the sequence of [cx cy bctyp bc nx ny sx sy], which corresponds to the coordinates of collocation points, BC type (0 for Dirichlet and 1 for Neumann), boundary value, directional cosines of outward normal, and coordinates of source points, shown as xy.csv below.

Using the code and the data set, we can calculate the approximate solution at any point in the domain as well as on the boundary by entering the coordinates. In Table 1.1, we present a comparison of the numerical solution with the exact solution at a few locations. We observe high accuracy of the MFS solution. The maximum error is found at the corner point $(0, 0)$, which is accurate up to 5 digits.

```
    % MFSLap2D.m: 2-D MFS code for Laplace equation
1   clear all, clc
    % Read input file:
    % cx,cy: Coordinates of collocation points
    % bctyp: BC type, 0 = Dirichlet, 1 = Neumann
    % bc: Boundary value
    % nx,ny: Directional cosines of boundary outward normal
    % sx, sy: Coordinates of source points
2   [cx,cy,bctyp,bc,nx,ny,sx,sy] = readvars('xy.csv');
    % Number of collocation points
3   nc = size(cx,1);
```

```
     % Form matrices for easy operation
4    nx = repmat(nx,1,nc);
5    ny = repmat(ny,1,nc);
6    bcmtx = repmat(bctyp,1,nc);
     % Form the distance matrix rij
7    dx = repmat(cx,1,nc)-repmat(sx,1,nc)';
8    dy = repmat(cy,1,nc)-repmat(sy,1,nc)';
9    rsqij = dx.^2+dy.^2;
     % Form the matrix A for A*b=c
10   A = (1-bcmtx).*log(rsqij)/2 + bcmtx.*(nx.*dx+ny.*dy)./rsqij;
     % Solve for coefficient c
11   c = A\bc;
     % Find approximate solution at (x,y)
12   xy = input('Enter [x y]\n');
13   phi = c'*log((xy(1)-sx).^2+(xy(2)-sy).^2)/2


% xy.csv in [cx,cy,bctyp,bc,nx,ny,sx,sy]
0 0 1 0 -0.7071068 -0.7071068 2.367161 1.216736
0.25 0 1 0 0 -1 2.009419 1.812118
0.5 0 1 0 0 -1 1.469619 2.249239
0.75 0 1 0 0 -1 0.8128689 2.475377
1 0 1 0 0 -1 0.118382 2.463254
1 0.25 0 2.318179 0 0 -0.5300761 2.214335
1 0.5 0 1.774257 0 0 -1.054292 1.758641
1 0.75 0 0.960221 0 0 -1.391037 1.151136
1 1 0 0 0 0 -1.499695 0.4650952
0.75 1 0 0 0 0 -1.367161 -0.2167359
0.5 1 0 0 0 0 -1.009419 -0.8121181
0.25 1 0 0 0 0 -0.4696192 -1.249239
0 1 1 0 -1 0 0.1871311 -1.475377
0 0.75 1 0 -1 0 0.881618 -1.463254
0 0.5 1 0 -1 0 1.530076 -1.214335
0 0.25 1 0 -1 0 2.054292 -0.7586408
1 0 0 2.509178 0 0 2.391037 -0.1511363
0 1 0 0 0 0 2.499695 0.5349048
```

In the above code `MFSLap2D.m`, Lines 2–6 are used for the reading and preparation of input data. Lines 7–9 compute the square of distance matrix $r_{ij} = \|x_i - x'_j\|$. Line 10 assigns the proper elements to the matrix A: $G = \ln r_{ij}$ for the Dirichlet type, and $\partial G/\partial n = [(x_i - x'_j)n_x + (y_i - y'_j)n_y]/r_{ij}^2$ for the Neumann type. The BC type `bcmtx` is used for the automatic assignment. Line 11 solves for the coefficients c_j as in (1.49). Line 13 computes the approximate solution given (x, y) coordinates.

Next, we solve a 3D BVP for the Laplace equation.

■ Example 1.2

We solve the exterior domain problem of a uniform flow of velocity U in the z-direction passing a sphere of radius a centered at the origin. The exact solution is

$$\phi = -Ua\left[\frac{r}{a} + \frac{1}{2}\left(\frac{a}{r}\right)^2\right]\cos\theta,$$

in the spherical coordinates (r, θ, φ), with $0 \leq \theta \leq \pi$. The solution is independent of the azimuth angle φ due to axisymmetry with respect to the z-axis.

As the solution needs to vanish at infinity, we shall drop the uniform flow part and solve for

$$\phi_1 = -\frac{Ua}{2}\left(\frac{a}{r}\right)^2\cos\theta$$

instead. Assuming $a = 1$ and $U = 1$, the original Neumann BC of zero flux on the surface of the sphere becomes

$$\frac{\partial\phi_1}{\partial n} = -\cos\theta, \quad \text{at } r = a.$$

To discretize the geometry, we use 40 collocation points uniformly spaced over the spherical surface using the MATLAB® function `SpiralSampleSphere` in the S^2 Sampling Toolbox. For the distribution of sources, a small sphere of radius $0.1a$ is used, with source points similarly distributed. These and other data are stored in a file `xyz.csv` in the sequence `[cx cy cz bctyp bc nx ny nz sx sy sz]` as shown below.

To solve the 3D problem, we present the code `MFSLap3D.m`. We notice that the extension of code from 2D to 3D is straightforward. This is one of the advantages of meshless methods. We also note that the code below can be used for exterior as well as interior domain problems, without modification. In Table 1.2, we compare the MFS approximation with the exact solution at a few selected locations. Again we observe high accuracy.

Table 1.2. Comparison of the MFS and exact solution for the potential flow around a sphere.

(r, θ, φ)	$(1, 0, 0)$	$(1, \pi/4, 0)$	$(1, \pi/2, 0)$	$(5, \pi/4, 0)$	$(5, \pi/2, 0)$
MFS	0.00000	-0.35353	-0.50000	-0.014142	-0.020000
Exact	0.00000	-0.35355	-0.50000	-0.014142	-0.020000

```
% MFSLap3D.m: 3-D MFS code for Laplace equation
1  clear all, clc
2  [cx,cy,cz,bctyp,bc,nx,ny,nz,sx,sy,sz] = readvars('xyz.csv');
3  nc = size(cx,1);
4  nx = repmat(nx,1,nc);
5  ny = repmat(ny,1,nc);
6  nz = repmat(nz,1,nc);
7  bcmtx = repmat(bctyp,1,nc);
8  dx = repmat(cx,1,nc)-repmat(sx,1,nc)';
9  dy = repmat(cy,1,nc)-repmat(sy,1,nc)';
10 dz = repmat(cz,1,nc)-repmat(sz,1,nc)';
11 rij = sqrt(dx.^2+dy.^2+dz.^2);
12 A=(1-bcmtx).*(1./rij)-bcmtx.*(nx.*dx+ny.*dy+nz.*dz)./rij.^3;
13 c = A\bc;
14 xyz = input('Enter [x y z]\n');
15 phi=c'*(1./sqrt((xyz(1)-sx).^2+(xyz(2)-sy).^2+(xyz(3)-sz).^2))
```

```
% xyz.csv in [cx,cy,cz,bctyp,bc,nx,ny,nz,sx,sy,sz]
0.00000 0.00000 1.00000 1 -1.00000 0.00000 0.00000 -1.00000 0.000000 0.000000
0.100000
-0.23310 0.21354 0.94872 1 -0.94872 0.23310 -0.21354 -0.94872 -0.023310
0.021354 0.094872
0.03857 -0.43946 0.89744 1 -0.89744 -0.03857 0.43946 -0.89744 0.003857
-0.043946 0.089744
0.32426 0.42294 0.84615 1 -0.84615 -0.32426 -0.42294 -0.84615 0.032426
0.042294 0.084615
-0.59750 -0.10569 0.79487 1 -0.79487 0.59750 0.10569 -0.79487 -0.059750
-0.010569 0.079487
0.56417 -0.35888 0.74359 1 -0.74359 -0.56417 0.35888 -0.74359 0.056417
-0.035888 0.074359
-0.18733 0.69686 0.69231 1 -0.69231 0.18733 -0.69686 -0.69231 -0.018733
0.069686 0.069231
-0.35376 -0.68113 0.64103 1 -0.64103 0.35376 0.68113 -0.64103 -0.035376
-0.068113 0.064103
0.75859 0.27703 0.58974 1 -0.58974 -0.75859 -0.27703 -0.58974 0.075859
0.027703 0.058974
-0.77890 0.32152 0.53846 1 -0.53846 0.77890 -0.32152 -0.53846 -0.077890
0.032152 0.053846
0.37015 -0.79098 0.48718 1 -0.48718 -0.37015 0.79098 -0.48718 0.037015
-0.079098 0.048718
```

```
0.26935 0.85874 0.43590 1 -0.43590 -0.26935 -0.85874 -0.43590 0.026935
0.085874 0.043590
-0.79866 -0.46284 0.38462 1 -0.38462 0.79866 0.46284 -0.38462 -0.079866
-0.046284 0.038462
0.92082 -0.20244 0.33333 1 -0.33333 -0.92082 0.20244 -0.33333 0.092082
-0.020244 0.033333
-0.55178 0.78485 0.28205 1 -0.28205 0.55178 -0.78485 -0.28205 -0.055178
0.078485 0.028205
-0.12504 -0.96494 0.23077 1 -0.23077 0.12504 0.96494 -0.23077 -0.012504
-0.096494 0.023077
0.75223 0.63398 0.17949 1 -0.17949 -0.75223 -0.63398 -0.17949 0.075223
0.063398 0.017949
-0.99090 0.04098 0.12821 1 -0.12821 0.99090 -0.04098 -0.12821 -0.099090
0.004098 0.012821
0.70673 -0.70329 0.07692 1 -0.07692 -0.70673 0.70329 -0.07692 0.070673
-0.070329 0.007692
-0.04618 0.99860 0.02564 1 -0.02564 0.04618 -0.99860 -0.02564 -0.004618
0.099860 0.002564
-0.64050 -0.76753 -0.02564 1 0.02564 0.64050 0.76753 0.02564 -0.064050
-0.076753 -0.002564
0.98813 0.13295 -0.07692 1 0.07692 -0.98813 -0.13295 0.07692 0.098813
0.013295 -0.007692
-0.81408 0.56642 -0.12821 1 0.12821 0.81408 -0.56642 0.12821 -0.081408
0.056642 -0.012821
0.21592 -0.95977 -0.17949 1 0.17949 -0.21592 0.95977 0.17949 0.021592
-0.095977 -0.017949
0.48376 0.84423 -0.23077 1 0.23077 -0.48376 -0.84423 0.23077 0.048376
0.084423 -0.023077
-0.91401 -0.29160 -0.28205 1 0.28205 0.91401 0.29160 0.28205 -0.091401
-0.029160 -0.028205
0.85587 -0.39544 -0.33333 1 0.33333 -0.85587 0.39544 0.33333 0.085587
-0.039544 -0.033333
-0.35636 0.85151 -0.38462 1 0.38462 0.35636 -0.85151 0.38462 -0.035636
0.085151 -0.038462
-0.30461 -0.84688 -0.43590 1 0.43590 0.30461 0.84688 0.43590 -0.030461
-0.084688 -0.043590
0.77304 0.40629 -0.48718 1 0.48718 -0.77304 -0.40629 0.48718 0.077304
0.040629 -0.048718
-0.81482 0.21478 -0.53846 1 0.53846 0.81482 -0.21478 0.53846 -0.081482
0.021478 -0.053846
0.43678 -0.67929 -0.58974 1 0.58974 -0.43678 0.67929 0.58974 0.043678
-0.067929 -0.058974
0.13000 0.75643 -0.64103 1 0.64103 -0.13000 -0.75643 0.64103 0.013000
0.075643 -0.064103
-0.57052 -0.44184 -0.69231 1 0.69231 0.57052 0.44184 0.69231 -0.057052
-0.044184 -0.069231
0.66635 -0.05521 -0.74359 1 0.74359 -0.66635 0.05521 0.74359 0.066635
-0.005521 -0.074359
-0.41205 0.44541 -0.79487 1 0.79487 0.41205 -0.44541 0.79487 -0.041205
0.044541 -0.079487
0.00260 -0.53293 -0.84615 1 0.84615 -0.00260 0.53293 0.84615 0.000260
-0.053293 -0.084615
0.29640 0.32674 -0.89744 1 0.89744 -0.29640 -0.32674 0.89744 0.029640
0.032674 -0.089744
-0.31477 -0.02917 -0.94872 1 0.94872 0.31477 0.02917 0.94872 -0.031477
-0.002917 -0.094872
0.00000 0.00000 -1.00000 1 1.00000 0.00000 0.00000 1.00000 0.000000 0.000000
-0.100000
```

1.7 A Literature Review

Since its modern revival, the MFS has found many applications. In the following we shall give a literature survey that is not intended to be exhaustive. Rather, it is a sampling of some earliest applications, the highly cited, and the more recent contributions.

The MFS has been implemented for a wide range of PDEs, such as the Laplace and Poisson equations [284, 285, 334, 430, 610, 744, 805], Helmholtz and modified Helmholtz equations [92, 445, 468], biharmonic equation [266, 443, 715, 725], polyharmonic equations [456, 853], diffusion equation [56, 134, 790], wave equation [351, 953], and fractal differential equations [117, 889, 936, 937]. The material constants can be isotropic or anisotropic [77, 78, 623], the coefficients of the equation can be constant or variable [143, 289], and the equation can be linear or nonlinear [57, 451, 619, 855].

The MFS has been applied to physical and engineering problems in elastostatics [35, 112, 749], beams and plates [525, 751, 864, 896, 909], fracture mechanics [79, 304, 447], functionally graded materials [88, 539, 895], thermoelasticity [586, 621, 852], poroelasticity [44], piezoelectricity [896], wave scattering [19, 286, 445, 496], elastodynamics [332, 495, 871], seismicity [267, 766, 767, 919], poroelastodynamics [596–598, 670], electromagnetics [142, 530, 942, 947], magnetoelectroelasticity [705], electrocardiography [102, 900], electroencephalography and magnetoencephalography [9, 10], electroporation [765], chemical vapor infiltration [607], Stokes flow [17, 948, 969], rarefied gas flow [389], water waves [785, 926, 927], and free-surface problems [131, 444, 728, 932]. Other applications include eigensolutions [23, 25, 31, 289], numerical conformal mapping [26, 27, 763], as well as hyper-dimensional problems [32].

The above applications involve direct and forward problems. The MFS has also been widely used as a tool to solve inverse and ill-posed problems [458]. The reason for using the MFS stems from that, not only it is an efficient direct solution tool, but also by its collocation nature, it can solve certain ill-posed problems without iteration [199, 203]. There have been extensive efforts utilizing the MFS to solve Cauchy problems, that is, problems with a part of the BCs missing, yet redundant conditions are found on other parts of the boundary. Some of the problems solved include the Laplace [796, 906, 951], Helmholtz [415, 617], and biharmonic [577, 797] equations, with applications in magnetics [237], heat transfer [614], and elasticity [613, 615, 959] problems. Other applications include the search for a heat source

[6, 416, 934, 935], inverse geometric problems [396, 454, 459, 639], Stefan problems [131, 426, 718], parameter determination [132, 642], as well as inverse problems of Stokes flow [149], Oseen flow [469], heat conduction [264, 395], thermoelasticity [464, 624], electrocardiography [66, 429, 900], and acoustic scattering [457, 460].

Since the early development of the MFS, there have been many efforts to improve its accuracy and efficiency, as well as to enhance its capability to solve different types of problems. The original MFS was limited to solving homogeneous linear PDEs. For inhomogeneous PDEs, such as the Poisson equation and other equations with body force terms, the method of particular solutions (MPS) has been employed to approximate and remove the RHS [144, 334, 335, 338, 955]. Transient problems are generally treated by a finite difference time-stepping scheme and solved using the stationary fundamental solutions. The inhomogeneous terms resulting from the previous time step are handled by the MPS. Such a scheme has been implemented for the diffusion and transport equations [20, 136, 714], wave and dynamic equations [483, 910, 953], Stokes flow [848], and Navier–Stokes equations [954]. Alternatively, time-dependent fundamental solutions can be directly applied, and time is treated as an extra dimension. Implementation has been found in diffusion and wave problems [569, 850, 944, 945, 952].

The accuracy of the MFS approximation has a strong dependence on the source locations. When the optimal locations are found, the error can be reduced by orders of magnitude without increasing the number of degrees of freedom in the approximation. Hence there has been a significant amount of effort devoted to finding the optimal locations of the source points. The MFS has an advantage over other methods in estimating the approximation error because the MFS approximation automatically satisfies the governing equation. Hence, one can determine the error estimate by merely comparing the computed approximate solution on the boundary with the Dirichlet BC. With the support of maximum principles for a class of elliptic partial differential operators [733], as well as the biharmonic [275, 721] and polyharmonic [722] operators, the boundary error gives the upper bound of the approximation error [203]. Such an *a posteriori* error estimate can be used to guide a search for optimal source placement to minimize the error.

In an optimization process, multiple objectives are generally lumped into a single one, quantified as a "cost function". The cost function is to be minimized with respect to a set of "decision variables". In the MFS, an overall error can be set as the target to be minimized. In the early implementation of the MFS, the least squares error on a set of boundary collocation nodes was chosen to be the cost function. All the source locations, the (x, y, z) coordinates, were taken as decision variables [285, 430, 630, 969] and a search was conducted in the multiple coordinates space. However, such a search is heavily dependent on the initial locations of the nodes. If a gradient method is used, the search will invariably settle on a nearest local minimum, and be trapped. To overcome this difficulty, metaheuristic search algorithms [697], such as the genetic algorithms [343, 431, 770], and simulated annealing [211], have been employed. In such methods, a more or less random search is conducted using biological evolution rules. Alternatively, one can narrow down the search space by using a finite set of fixed points, rather than floating points, to form an overdetermined system, and search for a best subset of these points using a greedy algorithm [558, 570, 573, 779]. In other methods, the node distribution follows certain multilevel geometric patterns, and the decision variables are limited to a few parameters defining the geometric shapes [536, 576, 798]. Alternatively, optimization can be conducted in two steps. The first step is to globally seek a best distance for the auxiliary boundary, and the second is to develop an adaptive scheme to nonuniformly distribute the nodes on the auxiliary boundary to even out the errors [686].

Other quality measures for the solution, such as the condition number, can be incorporated into the cost function [849]. Also assuming that the error is not available, an "energy gap function" is used as the cost function for minimization [888]. In instances that one does not wish to spend effort to compute the *a posteriori* error on the boundary, or an error bound is not available, an estimate of the error can be obtained through the cross validation technique, which is widely used in statistics for predicting regression error. For example, we can adopt the leave-one-out cross validation (LOOCV) method. An efficient LOOCV algorithm developed by Rippa [755] for the radial basis function collocation method (RBFCM) has been implemented for the MFS [10, 145, 367, 604].

In contrast to domain discretization methods, such as the FEM and the finite difference method (FDM), the MFS discretizes a boundary-like geometry and generates a smaller size discrete system, which is an important advantage. However, the MFS uses a global interpolation scheme, as compared to the local interpolation of the FEM and FDM. As a consequence, the MFS solution matrix is fully populated, which requires more effort to solve than a sparse one, and can be ill-conditioned. Various schemes have been devised to alleviate these difficulties. One method is to utilize parallel domain decomposition by subdividing the domain into a large number of sub-domains and use parallel processing to speed up the computation [406]. Another is to use a multilevel technique that builds up the approximation from a coarse grid to more refined ones [139], using for example a quadtree structure [322, 323]. For problems with symmetry in their geometry, such as axisymmetric or regular polygonal shapes, it is possible to use certain source and collocation point configurations to generate matrices that have circulant or block circulant properties. Matrix decomposition algorithms employing fast Fourier transforms can reduce the computational cost of solving a system based on an $N \times N$ grid to $\mathcal{O}(N^2 \log N)$ [86]. This technique has been employed to solve potential, biharmonic, and elasticity problems [448, 453, 810, 860]. In another approach, the adaptive cross approximation (ACA) method, used to approximate BEM and collocation matrices [67, 68], was incorporated into the MFS to speed up calculations [333, 907].

There are problems with multiplicity of geometry, such as a solid with a large number of heterogeneous inclusions, cavities, or fractures, and acoustic waves scattered from a large number of obstacles, in which the number of degrees of freedom can be overwhelming for any numerical method. This can be particularly critical for the MFS, because it generally yields a full matrix formed from these discrete nodes. One way of overcoming such an issue is to utilize fast algorithms [348, 349, 759] developed for particle interactions, which translate the point-to-point interactions to cell-to-cell ones by using multipole expansions and translations, where cells can have a hierarchical tree structure. The fast multipole method (FMM) has been successfully implemented with the MFS to solve potential [761], heat conduction [537, 593], acoustic [413, 595], and Stokes flow [101] problems.

Another important development in the MFS to overcome the large size full matrix issue is to perform collocation on local regions that overlap with each other, rather than globally. The local systems are tied together to form a global but sparse matrix system that can be solved more efficiently. The method then becomes similar to many other locally based meshless methods, such as the local radial basis function collocation method (LRBFCM) [261, 773, 876], the local radial basis function differential quadrature (RBF-DQ) method [245, 801, 803], the radial basis function finite difference (RBF-FD) method [308, 760, 920], the meshless local Petrov–Galerkin method (MLPG) [38, 40, 42], the element free Galerkin (EFG) method [74, 303, 602], etc. The differences among these methods are in the trial functions (the solution space) and test functions (space for weighted residuals) adopted, and the method used to minimize an error functional, such as the strong formulation (collocation), the weak formulation, and least squares, leading to the convergence of solution.

The localized method of fundamental solutions (LMFS) uses fundamental solutions as trial functions, and collocation to enforce the governing equation and BCs. The recently developed LMFS has been applied to the Laplace and biharmonic equations [291], inhomogeneous elliptic equations [357], Helmholtz equation [739] and acoustic problems [182, 740], anisotropic heat conduction [356], heat conduction in functionally-graded materials [931], convection–diffusion–reaction [589], isotropic [358, 359] and anisotropic elasticity [588], plates [742], elastic wave [541], fracture [360], piezoelectric structure [361], Signorini [292], and inverse problems [181, 890].

To conclude this section, we shall mention a few related numerical methods. The Trefftz collocation method (TCM) [486, 487, 550, 556] has a similar root as the MFS, but uses the general solutions of the governing equation, rather than the fundamental solutions, as the basis functions. To ensure solution convergence, the set of basis functions needs to be a complete set, known as T-complete (Trefftz-complete) [376, 378]. A similar collocation procedure is applied to satisfy the BCs. In fact, if we take a broader view of the general solutions, to include singular (fundamental) solutions, which satisfy the governing equation except at the singular point, then the MFS may be viewed as a subset of the TCM. The TCM has been applied to potential problems [526], elliptic and parabolic PDEs [206], the Helmholtz equation [116, 326, 723], heat conduction [656], the biharmonic equation and plate vibration [240, 249, 873], as well as free surface water wave [180], elasticity [265], inverse [575] and boundary detection problems [290].

We mentioned above that the general solutions can incorporate nonsingular as well as singular solutions. The singular solutions naturally contain a center, which is the singularity point. In the MFS, the same singular solution is placed on a number of locations to form the approximation bases. The solution at each center is an independent basis, as it cannot be represented by a collection of others. The nonsingular solutions, on the other hand, have two types — those with a center and those without one. Take, for example, the general solutions of the Laplace equation, which are the harmonic polynomials. Although each polynomial is defined with an origin, a translation or rotation does not generate a new basis. It is merely a linear combination of other polynomials in the family. Hence the full set is needed for the convergence of solution. There are also nonsingular general solutions that possess a center such as the Bessel function $J_0(kr)$. Paired with the Bessel function of the second kind $Y_0(kr)$, they are solutions of the 2D Helmholtz equation. The former is nonsingular and the latter singular. Both are radial functions that have a center, and can be placed at different locations to form a complete approximation set. The difference is that the singular one must be placed outside the solution domain, while the nonsingular one can be located on the boundary, leading to a simpler numerical implementation. Kang *et al.* [434] used the nonsingular radial solution, called the non-dimensional dynamic influence function (NDIF), for solving the Helmholtz and polymetaharmonic equations [435, 436]. This technique has been applied to membrane and plate eigenvalue and vibration problems [165, 437, 438].

The availability of a nonsingular radial solution also inspired the boundary knot method (BKM) [172, 173]. In addition to solving the Helmholtz and modified Helmholtz equations [174], the BKM attempts to solve other governing equations by lumping the non-Helmholtz parts to the RHS, as the inhomogeneous terms. The MPS and other techniques are then used to approximate the RHS. The method has been applied to heat conduction [316], convective-diffusion [392], acoustic [891], nonlinear wave [242, 244], elastic and viscoelastic [118], and inverse problems [417].

As mentioned above, the traditional MFS requires the use of an auxiliary boundary surrounding the physical one. Its creation may pose some difficulty in handling complex geometries whereas the NDIF and BKM can avoid it. However, nonsingular radial solutions appear to exist only for a few special governing equations. Hence there is incentive to develop schemes that can allow the singular solutions to be directly placed on the boundary. An obvious consequence of it is that the coefficient becomes unbounded, when a collocation node coincides with a center. In the matrix form, this occurs in the diagonal elements. A few methods have been devised to regularize the system. In the modified method of fundamental solutions (MMFS) [774, 950], which stems from boundary integral equations, such as (1.14), (1.30), and (1.32), local boundary elements were created, and analytical integration was performed to treat the integrable and non-integrable singularities, in the Cauchy principal value and Hadamard finite part senses. The nonsingular resultants were then used as the regularized diagonal terms.

In the regularized meshless method (RMM) [168, 947], a technique mimicking the "rigid body motion removal" in the integral equation is applied. In a standard BEM implementation, an integral equation for the constant potential or constant displacement (rigid body motion) is created. By subtracting this equation from the original integral equation, the resultant is no longer singular. A similar procedure has been developed for the MFS.

In yet another approach, the singular boundary method (SBM) [175, 176, 317], extra terms were added to the summation of fundamental solutions to represent the diagonal elements. These terms, called the "origin intensity factors", were obtained using an inverse interpolation technique, and are nonsingular. These methods have been successfully applied to solve potential [354], Helmholtz and modified Helmholtz [178, 534], heat conduction [353], convective diffusion [887], Stokes flow [738], water wave [535], isotropic and anisotropic elasticity [352, 355], acoustic wave [316, 536], seismic wave [571, 828], viscoelastic wave [821], and piezoelectricity problems [169].

Chapter 2

Fundamental Solutions

2.1 Dirac Delta Function

We shall begin by a discussion on the Dirac delta function. As shown in (1.7), the symbol δ and its operations were first introduced by Paul Dirac [260], originally without rigorous mathematical support. Its theoretical foundation as a generalized function was provided by the theory of distributions of Schwartz [787]. In the following presentation, we focus on its mathematical operations only.

A one-dimensional Dirac delta function is defined as a function that has the value

$$\delta(x, x') = \begin{cases} +\infty, & x = x', \\ 0, & \text{elsewhere.} \end{cases} \tag{2.1}$$

It is also constrained to satisfy

$$\int_a^b \delta(x, x')\, dx = \begin{cases} 1, & \text{if } a < x' < b, \\ 0, & \text{if } x' \notin [a, b], \end{cases} \tag{2.2}$$

and it is obvious that

$$\delta(x, x') = \delta(x', x). \tag{2.3}$$

Sometimes a different notation is used

$$\delta(x, x') \equiv \delta(x - x'), \quad \text{and} \quad \delta(x, 0) \equiv \delta(x). \tag{2.4}$$

The delta function picks up the value of a function at the singularity point as

$$f(x)\delta(x, x') = f(x')\delta(x, x'), \tag{2.5}$$

and by the integration

$$\int_a^b f(x)\delta(x, x')\, dx = f(x'), \quad a < x' < b, \tag{2.6}$$

assuming that $f(x)$ is continuous at x'. It can be scaled as

$$\delta(ct) = \frac{\delta(t)}{|c|}, \tag{2.7}$$

and it is related to the Heaviside unit step function, defined as

$$H(x, x') = \begin{cases} 0, & x' < x, \\ 1, & x' > x, \end{cases} \tag{2.8}$$

as its derivative

$$\frac{d\,H(x, x')}{dx} = \delta(x, x'). \tag{2.9}$$

The above can be integrated to give

$$\int_{-\infty}^x \delta(x, x')\, dx = H(x, x'). \tag{2.10}$$

We can distribute the singularity along the line by integrating with respect to x' to obtain the top hat function

$$\int_a^b \delta(x, x')\, dx' = H(x, a) - H(x, b), \quad b > a. \tag{2.11}$$

The Dirac delta function definition can be extended to 2D and 3D as

$$\delta(\boldsymbol{x}, \boldsymbol{x}') = \begin{cases} \delta(x, x')\delta(y, y'), & \text{(2D)}, \\ \delta(x, x')\delta(y, y')\delta(z, z'), & \text{(3D)}. \end{cases} \tag{2.12}$$

We can also formally consider that the delta function has derivatives. For example, when the solution of

$$\nabla^2 G^L(\boldsymbol{x}, \boldsymbol{x}') = \delta(\boldsymbol{x}, \boldsymbol{x}'), \tag{2.13}$$

is

$$G^L = -\frac{1}{4\pi r},\tag{2.14}$$

where $r = \|x - x'\|$, then the solution of

$$\nabla^2 G_x^L(x, x') = \frac{\partial\delta(x, x')}{\partial x},\tag{2.15}$$

is

$$G_x^L = \frac{\partial G^L}{\partial x} = -\frac{1}{4\pi}\frac{\partial(1/r)}{\partial x} = \frac{x - x'}{4\pi r^3}.\tag{2.16}$$

If the delta function represents a point electrical charge or a fluid mass source, then its derivative is called a dipole or a doublet, respectively. We may also distribute the point source over a line, an area, or a volume, with constant or variable intensity, by integration. For example, a line source of unit strength and length $2a$ on the x-axis, centered at the origin, can be obtained based on (2.11), as

$$\begin{aligned}
G_l^L &= -\frac{1}{4\pi}\int_{-a}^{a}\frac{1}{\sqrt{(x - x')^2 + y^2 + z^2}}\,dx' \\
&= \frac{1}{4\pi}\ln\left[\frac{x - a + \sqrt{(x - a)^2 + y^2 + z^2}}{x + a + \sqrt{(x + a)^2 + y^2 + z^2}}\right].
\end{aligned}\tag{2.17}$$

Other types of distributed sources can be obtained by similar integration.

2.2 Existence of Fundamental Solutions

With the properties and operations of the delta function clarified, we can define the fundamental solution as a particular solution of the equation

$$\mathcal{L}G(x, x') = \delta(x, x'),\tag{2.18}$$

where $\mathcal{L}$ is a linear partial differential operator. Based on the Malgrange-Ehrenpreis theorem [278, 279, 609], every linear partial differential operator with constant coefficients admits a fundamental solution. This means that the PDE

$$P\left(\frac{\partial}{\partial x_1}, \frac{\partial}{\partial x_2}, \ldots, \frac{\partial}{\partial x_n}\right)G(x, x') = \delta(x, x'), \quad x \in \mathbb{R}^n,\tag{2.19}$$

in which P is a polynomial in n variables, has a distributional solution G. The existence of G also guarantees the existence of a particular solution for the inhomogeneous equation

$$P\left(\frac{\partial}{\partial x_1}, \frac{\partial}{\partial x_2}, \ldots, \frac{\partial}{\partial x_n}\right)\phi(x) = f(x), \qquad (2.20)$$

in which $f(x)$ is a compactly supported distribution. The same statement, however, cannot be made when the coefficients are variable. Particularly, Lewy's example [531] showed that the Malgrange–Ehrenpreis theorem cannot be extended to linear PDEs with polynomial coefficients. We also note that the existence of a fundamental solution for the constant coefficient case does not necessarily mean that a closed form solution can be found, though an algebraic solution can always be found in the integral transform space.

2.3 Laplace Equation

The Laplace equation is

$$\nabla^2 \phi = 0, \qquad (2.21)$$

where the operator in d-dimensions is

$$\mathcal{L} \equiv \nabla^2 \equiv \Delta = \frac{\partial^2}{\partial x_1^2} + \frac{\partial^2}{\partial x_2^2} + \cdots + \frac{\partial^2}{\partial x_d^2}. \qquad (2.22)$$

In this book, we shall mix the use of the two notations, ∇^2 and Δ, whichever is more convenient.

It was mentioned in Section 1.2 that the fundamental solution $1/r$ for the 3D case was attributed to Lagrange. It was Carl Neumann [673] who embraced Green's idea and presented the fundamental solution in 2D as the logarithmic function [276]. In summary, the fundamental solutions of (2.13) are

$$G^L(x, x') = \frac{1}{2\pi}\ln r, \qquad \text{(2D)}, \qquad (2.23a)$$

$$= -\frac{1}{4\pi}\frac{1}{r}, \qquad \text{(3D)}, \qquad (2.23b)$$

$$= -\frac{1}{d(d-2)\alpha_d}\frac{1}{r^{d-2}}, \qquad (d\text{D}, \ d \ge 3). \qquad (2.23c)$$

In the above, the coefficient α_d is the volume of a d-dimensional unit ball, given by

$$\alpha_d = \frac{\pi^{d/2}}{\Gamma\left(\frac{d}{2} + 1\right)}, \tag{2.24}$$

where Γ is the Gamma function. We note that $G^L(x, x')$ in the above can also be expressed as $G^L(x - x')$, because the function is dependent on the Euclidean distance $\|x - x'\|$ only.

To prove that the expressions in (2.23) are fundamental solutions, we can substitute them into the LHS of (2.13) to show that they indeed vanish everywhere, except at $x = x'$, where a derivative cannot be taken due to the singularity. To demonstrate that the solutions produce a Dirac delta function on the RHS, we can observe the property

$$\int_\Omega \delta(x, x')\, dx = 1, \tag{2.25}$$

where Ω is an arbitrary domain containing x'. Take, for example, the 3D case with x' located at the origin. We can integrate (2.23b) over a sphere centered at the origin with radius a and apply the divergence theorem to obtain

$$\int_\Omega \nabla \cdot \nabla \left(-\frac{1}{4\pi r}\right) dx = \int_A \frac{\partial}{\partial r}\left(-\frac{1}{4\pi r}\right)\Big|_{r=a} dx$$
$$= \int_0^\pi 2\pi a^2 \sin\theta \frac{1}{4\pi a^2}\, d\theta = \int_0^\pi \frac{\sin\theta}{2}\, d\theta = 1. \tag{2.26}$$

where A is the surface of the sphere, and θ is the zenith angle made with the z-axis. The corresponding 2D case can be similarly proven.

2.4 Helmholtz-Type Equations

2.4.1 *Helmholtz equation*

The Helmholtz equation was first introduced by Helmholtz [375] in his study of acoustic problems. It takes the form

$$\nabla^2 \phi + k^2 \phi = 0, \tag{2.27}$$

in which k^2 may be viewed as an eigenvalue of the Laplacian operator. It is a result of the wave equation solved in the frequency domain, or by separation of variables. It is also related to the quantum mechanics wave equations, such as the Schrödinger equation, and the Klein–Gordon equation.

For example, the acoustic wave equation is

$$\nabla^2 p - \frac{1}{c}\frac{\partial^2 p}{\partial t^2} = 0, \tag{2.28}$$

in which p is the acoustic pressure, and c the wave speed. We can subject the above equation to a Fourier transform in time, or assume that the loading is sinusoidal and the solution has reached a steady state with a time factor as

$$p(\boldsymbol{x}, t) = \Re\left\{\tilde{p}(\boldsymbol{x})e^{\mathrm{i}\omega t}\right\}, \tag{2.29}$$

where $\mathrm{i} = \sqrt{-1}$ is the imaginary unit, and ω is the frequency. The above expression allows (2.28) to be written as

$$\nabla^2\tilde{p} + k^2\tilde{p} = 0, \tag{2.30}$$

where $k = \omega/c$ is the wave number.

The fundamental solution of the Helmholtz operator in (2.27) in dD is

$$G^H(\boldsymbol{x}, \boldsymbol{x}') = \frac{1}{4}\left(\frac{k}{2\pi r}\right)^{\frac{d}{2}-1} \mathrm{Y}_{\frac{d}{2}-1}(kr), \tag{2.31}$$

where Y_ν is the Bessel function of the second kind of order ν. In 2D and 3D, it becomes, respectively,

$$G^H(\boldsymbol{x}, \boldsymbol{x}') = \frac{1}{4}\mathrm{Y}_0(kr), \tag{2D} \tag{2.32a}$$

$$= \frac{1}{4}\sqrt{\frac{k}{2\pi r}}\,\mathrm{Y}_{\frac{1}{2}}(kr) = -\frac{1}{4\pi r}\cos(kr), \quad \text{(3D)}. \tag{2.32b}$$

The Helmholtz equation is generally associated with wave phenomena, such as acoustic, seismic, and electromagnetic waves. The time variable in the wave equation is removed by solving it in the frequency domain, or through separation of variables.

2.4.2 *Modified Helmholtz equation*

A variant of the Helmholtz equation is the modified Helmholtz equation

$$\nabla^2\phi - k^2\phi = 0. \tag{2.33}$$

The fundamental solution of the modified Helmholtz equation (2.33) in dD is

$$G^M(x, x') = -\frac{1}{2\pi}\left(\frac{k}{2\pi r}\right)^{\frac{d}{2}-1} K_{\frac{d}{2}-1}(kr), \tag{2.34}$$

where K_ν is the modified Bessel function of the second kind of order ν. In 2D and 3D, it becomes, respectively,

$$G^M(x, x') = -\frac{K_0(kr)}{2\pi}, \qquad (2D), \tag{2.35a}$$

$$= -\frac{1}{2\pi}\sqrt{\frac{k}{2\pi r}}\, K_{\frac{1}{2}}(kr) = -\frac{1}{4\pi r}e^{-kr}, \quad (3D). \tag{2.35b}$$

It is easy to show that the solutions presented above and in Section 2.4.1, respectively, satisfy the modified Helmholtz and the Helmholtz equations when $r \neq 0$, by a simple substitution. It takes additional effort to show that they generate the Dirac delta function on the RHS at $r = 0$. As the delta function is generated by the singular part of the solution, we can take the limit of these functions when $r \to 0$ in the 2D case as

$$Y_0(kr) \sim \frac{2}{\pi}\ln r, \tag{2.36}$$

$$K_0(kr) \sim -\ln r. \tag{2.37}$$

Comparing the above with the fundamental solution of the Laplacian operator, we confirm that a delta function is generated. We can show the same for the 3D case.

As an illustration, we plot the 2D fundamental solutions (2.32a) and (2.35a) versus r, alongside that of the Laplace equation (2.23a), in Fig. 2.1. In addition to the asymptotic behaviors, we observe that (2.32a) is a wavy function with multiple zeros. The same is true for the 3D solution (2.32b). The existence of zeros means that the solution of the Helmholtz equation using an integral equation can be prone to nonuniqueness, when encountering the eigenvalues of the system. This issue is more serious than that of the 2D Laplace equation, in which only one zero exists. The Helmholtz equation is typically solved with a fixed geometry and a range of wave numbers k, which effectively scale the geometry to different sizes. It is likely that the range can cover some degenerate scales, where the matrix becomes singular. Indeed, it was first reported by Copley [218, 219], and is now well

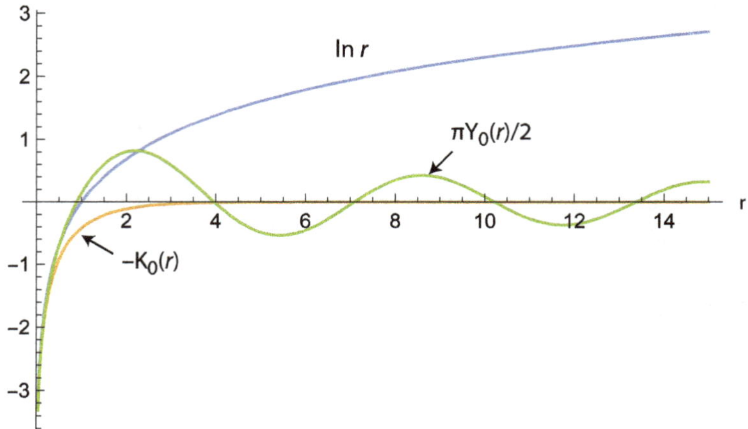

Figure 2.1. Plot of 2D Laplace and Helmholtz-type fundamental solutions.

known, that in the integral equation solution of the Helmholtz equation, there exist "spurious frequencies" where artificial resonances are observed. The same phenomenon can manifest itself in the MFS. There have been many proposed remedies, such as the Burton–Miller formulation [115], and the CHIEF method [782, 789]. Further discussion on this phenomenon will be provided in Section 6.5.

2.4.3 *Complex variable form*

In a wave problem, energy dissipation may exist, causing the damping of wave amplitude. For an elastic wave, we may write

$$c^2 \nabla^2 \phi - \frac{\partial^2 \phi}{\partial t^2} - 2D \frac{\partial \phi}{\partial t} - P\phi = 0, \tag{2.38}$$

where D is a viscous damping coefficient, and P a material damping coefficient. In the frequency domain, we observe that the wave number k in (2.27) is

$$k = \frac{1}{c}\sqrt{\omega^2 - P - 2i\omega D}, \tag{2.39}$$

which is a complex number. The solution ϕ is also complex.

Even with k a complex number, the fundamental solutions presented in Sections 2.4.1 and 2.4.2 remain valid. In this regard, we observe that the modified Helmholtz equation (2.33) can be obtained from the Helmholtz equation (2.27) by the substitution of k by ik. For the fundamental solutions, we can do the same. In the 2D case, we notice the relation

$$Y_0(iz) = -\frac{2}{\pi}K_0(z) + iI_0(z), \tag{2.40}$$

where z is a complex variable, and I_0 is the modified Bessel function of the first kind of order zero. Comparing the above with the two fundamental solutions (2.32a) and (2.35a), we observe that the equivalence is not complete, as they differ by a term $iI_0(kr)$, which is a general solution of the modified Helmholtz equation. We recognize, however, that fundamental solutions are particular solutions and are not unique. They can incorporate any general solution and still serve the same purpose for the MFS.

For consistent results, we may seek fundamental solutions of the Helmholtz equation (2.27) in the complex domain as

$$G^H(\boldsymbol{x}, \boldsymbol{x}') = -\frac{i}{4}H_0^{(1)}(kr), \qquad \text{(2D)}, \tag{2.41a}$$

$$= -\frac{1}{4\pi r}e^{ikr}, \qquad \text{(3D)}, \tag{2.41b}$$

$$= -\frac{i}{4}\left(\frac{k}{2\pi r}\right)^{\frac{d}{2}-1} H_{\frac{d}{2}-1}(kr), \quad \text{(}d\text{D)}, \tag{2.41c}$$

where

$$H_\nu^{(1)} = J_\nu + iY_\nu, \tag{2.42}$$

is the Hankel function of the first kind of order ν, J_ν is the Bessel function of the first kind of order ν, and k is a complex number. For fundamental solutions of the modified Helmholtz equation, we substitute of k by ik.

2.5 Diffusion Equation

The heat diffusion equation is given as follows:

$$\nabla^2 T - \frac{1}{\alpha_T}\frac{\partial T}{\partial t} = 0, \tag{2.43}$$

where T is the temperature, and α_T the thermal diffusivity, and it is subjected to both an initial condition (IC) and BCs. We note that the time dimension in PDEs, as well as in physics, is treated differently from the spatial dimensions. The concept of the "arrow of time" states that any disturbance introduced in time can only travel forward, and not backward. It is different from the spatial disturbances, such as a change in the BC, which can affect the entire domain. We will hence discuss several different time treatments.

Laplace Transform: We can apply the Laplace transform to (2.43) and obtain

$$\nabla^2 \tilde{T} - \frac{s}{\alpha_T} \tilde{T} = -\frac{T_o}{\alpha_T}, \tag{2.44}$$

where the tilde overbar denotes the Laplace transformed variable, s is the Laplace transform parameter, and T_o is the IC. As observed from (2.33), the above is an inhomogeneous modified Helmholtz equation. Its homogeneous solution can be found by utilizing the fundamental solutions introduced in Section 2.4.2. The solution in time can be obtained by performing a numerical inverse Laplace transform [136, 337].

Finite Difference Time Stepping: Another way to solve the diffusion equation is through the finite difference time stepping method. Assuming that the solution is known at the time step $n - 1$, we seek the solution at time step n by utilizing the weighted average scheme:

$$\frac{1}{\alpha_T} \frac{T^n - T^{n-1}}{\Delta t} = \theta \nabla^2 T^n + (1 - \theta) \nabla^2 T^{n-1}, \tag{2.45}$$

where T^{n-1} and T^n represent the solution at the respective time step, $\Delta t = t^n - t^{n-1}$, and $0 \leq \theta \leq 1$ is the weighting parameter. If we define

$$\phi^n = T^n + \frac{1 - \theta}{\theta} T^{n-1}, \tag{2.46}$$

we can transform (2.45) into

$$\nabla^2 \phi^n - k^2 \phi^n = -\frac{k^2}{\theta} T^{n-1}, \tag{2.47}$$

in which

$$k^2 = \frac{1}{\alpha_T \theta \Delta t}. \tag{2.48}$$

Equation (2.47) is an inhomogeneous modified Helmholtz equation, and the MFS can be employed to solve such problems [338, 406, 714].

Time Domain Fundamental Solution: The time domain fundamental solution for the diffusion equation satisfies

$$\nabla^2 G^D - \frac{1}{D}\frac{\partial G^D}{\partial t} = \delta(\boldsymbol{x} - \boldsymbol{x}')\delta(t - t'), \tag{2.49}$$

and is given by

$$G^D(\boldsymbol{x} - \boldsymbol{x}', t - t') = -\frac{D}{(\sqrt{\pi}\,r)^d}\xi^d e^{-\xi^2}\mathrm{H}(t - t'), \tag{2.50}$$

where

$$\xi = \frac{r}{2\sqrt{D(t - t')}}, \tag{2.51}$$

and $d \in \mathbb{N}$ is the number of dimensions. We notice in the fundamental solutions (2.50) that the time arrow is guaranteed by the arguments in the Heaviside unit step functions.

In the numerical solution of PDEs, the time dimension is either eliminated through the integral transform, or separately discretized from the spatial dimensions, as demonstrated in (2.45). While the spatially discretized unknowns at a given time are solved simultaneously, the unknowns in the time dimension are solved one step at a time. For the MFS, this means that the non-transient fundamental solutions are utilized. However, with the time-dependent fundamental solutions given above, it is of interest to explore a space–time scheme that treats time as if it were an extra spatial dimension.

For an easy graphic representation, we use a 1D diffusion equation as an example, which is defined within $0 \leq x \leq L$. The IC is given at $t = 0$ for all x, and BCs at $x = 0$ and L for all $t > 0$. We may distribute the sources surrounding the space-time boundary as shown in Fig. 2.2, which may be compared to Fig. 1.4 in spatial dimensions only. We notice that the top boundary is left open without sources, because sources do not travel back in time.

Assuming that there are N_d sources distributed below the base of the space–time domain, at $t = -\gamma\,\Delta t$, with $0 < \gamma < 1$, N_b sources surrounding the spatial boundary at a given time (for 1D problems, $N_b = 2$), and N_t sources in the time dimension (see Fig. 2.2), we can write an approximation formula as

$$\hat{T}(\boldsymbol{x}, t) = \sum_{i=1}^{N_d} c_i G^D(r_i, t + \gamma\,\Delta t) + \sum_{j=1}^{N_t}\sum_{i=1}^{N_b} d_{ij} G^D(r_i', t - (j - \gamma)\Delta t), \tag{2.52}$$

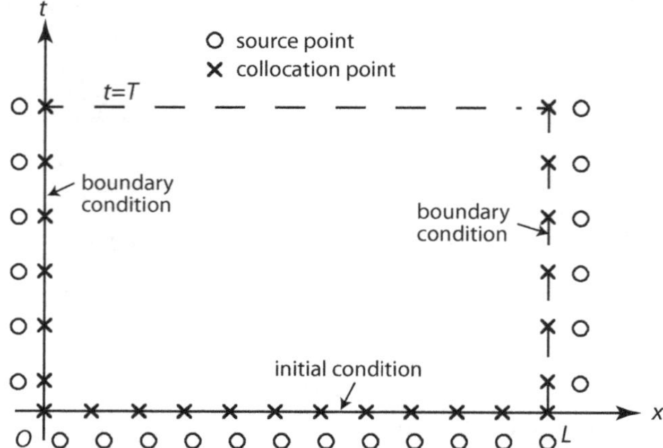

Figure 2.2. The MFS scheme that treats time similar to a space dimension for a 1D diffusion equation.

where $\hat{T}$ is the approximate solution, $r_i = \|x - x_i\|$, $r_i' = \|x - x_i'\|$, with x_i the source locations at the base, and x_i' the surrounding source locations, and c_i and d_{ij} are the undetermined coefficients. In the above we have used constant time spacings for the sources for convenience, which is not required. The above equation has $N_d + N_b \times N_t$ degrees of freedom. To form a solution system, we can collocate (2.52) at N_d nodes for the ICs at $t = 0$, and $N_b \times N_t$ nodes on the boundary and in time with a Δt increment. Once the c_i and d_{ij} are determined, the approximate solution is defined in the whole domain for $0 \le t \le N_t \times \Delta t$.

We note that (2.52) is constructed in an intuitive fashion without theoretical support. There have been several attempts to use the above methodology. Young *et al.* [945] solved 1D, 2D, and 3D diffusion problems. For the solution at $t = (n + 1)\Delta t$, the time dependent sources were placed in the domain at $t = (n - \gamma)\Delta t$, and outside the boundary at two time levels, $t = (n - \gamma)\Delta t$ and $(n + \gamma)\Delta t$. Time stepping was used to progress in time. Through several examples, they showed that the accuracy of the space-time scheme is better than that of the finite difference time-stepping scheme using stationary fundamental solutions. Chantasiriwan [130] solved these problems at multiple time levels simultaneously, based on (2.52). By further considering solution sensitivity to data perturbation, it was found that the modified Helmholtz equation based time-stepping scheme was more accurate. Hon and Wei [394, 395] and Mera [638] used (2.52) to solve inverse heat conduction problems. The sources, however, were distributed only below the base at a fixed negative time, and not around the boundary in the time direction. Johansson and his coworkers proposed a scheme to solve direct [423, 424] and inverse [425] problems, in which the sources were distributed outside the boundary along the time dimension, without using the domain sources at a negative time. They argued that the distribution of sources in the time dimension was required for the denseness of the solution.

Inspired by the work of Kupradze [505], in which a denseness result was proven if the source points are located outside the space domain only, based on the boundary integral equation formulation, it seems that we may extend such a result to the MFS. As an illustration, we introduce the boundary integral equation for the diffusion equation [655]

$$cT(\boldsymbol{x}, t) = \int_0^{t^+} \int_{\partial\Omega} \left[G^D(\boldsymbol{x} - \boldsymbol{x}', t - t') \frac{\partial T(\boldsymbol{x}')}{\partial n(\boldsymbol{x}')} - T(\boldsymbol{x}') \frac{\partial G^D(\boldsymbol{x} - \boldsymbol{x}', t - t')}{\partial n(\boldsymbol{x}')} \right] d\boldsymbol{x}' \, dt'. \qquad (2.53)$$

In the above equation, we ignored the IC and the inhomogeneous RHS, which require domain integration, but can be handled by the MPS, to be discussed in Section 4.1. Hence only a boundary integral is involved. As discussed in Section 1.3.2, discretizing the integral equation is equivalent to distributing discrete sources. Equation (2.53) is equivalent to distributing sources on or around the boundary in the time direction. Sources over the domain at a negative time are not needed, and may not be linearly independent. The linear independence and denseness of the MFS scheme by distributing fundamental solutions in the exterior of the domain and in the time dimension has been proven by Johansson [427, 428].

2.6 Wave Equation

For the wave equation (2.28), the fundamental solution is a particular solution of the following:

$$\nabla^2 G^W - \frac{1}{c^2} \frac{\partial^2 G^W}{\partial t^2} = \delta(\boldsymbol{x} - \boldsymbol{x}')\delta(t - t'). \qquad (2.54)$$

The fundamental solutions are [655]

$$G^W(x-x', t-t') = -\frac{c}{2}\mathrm{H}\left[(t-t') - \frac{|x-x'|}{c}\right], \qquad (1D),$$

$$(2.55a)$$

$$= -\frac{1}{2\pi\sqrt{(t-t')^2 - r^2/c^2}}\mathrm{H}\left[(t-t') - \frac{r}{c}\right], \quad (2D),$$

$$(2.55b)$$

$$= -\frac{1}{4\pi r}\delta\left[(t-t') - \frac{r}{c}\right], \qquad (3D).$$

$$(2.55c)$$

Wave problems are typically solved in the frequency domain, which transforms the time variable into a frequency [19, 286, 947]. The governing equation then becomes the Helmholtz equation, as described in Section 2.4.1. The MFS for the Helmholtz equation can then be utilized. The time domain responses can be obtained via the inverse Fourier transform, and typically by the fast Fourier transform (FFT).

Wave problems can also be solved in the time domain using a time stepping scheme [569, 953]. For example, the second time derivative in (2.28) can be approximated by the four level finite difference formula of Houbolt [399]

$$\frac{\partial^2 p^{n+1}}{\partial t^2} \approx \frac{1}{(\Delta t)^2}\left(2p^{n+1} - 5p^n + 4p^{n-1} - p^{n-2}\right). \qquad (2.56)$$

The spatial derivatives can be approximated as a weighted average at multiple time levels as in (2.45). Alternatively, in a fully implicit scheme, we can approximate (2.28) by

$$\nabla^2 p^{n+1} - \frac{2}{c(\Delta t)^2}p^{n+1} = \frac{1}{c(\Delta t)^2}\left(-5p^n + 4p^{n-1} - p^{n-2}\right). \qquad (2.57)$$

The above is an inhomogeneous modified Helmholtz equation, and can be solved by combining the MPS and the MFS.

2.7 Advection–Diffusion Equation

An advection–diffusion equation for a certain mass concentration c is given by

$$\frac{\partial c}{\partial t} = \nabla \cdot (D\nabla c) - \nabla \cdot (Vc), \tag{2.58}$$

where D is the diffusion coefficient, and V the advection velocity. Here, we only deal with cases with a constant diffusion coefficient and velocity. Fundamental solutions for variable coefficients are difficult, and some special cases are treated in Section 2.15.

Steady State: Under steady state and with constant coefficients, we can express (2.58) in the standard form

$$\nabla^2 c - 2\alpha \frac{\partial c}{\partial x} = 0, \tag{2.59}$$

in which $\alpha = V/2D$, and $V = |V|$. In the above equation, we have chosen the x-axis to coincide with the flow direction.

Equation (2.59) can also be the result of a diffusion equation in a moving coordinate system. Given a diffusion equation such as (2.43), if the loading of the system is moving at a constant speed V in the negative x-direction, and the domain is infinite, semi-infinite, or layered parallel to the x-axis, the field becomes steady at large time by an observer moving at the same speed. We therefore define a moving coordinate as

$$x' = x + Vt, \tag{2.60}$$

and express

$$\frac{\partial}{\partial t} = V \frac{\partial}{\partial x'}. \tag{2.61}$$

Equation (2.43) is then transformed into

$$\nabla'^2 T - \frac{V}{\alpha_T} \frac{\partial T}{\partial x'} = 0, \tag{2.62}$$

where ∇'^2 is the Laplacian in the transformed coordinate system. The above equation can also be expressed in the standard form (2.59).

The fundamental solutions of

$$\nabla^2 G^c - 2\alpha \frac{\partial G^c}{\partial x} = \delta(\boldsymbol{x} - \boldsymbol{x}'), \qquad (2.63)$$

are the following [185]

$$G^c = -\frac{1}{2\pi} \exp\left(\alpha x\right) \mathrm{K}_0\left(|\alpha|r\right), \qquad \text{(2D)}, \qquad (2.64\text{a})$$

$$= -\frac{1}{4\pi r} \exp\left(\alpha x\right) \exp\left(-|\alpha|r\right), \quad \text{(3D)}. \qquad (2.64\text{b})$$

When the velocity is not aligned with the x-axis, the fundamental solutions of

$$D\nabla^2 G^c - \nabla \cdot (\boldsymbol{V} G^c) = \delta(\boldsymbol{x} - \boldsymbol{x}'), \qquad (2.65)$$

are

$$G^c = -\frac{1}{2\pi D} \exp\left(\frac{\boldsymbol{V} \cdot \boldsymbol{r}}{2D}\right) \mathrm{K}_0\left(\frac{Vr}{2D}\right), \qquad \text{(2D)}, \qquad (2.66\text{a})$$

$$= -\frac{1}{4\pi D r} \exp\left(\frac{\boldsymbol{V} \cdot \boldsymbol{r}}{2D}\right) \exp\left(-\frac{Vr}{2D}\right), \quad \text{(3D)}, \qquad (2.66\text{b})$$

in which $\boldsymbol{r} = \boldsymbol{x} - \boldsymbol{x}'$.

Unsteady State: When the time derivative term in (2.58) exists, we can treat it similar to that of the diffusion equation case in Section 2.5, by expressing it into a two time level finite difference approximation, or by performing the Laplace transform. For the Laplace transform case, we can write (2.58) in a standard form, assuming that the velocity is aligned with the x-axis,

$$\nabla^2 \tilde{c} - 2\alpha \frac{\partial \tilde{c}}{\partial x} - \beta \tilde{c} = f(\boldsymbol{x}), \qquad (2.67)$$

where α can be positive or negative, β is positive, and the RHS $f(\boldsymbol{x})$ corresponds to the IC, or the solution of the previous time step for the finite difference case, which can be separately treated by the MPS.

To determine the fundamental solution, we seek a particular solution of

$$\nabla^2 G^c - 2\alpha \frac{\partial G^c}{\partial x} - \beta G^c = \delta(\boldsymbol{x} - \boldsymbol{x}'). \qquad (2.68)$$

In the following, we give a brief demonstration of its derivation, as it is not normally found in the literature. For 3D problems, we perform a triple

Fourier transform to (2.68) with respect to (x, y, z), with transform parameters (ξ, η, μ), to obtain

$$\overline{\overline{\overline{G}}} = -\frac{1}{\eta^2 + \mu^2 + \xi^2 - 2i\alpha\xi + \beta}. \tag{2.69}$$

The inverse transform with respect to η produces [683]

$$\overline{\overline{G}} = -\frac{\exp\left(-\sqrt{\mu^2 + \xi^2 - 2i\alpha\xi + \beta}\,|y|\right)}{2\sqrt{\mu^2 + \xi^2 - 2i\alpha\xi + \beta}}, \tag{2.70}$$

while another inverse transform with respect to μ gives

$$\overline{G} = -\frac{1}{2\pi}\mathrm{K}_0\left(\sqrt{y^2 + z^2}\sqrt{\xi^2 - 2i\alpha\xi + \beta}\right). \tag{2.71}$$

If we define

$$\zeta = \xi - i\alpha, \tag{2.72}$$

expression (2.71) then becomes

$$\overline{G} = -\frac{1}{2\pi}\mathrm{K}_0\left(\sqrt{y^2 + z^2}\sqrt{\zeta^2 + \alpha^2 + \beta}\right). \tag{2.73}$$

Performing the final inverse Fourier transform with respect to ζ, and applying the rule of translation of variable, we obtain

$$G^c = -\frac{1}{4\pi r}\exp\left(\alpha x\right)\exp\left(-\sqrt{\alpha^2 + \beta}\,r\right), \qquad \text{(3D)}. \tag{2.74}$$

Similarly, we can obtain for the 2D case

$$G^c = -\frac{1}{2\pi}\exp\left(\alpha x\right)\mathrm{K}_0\left(\sqrt{\alpha^2 + \beta}\,r\right), \qquad \text{(2D)}. \tag{2.75}$$

When $\boldsymbol{V}$ is not aligned with the x-axis, we seek the solution of

$$D\nabla^2 G^c - \nabla \cdot (\boldsymbol{V}G^c) - sG^c = \delta(\boldsymbol{x} - \boldsymbol{x}'), \tag{2.76}$$

and we obtain

$$G^c = -\frac{1}{2\pi D}\exp\left(\frac{\boldsymbol{V} \cdot \boldsymbol{r}}{2D}\right)\mathrm{K}_0\left(\sqrt{\frac{V^2}{4D^2} + \frac{s}{D}}\,r\right), \qquad \text{(2D)}, \tag{2.77a}$$

$$= -\frac{1}{4\pi D r}\exp\left(\frac{\boldsymbol{V} \cdot \boldsymbol{r}}{2D}\right)\exp\left(-\sqrt{\frac{V^2}{4D^2} + \frac{s}{D}}\,r\right), \qquad \text{(3D)}. \tag{2.77b}$$

2.8   Elastostatics (Cauchy–Navier Equations)

2.8.1   *Point force solution*

The inhomogeneous Cauchy–Navier equations of linear elasticity are

$$\mu \nabla^2 \boldsymbol{u} + \frac{\mu}{1 - 2v} \nabla \nabla \cdot \boldsymbol{u} = -\boldsymbol{F}, \tag{2.78}$$

where $\boldsymbol{u}$ is the displacement vector, μ the shear modulus, v the Poisson ratio, and $\boldsymbol{F}$ is the body force. We notice that $\boldsymbol{F}$ is a vector; hence it creates multiple fundamental solutions, one for each vector component. As mentioned earlier, fundamental solutions often have a correspondence with physical problems.

The physical solution of a point force located at the origin and in the k-direction is given by the following substitution:

$$F_{ik} = \delta_{ik}\,\delta(\boldsymbol{x}). \tag{2.79}$$

The solution of (2.78) with (2.79) then gives

$$u_{ik}^F = \frac{1}{8\pi\mu(1-v)}\left[\frac{x_i x_k}{r^2} - (3 - 4v)\delta_{ik}\ln r\right], \quad (2\text{D}), \tag{2.80a}$$

$$= \frac{1}{16\pi\mu(1-v)}\frac{1}{r}\left[\frac{x_i x_k}{r^2} + (3 - 4v)\delta_{ik}\right], \quad (3\text{D}), \tag{2.80b}$$

where the first subscript of u_{ik}^F indicates the vector component of the displacement, and the second corresponds to the direction of the point force. For simplicity of notation, we have placed the singularity at the origin, $\boldsymbol{x}' = \boldsymbol{0}$. For $\boldsymbol{x}' \neq \boldsymbol{0}$, we need to modify $r = \sqrt{x^2 + y^2 + z^2}$ to $r = \|\boldsymbol{x} - \boldsymbol{x}'\|$, and x_i to $x_i - x_i'$. The 3-D solution was presented by William Thompson (Lord Kelvin) [834], and is known as the Kelvin solution. Although Thompson did not shed light on its derivation, there have been many attempts to derive it [299]. In the following, we shall give a heuristic derivation for both the 2D and 3D cases.

There are a number of ways that a displacement vector can be decomposed into scalar and vector potentials, which satisfy (2.78). These include the Galerkin, Boussinesq, and Papkovich–Neuber functions [99, 811]. We shall utilize the Papkovich–Neuber functions here. Mindlin [647] gave the following result:

$$u_i = (\psi + x_j \Psi_j)_{,i} - 4(1 - v)\Psi_i, \tag{2.81}$$

in which ψ and Ψ_i are Papkovich–Neuber functions satisfying

$$\nabla^2 \Psi_i = \frac{1}{4\mu(1-v)} F_i, \tag{2.82}$$

$$\nabla^2 \psi = -\frac{1}{4\mu(1-v)} x_i F_i. \tag{2.83}$$

The above is the complete solution of (2.78) [645]. We can substitute (2.79) into (2.82) and (2.83), and obtain

$$\nabla^2 \Psi_{ik} = \frac{1}{4\mu(1-v)} \delta_{ik} \delta(\boldsymbol{x}), \tag{2.84}$$

$$\nabla^2 \psi_k = 0, \tag{2.85}$$

and in (2.85) we have used the property of the delta function that $x_i \delta(\boldsymbol{x}, 0) = 0$. As the fundamental solution is a particular solution, we can simply set

$$\psi_k = 0. \tag{2.86}$$

Following (2.23), the solution of (2.84) is

$$\Psi_{ik} = \frac{1}{8\pi\mu(1-v)} \delta_{ik} \ln r, \quad (2\text{D}), \tag{2.87a}$$

$$= -\frac{1}{16\pi\mu(1-v)} \delta_{ik} \frac{1}{r}, \quad (3\text{D}). \tag{2.87b}$$

Substituting the above into (2.81), we find the point force solutions given in (2.80).

To find the stresses, we utilize the constitutive equations

$$\sigma_{ij} = 2\mu e_{ij} + \frac{2\mu v}{1-2v} \delta_{ij} e, \tag{2.88}$$

where $e = e_{ii}$, and $e_{ij} = (u_{i,j} + u_{j,i})/2$ is the strain tensor. The stresses due to the point forces are

$$\sigma_{ijk}^F = \frac{1}{4\pi(1-v)} \frac{1}{r} \left[(1-2v)\left(\delta_{ij} r_{,k} - \delta_{jk} r_{,i} - \delta_{ik} r_{,j}\right) - 2r_{,i} r_{,j} r_{,k} \right], \quad (2\text{D}), \tag{2.89a}$$

$$= \frac{1}{8\pi(1-v)} \frac{1}{r^2} \left[(1-2v)\left(\delta_{ij} r_{,k} - \delta_{jk} r_{,i} - \delta_{ik} r_{,j}\right) - 3r_{,i} r_{,j} r_{,k} \right], \quad (3\text{D}), \tag{2.89b}$$

where $r_{,i} = \partial r / \partial x_i = x_i / r$.

2.8.2 *Double force, quadrupole, and hexapole*

By making use of the above results, we can derive other solutions, such as a double force, which is created by aligning two forces of equal but opposite magnitude on an axis, and bringing them together. To avoid exact cancellation, their magnitudes are increased inversely proportionally to the distance between them. This limiting process is equivalent to taking the derivative of the point force solution along the axis of its application. This concept is equivalent to the dipole in potential theory. A fundamental solution of the double force is then

$$u_{ik}^b = \frac{\partial u_{ik}^F}{\partial x_k},$$ (2.90)

in which the second subscript indicates the axis of its application.

We can bring two and three pairs of the double forces, orthogonal to each other, together at the origin, respectively for 2D and 3D problems. We then get the quadrupole and hexapole solutions [202]. This is equivalent to introducing the force term as

$$F_i = \delta_{ik} \frac{\partial \delta(\boldsymbol{x})}{\partial x_k} = \frac{\partial \delta(\boldsymbol{x})}{\partial x_i}.$$ (2.91)

The quadrupole and hexapole solutions are then the divergence of the force solution:

$$u_i^q = u_{ik,k}^F.$$ (2.92)

2.8.3 *Center of dilatation*

Another solution of interest is the center of dilatation. We shall take a different approach by borrowing the idea of an eigenstrain, used for the modeling of nonelastic strains such as thermal expansion, initial strain, plastic strain, etc. [661]. We shall write the total strain e_{ij} as the summation of a regular strain $\bar{e}_{ij}$, which satisfies the homogeneous Cauchy–Navier equations, and an eigenstrain E_{ij}, which corresponds to the singular part of the solution, as

$$e_{ij} = \bar{e}_{ij} + E_{ij}.$$ (2.93)

Substituting the above into the Cauchy–Navier equations (2.78) without the body force term, we obtain

$$\mu u_{i,jj} + \frac{\mu}{1-2v} u_{j,ji} = 2\mu E_{ij,j} + \frac{2\mu v}{1-2v} E_{jj,i}. \qquad (2.94)$$

To find the solution, we introduce the singular term [202]

$$E_{ij} = \left(\frac{\ln r}{2\pi}\right)_{,ij}, \qquad (2\text{D}), \qquad (2.95\text{a})$$

$$= \left(-\frac{1}{4\pi r}\right)_{,ij}, \qquad (3\text{D}). \qquad (2.95\text{b})$$

It is easy to show that the above expressions lead to the volumetric strain of

$$E = E_{ii} = \delta(x), \qquad (2.96)$$

hence fulfilling the description that it represents a unit volume expansion at the origin. Using (2.95) for the RHS of (2.94), we find that it is equivalent to the introduction of the body force term

$$F_i = -\frac{2\mu(1-v)}{1-2v} \frac{\partial \delta(x)}{\partial x_i}. \qquad (2.97)$$

Comparing the above with (2.91) and (2.92), it is clear that the center of dilatation solution is

$$u_i^c = -\frac{2\mu(1-v)}{1-2v} u_i^q = -\frac{2\mu(1-v)}{1-2v} u_{ik,k}^F. \qquad (2.98)$$

Rather than carrying out the differentiation of (2.80), we seek a different route to find the expression for u_i^c. We argue that the center of dilation must possess axial symmetry; that is, the displacement can be a function of the radial distance r only. This also means that the displacement field is irrotational and can be expressed as the gradient of a scalar function,

$$u_i^c = \Phi_{,i}. \qquad (2.99)$$

Substituting the above and (2.97) into (2.78), and removing a gradient operator from both sides, we find

$$\nabla^2 \Phi = \delta(\boldsymbol{x}). \tag{2.100}$$

We then easily obtain Φ from (2.23). The displacements in polar and spherical coordinates, respectively for the 2D and 3D cases, are

$$u_r^c = \frac{1}{2\pi} \frac{1}{r}, \quad \text{(2D)}, \tag{2.101a}$$

$$= \frac{1}{4\pi} \frac{1}{r^2}, \quad \text{(3D)}, \tag{2.101b}$$

which is the classical center of dilatation result. Alternatively, in Cartesian coordinates, we have

$$u_i^c = \frac{1}{2\pi} \frac{x_i}{r^2}, \quad \text{(2D)}, \tag{2.102a}$$

$$= \frac{1}{4\pi} \frac{x_i}{r^3}, \quad \text{(3D)}. \tag{2.102b}$$

2.8.4 *Displacement discontinuity*

Based on the work of Nedelec [69, 669], a displacement discontinuity tensor can be generated by the introduction of an eigenstrain as follows:

$$E_{ijkl} = -\frac{1}{2}(\delta_{ik}\delta_{jl} + \delta_{il}\delta_{jk})\delta(\boldsymbol{x}). \tag{2.103}$$

Substituting the above into (2.94), we obtain the RHS as

$$-\mu\left(\delta_{ik}\delta_{,l} + \delta_{il}\delta_{,k}\right) - \frac{2\mu v}{(1-2v)}\delta_{kl}\delta_{,i}$$

$$= -\mu\left(F_{ki,l} + F_{li,k}\right) - \frac{2\mu v}{(1-2v)}\delta_{kl}F_{ji,j}, \tag{2.104}$$

where $\delta_{,i} = \partial\delta(\boldsymbol{x})/\partial x_i$, etc. In the above, we have also substituted in (2.79) for the force solution, and flipped its indices, as $F_{ij} = F_{ji}$. We observe that the RHS is exactly the constitutive equation (2.88) with F_{ij} playing the role of displacement. As we can make the LHS and RHS of (2.94) go through the same operation, we then deduce the following correspondence [193]:

$$u_{ikl}^{d} = \sigma_{lki}^{F},$$

(2.105)

where the superscript d indicates a displacement discontinuity solution. We should particularly note the swapping of indices i and l in the displacement and the force expressions, and that these expressions are not symmetrical to i and l. We then obtain fundamental solutions for the displacement discontinuity as

$$u_{ijk}^{d} = \frac{1}{4\pi (1-v)} \frac{1}{r} \left[(1-2v) \left(\delta_{jk} r_{,i} - \delta_{ij} r_{,k} - \delta_{ik} r_{,j} \right) - 2 r_{,i} r_{,j} r_{,k} \right], \quad (2\text{D}),$$

(2.106a)

$$= \frac{1}{8\pi (1-v)} \frac{1}{r^2} \left[(1-2v) \left(\delta_{jk} r_{,i} - \delta_{ij} r_{,k} - \delta_{ik} r_{,j} \right) - 3 r_{,i} r_{,j} r_{,k} \right], \quad (3\text{D}).$$

(2.106b)

2.9 Stokes Flow

The governing equations for the Stokes flow, also known as creeping flow, are the following:

$$\mu \nabla^2 \boldsymbol{v} - \nabla p = \boldsymbol{0},$$

(2.107)

$$\nabla \cdot \boldsymbol{v} = 0,$$

(2.108)

in which $\boldsymbol{v}$ is the velocity, p the pressure, and μ the dynamic viscosity. Its fundamental solution, known as the stokeslet, was first derived by Lorentz [601]. We actually can deduce the fundamental solution from the elasticity one by noticing the full analogy between Stokes flow and an incompressible elastic material [511]. If we set the Poisson ratio of elasticity to $v = 1/2$, and equate $\boldsymbol{v}$ with $\boldsymbol{u}$, p with $-\sigma_{ii}/3$ for 3D, and $-\sigma_{ii}/2$ for 2D, then the two sets of governing equations become identical. The fundamental solutions due to a point force in the k-direction can be obtained from (2.80) and (2.89), as [2]

$$v_{ik}^{F} = \frac{1}{4\pi \mu} \left[\frac{x_i x_k}{r^2} - \delta_{ik} \ln r \right], \quad (2\text{D}),$$

(2.109a)

$$= \frac{1}{8\pi \mu} \frac{1}{r} \left[\frac{x_i x_k}{r^2} + \delta_{ik} \right], \quad (3\text{D}),$$

(2.109b)

and

$$p_k^F = \frac{1}{2\pi} \frac{x_k}{r^2}, \quad \text{(2D)}, \tag{2.110a}$$

$$= \frac{1}{4\pi} \frac{x_k}{r^3}, \quad \text{(3D)}. \tag{2.110b}$$

The constitutive equation is

$$\sigma_{ij} = 2\mu\epsilon_{ij} - \frac{2}{3}\mu\,\delta_{ij}\epsilon - \delta_{ij}\,p, \tag{2.111}$$

where $\epsilon_{ij} = (v_{i,j} + v_{j,i})/2$ is the strain rate tensor, and $\epsilon = \epsilon_{ii}$. The stresses for the stokeslet are then

$$\sigma_{ijk}^F = -\frac{x_i x_j x_k}{\pi r^4}, \quad \text{(2D)}, \tag{2.112a}$$

$$= -\frac{3 x_i x_j x_k}{4\pi r^5}, \quad \text{(3D)}. \tag{2.112b}$$

Other types of fundamental solutions can be found following the same technique as in elasticity. For example, following the definition of displacement discontinuity, we can obtain the fundamental solution known as the "stresslet" as

$$v_{ijk}^d = \sigma_{ijk}^F. \tag{2.113}$$

We notice that no switching of indices is necessary as they are symmetrical.

2.10   Elastodynamics

The equation for elastodynamics is formed by adding an inertia term to the elastostatics equation (2.78), and we obtain

$$\mu\nabla^2 \boldsymbol{u} + \frac{\mu}{1-2v}\nabla\nabla\cdot\boldsymbol{u} - \rho\frac{\partial^2 \boldsymbol{u}}{\partial t^2} = -\boldsymbol{F}, \tag{2.114}$$

where ρ is the mass density. In the frequency domain, it is expressed as

$$\mu\nabla^2 \tilde{\boldsymbol{u}} + \frac{\mu}{1-2v}\nabla\nabla\cdot\tilde{\boldsymbol{u}} + \rho\omega^2\tilde{\boldsymbol{u}} = -\tilde{\boldsymbol{F}}. \tag{2.115}$$

Following the Helmholtz theorem, the displacement vector can be decomposed into a rotational (distortional) and an irrotational (dilatational) part, as

$$\tilde{\boldsymbol{u}} = \nabla\tilde{\phi} + \nabla \times \tilde{\boldsymbol{\psi}}, \tag{2.116}$$

where $\tilde{\phi}$ is a scalar potential representing the irrotational field, and $\tilde{\boldsymbol{\psi}}$ a vector potential for the rotational field. For convenience, from this point on we shall drop the tilde sign above the variables, as we will be solving the equations only in the frequency domain.

Substituting (2.116) into (2.115) with $\boldsymbol{F} = \boldsymbol{0}$, and separating the rotational and irrotational parts, we obtain the following Helmholtz equations:

$$\nabla^2\phi + k_p^2\phi = 0, \tag{2.117}$$

$$\nabla^2\boldsymbol{\psi} + k_s^2\boldsymbol{\psi} = \boldsymbol{0}, \tag{2.118}$$

where

$$k_p = \frac{\omega}{c_p}, \quad k_s = \frac{\omega}{c_s}, \tag{2.119}$$

are the wave numbers, and

$$c_p = \sqrt{\frac{E_d}{\rho}}, \tag{2.120}$$

$$c_s = \sqrt{\frac{\mu}{\rho}}, \tag{2.121}$$

are, respectively, the compressional (longitudinal) and shear (transverse) wave speeds, with

$$E_d = \frac{2\mu(1 - v)}{1 - 2v}, \tag{2.122}$$

the dynamic modulus. We observe that the system represented by ϕ and $\boldsymbol{\psi}$ has one more variable than that of the original system for $\boldsymbol{u}$; hence a constraint is needed to make the representation unique, which is typically chosen as

$$\nabla \cdot \boldsymbol{\psi} = 0. \tag{2.123}$$

2.10.1   *Force fundamental solution*

We may consider the response of an elastodynamic system subject to an impulse concentrated force in the k-direction

$$F_{ik} = \delta_{ik}\delta(\boldsymbol{x})\delta(t), \tag{2.124}$$

where $i, k = 1, 2,$ and $1, 2, 3$, respectively, for 2D and 3D cases. In the frequency domain, we can express (2.115) as

$$\mu u_{ik,jj}^F + \frac{\mu}{1-2v} u_{jk,ji}^F + \rho\,\omega^2 u_{ik}^F = -\delta_{ik}\delta(\boldsymbol{x}). \tag{2.125}$$

To derive the solution, we shall separate it into an irrotational part u_{ik}^p (compressional wave), and a rotational part u_{ik}^s (shear wave), as

$$u_{ik}^F = u_{ik}^p + u_{ik}^s. \tag{2.126}$$

We can compare the two parts on the RHS with (2.116) for their correspondences.

Likewise, we seek to separate the force term on the RHS of (2.125) into two parts, $\boldsymbol{F}^p$ and $\boldsymbol{F}^s$. To do so, we write the body force term as

$$F_{ik} = \delta_{ik}\delta(\boldsymbol{x})\boldsymbol{e}_k, \tag{2.127}$$

where $\boldsymbol{e}_k$ is the unit vector giving the direction of the force, and the Einstein summation is not applied to the index k to retain the tensor form.

3D Case: Observing that the Dirac delta function can be generated by the Laplacian and its fundamental solution (2.23b), we can express (2.127) as [4]

$$\begin{aligned}
F_{ik} &= \delta_{ik}\nabla^2\left(-\frac{1}{4\pi r}\boldsymbol{e}_k\right) \\
&= \delta_{ik}\nabla\times\left(\nabla\times\frac{1}{4\pi r}\boldsymbol{e}_k\right) - \delta_{ik}\nabla\left(\nabla\cdot\frac{1}{4\pi r}\boldsymbol{e}_k\right) \\
&= F_{ik}^s + F_{ik}^p.
\end{aligned} \tag{2.128}$$

In the above, the transformation from the first to the second line is a vector identity. We notice that the first part is rotational, and the second part irrotational.

Substituting (2.126) and (2.128) into (2.125), and separating the rotational and irrotational parts, we obtain two wave equations:

$$\nabla^2 u_{ik}^p + k_p^2 u_{ik}^p = \frac{\delta_{ik}}{\rho c_p^2} \nabla \left(\nabla \cdot \frac{1}{4\pi r} \boldsymbol{e}_k \right), \tag{2.129}$$

$$\nabla^2 u_{ik}^s + k_s^2 u_{ik}^s = -\frac{\delta_{ik}}{\rho c_s^2} \nabla \times \left(\nabla \times \frac{1}{4\pi r} \boldsymbol{e}_k \right). \tag{2.130}$$

Being aware of the relation (2.116), we can express the above in terms of the potentials

$$\nabla^2 \phi_k + k_p^2 \phi_k = \frac{1}{\rho c_p^2} \left(\frac{1}{4\pi r} \right)_{,k}, \tag{2.131}$$

$$\nabla^2 \psi_{ik} + k_s^2 \psi_{ik} = -\frac{\delta_{ik}}{\rho c_s^2} \left(\nabla \times \frac{1}{4\pi r} \boldsymbol{e}_k \right). \tag{2.132}$$

In the above, we have respectively removed a gradient and a curl operator from both sides of the equations. By inspection, it is easy to find that the solution of (2.131) is

$$\phi_k = -\frac{1}{\rho \omega^2} \left(\frac{e^{ik_p r}}{4\pi r} - \frac{1}{4\pi r} \right)_{,k}. \tag{2.133}$$

We observe that the terms in the parentheses are the difference between the fundamental solutions of the Helmholtz and the Laplace equations. Similarly by observation, we can show that

$$\psi_{ik} = \frac{\delta_{ik}}{\rho \omega^2} \nabla \times \left(\frac{e^{ik_s r}}{4\pi r} - \frac{1}{4\pi r} \right) \boldsymbol{e}_k. \tag{2.134}$$

From the above, we can recover the displacements

$$u_{ik}^p = \phi_{k,i} = -\frac{1}{\rho \omega^2} \left(\frac{e^{ik_p r}}{4\pi r} - \frac{1}{4\pi r} \right)_{,ik}, \tag{2.135}$$

and

$$u_{ik}^s = \nabla \times \boldsymbol{\psi}_k = \frac{\delta_{ik}}{\rho\,\omega^2} \nabla \times \nabla \times \left(\frac{e^{ik_s r}}{4\pi r} - \frac{1}{4\pi r} \right) \boldsymbol{e}_k$$

$$= \frac{\delta_{ik}}{\rho\,\omega^2} \left\{ \nabla \left[\nabla \cdot \left(\frac{e^{ik_s r}}{4\pi r} - \frac{1}{4\pi r} \right) \boldsymbol{e}_k \right] - \nabla^2 \left(\frac{e^{ik_s r}}{4\pi r} - \frac{1}{4\pi r} \right) \boldsymbol{e}_k \right\}$$

$$= \frac{1}{\rho\,\omega^2} \left[\left(\frac{e^{ik_s r}}{4\pi r} - \frac{1}{4\pi r} \right)_{,ik} + \delta_{ik} k_s^2 \frac{e^{ik_s r}}{4\pi r} \right]. \tag{2.136}$$

Summing (2.135) and (2.136), we obtain the force fundamental solution as

$$u_{ik}^F = \frac{1}{4\pi\rho\,\omega^2 r} \left[\left(e^{ik_s r} - e^{ik_p r} \right)_{,ik} + \delta_{ik} k_s^2 e^{ik_s r} \right], \quad (3\text{D}). \tag{2.137}$$

2D Case: Following the same procedure, we obtain the fundamental solution for the 2D case as

$$u_{ik}^F = \frac{i}{4\rho\,\omega^2} \left\{ \left[\mathrm{H}_0^{(1)}(k_s r) - \mathrm{H}_0^{(1)}(k_p r) \right]_{,ik} + \delta_{ik} k_s^2 \mathrm{H}_0^{(1)}(k_s r) \right\}, \quad (2\text{D}). \tag{2.138}$$

The stress expressions can be found by using the constitutive equation (2.88).

2.10.2 *Spherical wave fundamental solution*

From the previous section, it is clear that the fundamental solutions corresponding to the physical application of forces take some effort to derive, due to the nature of the vector variable and the vector forcing function. We also observe that the vector governing equation can be decomposed into scalar wave (Helmholtz) equations. We can take advantage of these scalar equations to obtain an easy derivation of different fundamental solutions. To do so, we introduce the Dirac delta function to the RHS of (2.117) and (2.118) to obtain

$$\nabla^2 \phi^* + k_p^2 \phi^* = -\delta(\boldsymbol{x}), \tag{2.139}$$

$$\nabla^2 \psi_{ik}^* + k_s^2 \psi_{ik}^* = -\delta_{ik}\,\delta(\boldsymbol{x}). \tag{2.140}$$

The fundamental solutions of these Helmholtz equations are readily available in (2.41), with the proper substitution of the wave numbers.

3D Case: Written out explicitly for the 3D case, we have

$$\phi^*(r) = \frac{1}{4\pi r} e^{ik_p r}, \tag{2.141}$$

$$\psi_{ik}^*(r) = \frac{\delta_{ik}}{4\pi r} e^{ik_s r}. \tag{2.142}$$

Recalling that these are Fourier transformed quantities, we can perform an inverse transform to (2.141) and obtain

$$\phi^*(r, t) = \frac{1}{4\pi r} \delta\left(t - \frac{r}{c_p}\right), \tag{2.143}$$

and similarly for (2.142). Hence (2.141) and (2.142) represent waves resulting from an impulse source at the origin and propagating outwardly with a spherical front at velocities c_p and c_s, with an amplitude decaying as $1/r$.

Separately substituting (2.141) and (2.142) into (2.116), we obtain the following: for the compressional wave

$$u_{i0}^* = \frac{x_i}{4\pi r^3}(-1 + ik_p r) e^{ik_p r}, \tag{2.144}$$

and for the shear waves

$$u_{11}^* = u_{22}^* = u_{33}^* = 0, \tag{2.145}$$

$$u_{21}^* = -u_{12}^* = \frac{z}{4\pi r^3}(-1 + ik_s r) e^{ik_s r}, \tag{2.146}$$

$$u_{13}^* = -u_{31}^* = \frac{y}{4\pi r^3}(-1 + ik_s r) e^{ik_s r}, \tag{2.147}$$

$$u_{32}^* = -u_{23}^* = \frac{x}{4\pi r^3}(-1 + ik_s r) e^{ik_s r}. \tag{2.148}$$

Consolidating the above notations, we can express these solutions as u_{ik}^*, with $i = 1, 2, 3$, denoting the vector component, and $k = 0, 1, 2, 3$, representing the wave type: 0 for the compressional wave, and 1, 2, and 3 for the first, second, and third shear waves. In comparison with the force fundamental solutions u_{ik}^F, in which $k = 1, 2, 3$, the spherical ones have one set more than the number of modeling variables, u_i, $i = 1, 2, 3$. The reason is that these shear wave solutions are not independent, as they satisfy a divergence free condition. Its implication in the numerical solution is discussed in Section 6.1.3.

2D Case: The 2D case follows similarly with

$$\phi^*(r) = -\frac{i}{4} H_0^{(1)}(k_p r), \tag{2.149}$$

$$\psi_{ik}^*(r) = -\frac{i}{4} \delta_{ik} H_0^{(1)}(k_s r). \tag{2.150}$$

We then obtain

$$u_{i0}^* = -\frac{i k_p x_i}{4r} H_1^{(1)}(k_p r), \tag{2.151}$$

and

$$u_{11}^* = u_{22}^* = 0, \tag{2.152}$$

$$u_{21}^* = \frac{i k_s y}{4r} H_1^{(1)}(k_s r), \tag{2.153}$$

$$u_{12}^* = -\frac{i k_s x}{4r} H_1^{(1)}(k_s r). \tag{2.154}$$

Comparison with Force Fundamental Solutions: Similar to force fundamental solutions, spherical wave solutions can be used in the MFS procedure to solve BVPs. Based on the above presentation, we observe that spherical wave solutions are easier to derive and shorter in their expressions. This is even more so in the cases of coupled physical systems, such as the coupling of elasticity with the heat effect in dynamic thermoelasticity [681], with fluid effect in poroelasticity, and with both effects in porothermoelasticity [202]. In these cases, the various physically based fundamental solutions, such as those due to forces, are difficult to derive. On the other hand, without being committed to a certain physical quantity, the spherical wave potential solutions are easy to derive. The MFS only requires a complete set of singular solutions, and not a physical association. The spherical wave potential based MFS has been successfully applied to wave scattering problems in elastic [267], poroelastic [598], and double-porosity dual-permeability poroelastic [599] materials.

2.11 Biharmonic and Polyharmonic Equations

The biharmonic equation is given by

$$\Delta^2 \phi = \nabla^4 \phi = 0, \tag{2.155}$$

where $\Delta \equiv \nabla^2$ is the Laplacian operator. The fundamental solutions of the biharmonic operator are

$$G^B(\boldsymbol{x}, \boldsymbol{x}') = \frac{1}{8\pi} r^2 \ln r, \quad \text{(2D)}, \tag{2.156a}$$

$$= -\frac{1}{8\pi} r, \quad \text{(3D)}. \tag{2.156b}$$

Such an equation may be associated with physical problems such as elasticity involving the Airy stress function, the stream function of Stokes flow, the torsion of a prismatic bar, plate theories, among other applications.

The products of the Laplacian operator Δ give the polyharmonic equation of degree n

$$\Delta^n \phi = 0. \tag{2.157}$$

The case $n = 2$ corresponds to the biharmonic equation given in (2.155). We may obtain its fundamental solutions as follows [191]. For the nth degree (in Δ) operator, we define the fundamental solution G_n^P as the solution of

$$\Delta^n G_n^P = \delta(\boldsymbol{x}, \boldsymbol{x}'). \tag{2.158}$$

We observe from the above definition the following recurrence relations among the solutions:

$$\Delta G_1^P = \delta(\boldsymbol{x}, \boldsymbol{x}'), \tag{2.159}$$

$$\Delta G_n^P = G_{n-1}^P. \tag{2.160}$$

Such a relation allows us to obtain the higher order fundamental solutions by the integration of lower order ones. For example, we may start from the relation

$$\Delta G_2^P = G_1^P, \tag{2.161}$$

and move upward, as follows.

2D Case: In 2D, the above equation can be written as

$$\frac{1}{r} \frac{\partial}{\partial r} \left(r \frac{\partial G_2^P}{\partial r} \right) = \frac{1}{2\pi} \ln r. \tag{2.162}$$

Integrating twice, we obtain

$$G_2^P = \frac{r^2}{8\pi}(\ln r - 1). \tag{2.163}$$

We may compare the above expression with (2.156a), and find that they differ by a term $-r^2/8\pi$, which is a general solution of the biharmonic equation. As fundamental solutions are particular solutions, which are not unique, both expressions are acceptable.

By successive integration, we obtain a general formula as [191, 195]

$$G_n^P = \frac{r^{2(n-1)}}{2^{2n-1}\pi (n-1)!(n-1)!}(\ln r - H_{n-1}), \quad n = 1, 2, 3, \ldots \tag{2.164}$$

in which H_n is the harmonic number, defined as

$$H_n = 1 + \frac{1}{2} + \frac{1}{3} + \cdots + \frac{1}{n}. \tag{2.165}$$

Clearly, we may drop the general solution and retain only the singular part, as

$$G_n^P = \frac{r^{2(n-1)}}{2^{2n-1}\pi (n-1)!(n-1)!}\ln r. \tag{2.166}$$

We note, however, that the above fundamental solution does not possess the recursive relation (2.160), which is a convenient relation to use when we later seek the particular solutions of the polyharmonic splines in Section 4.5.

The first few of the fundamental solutions are

$$G_1^P \equiv G^L, \tag{2.167}$$

$$G_2^P \equiv G^B, \tag{2.168}$$

$$G_3^P = \frac{r^4}{128\pi}\left(\ln r - \frac{3}{2}\right), \tag{2.169}$$

$$G_4^P = \frac{r^6}{4608\pi}\left(\ln r - \frac{11}{6}\right). \tag{2.170}$$

3D Case: For the 3D case, we can follow a similar procedure and obtain the following general formula

$$G_n^P = -\frac{1}{4\pi (2n-2)!}r^{2n-3}. \tag{2.171}$$

The first few terms are: $G_1^P \equiv G^L$, $G_2^P \equiv G^B$, and

$$G_3^P = -\frac{r^3}{96\pi}, \tag{2.172}$$

$$G_4^P = -\frac{r^4}{2880\pi}. \tag{2.173}$$

These solutions satisfy the recurrence relation (2.160).

2.12 Polymetaharmonic Equations

A polymetaharmonic equation of degree n is given by [874]

$$P_n(\Delta)\phi \equiv \Delta^n\phi + a_1\Delta^{n-1}\phi + \cdots + a_n\phi = 0, \tag{2.174}$$

where $P_n(\Delta)$ is a polynomial of degree n in Δ. When $n = 1$, it is the Helmholtz or modified Helmholtz equation, also called the metaharmonic equation.

Many problems in engineering and applied science are governed by a system of coupled, linear, constant coefficient PDEs. Examples include multiple layered aquifers [189, 190], multiple porosity media [194], and multiple diffusion processes [7]. These coupled systems are polynomials in differential operators, and can be manipulated using algebraic rules [189, 190, 397, 433]. The uncoupled systems are polynomials of operators in higher degrees, such as (2.157) and (2.174). A similar situation exists in the technique known as the multiple reciprocity boundary element method (MRBEM) [682], which converts a domain integral into a boundary one. The procedure requires iterated fundamental solutions, and has been applied to the Laplace [682], metaharmonic [407, 408], and elasticity [674] operators. These high order PDEs have been studied extensively in the mathematics literature [319, 320, 694, 695, 787, 844].

2.12.1 *Product of Helmholtz operators*

We shall investigate the special case when (2.174) can be factored into the following form [191]:

$$(\Delta + \lambda_1)^{m_1}(\Delta + \lambda_2)^{m_2}\cdots(\Delta + \lambda_n)^{m_n}G_n^H = \delta(\boldsymbol{x}, \boldsymbol{x}'), \tag{2.175}$$

where λ_i are the roots of the polynomial equation in Δ, and are positive real numbers, $\lambda_1 \neq \lambda_2 \neq \cdots \neq \lambda_n$, and m_1, m_2, ... indicate their multiplicity. The fundamental solution is given by [320, 695, 844]

$$G_n^H = \sum_{i=1}^{n} \frac{(-1)^{m_i-1}}{(m_i - 1)!} \frac{\partial^{m_i-1} c_i \, G_1^H(\lambda_i, x, x')}{\partial \lambda_i^{m_i-1}}, \tag{2.176}$$

in which

$$c_i = \prod_{j=1, j \neq i}^{n} \frac{1}{(\lambda_j - \lambda_i)^{m_j}}, \tag{2.177}$$

and $G_1^H(\lambda, x, x')$ satisfies

$$(\Delta + \lambda) G_1^H = \delta(x, x'). \tag{2.178}$$

We shall present a few special cases as follows.

Distinct Roots: When all the roots are distinct, (2.175) becomes [191]

$$(\Delta + \lambda_1)(\Delta + \lambda_2) \cdots (\Delta + \lambda_n) G_n^H = \delta(x, x'). \tag{2.179}$$

It can be shown that these fundamental solutions satisfy the recurrence relation:

$$(\Delta + \lambda_n) G_n^H = G_{n-1}^H. \tag{2.180}$$

For $m_i = 1$, (2.176) simplifies to

$$G_n^H = \sum_{i=1}^{n} \frac{G_1^H(\lambda_i, x, x')}{\prod_{j=1, j \neq i}^{n}(\lambda_j - \lambda_i)}. \tag{2.181}$$

Substituting the fundamental solutions in (2.32), we obtain:

$$G_n^H = \frac{1}{4} \sum_{i=1}^{n} \frac{Y_0(\sqrt{\lambda_i} r)}{\prod_{j=1, j \neq i}^{n}(\lambda_j - \lambda_i)}, \quad (2D), \tag{2.182a}$$

$$= -\frac{1}{4\pi r} \sum_{i=1}^{n} \frac{\cos(\sqrt{\lambda_i} r)}{\prod_{j=1, j \neq i}^{n}(\lambda_j - \lambda_i)}, \quad (3D). \tag{2.182b}$$

The first few expressions are: for the 2D case

$$G_1^H \equiv G^H, \tag{2.183}$$

$$G_2^H = \frac{1}{4}\left[\frac{Y_0(\sqrt{\lambda_1}r)}{\lambda_2 - \lambda_1} + \frac{Y_0(\sqrt{\lambda_2}r)}{\lambda_1 - \lambda_2}\right], \tag{2.184}$$

$$G_3^H = \frac{1}{4}\left[\frac{Y_0(\sqrt{\lambda_1}r)}{(\lambda_2 - \lambda_1)(\lambda_3 - \lambda_1)} + \frac{Y_0(\sqrt{\lambda_2}r)}{(\lambda_1 - \lambda_2)(\lambda_3 - \lambda_2)}\right.$$
$$\left. + \frac{Y_0(\sqrt{\lambda_3}r)}{(\lambda_1 - \lambda_3)(\lambda_2 - \lambda_3)}\right], \tag{2.185}$$

and for the 3D case, $G_1^H \equiv G^H$, and

$$G_2^H = -\frac{1}{4\pi r}\left[\frac{\cos(\sqrt{\lambda_1}r)}{\lambda_2 - \lambda_1} + \frac{\cos(\sqrt{\lambda_2}r)}{\lambda_1 - \lambda_2}\right], \tag{2.186}$$

$$G_3^H = -\frac{1}{4\pi r}\left[\frac{\cos(\sqrt{\lambda_1}r)}{(\lambda_2 - \lambda_1)(\lambda_3 - \lambda_1)} + \frac{\cos(\sqrt{\lambda_2}r)}{(\lambda_1 - \lambda_2)(\lambda_3 - \lambda_2)}\right.$$
$$\left. + \frac{\cos(\sqrt{\lambda_3}r)}{(\lambda_1 - \lambda_3)(\lambda_2 - \lambda_3)}\right]. \tag{2.187}$$

Repeated Roots: Next, we examine the case of m repeated roots. In the MRBEM procedure, the solutions of the following iterative metaharmonic equations are needed [407, 408]:

$$(\Delta + \lambda)G_1^H = \delta(\boldsymbol{x}, \boldsymbol{x}'), \tag{2.188}$$

$$(\Delta + \lambda)G_m^H = G_{m-1}^H, \tag{2.189}$$

for $m \geq 2$. By successive substitution, it is clear that they satisfy

$$(\Delta + \lambda)^m G_m^H = \delta(\boldsymbol{x}, \boldsymbol{x}'), \tag{2.190}$$

and (2.176) becomes

$$G_m^H = \frac{(-1)^{m-1}}{(m-1)!}\frac{\partial^{m-1}G_1^H(\lambda, \boldsymbol{x}, \boldsymbol{x}')}{\partial\lambda^{m-1}}$$
$$= -\frac{1}{(m-1)}\frac{\partial G_{m-1}^H(\lambda, \boldsymbol{x}, \boldsymbol{x}')}{\partial\lambda}. \tag{2.191}$$

Substituting (2.32) into the above, we find the fundamental solution in 2D as [787]

$$G_m^H = \frac{1}{2^{m+1}(m-1)!} \left(\frac{r}{\sqrt{\lambda}}\right)^{m-1} Y_{m-1}(\sqrt{\lambda}\,r), \tag{2.192}$$

and in 3D

$$\begin{aligned} G_m^H &= \frac{(-1)^{m-1}}{2^{m+\frac{3}{2}}\,\pi^{\frac{1}{2}}\,(m-1)!} \left(\frac{r}{\sqrt{\lambda}}\right)^{m-\frac{3}{2}} Y_{\frac{3}{2}-m}(\sqrt{\lambda}\,r) \\ &= \frac{1}{2^{m+1}\,\pi\,(m-1)!\,r} \left(\frac{r}{\sqrt{\lambda}}\right)^{m-1} \\ &\quad \times \Bigg[\sin\left(\sqrt{\lambda}\,r - \frac{m\pi}{2}\right) \sum_{i=0}^{\lfloor\frac{m}{2}-1\rfloor} \frac{(-1)^i(m+2i-2)!}{2^{2i}(2i)!(m-2i-2)!(\sqrt{\lambda}\,r)^{2i}} \\ &\quad\quad + \cos\left(\sqrt{\lambda}\,r - \frac{m\pi}{2}\right) \sum_{i=0}^{\lfloor\frac{m}{2}-\frac{3}{2}\rfloor} \frac{(-1)^i(m+2i-1)!}{2^{2i+1}(2i+1)!(m-2i-3)!(\sqrt{\lambda}\,r)^{2i+1}} \Bigg], \end{aligned} \tag{2.193}$$

for $m \geq 2$. In the above, $\lfloor\cdot\rfloor$ is the floor function that takes the greatest integer less than or equal to its argument. Again, we list the first few fundamental solutions below: in 2D

$$G_2^H = \frac{r}{8\sqrt{\lambda}} Y_1(\sqrt{\lambda}\,r), \tag{2.194}$$

$$G_3^H = \frac{r^2}{32\lambda} Y_2(\sqrt{\lambda}\,r), \tag{2.195}$$

$$G_4^H = \frac{r^3}{192\lambda^{3/2}} Y_3(\sqrt{\lambda}\,r), \tag{2.196}$$

and in 3D

$$G_2^H = -\frac{1}{8\pi\sqrt{\lambda}} \sin(\sqrt{\lambda}\,r), \tag{2.197}$$

$$G_3^H = \frac{1}{32\pi\,\lambda^{3/2}} \left[\sqrt{\lambda}\,r\,\cos(\sqrt{\lambda}\,r) - \sin(\sqrt{\lambda}\,r)\right], \tag{2.198}$$

$$G_4^H = -\frac{1}{192\pi \, \lambda^{5/2}} \left[3\sin(\sqrt{\lambda}\,r) - 3\sqrt{\lambda}\,r\,\cos(\sqrt{\lambda}\,r) - \lambda r^2 \sin(\sqrt{\lambda}\,r) \right].$$

$$(2.199)$$

2.12.2 *Product of modified Helmholtz operators*

Corresponding to (2.175), we shall examine the case

$$(\Delta - \lambda_1)^{m_1}(\Delta - \lambda_2)^{m_2} \cdots (\Delta - \lambda_n)^{m_n} G_n^M = \delta(\boldsymbol{x}, \boldsymbol{x}'), \qquad (2.200)$$

where the λ_i are positive real numbers, and m_i are positive integers. The fundamental solution is given by [695, 844]

$$G_n^M = \sum_{i=1}^{n} \frac{1}{(m_i - 1)!} \frac{\partial^{m_i-1} c_i\, G_1^M(\lambda_i, \boldsymbol{x}, \boldsymbol{x}')}{\partial \lambda_i^{m_i-1}}, \qquad (2.201)$$

in which

$$c_i = \prod_{j=1, j\neq i}^{n} \frac{1}{(\lambda_i - \lambda_j)^{m_j}}, \qquad (2.202)$$

and $G_1^M(\lambda, \boldsymbol{x}, \boldsymbol{x}')$ is the fundamental solution of the simple metaharmonic operator

$$(\Delta - \lambda) G_1^M = \delta(\boldsymbol{x}, \boldsymbol{x}'). \qquad (2.203)$$

Two cases of special interest are discussed as follows.

Distinct Roots: An equation that arises from an n-layered aquifer system [189, 190] or an n-porosity medium [194] is

$$(\Delta - \lambda_1)(\Delta - \lambda_2) \cdots (\Delta - \lambda_n) G_n^M = \delta(\boldsymbol{x}, \boldsymbol{x}'), \qquad (2.204)$$

where G_n^M satisfies the recurrence relation

$$(\Delta - \lambda_n) G_n^M = G_{n-1}^M. \qquad (2.205)$$

In this case, (2.201) simplifies to

$$G_n^M = \sum_{i=1}^{n} \frac{G_1^M(\lambda_i, \boldsymbol{x}, \boldsymbol{x}')}{\prod_{j=1, j\neq i}^{n}(\lambda_i - \lambda_j)}. \qquad (2.206)$$

Substituting the fundamental solutions given in (2.35) into the above, we obtain

$$G_n^M = -\frac{1}{2\pi} \sum_{i=1}^{n} \frac{K_0(\sqrt{\lambda_i}\, r)}{\prod_{j=1, j\neq i}^{n}(\lambda_i - \lambda_j)}, \quad (2D), \tag{2.207a}$$

$$= -\frac{1}{4\pi r} \sum_{i=1}^{n} \frac{e^{-\sqrt{\lambda_i}\, r}}{\prod_{j=1, j\neq i}^{n}(\lambda_i - \lambda_j)}, \quad (3D). \tag{2.207b}$$

The first few solutions are as follows: for the 2D case

$$G_1^M \equiv G^M, \tag{2.208}$$

$$G_2^M = -\frac{1}{2\pi}\left[\frac{K_0(\sqrt{\lambda_1}\, r)}{\lambda_1 - \lambda_2} + \frac{K_0(\sqrt{\lambda_2}\, r)}{\lambda_2 - \lambda_1}\right], \tag{2.209}$$

$$G_3^M = -\frac{1}{2\pi}\left[\frac{K_0(\sqrt{\lambda_1}\, r)}{(\lambda_1 - \lambda_2)(\lambda_1 - \lambda_3)} + \frac{K_0(\sqrt{\lambda_2}\, r)}{(\lambda_2 - \lambda_1)(\lambda_2 - \lambda_3)}\right.$$
$$\left. + \frac{K_0(\sqrt{\lambda_3}\, r)}{(\lambda_3 - \lambda_1)(\lambda_3 - \lambda_2)}\right], \tag{2.210}$$

and for 3D, $G_1^M \equiv G^M$ and

$$G_2^M = -\frac{1}{4\pi r}\left[\frac{e^{-\sqrt{\lambda_1}\, r}}{\lambda_1 - \lambda_2} + \frac{e^{-\sqrt{\lambda_2}\, r}}{\lambda_2 - \lambda_1}\right], \tag{2.211}$$

$$G_3^M = -\frac{1}{4\pi r}\left[\frac{e^{-\sqrt{\lambda_1}\, r}}{(\lambda_1 - \lambda_2)(\lambda_1 - \lambda_3)} + \frac{e^{-\sqrt{\lambda_2}\, r}}{(\lambda_2 - \lambda_1)(\lambda_2 - \lambda_3)}\right.$$
$$\left. + \frac{e^{-\sqrt{\lambda_3}\, r}}{(\lambda_3 - \lambda_1)(\lambda_3 - \lambda_2)}\right]. \tag{2.212}$$

Repeated Roots: For the case of m repeated roots, (2.200) becomes

$$(\Delta - \lambda)^m G_m^M = \delta(\boldsymbol{x}, \boldsymbol{x}'), \tag{2.213}$$

where G_m^M satisfies the iterative relation

$$(\Delta - \lambda) G_1^M = \delta(\boldsymbol{x}, \boldsymbol{x}'), \tag{2.214}$$

$$(\Delta - \lambda) G_m^M = G_{m-1}^M. \tag{2.215}$$

The solution is given by

$$G_m^M = \frac{1}{(m-1)!} \frac{\partial^{m-1} G_1^M(\lambda, \boldsymbol{x}, \boldsymbol{x}')}{\partial \lambda^{m-1}}$$

$$= \frac{1}{(m-1)} \frac{\partial G_{m-1}^M(\lambda, \boldsymbol{x}, \boldsymbol{x}')}{\partial \lambda}. \tag{2.216}$$

Substituting (2.35) into the above, the following solutions are obtained: in 2D

$$G_m^M = \frac{(-1)^m}{2^m \pi (m-1)!} \left(\frac{r}{\sqrt{\lambda}}\right)^{m-1} \mathrm{K}_{m-1}(\sqrt{\lambda}\, r), \tag{2.217}$$

and in 3D

$$G_m^M = \frac{(-1)^m}{2^{m+\frac{1}{2}} \pi^{\frac{3}{2}} (m-1)!} \left(\frac{r}{\sqrt{\lambda}}\right)^{m-\frac{3}{2}} \mathrm{K}_{m-\frac{3}{2}}(\sqrt{\lambda}\, r)$$

$$= \frac{(-1)^m}{2^{m+1} \pi (m-1)!\, r} \left(\frac{r}{\sqrt{\lambda}}\right)^{m-1} e^{-\sqrt{\lambda}\, r}$$

$$\times \sum_{i=0}^{m-2} \frac{(m+i-2)!}{2^i\, i!\, (m-i-2)!\, (\sqrt{\lambda}\, r)^i}, \tag{2.218}$$

for $m \geq 2$. The first few solutions are: in 2D

$$G_2^M = \frac{r}{4\pi \sqrt{\lambda}} \mathrm{K}_1(\sqrt{\lambda}\, r), \tag{2.219}$$

$$G_3^M = -\frac{r^2}{16\pi \lambda} \mathrm{K}_2(\sqrt{\lambda}\, r), \tag{2.220}$$

$$G_4^M = \frac{r^3}{96\pi \lambda^{3/2}} \mathrm{K}_3(\sqrt{\lambda}\, r), \tag{2.221}$$

and in 3D

$$G_2^M = \frac{1}{8\pi \sqrt{\lambda}} e^{-\sqrt{\lambda}\, r}, \tag{2.222}$$

$$G_3^M = -\frac{1 + \sqrt{\lambda}\, r}{32\pi \lambda^{3/2}} e^{-\sqrt{\lambda}\, r}, \tag{2.223}$$

$$G_4^M = \frac{3 + 3\sqrt{\lambda}\, r + \lambda r^2}{192\pi \lambda^{5/2}} e^{-\sqrt{\lambda}\, r}. \tag{2.224}$$

2.12.3 *Product of mixed metaharmonic operators*

In Sections 2.12.1 and 2.12.2, we separately treated the Helmholtz and modified Helmholtz operators to obtain explicit results. As these equations are similar, we can in fact obtain one set of results from the other. This is done by allowing the coefficients λ_i in (2.175) and (2.200) to take negative values. In the 2D cases, the factor $\sqrt{\lambda_i}$ contained in the Bessel functions Y_0 and K_0 becomes imaginary, and the functions become complex. As only the real part is needed, this action is equivalent to replacing the terms $(1/4)Y_0(\sqrt{\lambda_i}\,r)$ by $-(1/2\pi)K_0(\sqrt{\lambda_i}\,r)$, and vice versa. For 3D, we replace $\cos(\sqrt{\lambda_i}\,r)$ by $-\exp(-\sqrt{\lambda_i}\,r)$. We also need to switch the sign of λ_i in the conversion. We shall give an example below.

Given the mixed product

$$(\Delta + \lambda_1)(\Delta - \lambda_2)G_2 = \delta(\boldsymbol{x}, \boldsymbol{x}'), \tag{2.225}$$

we can find the 2D fundamental solution by converting (2.184) or (2.209). Either way, we obtain [949]

$$G_2 = -\frac{1}{4}\frac{Y_0(\sqrt{\lambda_1}\,r)}{\lambda_1 + \lambda_2} - \frac{1}{2\pi}\frac{K_0(\sqrt{\lambda_2}\,r)}{\lambda_1 + \lambda_2}, \tag{2.226}$$

and for 3D, we have

$$G_2 = \frac{1}{4\pi r}\left[\frac{\cos(\sqrt{\lambda_1}\,r)}{\lambda_1 + \lambda_2} - \frac{e^{-\sqrt{\lambda_2}\,r}}{\lambda_1 + \lambda_2}\right]. \tag{2.227}$$

All results in Sections 2.12.1 and 2.12.2 can be converted this way for mixed operators.

2.12.4 *Product of mixed harmonic and metaharmonic operators*

When the product is a combination of the Laplacian and Helmholtz operators, we may deduce the fundamental solution from the results in Sections 2.12.1 and 2.12.2, by considering the Laplacian operator a limiting case of the Helmholtz one with $\lambda \to 0$. Here we present a special case that becomes useful in the derivation of the particular solutions involving the polyharmonic radial basis functions (RBFs) in Section 4.5.2. Consider the equation

$$\Delta^n(\Delta \pm \lambda)G_n^X = \delta(\boldsymbol{x}, \boldsymbol{x}'), \quad n = 0, 1, 2, \ldots \tag{2.228}$$

We may obtain such a solution by utilizing (2.176) and (2.201), and taking the limit with respect to λ. However, we shall give a heuristic derivation as follows.

First, we write (2.228) as

$$\Delta^n(\Delta \pm \lambda)\, G_n^X = \Delta^n G_n^P, \tag{2.229}$$

where G_n^P is the fundamental solution of the polyharmonic operator defined in (2.158), and given by (2.164) and (2.171). The above is an equality because both sides can be equated to a Dirac delta function. As we are seeking a particular solution, it is safe to remove the operator Δ^n from both sides to obtain

$$(\Delta \pm \lambda)\, G_n^X = G_n^P. \tag{2.230}$$

Similarly, we can write from (2.160)

$$\Delta^n(\Delta \pm \lambda)\, G_n^X = \Delta^{n-1}(\Delta \pm \lambda)\, G_{n-1}^X, \tag{2.231}$$

which leads to

$$\Delta G_n^X = G_{n-1}^X. \tag{2.232}$$

Substituting the above into (2.230) produces the recurrence relation

$$G_n^X = \frac{\pm 1}{\lambda}\left(G_n^P - G_{n-1}^X\right), \quad n = 1, 2, \dots \tag{2.233}$$

The above equation can be seeded by

$$G_0^X \equiv G^H \quad \text{and} \quad G_0^X \equiv G^M, \tag{2.234}$$

respectively for the plus and minus sign in (2.228), where G^H is the fundamental solution of the Helmholtz equation, given by (2.32), and G^M is that for the modified Helmholtz equation, given by (2.35). Using the above, and through successive substitutions, we obtain the following direct formulas:

$$G_n^X = \frac{(-1)^n}{\lambda^n}\, G^H + \sum_{i=1}^{n} \frac{(-1)^{n-i}}{\lambda^{n+1-i}}\, G_i^P, \quad n = 1, 2, \dots \tag{2.235}$$

for the plus sign in (2.228), and

$$G_n^X = \frac{1}{\lambda^n} G^M - \sum_{i=1}^{n} \frac{1}{\lambda^{n+1-i}} G_i^P, \quad n = 1, 2, \ldots \tag{2.236}$$

for the minus sign. For the explicit result, we can write (2.235) in 2D as

$$G_n^X = \frac{(-1)^n}{4\lambda^n} Y_0(\sqrt{\lambda}\, r) + \frac{1}{2\pi} \sum_{i=1}^{n} \frac{(-1)^{n-i}}{4^{i-1}\lambda^{n+1-i}(i-1)!(i-1)!}$$

$$\times r^{2(i-1)} (\ln r - H_{i-1}). \tag{2.237}$$

In the above, we can drop the term with the harmonic number H_{i-1}, and the expression is still a valid fundamental solution; though this solution does not possess the recursive relation (2.232). In 3D, we have

$$G_n^X = \frac{(-1)^{n+1}}{4\pi\,\lambda^n} \frac{\cos(\sqrt{\lambda}\, r)}{r} + \frac{1}{4\pi} \sum_{i=1}^{n} \frac{(-1)^{n+1-i}}{\lambda^{n+1-i}(2i-2)!} r^{2i-3}. \tag{2.238}$$

The explicit expressions for (2.236) can be similarly obtained.

We can show some specific results in the following. For the product of harmonic and Helmholtz operators, that is, taking the plus sign in (2.228), we have

$$G_1^X = -\frac{1}{4\lambda} Y_0(\sqrt{\lambda}\, r) + \frac{1}{2\pi\,\lambda} \ln r, \quad \text{(2D)}, \tag{2.239a}$$

$$= \frac{1}{4\pi\,\lambda r} \cos(\sqrt{\lambda}\, r) - \frac{1}{4\pi\,\lambda r}, \quad \text{(3D)}, \tag{2.239b}$$

and

$$G_2^X = \frac{1}{4\lambda^2} Y_0(\sqrt{\lambda}\, r) + \frac{\lambda r^2 - 4}{8\pi\,\lambda^2} \ln r, \quad \text{(2D)}, \tag{2.240a}$$

$$= -\frac{1}{4\pi\,\lambda^2 r} \cos(\sqrt{\lambda}\, r) - \frac{\lambda r^2 - 2}{8\pi\,\lambda^2 r}, \quad \text{(3D)}. \tag{2.240b}$$

For the modified Helmholtz operator, with the minus sign in (2.228), we have

$$G_1^X = -\frac{1}{2\pi\,\lambda} K_0(\sqrt{\lambda}\, r) - \frac{1}{2\pi\,\lambda} \ln r, \quad \text{(2D)}, \tag{2.241a}$$

$$= -\frac{1}{4\pi\,\lambda r} e^{-\sqrt{\lambda}\, r} + \frac{1}{4\pi\,\lambda r}, \quad \text{(3D)}, \tag{2.241b}$$

and

$$G_2^X = -\frac{1}{2\pi\lambda^2}K_0(\sqrt{\lambda}\,r) - \frac{\lambda r^2 + 4}{8\pi\lambda^2}\ln r, \quad (2D), \tag{2.242a}$$

$$= -\frac{1}{4\pi\lambda^2 r}e^{-\sqrt{\lambda}\,r} + \frac{\lambda r^2 + 2}{8\pi\lambda^2 r}, \qquad (3D). \tag{2.242b}$$

2.12.5 *Plate equations*

Many of the elastic plate equations [734] involve higher-order PDEs that can be classified into the polyharmonic and polymetaharmonic categories. Due to their importance in applications, we shall provide a special discussion in this section. The most basic is the Kirchhoff static plate equation:

$$\Delta^2 w = -\frac{q}{D}, \tag{2.243}$$

in 2D, where w is the deflection, q the load normal to the neutral plane, and D is the flexural rigidity given by

$$D = \frac{2h^3 E}{3(1 - \nu^2)}, \tag{2.244}$$

in which h is the thickness of the plate, and E is Young's modulus. The LHS of (2.243) is obviously the biharmonic operator, whose fundamental solution is given by (2.156a).

The dynamic version of (2.243) in the frequency domain is

$$\Delta^2 \tilde{w} - \frac{\rho h \omega^2}{D}\tilde{w} = -\frac{\tilde{q}}{D}, \tag{2.245}$$

where ρ is the mass density, and ω the frequency. We hence seek the fundamental solution satisfying

$$\Delta^2 G - \lambda^2 G = (\Delta + \lambda)(\Delta - \lambda)G = \delta(x, x'). \tag{2.246}$$

Following (2.226), we obtain

$$G = -\frac{1}{8\lambda}Y_0(\sqrt{\lambda}\,r) - \frac{1}{4\pi\lambda}K_0(\sqrt{\lambda}\,r). \tag{2.247}$$

In a more involved case, the plate is resting on a Pasternak foundation with a foundation modulus k, and a shear modulus μ, and we have [734]

$$\Delta^2 \tilde{w} - \frac{\mu}{D}\Delta\tilde{w} + \frac{k}{D}\tilde{w} + \frac{i\omega c}{D}\tilde{w} - \frac{\rho h \omega^2}{D}\tilde{w} = -\frac{\tilde{q}}{D}, \tag{2.248}$$

in which c is a viscous damping coefficient of the plate material. We then seek the fundamental solution satisfying

$$(\Delta - \lambda_1)(\Delta - \lambda_2)G = \delta(x, x'), \tag{2.249}$$

where

$$\lambda_{1,2} = \frac{\mu \pm \sqrt{\mu^2 + 4D(\rho h \omega^2 - i\omega c - k)}}{2D}, \tag{2.250}$$

in which λ_1 and λ_2 are complex numbers. The fundamental solution is given by (2.209), or in terms of Hankel functions as [521]

$$G = \frac{i}{4}\left[\frac{H_0^{(1)}(\sqrt{-\lambda_1}\,r)}{\lambda_2 - \lambda_1} + \frac{H_0^{(1)}(\sqrt{-\lambda_2}\,r)}{\lambda_1 - \lambda_2}\right]. \tag{2.251}$$

Similarly, the governing equations of a thick plate can be condensed into a polymetaharmonic equation of the third degree [11, 748]. We then seek the solution of

$$(\Delta - \lambda_1)(\Delta - \lambda_2)(\Delta - \lambda_3)G = \delta(x, x'), \tag{2.252}$$

with complex λ values. The solution is given by (2.210), or in the form of Hankel functions similar to (2.251).

2.13 Anisotropic Elliptic Equations

Many physical phenomena are modeled by two laws, a conservation law, which is associated with the divergence operator, and a flux law, which is given by the gradient operator. The proportionality constants of the flux law can be spatially varying and anisotropic. For the advection–diffusion transport under such conditions, we may write (2.58) in the following form:

$$\frac{\partial c}{\partial t} = \left(D_{ij}\,c_{,j}\right)_{,i} - (V_i\,c)_{,i}. \tag{2.253}$$

As the diffusion coefficient D_{ij} is derived based on physics laws, Onsager's reciprocal principle [692, 693] requires that it is a positive definite symmetric tensor.

2.13.1 *Quasi-harmonic equation*

For a steady state diffusion equation without advection, $V = 0$, and with constant diffusion coefficients, we obtain the following:

$$D_{ij}\, c_{,ij} = 0. \tag{2.254}$$

The above equation is called the quasi-harmonic equation. As presented in [128, 129], the fundamental solutions that satisfy

$$D_{ij}\, G_{,ij}(x, x') = \delta(x, x'), \tag{2.255}$$

are

$$G = \frac{|d_{ij}|^{1/2}}{2\pi}\, \ln \sqrt{d_{ij} x_i x_j}, \quad \text{(2D)}, \tag{2.256a}$$

$$= -\frac{|d_{ij}|^{1/2}}{4\pi \sqrt{d_{ij} x_i x_j}}, \quad \text{(3D)}. \tag{2.256b}$$

In the above,

$$d_{ij} = D_{ij}^{-1} \tag{2.257}$$

is the inverse of the conductivity tensor D_{ij}, and $|d_{ij}|$ is its determinant. The positive definiteness guarantees the invertibility of D_{ij}. In (2.256), for shorthand, we have assumed that x' is at the origin. If not, we should replace all x_i by $x_i - x_i'$.

The properties of D_{ij} ensure real and positive eigenvalues. Hence, there exist principal directions under which the off-diagonal terms of D_{ij} vanish. We can express (2.254) as

$$D_x \frac{\partial^2 c}{\partial x^2} + D_y \frac{\partial^2 c}{\partial y^2} = 0, \quad \text{(2D)}, \tag{2.258}$$

$$D_x \frac{\partial^2 c}{\partial x^2} + D_y \frac{\partial^2 c}{\partial y^2} + D_z \frac{\partial^2 c}{\partial z^2} = 0, \quad \text{(3D)}, \tag{2.259}$$

and the fundamental solutions reduce to

$$G = \frac{1}{2\pi \sqrt{D_x D_y}}\, \ln \sqrt{\frac{x^2}{D_x} + \frac{y^2}{D_y}}, \quad \text{(2D)}, \tag{2.260a}$$

$$= -\frac{1}{4\pi \sqrt{D_x D_y D_z}} \left(\frac{x^2}{D_x} + \frac{y^2}{D_y} + \frac{z^2}{D_z} \right)^{-\frac{1}{2}}, \quad \text{(3D)}. \tag{2.260b}$$

2.13.2 *Helmholtz-type equations*

As discussed in Section 2.5, the diffusion equation can be transformed into a modified Helmholtz equation via the Laplace transform, or a finite difference time stepping. For the Laplace transform case, (2.253) becomes

$$D_{ij}\,\tilde{c}_{,ij} - s\tilde{c} = 0, \tag{2.261}$$

ignoring the IC. Following the same coordinate transformation rules as in the previous section, we find the fundamental solutions

$$G = -\frac{|d_{ij}|^{1/2}}{2\pi}\,\mathrm{K}_0\left(\sqrt{s\,d_{ij}x_ix_j}\right), \qquad (2\mathrm{D}), \tag{2.262a}$$

$$= -\frac{|d_{ij}|^{1/2}}{4\pi\,\sqrt{d_{ij}x_ix_j}}\,\exp\left(-\sqrt{s\,d_{ij}x_ix_j}\right), \quad (3\mathrm{D}). \tag{2.262b}$$

2.13.3 *Advection–diffusion type equations*

As in Section 2.7, we shall examine two cases, the steady state and the unsteady state.

Steady State: We seek the fundamental solution of

$$D_{ij}\,G_{,ij} - V_i G_{,i} = \delta(\boldsymbol{x}, \boldsymbol{x}'). \tag{2.263}$$

The solution of the above equation in 2D has been derived by Azis [48] using complex variables for the coordinate transformation. Here we present the solutions both in 2D and 3D based on an inspection of solutions in Sections 2.7, 2.13.1, and 2.13.2, as the following:

$$G = -\frac{|d_{ij}|^{1/2}}{2\pi}\,\exp\left(\frac{1}{2}d_{ij}V_ix_j\right)\mathrm{K}_0\left(\frac{1}{2}\sqrt{d_{ij}V_iV_j}\sqrt{d_{k\ell}x_kx_\ell}\right), \qquad (2\mathrm{D}),$$

$$\tag{2.264a}$$

$$= -\frac{|d_{ij}|^{1/2}}{4\pi\,\sqrt{d_{ij}x_ix_j}}\,\exp\left(\frac{1}{2}d_{ij}V_ix_j\right)\exp\left(-\frac{1}{2}\sqrt{d_{ij}V_iV_j}\sqrt{d_{k\ell}x_kx_\ell}\right), \quad (3\mathrm{D}).$$

$$\tag{2.264b}$$

The validity of the above solutions can be verified by substitution into (2.263).

Unsteady State: We seek the fundamental solution of

$$D_{ij} G_{,ij} - V_i G_{,i} - sG = \delta(\boldsymbol{x}, \boldsymbol{x}'). \tag{2.265}$$

Following the same lead as in the steady state case, we obtain the following fundamental solutions

$$G = -\frac{|d_{ij}|^{1/2}}{2\pi} \exp\left(\frac{1}{2} d_{ij} V_i x_j\right) K_0\left(\sqrt{s + \frac{d_{ij} V_i V_j}{4}} \sqrt{d_{k\ell} x_k x_\ell}\right), \quad \text{(2D)}, \tag{2.266a}$$

$$= -\frac{|d_{ij}|^{1/2}}{4\pi \sqrt{d_{ij} x_i x_j}} \exp\left(\frac{1}{2} d_{ij} V_i x_j\right)$$

$$\times \exp\left(-\sqrt{s + \frac{d_{ij} V_i V_j}{4}} \sqrt{d_{k\ell} x_k x_\ell}\right), \quad \text{(3D)}. \tag{2.266b}$$

2.14 Anisotropic Elasticity

2.14.1 *3D general anisotropy*

The constitutive equation of elasticity for general anisotropy is

$$\sigma_{ij} = C_{ijpq} e_{pq}, \quad i, j, p, q = 1, 2, 3, \tag{2.267}$$

where σ_{ij} is the stress tensor, e_{pq} is the strain tensor, and the C_{ijpq} are the elastic moduli. Here we have assumed that the material is homogeneous in space; hence the C_{ijpq} are constants. These constants possess the following symmetry properties

$$C_{ijpq} = C_{jipq} = C_{ijqp} = C_{pqij}, \tag{2.268}$$

and only 21 of them are independent. By combining with the equilibrium equation

$$\sigma_{ij,j} = -F_i, \tag{2.269}$$

we obtain the Cauchy–Navier equations

$$C_{ijpq} u_{p,qj} = -F_i. \tag{2.270}$$

The thermodynamic equilibrium ensures the ellipticity of the system, i.e.

$$C_{ijpq} u_{i,j} u_{p,q} > 0. \tag{2.271}$$

For the fundamental solutions corresponding to the point forces, we solve the following

$$C_{ijpq} u^*_{pk,qj} = -\delta_{ik} \delta(\boldsymbol{x}, \boldsymbol{x}'). \tag{2.272}$$

Applying the Fourier transform and its inverse to the above, Lifshitz and Rosenzweig [567] obtained the solution

$$u^*_{ik}(\boldsymbol{x}, \boldsymbol{x}') = \frac{1}{8\pi^2 r} \oint_{|\boldsymbol{\xi}|=1} Q^{-1}_{ik}(\boldsymbol{\xi}) \, ds, \tag{2.273}$$

in which the line integral is taken along a unit circle $|\boldsymbol{\xi}| = 1$ on an oblique plane normal to $\boldsymbol{r}$ and passing through $\boldsymbol{x}$, with $\boldsymbol{r} = \boldsymbol{x} - \boldsymbol{x}'$ the distance vector. The integrand is given by

$$Q_{ik} = C_{ijk\ell} \xi_j \xi_\ell, \tag{2.274}$$

with Q^{-1}_{ik} denoting the elements of the inverse matrix. It is of interest to note that the integral in (2.273) is dependent only on the direction of $\boldsymbol{r}$, and not on its magnitude. To facilitate the integration, we may define two mutually orthogonal unit vectors $\boldsymbol{m}$ and $\boldsymbol{n}$ on the oblique plane. We then express $\boldsymbol{\xi}$ as

$$\boldsymbol{\xi} = \boldsymbol{m} \cos \theta + \boldsymbol{n} \sin \theta, \tag{2.275}$$

in which θ is the polar angle between $\boldsymbol{\xi}$ and $\boldsymbol{m}$. Equation (2.274) then becomes

$$Q_{ik}(\theta) = C_{ijk\ell} (m_j \cos \theta + n_j \sin \theta)(m_\ell \cos \theta + n_\ell \sin \theta), \tag{2.276}$$

and (2.273) is rewritten as [839]

$$u^*_{ik}(\boldsymbol{x}, \boldsymbol{x}') = \frac{1}{4\pi^2 r} \int_{-\pi/2}^{\pi/2} Q^{-1}_{ik}(\theta) \, d\theta. \tag{2.277}$$

In the above we have taken advantage of the symmetry of the integrand and halved the range of integration. A quadrature rule can be applied for the numerical integration of (2.277). Such an approach has been taken in the evaluation of the fundamental solution in the BEM solution of 3D anisotropic elasticity problems [917].

In another approach, the Radon transform and its inverse were utilized. Wang [885], as well as Ting and Lee [839], found an explicit expression of the line integral (2.273) by integrating around its poles, and obtained

$$u_{ik}^* = -\frac{1}{2\pi r}\Im\left\{\sum_{\alpha=1}^{3}\frac{\hat{\Gamma}_{ik}(\mu_\alpha)}{|\Gamma_{ik}(\mu_\alpha)|'}\right\},\tag{2.278}$$

where $\Im$ denotes the imaginary part. In the above, Γ_{ik} is the Stroh matrix given by [817, 831, 839]

$$\boldsymbol{\Gamma}(\mu) = \boldsymbol{Q} + \mu(\boldsymbol{R} + \boldsymbol{R}^T) + \mu^2\boldsymbol{T},\tag{2.279}$$

where

$$Q_{ik} = C_{ijk\ell}\,m_j m_\ell,\tag{2.280}$$

$$R_{ik} = C_{ijk\ell}\,m_j n_\ell,\tag{2.281}$$

$$T_{ik} = C_{ijk\ell}\,n_j n_\ell,\tag{2.282}$$

and $\hat{\boldsymbol{\Gamma}}$ is the adjugate matrix of $\boldsymbol{\Gamma}$, defined as the transpose of its cofactor matrix, the prime next to the determinant of $\boldsymbol{\Gamma}$ indicates differentiation with respect to μ. We note that the determinant $|\boldsymbol{\Gamma}|$ is a sextic polynomial equation in μ, which has three pairs of complex roots, μ_α and $\bar{\mu}_\alpha$, with $\alpha = 1, 2, 3$. Here we assume that these roots are distinct, and define the μ_α as the roots with positive imaginary part. These roots are known as the Stroh eigenvalues [817]. We can expand the determinant and factor for the roots as

$$|\boldsymbol{\Gamma}(\mu)| = \sum_{p=0}^{6} a_p\,\mu^p = a_6\prod_{\alpha=1}^{3}(\mu - \mu_\alpha)(\mu - \bar{\mu}_\alpha).\tag{2.283}$$

It has been proven by Head [372–374] that under general anisotropy, these eigenvalues cannot be found explicitly; hence have to be evaluated numerically. With these eigenvalues available, we can rewrite (2.278) as

$$u_{ik}^* = -\frac{1}{2\pi r}\Im\left\{\sum_{\alpha=1}^{3}\frac{\hat{\Gamma}_{ik}(\mu_\alpha)}{a_6(\mu_\alpha - \bar{\mu}_\alpha)\prod_{\beta=1,\,\beta\neq\alpha}^{3}(\mu_\alpha - \mu_\beta)(\mu_\alpha - \bar{\mu}_\beta)}\right\}.\tag{2.284}$$

A detailed procedure for the computation of (2.284) can be found in [840].

A third approach is to express the fundamental solution in terms of the Stroh eigenvectors associated with the oblique plane defined by the orthogonal unit vectors $\boldsymbol{m}$ and $\boldsymbol{n}$, as [62, 925]:

$$u_{ik}^* = -\frac{1}{2\pi r}\boldsymbol{A}\boldsymbol{A}^T = \frac{\mathrm{i}}{4\pi r}\boldsymbol{H}\,[\boldsymbol{r}], \qquad (2.285)$$

where $\boldsymbol{A}$ is the eigenvector matrix, and $\boldsymbol{A} = [\boldsymbol{a}_1, \boldsymbol{a}_2, \boldsymbol{a}_3]^T$, with $\boldsymbol{a}_\alpha$ satisfying

$$\boldsymbol{\Gamma}(\mu_\alpha)\boldsymbol{a}_\alpha = \boldsymbol{0}, \quad \alpha = 1, 2, 3, \qquad (2.286)$$

in which $\boldsymbol{\Gamma}$ is defined in (2.279), and μ_α are the eigenvalues. We also note that $\boldsymbol{H}$ is one of the Barnett–Lothe tensors [62, 838]. The notation $[\boldsymbol{r}]$ is used to indicate that $\boldsymbol{H}$ is dependent only on the direction of $\boldsymbol{r}$, and not on its magnitude.

The fundamental solutions expressed as (2.273), (2.278), and (2.285), need to be evaluated numerically. In a numerical method, such as the BEM or the MFS, not only the fundamental solution, but also its derivatives, are calculated many times; hence an efficient algorithm for their calculation is important. Xie *et al.* [933] tested these three expressions, and found that the one based on (2.278), which needed the eigenvalues, was the most efficient. The one based on (2.273) by numerical quadrature was the next most efficient. The method based on (2.285), which needed both the eigenvectors and eigenvalues, was by far the least efficient. In another method, Shiah *et al.* [794, 795] utilized Fourier series to approximately evaluate the anisotropic fundamental solution.

For special cases, particularly if there is material symmetry so that the characteristic equation degenerates, explicit results can be obtained. For example, for cubic materials, (2.283) can be expressed into a cubic equation in μ^2, and explicit expressions for fundamental solutions have been found [241, 565]. Similarly, explicit fundamental solutions have been derived for transversely isotropic materials [280, 663, 708].

2.14.2 *2D general anisotropy*

For 2D general anisotropy problems, we assume that all variables are functions of two spatial dimensions only; that is, $u_x = u_x(x, y)$, $u_y = u_y(x, y)$, and $u_z = u_z(x, y)$. In that case, (2.279) becomes [212, 882]

$$\boldsymbol{\Gamma}(\mu) = \Gamma_{ik}(\mu) = C_{i1k1} + (C_{i1k2} + C_{i2k1})\,\mu + C_{i2k2}\,\mu^2, \quad i, k = 1, 2, 3.$$
$$(2.287)$$

Similar to (2.279), its determinant is a sextic equation in μ. Clements and Rizzo [212] and Clements and Haselgrove [214] presented the fundamental solutions in terms of the eigenvectors, in a form similar to (2.285). The expression, however, is complicated. Here we adopt the form similar to (2.278), derived by Wang [882]

$$u_{ik}^* = \frac{1}{\pi} \Im \left\{ \sum_{\alpha=1}^{3} \frac{\hat{\Gamma}_{ik}(\mu_\alpha)}{|\Gamma_{ik}(\mu_\alpha)|'} \ln \left(z_\alpha - z_\alpha' \right) \right\}, \qquad (2.288)$$

where

$$z_\alpha = x + \mu_\alpha y, \quad z_\alpha' = x' + \mu_\alpha y'. \qquad (2.289)$$

The above defined 2D problem in fact has a 3D deformation field. It is of interest to examine two special cases:

- Plane strain: $u_x = u_x(x, y)$, $u_y = u_y(x, y)$, and $u_z = 0$.
- Anti-plane strain: $u_x = u_y = 0$, $u_z = u_z(x, y)$.

For plane strain problems, it is more convenient to switch to the engineering (Voigt) notation to shorten the expressions. The constitutive equation (2.267) can be expressed as

$$\sigma_i = C_{ij} e_j, \quad i, j = 1, \ldots, 6, \qquad (2.290)$$

where $\sigma_i = \{\sigma_{xx}, \sigma_{yy}, \sigma_{zz}, \sigma_{yz}, \sigma_{xz}, \sigma_{xy}\}$, and $e_i = \{e_{xx}, e_{yy}, e_{zz}, \gamma_{yz}, \gamma_{xz}, \gamma_{xy}\}$, with $\gamma_{xy} = 2e_{xy}$, etc. The correspondences between the C_{ij} and C_{ijpq} can be found in elasticity textbooks [99, 811]. For plane strain problems, the indices reduce to $i, j = 1, 2, 6$, with $\sigma_i = \{\sigma_{xx}, \sigma_{yy}, \sigma_{xy}\}$, $e_i = \{e_{xx}, e_{yy}, \gamma_{xy}\}$, and six independent constants, $\{C_{11}, C_{22}, C_{66}, C_{12}, C_{16}, C_{26}\}$. With this notation, we can explicitly write out (2.287) as [588]

$$\mathbf{\Gamma} = \begin{bmatrix} C_{11} + 2C_{16}\mu + C_{66}\mu^2 & C_{16} + (C_{12} + C_{66})\mu + C_{26}\mu^2 \\ C_{16} + (C_{12} + C_{66})\mu + C_{26}\mu^2 & C_{66} + 2C_{26}\mu + C_{22}\mu^2 \end{bmatrix}. \qquad (2.291)$$

The determinant of $\mathbf{\Gamma}$ is a quartic equation in μ with two pairs of complex roots

$$|\mathbf{\Gamma}(\mu)| = \sum_{p=0}^{4} a_p \mu^p = a_4 \prod_{\alpha=1}^{2} (\mu - \mu_\alpha)(\mu - \bar{\mu}_\alpha), \qquad (2.292)$$

where μ_1 and μ_2 are the two roots with positive imaginary part, and $\bar{\mu}_1$ and $\bar{\mu}_2$ their conjugates. Unlike the 3D case, these roots can be found explicitly with a computer algebra program, such as Mathematica® [918]. The expressions, however, are too lengthy to be presented. Hence we may rely on a numerical method to find these roots. Employing these results in (2.288), we obtain the fundamental solutions as

$$
u_{ik}^* = \frac{1}{\pi} \Im \left\{ \frac{\hat{\Gamma}_{ik}(\mu_1)}{a_4(\mu_1 - \bar{\mu}_1)(\mu_1 - \mu_2)(\mu_1 - \bar{\mu}_2)} \ln\left(z_1 - z_1'\right) \right.
$$

$$
\left. + \frac{\hat{\Gamma}_{ik}(\mu_2)}{a_4(\mu_2 - \bar{\mu}_2)(\mu_2 - \mu_1)(\mu_2 - \bar{\mu}_1)} \ln\left(z_2 - z_2'\right) \right\}, \quad i, k = 1, 2,
$$

$$\tag{2.293}$$

where

$$
a_4 = C_{22}C_{66} - C_{26}^2, \tag{2.294}
$$

and the adjugate matrix is given by

$$
\hat{\Gamma}_{ik} = \begin{bmatrix} C_{66} + 2C_{26}\mu + C_{22}\mu^2 & -C_{16} - (C_{12} + C_{66})\mu - C_{26}\mu^2 \\ -C_{16} + (C_{12} - C_{66})\mu - C_{26}\mu^2 & C_{11} + 2C_{16}\mu + C_{66}\mu^2 \end{bmatrix}.
$$

$$\tag{2.295}$$

We next examine the anti-plane case, with $u_z = u_z(x, y)$ which produces only the shear stresses σ_{xz} and σ_{yz}. Combining the equilibrium equation (2.269) and the constitutive equation (2.290) produces the governing equation

$$
C_{55}\frac{\partial^2 u_z}{\partial x^2} + 2C_{56}\frac{\partial^2 u_z}{\partial x \partial y} + C_{66}\frac{\partial^2 u_z}{\partial y^2} = 0, \tag{2.296}
$$

which is of the same type as (2.254), with the fundamental solution given by (2.256a).

2.14.3 *Transverse isotropy*

For materials with three axes of symmetry, called orthotropic, there are 13 independent constants; and for materials with one plane of isotropy, called transverse isotropic, there are five independent constants. In the following, we shall present the special case of transverse isotropy with explicit results.

Assuming that the plane of isotropy is the x–y plane, the independent constants are $\{C_{11}, C_{33}, C_{44}, C_{66}, C_{13}\}$. We seek the fundamental solution for the forcing function $F_{ik} = \delta_{ik}\delta(\boldsymbol{x})$, with $i, k = 1, 2, 3$. The solution was given by a few authors [527, 916], and in the following, we present the version by Pan and Chou [708].

For the force in the z-direction (perpendicular to the isotropy plane), we have

$$u_{13}^* = A\left(-\frac{xz_1}{\rho^2 R_1} + \frac{xz_2}{\rho^2 R_2}\right), \tag{2.297}$$

$$u_{23}^* = A\left(-\frac{yz_1}{\rho^2 R_1} + \frac{yz_2}{\rho^2 R_2}\right), \tag{2.298}$$

$$u_{33}^* = A\left[\frac{C_{11} - C_{44}v_1^2}{v_1(C_{13} + C_{44})}\frac{1}{R_1} - \frac{C_{11} - C_{44}v_2^2}{v_2(C_{13} + C_{44})}\frac{1}{R_2}\right]. \tag{2.299}$$

For forces parallel to the x–y plane, we find

$$u_{11}^* = \sum_{i=1}^{2} 2v_i A_i' \frac{y^2 R_i^2 - x^2 z_i^2}{\rho^4 R_i} + D\frac{x^2 R_3^2 - y^2 z_3^2}{\rho^4 R_3}, \tag{2.300}$$

$$u_{21}^* = u_{12}^* = \sum_{i=1}^{2} -2v_i A_i' \frac{xy(\rho^2 + 2z_i^2)}{\rho^4 R_i} + D\frac{xy(\rho^2 + 2z_3^2)}{\rho^4 R_3}, \tag{2.301}$$

$$u_{31}^* = \sum_{i=1}^{2} \frac{A_i'(C_{11} - C_{44}v_i^2)}{C_{13} + C_{44}}\frac{2xz_i}{\rho^2 R_i}, \tag{2.302}$$

$$u_{22}^* = \sum_{i=1}^{2} 2v_i A_i' \frac{x^2 R_i^2 - y^2 z_i^2}{\rho^4 R_i} + D\frac{y^2 R_3^2 - x^2 z_3^2}{\rho^4 R_3}, \tag{2.303}$$

$$u_{32}^* = \sum_{i=1}^{2} \frac{A_i'(C_{11} - C_{44}v_i^2)}{C_{13} + C_{44}}\frac{2yz_i}{\rho^2 R_i}. \tag{2.304}$$

In the above, we have defined

$$v_1 = \left[\frac{(\bar{C}_{13} - C_{13})(\bar{C}_{13} + C_{13} + 2C_{44})}{4C_{33}C_{44}}\right]^{\frac{1}{2}}$$

$$+ \left[\frac{(\bar{C}_{13} + C_{13})(\bar{C}_{13} - C_{13} - 2C_{44})}{4C_{33}C_{44}}\right]^{\frac{1}{2}}, \tag{2.305}$$

$$v_2 = \left[\frac{(\bar{C}_{13} - C_{13})(\bar{C}_{13} + C_{13} + 2C_{44})}{4C_{33}C_{44}} \right]^{\frac{1}{2}}$$

$$- \left[\frac{(\bar{C}_{13} + C_{13})(\bar{C}_{13} - C_{13} - 2C_{44})}{4C_{33}C_{44}} \right]^{\frac{1}{2}}, \qquad (2.306)$$

$$v_3 = \left(\frac{C_{66}}{C_{44}} \right)^{\frac{1}{2}}, \qquad (2.307)$$

$$\bar{C}_{13} = (C_{11}C_{33})^{\frac{1}{2}}, \qquad (2.308)$$

$$A = \frac{C_{13} + C_{44}}{4\pi\, C_{33}C_{44}(v_2^2 - v_1^2)}, \qquad (2.309)$$

$$A_1' = -\frac{C_{44} - C_{33}v_1^2}{8\pi\, C_{33}C_{44}(v_1^2 - v_2^2)v_1^2}, \qquad (2.310)$$

$$A_2' = \frac{C_{44} - C_{33}v_2^2}{8\pi\, C_{33}C_{44}(v_1^2 - v_2^2)v_2^2}, \qquad (2.311)$$

$$D = \frac{1}{4\pi\, C_{44}\, v_3}, \qquad (2.312)$$

$$\rho = \sqrt{x^2 + y^2}, \qquad (2.313)$$

$$z_i = v_i z, \quad i = 1, 2, 3, \qquad (2.314)$$

$$R_i = \sqrt{\rho^2 + z_i^2}, \quad i = 1, 2, 3. \qquad (2.315)$$

The above results are valid for $\bar{C}_{13} - C_{13} - 2C_{44} \neq 0$.

For $\bar{C}_{13} - C_{13} - 2C_{44} = 0$, we find $v_1 = v_2$, and the above expressions degenerate to

$$u_{13}^* = -v B x \left(\frac{z_1}{R_1^3} + \frac{z_2}{R_2^3} \right), \qquad (2.316)$$

$$u_{23}^* = -v B y \left(\frac{z_1}{R_1^3} + \frac{z_2}{R_2^3} \right), \qquad (2.317)$$

$$u_{33}^* = \sum_{i=1}^{2} \left[-\frac{C_{11}B}{C_{13} + C_{44}} \frac{1}{R_i} - \frac{Bv^2}{C_{13} + C_{44}} \frac{C_{44}\rho^2 + C_{11}z_i^2}{R_i^3} \right], \qquad (2.318)$$

$$u_{11}^* = \sum_{i=1}^{2} \left[v(A' - B') \left(\frac{1}{R_i} - \frac{x^2}{R_i^3} \right) + 2vB' \frac{y^2 R_i^2 - x^2 z_i^2}{\rho^4 R_i} \right]$$

$$+ D \frac{x^2 R_3^2 - y^2 z_3^2}{\rho^4 R_3}, \tag{2.319}$$

$$u_{21}^* = u_{12}^* = \sum_{i=1}^{2} \left[-v(A' - B') \frac{xy}{R_i^3} - 2vB' \frac{xy(\rho^2 + 2z_i^2)}{\rho^4 R_i} \right]$$

$$+ D \frac{xy(\rho^2 + 2z_3^2)}{\rho^4 R_3}, \tag{2.320}$$

$$u_{31}^* = \sum_{i=1}^{2} \left[\frac{C_{11}A' - C_{44}v^2 B'}{C_{13} + C_{44}} \frac{2xz_i}{\rho^2 R_i} - v^2 k(A' - B') \frac{xz_i}{R_i^3} \right], \tag{2.321}$$

$$u_{22}^* = \sum_{i=1}^{2} \left[v(A' - B') \left(\frac{1}{R_i} - \frac{y^2}{R_i^3} \right) + 2vB' \frac{x^2 R_i^2 - y^2 z_i^2}{\rho^4 R_i} \right]$$

$$+ D \frac{y^2 R_3^2 - x^2 z_3^2}{\rho^4 R_3}, \tag{2.322}$$

$$u_{32}^* = \sum_{i=1}^{2} \left[\frac{C_{11}A' - C_{44}v^2 B'}{C_{13} + C_{44}} \frac{2yz_i}{\rho^2 R_i} - v^2 k(A' - B') \frac{yz_i}{R_i^3} \right]. \tag{2.323}$$

In the above, we have defined

$$v = \left[\frac{C_{13} + 2C_{44}}{C_{33}} \right]^{\frac{1}{2}}, \tag{2.324}$$

$$k = \frac{C_{11} - v^2 C_{44}}{v^2 (C_{13} + C_{44})}, \tag{2.325}$$

$$B = -\frac{C_{13} + C_{44}}{16\pi C_{11} C_{44}}, \tag{2.326}$$

$$A' = \frac{1}{16\pi C_{11}}, \tag{2.327}$$

$$B' = \frac{1}{16\pi C_{44} v^2}. \tag{2.328}$$

2.14.4 *Plane strain orthotropy*

For plane strain orthotropy, there are four independent constants, $\{C_{11}, C_{22}, C_{66}, C_{12}\}$. The force fundamental solutions were given by Rizzo and Shippy [757] as

$$u_{xx}^* = D \left(\sqrt{\lambda_1} A_2^2 \ln \rho_1 - \sqrt{\lambda_2} A_1^2 \ln \rho_2 \right), \tag{2.329}$$

$$u_{xy}^* = u_{yx}^* = D A_1 A_2 \left(\tan^{-1} \frac{y}{\sqrt{\lambda_2}\, x} - \tan^{-1} \frac{y}{\sqrt{\lambda_1}\, x} \right), \tag{2.330}$$

$$u_{yy}^* = -D \left(\frac{A_1^2}{\sqrt{\lambda_1}} \ln \rho_1 - \frac{A_2^2}{\sqrt{\lambda_2}} \ln \rho_2 \right), \tag{2.331}$$

where

$$\lambda_{1,2} = \frac{B \pm \sqrt{B^2 - 4 C_{11} C_{22} C_{66}^2}}{2 C_{11} C_{66}}, \tag{2.332}$$

$$B = C_{11} C_{22} - C_{12}^2 - 2 C_{12} C_{66}, \tag{2.333}$$

$$D = \frac{C_{11}}{2\pi\, (\lambda_1 - \lambda_2)(C_{11} C_{22} - C_{12}^2)}, \tag{2.334}$$

$$A_i = -\frac{C_{12} + \lambda_i C_{11}}{C_{11} C_{22} - C_{12}^2}, \tag{2.335}$$

$$\rho_i = \sqrt{\lambda_i x^2 + y^2}. \tag{2.336}$$

2.14.5 *Elastodynamics*

The governing equation for anisotropic elastodynamics in the frequency domain is

$$C_{ijpq} \tilde{u}_{p,qj} + \rho \omega^2 \tilde{u}_i = -\tilde{F}_i. \tag{2.337}$$

For fundamental solutions due to an instantaneous point force, we solve the following:

$$C_{ijpq} u_{pk,qj}^* + \rho \omega^2 u_{ik}^* = -\delta_{ik}\, \delta(\boldsymbol{x}, \boldsymbol{x}'). \tag{2.338}$$

Utilizing the Radon transform, Wang and Achenbach [883, 884] presented the following fundamental solution

$$u_{ik}^*(\boldsymbol{x}, \boldsymbol{x}', \omega) = u_{ik}^s(\boldsymbol{x}, \boldsymbol{x}') + u_{ik}^d(\boldsymbol{x}, \boldsymbol{x}', \omega), \tag{2.339}$$

in which u_{ik}^s is the static fundamental solution presented in Section 2.14.1, and u_{ik}^d is the dynamic part given as follows:

$$u_{ik}^d(\boldsymbol{x}, \boldsymbol{x}', \omega) = \frac{\mathrm{i}}{4\pi^2} \int_{\substack{|\boldsymbol{\xi}| = 1 \\ \boldsymbol{\xi}\cdot\boldsymbol{r}>0}} \sum_{m=1}^{3} \frac{k_m E_{im} E_{km}}{2\rho c_m^2} \exp\left(\mathrm{i} k_m |\boldsymbol{\xi} \cdot \boldsymbol{r}|\right) dS(\boldsymbol{\xi}),$$

$$(2.340)$$

where the integration takes place over one half of the surface of a sphere (due to symmetry) with radius $|\boldsymbol{\xi}| = 1$, centered at $\boldsymbol{x}'$, and $\boldsymbol{r} = \boldsymbol{x} - \boldsymbol{x}'$. In the above, c_m are the phase velocities and k_m the wave numbers that can be obtained from the eigenvalue system

$$\Gamma_{ip}(\boldsymbol{\xi}) E_{pm} = \lambda_m E_{im}, \quad m = 1, 2, 3, \qquad (2.341)$$

in which

$$\Gamma_{ip}(\boldsymbol{\xi}) = C_{ijpq} \xi_j \xi_q, \qquad (2.342)$$

where λ_m are the eigenvalues, and E_{im} the eigenvectors, and no summation is performed over m in (2.341). We therefore obtain

$$c_m = \sqrt{\frac{\lambda_m}{\rho}}, \qquad (2.343)$$

$$k_m = \frac{\omega}{c_m}. \qquad (2.344)$$

We observe that c_m, k_m, and E_{im} are all functions of $\boldsymbol{\xi}$. The integral in (2.340) has to be evaluated numerically.

As eigenvectors are more difficult to obtain than eigenvalues, we would like to bypass their evaluation. If the eigenvalue system (2.341) has three distinct roots, we can find the product of eigenvectors in (2.340) as

$$E_{im} E_{km} = \frac{A_{ik}^m}{A_{\ell\ell}^m}, \qquad (2.345)$$

where

$$A_{ik}^m = A_{ik}(c_m, \boldsymbol{\xi}) = \mathrm{adj}\left\{\Gamma_{ik}(\boldsymbol{\xi}) - \rho c_m^2 \delta_{ik}\right\}, \qquad (2.346)$$

with adj denoting the adjugate matrix. An algorithm for the evaluation of the fundamental solution [268, 269] and its application in the BEM [677] have been attempted. The special cases when there are two or one distinct roots can be found in Wang and Achenbach [884].

2.15　Variable Coefficients

In the governing equations investigated so far we have assumed that the equation coefficients are constant in space and time. In applications, the coefficients are often material properties, which can vary in space. As there are an infinite number of possible variations, it is not feasible to seek their fundamental solutions. On the other hand, the detailed spatial variations of material constants are rarely known. Determining them requires a dense sampling. In practice, material constants are typically modeled as piecewise constant, or interpolated using available data points. In the former case, fundamental solutions can be found for each sub-domain. The sub-domains are then coupled through continuity conditions at the interfaces. In the latter, we shall demonstrate in this section that if the data are fitted in certain ways, fundamental solutions can be found for the fitted field.

In Section 2.2, we presented an existence of fundamental solutions theorem for linear PDEs with constant coefficients. Although the theorem cannot be extended to all linear PDEs with variable coefficients, the existence of a fundamental solution has been demonstrated for second order elliptic systems [236, 388]. In the following, we shall derive the explicit fundamental solutions for a few special variable coefficient cases.

2.15.1　*Quasi-harmonic equation*

We shall investigate the quasi-harmonic equation

$$\nabla \cdot [D(x)\nabla \phi(x)] = 0, \tag{2.347}$$

where D is a function of space. When D is a constant, it reduces to the harmonic equation (Laplace equation). In physics, D can be the thermal conductivity of heat flow, the hydraulic conductivity of porous medium flow, or the coefficient of other gradient laws.

1D Variation — Functionally graded material

We shall first investigate the case when D is a function of one spatial dimension. In modern industry applications, materials known as functionally graded materials are made to continuously vary in one direction, in order to optimize a certain physical outcome. One-dimensional varying properties are also found in nature. For example, geological formations often have their elastic moduli and permeabilities changing in the direction of depth, due to the consolidation by self weight. To model these spatial variations, material constants are typically expressed as linear, quadratic, or exponential functions.

For such problems, we seek a fundamental solution satisfying

$$\nabla \cdot \left[D(x) \nabla G(x, x') \right] = \delta(x, x').$$ (2.348)

Clements [213] derived a 2D fundamental solution for the case

$$D(x) = D_0 (1 + \lambda x)^\eta,$$ (2.349)

where D_0 is a constant, and λ and η are real numbers. The solution was a cumbersome infinite series. In the cases when $\eta = 2N$, where N is an integer, the series terminates in a finite number of terms. We give a simple case below. For

$$D(x) = D_0 (1 + \lambda x)^2,$$ (2.350)

the fundamental solutions for the 2D and 3D cases are [186]

$$G = \frac{1}{2\pi D_0 (1 + \lambda x)(1 + \lambda x')} \ln r, \quad \text{(2D)},$$ (2.351a)

$$= -\frac{1}{4\pi D_0 (1 + \lambda x)(1 + \lambda x')} \frac{1}{r}, \quad \text{(3D)}.$$ (2.351b)

In the above, the 3D solution was obtained by a methodology presented below.

Another popular fitting function is the exponential function

$$D(x) = D_0 e^{\pm \lambda x}, \quad \lambda > 0,$$ (2.352)

in which case the fundamental solutions are

$$G = -\frac{1}{2\pi D_0} \exp \left[\frac{\mp \lambda (x + x')}{2} \right] K_0 \left(\frac{\lambda r}{2} \right), \quad \text{(2D)},$$ (2.353a)

$$= -\frac{1}{4\pi D_0} \left[\frac{\mp \lambda (x + x')}{2} \right] \frac{\exp(-\lambda r / 2)}{r}, \quad \text{(3D)}.$$ (2.353b)

For

$$D(x) = D_0 \cos^2(\lambda x), \quad \lambda > 0,$$ (2.354)

we find

$$G = \frac{1}{4D_0 \cos(\lambda x)\cos(\lambda x')} Y_0(\lambda r), \qquad (2D), \qquad (2.355a)$$

$$= -\frac{1}{4\pi D_0 \cos(\lambda x)\cos(\lambda x')}\frac{\cos(\lambda r)}{r}, \qquad (3D). \qquad (2.355b)$$

Other fundamental solutions for 1D varying coefficients can be found in [186].

2D and 3D variations

When the material is heterogeneous in space, $D = D(x)$, we seek a fundamental solution satisfying

$$\nabla \cdot \left[D(x)\nabla G(x, x')\right] = \delta(x, x'). \qquad (2.356)$$

Cheng [183, 186] introduced the transformation:

$$G(x, x') = D^{-\frac{1}{2}}(x')\, D^{-\frac{1}{2}}(x)\, G^*(x, x'). \qquad (2.357)$$

Substituting the above into the LHS of (2.356), we obtain

$$\nabla \cdot \left[D(x)\nabla D^{-\frac{1}{2}}(x')\, D^{-\frac{1}{2}}(x)\, G^*(x, x')\right]$$

$$= D^{-\frac{1}{2}}(x')\nabla \cdot \left[D^{\frac{1}{2}}(x)\nabla G^*(x, x') - G^*(x, x')\nabla D^{\frac{1}{2}}(x)\right]$$

$$= D^{-\frac{1}{2}}(x')\, D^{\frac{1}{2}}(x)\,\nabla^2 G^*(x, x') - G^*(x, x')\, D^{-\frac{1}{2}}(x')\,\nabla^2 D^{\frac{1}{2}}(x).$$

$$(2.358)$$

We note that in the second line of the above equation, we used the identity

$$D(x)\,\nabla D^{-\frac{1}{2}}(x) = -\nabla D^{\frac{1}{2}}(x). \qquad (2.359)$$

If we require G^* to be the fundamental solution of the Laplace equation, i.e.

$$\nabla^2 G^*(x, x') = \delta(x, x'), \qquad (2.360)$$

we can show that

$$D^{-\frac{1}{2}}(x')\,D^{\frac{1}{2}}(x)\,\nabla^2 G^*(x,x') = D^{-\frac{1}{2}}(x')\,D^{\frac{1}{2}}(x)\,\delta(x,x') = \delta(x,x'),$$
$$(2.361)$$

because $\delta(x,x')$ is zero everywhere except at $x = x'$. If we further require the condition

$$\nabla^2 D^{\frac{1}{2}}(x) = 0, \tag{2.362}$$

in (2.358), we recover (2.356). Then (2.357) is indeed a fundamental solution of (2.356). Using (2.23), the fundamental solutions are

$$G(x,x') = \frac{1}{2\pi} D^{-\frac{1}{2}}(x')\,D^{-\frac{1}{2}}(x)\,\ln r, \quad \text{(2D)}, \tag{2.363a}$$

$$= -\frac{1}{4\pi r} D^{-\frac{1}{2}}(x')\,D^{-\frac{1}{2}}(x), \quad \text{(3D)}, \tag{2.363b}$$

which are exact and easy to construct.

Equation (2.362) shows that any harmonic function can be used to interpolate $D^{\frac{1}{2}}$. For example, $D^{\frac{1}{2}}$ in (2.350) satisfies (2.362). Based on (2.363), the fundamental solutions are found to be the ones given in (2.351). For D varying in 2D, we can have

$$D(x,y) = (c_0 + c_1 x + c_2 y + c_3 xy)^2, \tag{2.364}$$

and in 3D

$$D(x,y,z) = (c_0 + c_1 x + c_2 y + c_3 z + c_4 xy + c_5 xz + c_6 yz + c_7 xyz)^2, \tag{2.365}$$

with the fundamental solutions given by (2.363). The coefficients c_i in (2.364) and (2.365) can be determined by collocating with the available data of D. We also note that (2.350) is a 1D version of the above.

In another transformation [183, 186], we can add and subtract a term in the third line of (2.358), and obtain

$$\nabla \cdot \left[D(x)\nabla G(x,x') \right] = D^{-\frac{1}{2}}(x')\,D^{\frac{1}{2}}(x)\left[\nabla^2 G^*(x,x') \pm \lambda^2 G^*(x,x') \right]$$
$$- G^*(x,x')\,D^{-\frac{1}{2}}(x')\left[\nabla^2 D^{\frac{1}{2}}(x) \pm \lambda^2 D^{\frac{1}{2}}(x) \right]. \tag{2.366}$$

Hence, if $D^{\frac{1}{2}}$ satisfies

$$\nabla^2 D^{\frac{1}{2}}(\boldsymbol{x}) \pm \lambda^2 D^{\frac{1}{2}}(\boldsymbol{x}) = 0, \tag{2.367}$$

then the fundamental solutions are, for the plus sign,

$$G(\boldsymbol{x}, \boldsymbol{x}') = \frac{1}{4} D^{-\frac{1}{2}}(\boldsymbol{x}') D^{-\frac{1}{2}}(\boldsymbol{x}) \, Y_0(\lambda r), \qquad \text{(2D)}, \tag{2.368a}$$

$$= -\frac{1}{4\pi} D^{-\frac{1}{2}}(\boldsymbol{x}') D^{-\frac{1}{2}}(\boldsymbol{x}) \frac{\cos(\lambda r)}{r}, \quad \text{(3D)}, \tag{2.368b}$$

and for the minus sign,

$$G(\boldsymbol{x}, \boldsymbol{x}') = -\frac{1}{2\pi} D^{-\frac{1}{2}}(\boldsymbol{x}') D^{-\frac{1}{2}}(\boldsymbol{x}) \, K_0(\lambda r), \qquad \text{(2D)}, \tag{2.369a}$$

$$= -\frac{1}{4\pi} D^{-\frac{1}{2}}(\boldsymbol{x}') D^{-\frac{1}{2}}(\boldsymbol{x}) \frac{\exp(-\lambda r)}{r}, \quad \text{(3D)}. \tag{2.369b}$$

Using the above formulas, we can derive (2.352)–(2.355).

We may argue that the fundamental solutions (2.363), (2.368), and (2.369) are special cases, as condition (2.362) or (2.367) must be satisfied, which is uncommon. However, their applicability could be broader than we anticipate. In real life, it is rare that the material properties are known as an exact function, or controlled to be so. In geophysics, and many other applications, properties are known only at a small number of sampled points. To find the missing data, interpolation is practiced. The Laplace interpolation that requires the interpolated data $f(\boldsymbol{x})$ to satisfy the Laplace equation

$$\nabla^2 f(\boldsymbol{x}) = 0, \tag{2.370}$$

is a method that has been used in terrain data contouring [224, 679] in early days. In modern days, it has been used in image compression [114], image smoothing [863], shape interpolation, and surface smoothing [71]. The reason for choosing the Laplace equation is that its solution yields some of the smoothest interpolations [732]. In particular, its solution minimizes the integrated square of the gradient $\int_\Omega |\nabla f(\boldsymbol{x})|^2 \, d\boldsymbol{x}$. These interpolations, however, have a serious limitation: the solutions of the Laplace and the modified Helmholtz equation are subject to the maximum principle. As a consequence, the data set must not contain extrema inside the interpolated region. For a complex field with a sparse set of data, it is recommended that the interpolation be conducted on the vertices of a convex polygon in 2D, and of similar convex shapes in 3D that do not contain internal data. Within the interpolated region, the localized method of fundamental solutions (LMFS) can be implemented (see Section 6.8).

■ Example 2.1

As an illustration, we consider the interpolation of data located at the corners of a quadrilateral. For convenience, we use a unit square with corners located at $(0, 0)$, $(0, 1)$, $(1, 1)$, $(0, 1)$. On these nodes, the values $D = 4, 5, 6, 3$ are respectively assigned.

We can perform a fit of $\sqrt{D}$ using the bilinear function as in (2.364). The result of the interpolation of $D(x, y)$ is plotted in Fig. 2.3, and the fundamental solution within this region is given by (2.363a).

2.15.2 *Helmholtz-type equations*

The governing equation for wave propagation in an inhomogeneous medium can be expressed as

$$\nabla \cdot [\mu(x)\nabla\phi(x)] + \rho(x)\omega^2\phi = 0, \tag{2.371}$$

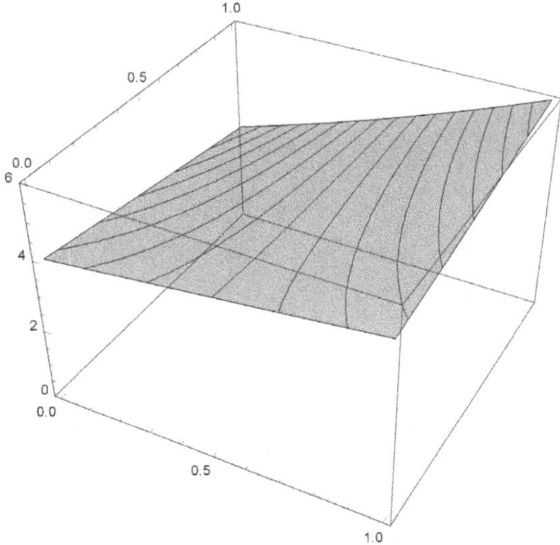

Figure 2.3. $D(x, y)$ interpolated by (2.364).

where μ is the elastic modulus for the compressional or shear wave, ρ the density, and ω the wave frequency. We may introduce the transformation

$$\phi = \mu^{-\frac{1}{2}}\phi^*. \tag{2.372}$$

Substitution of the above into the first term of (2.371) yields

$$\nabla \cdot \left[\mu \nabla \mu^{-\frac{1}{2}}\phi^* \right] = \nabla \cdot \left[\mu^{\frac{1}{2}} \nabla \phi^* - \phi^* \nabla \mu^{\frac{1}{2}} \right] = \mu^{\frac{1}{2}} \nabla^2 \phi^* - \phi^* \nabla^2 \mu^{\frac{1}{2}}. \tag{2.373}$$

This enables us to transform (2.371) into a Helmholtz equation with variable wave number

$$\nabla^2 \phi^*(x) + k^2(x)\phi^*(x) = 0, \tag{2.374}$$

where

$$k^2 = \frac{\rho \omega^2}{\mu} - \mu^{-\frac{1}{2}} \nabla^2 \mu^{\frac{1}{2}}. \tag{2.375}$$

Our interest is to find a fundamental solution of (2.374) satisfying

$$\nabla^2 G(x, x') + k^2(x)\, G(x, x') = \delta(x, x'). \tag{2.376}$$

Unfortunately, closed form fundamental solutions for a variable $k(x)$ are hard to come by. Li *et al.* [547] derived the 3D fundamental solution for the case of a 1D profile

$$k^2 = k_0^2(1 + az), \tag{2.377}$$

where the fundamental solution is given as

$$G(x, x') = \frac{e^{i\pi/6}}{2r} \left[\mathrm{Ai}'(F_1)\,\mathrm{Ai}(F_2) + e^{i\pi/3}\,\mathrm{Ai}(F_1)\,\mathrm{Ai}'(F_2) \right], \quad (3\mathrm{D}), \tag{2.378}$$

where Ai is the Airy function, Ai$'$ is its derivative [3], and

$$F_1 = -\left(\frac{k_0^2}{a^2} \right)^{1/3} \left[1 + \frac{a}{2}(z + z') + \frac{|a|}{2}r \right], \tag{2.379}$$

$$F_2 = -e^{-2i\pi/3} \left(\frac{k_0^2}{a^2} \right)^{1/3} \left[1 + \frac{a}{2}(z + z') - \frac{|a|}{2}r \right]. \tag{2.380}$$

This result can be used to model the acoustic wave in a material with a 1D variation of wave speed, such as a functionally graded material, or a temperature and salinity stratified ocean.

Other closed form fundamental solutions were presented by Shaw and Makris [792] and Clements and Larsson [215]. However, the conditions required for the coefficients μ and ρ can be too restrictive to be of practical interest; hence they are not presented here.

2.16 Variable Anisotropic Coefficients

2.16.1 *Quasi-harmonic equation*

We shall examine cases when the elliptic operator has anisotropic and variable coefficients. In particular, we seek a fundamental solution $G(x, x')$ satisfying

$$\left[D_{ij}(x) G_{,j}(x, x')\right]_{,i} = \delta(x, x'). \tag{2.381}$$

Following the idea in [30, 46, 764], we shall assume that all components of the coefficient tensor have the same spatial variation:

$$D_{ij}(x) = D_{ij}^o\, g(x), \tag{2.382}$$

where $D_{ij}^o = D_{ij}(0)$ is a constant, and $g(x)$ is a twice differentiable function with $g(0) = 1$. Utilizing the same idea as Cheng [183, 186], we can express G as

$$G(x, x') = g^{-\frac{1}{2}}(x') g^{-\frac{1}{2}}(x) G^*(x, x'), \tag{2.383}$$

and substituting the above into the LHS of (2.381), we obtain

$$\{D_{ij}^0\, g(x)[g^{-\frac{1}{2}}(x') g^{-\frac{1}{2}}(x) G^*(x, x')]_{,j}\}_{,i}$$

$$= D_{ij}^0\, g^{-\frac{1}{2}}(x')[g^{\frac{1}{2}}(x) G_{,ij}^*(x, x') - g_{,ij}^{\frac{1}{2}}(x) G^*(x, x')]. \tag{2.384}$$

In the above, we have used the relation

$$g\, g_{,j}^{-\frac{1}{2}} = -g_{,j}^{\frac{1}{2}}, \tag{2.385}$$

and canceled terms like

$$D_{ij}^0\, g_{,i}^{\frac{1}{2}}\, G_{,j}^* = D_{ij}^0\, g_{,j}^{\frac{1}{2}}\, G_{,i}^*, \tag{2.386}$$

due to the symmetry of $D_{ij}^0 = D_{ji}^0$. We now observe that if G^* is the fundamental solution in the constant coefficient case, that is

$$D_{ij}^0\, G_{,ij}^*(x, x') = \delta(x, x'), \tag{2.387}$$

and the square root of $g(x)$ satisfies the same governing equation

$$D_{ij}^0 \, g_{,ij}^{\frac{1}{2}}(x) = 0, \tag{2.388}$$

then (2.383) is the fundamental solution in (2.381).

We note that G^* in (2.383) is given by (2.256), with $D_{ij} = D_{ij}^0$. The function $g^{\frac{1}{2}}(x)$ needs to satisfy (2.388), and is no longer given by any harmonic function as in the case of (2.362). Nevertheless, the simple function

$$g(x) = (1 + ax + by + cz)^2, \tag{2.389}$$

is still valid. To have more options, we seek the general solutions of

$$D_{ij}^0 \, f_{,ij}(x) = 0. \tag{2.390}$$

As a first step, we find the principal axes of D_{ij}^0, in which only the diagonal terms exist. Performing a rotation around the principal axis system $\bar{x}$, by the relations

$$\bar{x}_i = \ell_{ij} x_j, \tag{2.391}$$

where the rotation matrix ℓ_{ij} is given by the eigenvectors of D_{ij}^0, (2.390) then transforms into

$$D_{\bar{x}\bar{x}}^0 \frac{\partial^2 f}{\partial \bar{x}^2} + D_{\bar{y}\bar{y}}^0 \frac{\partial^2 f}{\partial \bar{y}^2} + D_{\bar{z}\bar{z}}^0 \frac{\partial^2 f}{\partial \bar{z}^2} = 0, \tag{2.392}$$

where $D_{\bar{x}\bar{x}}^0$, $D_{\bar{y}\bar{y}}^0$, and $D_{\bar{z}\bar{z}}^0$ are the eigenvalues of the matrix D_{ij}^0. We next rescale the axes to

$$\bar{\bar{x}} = \frac{\bar{x}}{\sqrt{D_{\bar{x}\bar{x}}^0}}, \quad \bar{\bar{y}} = \frac{\bar{y}}{\sqrt{D_{\bar{y}\bar{y}}^0}}, \quad \bar{\bar{z}} = \frac{\bar{z}}{\sqrt{D_{\bar{z}\bar{z}}^0}}, \tag{2.393}$$

and (2.392) becomes the Laplace equation

$$\frac{\partial^2 f}{\partial \bar{\bar{x}}^2} + \frac{\partial^2 f}{\partial \bar{\bar{y}}^2} + \frac{\partial^2 f}{\partial \bar{\bar{z}}^2} = 0. \tag{2.394}$$

Hence, if $f(\bar{\bar{x}})$ is a harmonic function in $\bar{\bar{x}}$, such as a harmonic polynomial, then

$$g(x) = f^2(x), \tag{2.395}$$

fulfills the required condition (2.388), and (2.383) is the fundamental solution of (2.381).

2.16.2 *Helmholtz-type equations*

Helmholtz-type equations with variable coefficients in the following form:

$$\left[D_{ij}(x)\, G_{,j}(x, x') \right]_{,i} + \beta^2(x) G(x, x') = \delta(x, x'), \tag{2.396}$$

were investigated by Aziz *et al.* [49, 50]. The fundamental solutions, however, were derived under a rather restrictive assumption, namely that the spatial variability of both the diffusion coefficients and the wave number are defined by the same function, in the following manner

$$D_{ij}(x) = D_{ij}^{o}\, g(x), \tag{2.397}$$

$$\beta^2(x) = \beta_o^2\, g(x). \tag{2.398}$$

Details can be found in the above cited papers.

2.16.3 *Cauchy–Navier equations*

If the constitutive constants C_{ijpq} in (2.267) are functions of space, the Cauchy–Navier equations of elasticity become

$$\left(C_{ijpq}\, u_{p,q} \right)_{,j} = 0, \tag{2.399}$$

and we seek a fundamental solution that satisfies

$$\left(C_{ijpq}\, u^{*}_{pk,q} \right)_{,j} = \delta_{ik}\, \delta(x, x'). \tag{2.400}$$

Following Azis and Clements [45, 47], we shall treat the special case when

$$C_{ijpq}(x) = C_{ijpq}^{0}\, g(x), \tag{2.401}$$

where $C_{ijpq}^{0} = C_{ijpq}(\mathbf{0})$ and $g(\mathbf{0}) = 1$.
  Similar to (2.383), we designate

$$u^{*}_{pk}(x, x') = g^{-\frac{1}{2}}(x')\, g^{-\frac{1}{2}}(x)\, u^{0}_{pk}(x, x'). \tag{2.402}$$

Substituting the above into the LHS of (2.400), we obtain

$$\{ C_{ijpq}^{0}\, g(x) [g^{-\frac{1}{2}}(x')\, g^{-\frac{1}{2}}(x)\, u^{0}_{pk}(x, x')]_{,q} \}_{,j}$$

$$= C_{ijpq}^{0}\, g^{-\frac{1}{2}}(x') [g^{\frac{1}{2}}(x)\, u^{0}_{pk,qj}(x, x') - g^{\frac{1}{2}}_{,qj}(x)\, u^{0}_{pk}(x, x')]. \tag{2.403}$$

Hence, if u_{pk}^0 is a fundamental solution satisfying

$$C_{ijpq}^0 u_{pk,qj}^0 (x, x') = \delta_{ik} \delta(x, x'), \qquad (2.404)$$

and the square root of $g(x)$ is a general solution of

$$C_{ijpq}^0 g_{,qj}^{\frac{1}{2}}(x) = 0, \qquad (2.405)$$

then (2.402) is a fundamental solution of (2.400). We note that u_{pk}^0 in (2.404) can be obtained following Section 2.14, and the simplest function for g is given by (2.389). We should also mention that Martin *et al.* [627] presented fundamental solutions for isotropic elasticity with the Lamé constants varying as an exponential function in space.

For elastodynamics, fundamental solutions in the frequency domain for certain inhomogeneous anisotropic materials have been derived in [234, 611, 745].

2.17   Half Plane and Half Space Solutions

As commented in Sections 1.1 and 1.2, Green's functions are singular solutions corresponding to given geometries with a null BC. As there are infinitely many possible geometries, and their evaluation generally involves numerical effort, it may not be beneficial to construct the various Green's functions. However, there are certain frequently encountered geometries, such as the half plane in 2D, the half space in 3D, layered domains, and spherical geometries, for which analytical Green's functions are available. We shall examine some of these cases in this section.

2.17.1   *Laplace equation*

Using the method of images, we can easily construct the half plane and half space Green's functions. In a 2D problem, we assume that the domain is defined in the upper half plane, that is, $y \geq 0$, and that a Dirac delta function is located at (a, b). On the boundary, $y = 0$, we may prescribe two types of BCs, either a Dirichlet condition $\phi = 0$, or a Neumann condition $\partial \phi / \partial y = 0$.

The solution of the Dirichlet problem is easily obtained by placing a free-space Green's function (fundamental solution) centered at (a, b), and adding to it a negative image at $(a, -b)$. In other words, the Green's function is

$$G = \frac{1}{2\pi} \, (\ln r_1 - \ln r_2), \tag{2.406}$$

where $r_1 = \sqrt{(x - a)^2 + (y - b)^2}$ and $r_2 = \sqrt{(x - a)^2 + (y + b)^2}$. It is easy to show that the condition $G = 0$ is satisfied at $y = 0$. For the Neumann type, we make the image a positive one, and obtain

$$G = \frac{1}{2\pi} \, (\ln r_1 + \ln r_2). \tag{2.407}$$

We can show that the no-flux condition is satisfied at $y = 0$. The 3D cases can be constructed in a similar way.

For solving BVPs, in which a part of the boundary is bounded by a half plane or half space with the null BCs, the use of such Green's functions in the MFS, instead of the free space ones, can eliminate the need of collocation of BCs on these surfaces, as they are automatically satisfied. This can lead to significant savings in the numerical effort.

2.17.2 *Helmholtz-type equations*

Following the above section, the half plane and half space solutions for Helmholtz-type equations can be similarly constructed. For wave problems formulated as the Helmholtz equation, there are two types of BCs on the surface: a fully absorbing condition, given as $\phi = 0$, and a fully reflective condition $\partial\phi/\partial y = 0$. For the 2D Helmholtz equation, the solutions are, respectively,

$$G = \frac{1}{4} \, [Y_0(k r_1) - Y_0(k r_2)], \tag{2.408}$$

and

$$G = \frac{1}{4} \, [Y_0(k r_1) + Y_0(k r_2)]. \tag{2.409}$$

The 3D solution, and those for the modified Helmholtz equation, follow the same principle.

2.17.3 *Cauchy–Navier equations*

The half plane and half space solutions for elasticity are more difficult to construct, because they are vectors. While there are multiple BCs to satisfy, a mirror image provides only one degree of freedom. For example, we may use an image to eliminate the normal stress on the half space boundary, but we cannot simultaneously eliminate the shear stresses. Nevertheless, these classical solutions have been obtained.

Half Space (Mindlin) Solution: The 3D half space solution with a free stress boundary was obtained by Mindlin [645]. Referring to Fig. 2.4, a concentrated force F of unit magnitude is acting on the point $(0, 0, c)$ located in the half space $z \geq 0$. The bounding plane $z = 0$ is free from normal and shear stresses; i.e. $\sigma_{zz} = \sigma_{xz} = \sigma_{yz} = 0$. We are thus aiming to solve two problems: one is due to a force normal to the plane, in the z-direction, and the second due to a force parallel to the plane, which we have conveniently chosen to be the x-direction. We shall present the normal force solution first.

Following Mindlin [645], we first place at $(0, 0, c)$ in the z-direction a full space point force solution (Kelvin solution), given by (2.89b). This will create normal and shear stresses on the plane $z = 0$. To cancel these stresses, Mindlin in a heuristic way added at the image location $(0, 0, -c)$ a concentrated force and a double force in the z-direction, a center of dilatation, a doublet (derivative of center of dilatation) in the z-direction, and a line of dilation along the z-axis from $z = 0$ to $z = -\infty$, with proper magnitudes.

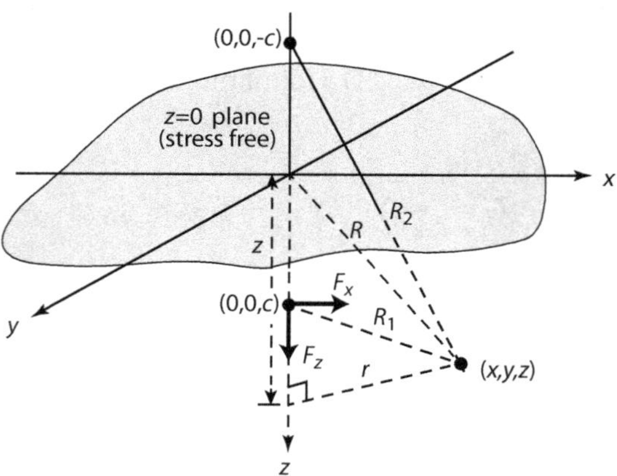

Figure 2.4. Mindlin solution: Point force in a half space.

The resultant, expressed in terms of the Galerkin vector V [645, 762], only has a nonzero z-component, namely

$$V_z = \frac{1}{8\pi(1-v)}\left\{ R_1 + [8v(1-v)-1]R_2 - \frac{2cz}{R_2} \right.$$

$$\left. + 4(1-2v)[(1-v)z - cv]\ln(R_2 + z + c) \right\}, \qquad (2.410)$$

with

$$R_1 = \sqrt{r^2 + (z-c)^2}, \qquad (2.411)$$

$$R_2 = \sqrt{r^2 + (z+c)^2}, \qquad (2.412)$$

$$r = \sqrt{x^2 + y^2}. \qquad (2.413)$$

The displacement vector can be obtained from the Galerkin vector as

$$u = \frac{1}{2\mu}\left[2(1-v)\nabla^2 V - \nabla(\nabla \cdot V) \right]. \qquad (2.414)$$

Substituting in (2.410), we obtain

$$u_{xz}^* = \frac{x}{16\pi\mu(1-v)}\left[\frac{z-c}{R_1^3} + \frac{(3-4v)(z-c)}{R_2^3} \right.$$

$$\left. - \frac{4(1-v)(1-2v)}{R_2(R_2 + z + c)} + \frac{6cz(z+c)}{R_2^5} \right], \qquad (2.415)$$

$$u_{yz}^* = \frac{y}{16\pi\mu(1-v)}\left[\frac{z-c}{R_1^3} + \frac{(3-4v)(z-c)}{R_2^3} \right.$$

$$\left. - \frac{4(1-v)(1-2v)}{R_2(R_2 + z + c)} + \frac{6cz(z+c)}{R_2^5} \right], \qquad (2.416)$$

$$u_{zz}^* = \frac{1}{16\pi\mu(1-v)}\left[\frac{3-4v}{R_1} + \frac{8v^2 - 12v + 5}{R_2} + \frac{(z-c)^2}{R_1^3} \right.$$

$$\left. + \frac{(3-4v)(z+c)^2 - 2cz}{R_2^3} + \frac{6cz(z+c)^2}{R_2^5} \right]. \qquad (2.417)$$

In the above, the second index of the displacements indicates the direction of the force.

For a force in the x-direction, Mindlin [646] added images of a variety of nuclei of strain to create the Galerkin vector

$$V_x = \frac{1}{8\pi(1-v)}\left\{ R_1 + R_2 - \frac{2c^2}{R_2} \right.$$

$$\left. + 4(1-v)(1-2v)\left[(z+c)\ln(R_2+z+c) - R_2 \right] \right\}, \quad (2.418)$$

$$V_z = \frac{1}{8\pi(1-v)}\left[\frac{2cx}{R_2} + 2(1-2v)x\ln(R_2+z+c) \right]. \quad (2.419)$$

The displacements are

$$u_{xx}^* = \frac{1}{16\pi\mu(1-v)}\left\{ \frac{3-4v}{R_1} + \frac{1}{R_2} + \frac{x^2}{R_1^3} + \frac{(3-4v)x^2}{R_2^3} \right.$$

$$\left. + \frac{2cz}{R_2^3}\left(1 - \frac{3x^2}{R_2^2}\right) + \frac{4(1-v)(1-2v)}{R_2+z+c}\left[1 - \frac{x^2}{R_2(R_2+z+c)} \right] \right\},$$

$$(2.420)$$

$$u_{yx}^* = \frac{xy}{16\pi\mu(1-v)}\left[\frac{1}{R_1^3} + \frac{3-4v}{R_2^3} - \frac{6cz}{R_2^5} - \frac{4(1-v)(1-2v)}{R_2(R_2+z+c)^2} \right],$$

$$(2.421)$$

$$u_{zx}^* = \frac{x}{16\pi\mu(1-v)}\left[\frac{z-c}{R_1^3} + \frac{(3-4v)(z-c)}{R_2^3} - \frac{6cz(z+c)}{R_2^5} \right.$$

$$\left. + \frac{4(1-v)(1-2v)}{R_2(R_2+z+c)} \right]. \quad (2.422)$$

For the case of a force in the y-direction, we can interchange the roles of x and y in the above, to obtain u_{xy}^*, u_{yy}^*, and u_{zy}^*.

The above results are presented with the force located on the z-axis, as shown in Fig. 2.4, for an easy illustration. For a force acting at a general point (x', y', z') in the half space, we can replace x by $x - x'$, y by $y - y'$, and c by z', in the above expressions to get the desired result. The stresses can be found using the constitutive equations, which were derived in [646].

Half Plane (Melan) Solution: In 2D, the solution in terms of stresses of a force acting at the point $(0, c)$ in the half plane $y \geq 0$, with a free

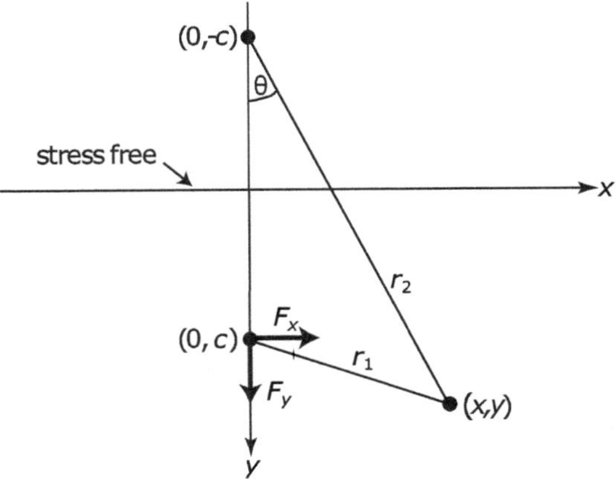

Figure 2.5. Melan solution: Point force in a half plane.

stress boundary at $y = 0$ (see Fig. 2.5), was obtained by Melan [635]. The displacements were found as [833]

$$u_{xx}^* = \frac{1}{8\pi\mu(1-v)}\left[\frac{x^2}{r_1^2} - (3-4v)\ln r_1 - (8v^2 - 12v + 5)\ln r_2\right.$$
$$\left. + \frac{(3-4v)x^2 + 2cy}{r_2^2} - \frac{4cx^2y}{r_2^4}\right], \tag{2.423}$$

$$u_{yx}^* = \frac{1}{8\pi\mu(1-v)}\left[\frac{x(y-c)}{r_1^2} + \frac{(3-4v)(y-c)x}{r_2^2} - \frac{4cxy(y+c)}{r_2^4}\right.$$
$$\left. + 4(1-v)(1-2v)\theta\right], \tag{2.424}$$

$$u_{xy}^* = \frac{1}{8\pi\mu(1-v)}\left[\frac{x(y-c)}{r_1^2} + \frac{(3-4v)(y-c)x}{r_2^2} + \frac{4cxy(y+c)}{r_2^4}\right.$$
$$\left. - 4(1-v)(1-2v)\theta\right], \tag{2.425}$$

$$u_{yy}^* = \frac{1}{8\pi\mu(1-v)}\left[\frac{(y-c)^2}{r_1^2} - (3-4v)\ln r_1 - (8v^2 - 12v + 5)\ln r_2\right.$$
$$\left. + \frac{(3-4v)(y+c)^2 - 2cy}{r_2^2} + \frac{4cy(y+c)^2}{r_2^4}\right], \tag{2.426}$$

in which

$$r_1 = \sqrt{x^2 + (y-c)^2}, \tag{2.427}$$

$$r_2 = \sqrt{x^2 + (y+c)^2}, \tag{2.428}$$

$$\theta = \tan^{-1} \frac{x}{y+c}. \tag{2.429}$$

In the above, the second index of the displacement denotes the direction of the force. The stresses were obtained by Melan [635], and corrected by Telles and Brebbia [833] as

$$
\begin{aligned}
\sigma_{xxx} = \frac{-1}{4\pi(1-v)} \Bigg\{ & \frac{(1-2v)x}{r_1^2} + \frac{2x^3}{r_1^4} + \frac{3(1-2v)x}{r_2^2} \\
& + \frac{2x\left[x^2 - 4cy - 2c^2 - 2(1-2v)y(y+c)\right]}{r_2^4} \\
& + \frac{16cxy(y+c)^2}{r_2^6} \Bigg\},
\end{aligned}
\tag{2.430}
$$

$$
\begin{aligned}
\sigma_{xyx} = \sigma_{yxx} = \frac{-1}{4\pi(1-v)} \Bigg\{ & \frac{(1-2v)(y-c)}{r_1^2} + \frac{2x^2(y-c)}{r_1^4} \\
& + \frac{(1-2v)(3y+c)}{r_2^2} \\
& + \frac{2\left[(x^2+2cy)(y+c) - 2(1-2v)y(y+c)^2\right]}{r_2^4} \\
& - \frac{16cx^2y(y+c)}{r_2^6} \Bigg\},
\end{aligned}
\tag{2.431}
$$

$$
\begin{aligned}
\sigma_{yyx} = \frac{1}{4\pi(1-v)} \Bigg\{ & \frac{(1-2v)x}{r_1^2} - \frac{2x(y-c)^2}{r_1^4} - \frac{(1-2v)x}{r_2^2} \\
& - \frac{2x\left[y^2 - 6cy - c^2 + 2(1-2v)y(y+c)\right]}{r_2^4} - \frac{16cx^3y}{r_2^6} \Bigg\}, \tag{2.432}
\end{aligned}
$$

$$\sigma_{xxy} = \frac{1}{4\pi(1-v)} \left\{ \frac{(1-2v)(y-c)}{r_1^2} - \frac{2x^2(y-c)}{r_1^4} - \frac{(1-2v)(y+3c)}{r_2^2} \right.$$

$$- \frac{2\left[(x^2+2c^2)(y+c) - 2cx^2 + 2(1-2v)x^2y\right]}{r_2^4}$$

$$\left. - \frac{16cx^2y(y+c)}{r_2^6} \right\}, \tag{2.433}$$

$$\sigma_{xyy} = \sigma_{yxy} = \frac{-1}{4\pi(1-v)} \left\{ \frac{(1-2v)x}{r_1^2} + \frac{2x(y-c)^2}{r_1^4} - \frac{(1-2v)x}{r_2^2} \right.$$

$$+ \frac{2x\left[y^2 - 2cy - c^2 + 2(1-2v)y(y+c)\right]}{r_2^4}$$

$$\left. + \frac{16cxy(y+c)^2}{r_2^6} \right\}, \tag{2.434}$$

$$\sigma_{yyy} = \frac{-1}{4\pi(1-v)} \left\{ \frac{(1-2v)(y-c)}{r_1^2} + \frac{2(y-c)^3}{r_1^4} + \frac{(1-2v)(3y+c)}{r_2^2} \right.$$

$$+ \frac{2\left[(y^2 + 4cy + c^2)(y+c) - 2(1-2v)x^2y\right]}{r_2^4}$$

$$\left. - \frac{16cx^2y(y+c)}{r_2^6} \right\}. \tag{2.435}$$

Again, the last index of the stresses denotes the force direction. For convenience, the force was placed on the y-axis (see Fig. 2.5). For a force applied at a general location (x', y') in the $y > 0$ plane, we can replace x by $x - x'$, and c by y' in the above displacement and stress expressions.

2.17.4 *Layered geometry*

Another geometry that has a wide range of applications in engineering and physical sciences is the layered geometry. In layered media, the adjacent layers have different material constants and there are interfacial boundary conditions between them. Due to the need to satisfy the boundary and interfacial conditions, the derivation of these Green's functions generally involved integral transforms that needed to be inverted via numerical means.

In elasticity, an important solution is the generalized Mindlin problem, in which a point force is applied to a layered system [123, 699]. Other half plane/space solutions include those for elastodynamics [321, 514, 707, 769], poroelasticity [171, 701, 735, 743], and high-tech applications such as piezoelectric [704], microelectronic [943], and electromagnetic [522, 705] materials.

Another important geometry is the spherically layered system, such as Earth's mantle. These Green's functions have important applications in earth sciences such as the study of earth tide and seismic wave propagation [295, 706, 716].

2.18 Coupled Equations

In many engineering applications, there exist coupled physical effects, known as multiphysics. For example, it is well known that heating a solid can generate thermal stresses. The phenomenon has been modeled as the theory of thermal stresses [96]. The theory is uncoupled, because one can solve the heat conduction problem to obtain the temperature distribution in a solid, and then solve the elasticity problem with the gradient of temperature as a body force. However, one should be aware of the reciprocal effect, that is, a solid body stressed will generate heat. Hence, the time rate of stress change should enter as a source term in the heat diffusion equation. Mathematically, this causes the two governing equations to be coupled, and they must be solved simultaneously. This leads to the coupled theory of thermoelasticity [681].

In porous solids, such as rocks, soils, and biological tissues, saturated with a fluid, not only can the solid deformation generate a fluid pressure, but also a fluid pressure can affect the solid stress in the form of an "effective stress". There is a full equivalence between the mathematical theory of poroelasticity [89] with that of thermoelasticity, where the pore pressure plays the same role as the temperature [325]. For example, we may write the equations of poroelasticity in the following form [251]:

$$\mu \nabla^2 \boldsymbol{u} + \frac{\mu}{1-2v} \nabla \nabla \cdot \boldsymbol{u} - \alpha \nabla p = \boldsymbol{0}, \qquad (2.436)$$

$$\nabla^2 p - \frac{1}{\kappa M} \frac{\partial p}{\partial t} - \frac{\alpha}{\kappa} \frac{\partial (\nabla \cdot \boldsymbol{u})}{\partial t} = 0, \qquad (2.437)$$

in which α is the Biot effective stress coefficient, M the Biot modulus, and κ the mobility coefficient. The above equations can be compared to (2.43) and (2.78).

Fundamental solutions to the above operators correspond to the introduction of Dirac delta functions in the RHS, in turn, to the first and the second equation. The solutions in the time domain have been derived in [193], and in the Laplace transform domain in [184, 187]. Those for poroelastodynamics have been presented in the frequency domain [188, 255].

There are also other physicochemical mechanisms, such as chemical (osmotic), electrical, and magnetic effects. According to Onsager's reciprocity principle [692, 693], when one effect exerts an influence on the other, there is a reciprocal effect. As a consequence, the governing equations for multiphysics modeling are coupled. Closed form fundamental solutions for porothermoelasticity and porochemoelasticity have been derived in [202]. The coupled electric and elastic effects lead to piezoelectricity, and fundamental solutions are available for isotropic and anisotropic materials [256, 257, 277, 523, 700, 702]. Furthermore, fundamental solutions for magnetoelectroelastic [179, 258, 703, 705] and electromagnetothermoelastic [398] materials have been developed.

2.19 Nonlinear PDEs

The MFS and TCM are based on the superposition of basis functions that satisfy a linear PDE with undetermined coefficients. Such a principle, however, does not work for nonlinear PDEs; i.e. if u_1 and u_2 are each a solution of the governing equation, $u_1 + u_2$ is generally not. Consequently, the MFS cannot be directly applied to solve nonlinear PDEs. However, like all other numerical methods, we can seek indirect ways to circumvent the difficulty.

One technique that allows the MFS to treat nonlinear PDEs is by variable transformation. For example, the Kirchhoff transformation can render a nonlinear diffusion equation linear. The resultant linear PDE is then amenable for treatment by the BEM [87, 419, 484, 922] and the MFS [315, 451, 619, 892].

Another method is by the iteration of trial solutions, accelerated by techniques such as the Picard iteration method (PIM). For a nonlinear PDE, we may extract a linear operator and place it on the LHS. The remainder is lumped to the RHS as inhomogeneous parts of a linear PDE. An initial guess of the solution is attempted, which makes the RHS known. The MPS-MFS presented in Section 4.1 and Chapter 8 can be employed for the solution. The obtained trial solutions are used to successively update the RHS, until convergence is achieved. This technique has been used to solve nonlinear elliptic equations [55, 57, 133], as well as other nonlinear PDEs [493, 494, 771], by the MFS. The deficiency of this approach is that the rearrangement of the equations is largely an ad-hoc act, and there is no guarantee that the solution will converge.

Other techniques involved the expansion of the nonlinear PDEs into a series of linear ones. The asymptotic numerical method (ANM) [216, 232] has been combined with the MFS to solve nonlinear Poisson [845], biharmonic [847], and Lagrangian elasticity [35] problems. The analog equation method (AEM) [474, 671] was employed in the MFS [544, 846, 894] and the LMFS [966] to solve nonlinear elliptic equations. The AEM has also been combined with the BKM to solve the regularized long wave equation [242], hyperbolic telegraph equation [243], and with the BEM for plates and other nonlinear equations [475, 476]. The MFS can be similarly applied to these problems.

Another useful method is the Homotopy Analysis Method (HAM) [562–564]. Originally developed for the analytical solution of nonlinear ODEs, it has been combined with the FEM [227, 572], the FDM [226, 970], and the BEM [560, 561, 930] to solve nonlinear PDEs. The advantage of the HAM over other perturbation based methods is that it does not rely on the existence of a small physical parameter to perform the perturbation. Also, parameters can be introduced to control the convergence radius such that it can be applied to highly nonlinear problems. It has been combined with the MFS to solve a nonlinear Poisson equation [855], diffusion equation [126], and fourth order coupled plate equations [867].

In the following, we shall give the Kirchhoff transformation a brief description.

Kirchhoff Transformation: In heat conduction, the thermal conductivity coefficient is generally a property of the material. However, when the range of temperature is large, the coefficient can be a function of the temperature as well [121]. Similarly, for the mass diffusion in a liquid, the diffusion coefficient can be dependent on the solute concentration. In steady state, we may rewrite (2.347) into the following form:

$$\nabla \cdot [D(\phi)\nabla\phi(x)] = 0. \tag{2.438}$$

The above equation is nonlinear and can be subjected to the well known Kirchhoff transformation. We may introduce a functional $\Phi(\phi)$ as

$$\Phi(\phi) = \int_{\phi_o}^{\phi} D(\phi)\, d\phi, \tag{2.439}$$

where ϕ_o is a reference value and it is obvious that

$$D(\phi) = \frac{d\Phi}{d\phi}. \tag{2.440}$$

Substituting the above into (2.438), and being aware of the relation

$$\nabla \Phi(\phi) = \frac{d\Phi}{d\phi} \nabla \phi(x), \tag{2.441}$$

we can transform (2.438) into the Laplace equation

$$\nabla^2 \Phi = 0. \tag{2.442}$$

To solve the above equation by the MFS as a BVP, we need the BCs also expressed in terms of Φ. The Dirichlet and Neumann BCs can generally be resolved into the same type of conditions. Robin type BCs, however, become nonlinear. Iteration can be performed for the convergence of the BCs.

For a demonstration, we treat the simplest case of a nonlinear conductivity coefficient as

$$D(\phi) = D_o[1 + \beta(\phi - \phi_o)] = D_o(1 + \beta\bar{\phi}), \tag{2.443}$$

where we have defined a new variable $\bar{\phi}(x) = \phi(x) - \phi_o$, and ϕ_o and $D_o = D(\phi_o)$ are reference values. Based on the above, we can find Φ using (2.439), and obtain

$$\Phi = \frac{D_o \beta}{2} \bar{\phi}^2. \tag{2.444}$$

The Dirichlet condition

$$\bar{\phi} = g(x), \quad x \in \partial\Omega_D, \tag{2.445}$$

transforms into another Dirichlet condition

$$\Phi = \frac{D_o \beta}{2} g^2(x), \quad x \in \partial\Omega_D. \tag{2.446}$$

The Neumann condition is given by the normal flux

$$-D(\bar{\phi})\frac{\partial \bar{\phi}}{\partial n} = q(x), \quad x \in \partial\Omega_N. \tag{2.447}$$

We can show that

$$D(\bar{\phi})\frac{\partial \bar{\phi}}{\partial n} = D(\bar{\phi})\frac{d\bar{\phi}}{d\Phi}\frac{\partial \Phi}{\partial n} = \frac{\partial \Phi}{\partial n}, \tag{2.448}$$

in which we have utilized (2.440), and (2.447), becomes

$$\frac{\partial \Phi}{\partial n} = -q(x), \quad x \in \partial\Omega_N. \tag{2.449}$$

With BCs (2.446) and (2.449), we are in a position to solve (2.442) by the MFS. Once Φ is determined, we can obtain $\bar{\phi}$ from (2.444).

Chapter 3

Basis Functions

In Chapter 2, we presented a range of fundamental solutions, which can be used as the basis functions in the MFS for the approximate solution of BVPs. The advantage of a fundamental solution is that it automatically satisfies the homogeneous PDE governed by the corresponding partial differential operator, and we anticipate a highly accurate and efficient solution. Another type of solutions that has this property are the general solutions of a PDE and these have been used as the basis functions in the TCM for solving BVPs.

When the PDEs are homogeneous, that is, their RHSs are null, the MFS and the TCM are among the most efficient methods for solving BVPs. On the other hand, when the governing equations are inhomogeneous, the fundamental and general solutions are not sufficient to approximate the solutions. The success of these methods relies on an efficient treatment of the RHS, using the MPS. In the MPS, the RHS function is approximated using bases other than the fundamental and general solutions. In this section, we shall discuss these other basis functions, as well as the general solutions, used in the approximation.

3.1 T-Complete General Solutions

A general solution is a solution that satisfies a homogeneous PDE without necessarily satisfying the associated BCs. It is similar to a fundamental solution, when it is used as a basis function in approximating the solution of a BVP. In general, a complete family of solutions is required to ensure the convergence of solution in a boundary solution procedure, such as the BIEM, the MFS, and the TCM. There are two possible ways in which such

117

a family can be constructed: one is a denumerable family of regular general solutions, and the other is by means of a fundamental singular solution [281, 376]. These sets have been called Trefftz-complete (T-complete) by Herrera [378, 974] to honor Trefftz [842, 843], who first proposed the use of such sets as basis functions for the solution of BVPs.

We note that the T-complete families are not unique to a PDE. They can take many forms. We shall introduce a few that are convenient to use. T-complete sets of functions have been developed for many governing equations, such as the Laplace, biharmonic, Helmholtz, modified Helmholtz, elasticity, elastodynamics, Stokes flow, and diffusion equations [217, 344, 377, 492, 500, 768].

3.1.1 *Laplacian operator*

For the Laplace equation (2.21), a T-complete set for interior domain problems is given by

$$\psi_n = \left\{1,\, r^n \cos(n\theta),\, r^n \sin(n\theta)\right\}, \qquad\qquad \text{(2D)}, \quad (3.1a)$$

$$= \left\{1,\, r^n \mathrm{P}_n^m(\cos\theta)\cos(m\varphi),\, r^n \mathrm{P}_n^m(\cos\theta)\sin(m\varphi)\right\}, \quad \text{(3D)}, \quad (3.1b)$$

where $n \in \mathbb{N}$, $0 \le m \le n$, the P_n^m are the associated Legendre polynomials of the first kind [688], θ is the polar angle for 2D ($0 \le \theta < 2\pi$), and the zenith (polar) angle for 3D ($0 \le \theta < \pi$), and φ is the azimuth angle for 3D ($0 \le \varphi < 2\pi$). The 2D case can also be expressed in the complex variable form as

$$\psi_n = \left\{1,\, \Re(z^n),\, \Im(z^n)\right\}, \quad n \in \mathbb{N}, \qquad (3.2)$$

where $z = x + iy$, and $\Re$ and $\Im$, respectively, denote the real and the imaginary part.

A closer examination of (3.1) reveals that they are polynomials, and are called harmonic polynomials,

$$\psi_n = \left\{1,\, x,\, y,\, xy,\, x^2 - y^2,\, x^3 - 3xy^2,\, y^3 - 3x^2y, \ldots\right\}, \qquad \text{(2D)}, \tag{3.3a}$$

$$= \left\{1,\, x,\, y,\, z,\, xy,\, xz,\, yz,\, x^2 - y^2,\, x^2 + y^2 - 2z^2, \ldots\right\}, \quad \text{(3D)}. \tag{3.3b}$$

The above listings give polynomials complete only up to the third and the second degree, respectively. Higher degree polynomials can be generated using recursive formulas [791]. In the actual numerical implementation, the polar and spherical coordinate expressions in (3.1) are preferred because they are easily programmed, assuming that the special functions can be automatically generated. We note that the above solutions are all regular; so the infinite set is needed for the full convergence. Clearly, in a numerical implementation, a truncated set is used.

For exterior domain problems, the sets in (3.1) become unbounded as $r \to \infty$; hence we need to choose different ones. These are given by

$$\psi_n = \left\{\ln r, \, r^{-n}\cos(n\theta), \, r^{-n}\sin(n\theta)\right\}, \qquad\qquad \text{(2D)}, \tag{3.4a}$$

$$= \left\{1/r, \, r^{-n-1}P_n^m(\cos\theta)\cos(m\varphi), \, r^{-n-1}P_n^m(\cos\theta)\sin(m\varphi)\right\}, \quad \text{(3D)}. \tag{3.4b}$$

In complex variable form, the 2D case can be expressed as

$$\psi_n = \left\{\ln r, \, \Re(z^{-n}), \, \Im(z^{-n})\right\}, \quad n \in \mathbb{N}. \tag{3.5}$$

The above sets are unbounded at $r = 0$, so the origin must be excluded from the domain. We also notice that the term $\ln r$ in (3.4a) is unbounded at infinity. If the solution sought is bounded, such as the potential flow around an airfoil discussed in Section 1.4, the term needs to be excluded. Alternatively, there should be a sufficient number of such terms at different locations to allow an auxiliary condition ensuring that their strengths sum to zero.

We in fact observe that the first terms in (3.4) are the fundamental solutions. The subsequent terms are solutions with increasingly stronger singularities. These terms can be obtained by differentiating the lower-order terms. For example, in the 2D case with $n = 1$, the second and the third term in the set are, respectively,

$$\frac{\cos\theta}{r} = \frac{x}{r^2} = \frac{\partial \ln r}{\partial x}, \quad \frac{\sin\theta}{r} = \frac{y}{r^2} = \frac{\partial \ln r}{\partial y}. \tag{3.6}$$

These are just dipoles in the x- and y-direction, respectively. The sets in (3.4) can be classified as RBFs, and their spatial derivatives. Each of these functions has a "center", where the singularity is located. For RBFs, a single function can be a T-complete set by placing it at an infinite number of locations. This is the foundation of the MFS — solutions are approximated by summing discrete sources, dipoles, or other singularities, outside of the solution domain. Each term in (3.4) can be used for this purpose. By this methodology, these bases can be used for both interior and exterior domain problems.

We should note that the above statement does not apply to the regular solutions in (3.1). These terms may have the appearance of RBFs; but they are not. This is evident by observing the set in (3.3), which are polynomials in Cartesian coordinates. Although each term has an origin, they do not possess a "center" to be translated to form an independent basis. A translation only generates a different harmonic polynomial that is a linear combination of a finite number of terms in the family.

3.1.2 *Helmholtz-type operators*

For the Helmholtz equation (2.27), T-complete sets for interior domain problems are given by

$$\psi_n = \{J_0(kr), \ J_n(kr)\cos(n\theta), \ J_n(kr)\sin(n\theta)\}, \qquad \text{(2D)}, \quad \text{(3.7a)}$$

$$= \{j_0(kr), \ j_n(kr)P_n^m(\cos\theta)\cos(m\varphi), \ j_n(kr)P_n^m(\cos\theta)\sin(m\varphi)\},$$

$$\text{(3D)}, \quad \text{(3.7b)}$$

where the j_n are the spherical Bessel functions of the first kind [688], and m and n are the same as those defined below (3.1). For the modified Helmholtz equation (2.33), we have

$$\psi_n = \{I_0(kr), \ I_n(kr)\cos(n\theta), \ I_n(kr)\sin(n\theta)\}, \quad \text{(2D)}, \qquad \text{(3.8a)}$$

$$= \{i_0^{(1)}(kr), \ i_n^{(1)}(kr)P_n^m(\cos\theta)\cos(m\varphi),$$

$$i_n^{(1)}(kr)P_n^m(\cos\theta)\sin(m\varphi)\}, \qquad \text{(3D)}, \qquad \text{(3.8b)}$$

where the $i_n^{(1)}$ are the modified spherical Bessel functions [688]. We notice that the functions in (3.7)–(3.8) are regular.

For exterior domain problems, we have, for the Helmholtz equation,

$$\psi_n = \{Y_0(kr), \ Y_n(kr)\cos(n\theta), \ Y_n(kr)\sin(n\theta)\}, \quad \text{(2D)}, \qquad \text{(3.9a)}$$

$$= \{y_0(kr), \ y_n(kr)P_n^m(\cos\theta)\cos(m\varphi),$$

$$y_n(kr)P_n^m(\cos\theta)\sin(m\varphi)\}, \qquad \text{(3D)}, \qquad \text{(3.9b)}$$

where the y_n are the spherical Bessel functions of the second kind; and for the modified Helmholtz equation,

$$\psi_n = \{K_0(kr), K_n(kr)\cos(n\theta), K_n(kr)\sin(n\theta)\}, \quad (2D), \qquad (3.10a)$$

$$= \{k_0(kr), k_n(kr)P_n^m(\cos\theta)\cos(m\varphi),$$

$$k_n(kr)P_n^m(\cos\theta)\sin(m\varphi)\}, \qquad (3D), \qquad (3.10b)$$

where the k_n are the modified spherical Bessel functions. These functions are singular.

In the above equations, we observe the following special relations for the spherical Bessel functions:

$$i_0^{(1)}(z) = \frac{\sinh z}{z}, \qquad (3.11)$$

$$j_0(z) = \frac{\sin z}{z}, \qquad (3.12)$$

$$y_0(z) = -\frac{\cos z}{z}, \qquad (3.13)$$

$$k_0(z) = \frac{\pi}{2}\frac{e^{-z}}{z}. \qquad (3.14)$$

With the above, we recognize that the first terms of (3.9)–(3.10) are the fundamental solutions given in (2.32) and (2.35), differing by a constant factor. The remaining terms involve stronger singularities. Similar to the Laplacian operator case, any such term can be used as an RBF, and a T-complete basis, by distributing it at different locations.

We further observe, with curiosity, that the regular solutions presented in (3.7)–(3.8) are also RBFs or their derivatives. By distributing a single term at different locations, we can create a T-complete basis. Being nonsingular, their centers need not be exterior to the domain, and can be placed on the boundary. This property has inspired the BKM, which utilized the nonsingular solutions I_0 and J_0 for the basis functions, as reviewed in Section 1.7.

3.1.3 *Biharmonic operator*

For the biharmonic equation (2.155), Goursat in 1898 [345] demonstrated that a biharmonic function ψ can be represented by two harmonic functions as

$$\psi = \Re\{v(z) + \bar{z}\,w(z)\}, \tag{3.15}$$

where z is a complex variable, $\bar{z}$ is its conjugate, and v and w are complex analytic functions. The above is a complete representation. As (3.15) is in complex variable form, the theory applies only to 2D. One year later, Almansi [13] gave the following representation for 3D biharmonic functions in the real domain:

$$\psi = v + r^2 w, \tag{3.16}$$

where v and w are harmonic functions. Almansi's treatment of the completeness question, however, was valid only for regions which possess certain directional convexities [311]. Timoshenko and Goodier [837] showed that (3.15) can be written as

$$\psi = \Re\left\{v(z) + z\bar{z}\,\frac{w(z)}{z}\right\}, \tag{3.17}$$

where $z\bar{z} = r^2$. Hence (3.16) is equivalent to (3.15), and is a complete representation in 2D. Utilizing (3.1a), we find the T-complete set for 2D biharmonic interior domain problems [344]:

$$\psi_n = \left\{1,\, r^2,\, r^n\cos(n\theta),\, r^{n+2}\cos(n\theta),\, r^n\sin(n\theta),\, r^{n+2}\sin(n\theta)\right\}, \quad (2D), \tag{3.18}$$

where $n \in \mathbb{N}$. For the 3D case, Fosdick [310] showed that a complete set requires four functions, given by

$$\psi = v + \boldsymbol{x} \cdot \boldsymbol{w}, \tag{3.19}$$

where v and $\boldsymbol{w}$ are harmonic functions, with $\boldsymbol{w}$ a vector. The above result, however, is valid only for a domain that is nonperiphractic (that is, without cavities). For periphractic domains, special functions need to be added [311]. Bergman and Schiffer [81], on the other hand, showed that for a star-shaped domain (a star convex set), only two harmonic functions were needed for completeness, which essentially leads to Almansi's result.

With the above restrictions in mind, we can form a 3D T-complete set for interior problems as

$$\psi_n = \left\{1,\, r^2,\, r^n \, P_n^m(\cos\theta)\cos(m\varphi),\, r^{n+2} P_n^m(\cos\theta)\cos(m\varphi),\right.$$

$$\left. r^n\, P_n^m(\cos\theta)\sin(m\varphi),\, r^{n+2} P_n^m(\cos\theta)\sin(m\varphi)\right\},\quad (3D), \quad (3.20)$$

with $0 \le m \le n$. The functions in (3.18) and (3.20) are all regular.

For exterior domain problems, we have the following T-complete sets,

$$\psi_n = \left\{\ln r,\, r^2 \ln r,\, r^{-n}\cos(n\theta),\, r^{-n+2}\cos(n\theta),\, r^{-n}\sin(n\theta),\right.$$

$$\left. r^{-n+2}\sin(n\theta)\right\},\quad (2D), \tag{3.21a}$$

$$= \left\{1/r,\, r,\, r^{-n-1}P_n^m(\cos\theta)\cos(m\varphi),\, r^{-n+1}P_n^m(\cos\theta)\cos(m\varphi),\right.$$

$$\left. r^{-n-1}P_n^m(\cos\theta)\sin(m\varphi),\, r^{-n+1}P_n^m(\cos\theta)\sin(m\varphi)\right\},\quad (3D). \tag{3.21b}$$

We note that some of the above terms are not singular, such as $r^2 \ln r$ and r. However, their derivatives, particularly in dealing with second derivative BCs, such as the Laplacian, become singular. Hence all the above terms are regarded as singular. We also observe in the above sets that $r^2 \ln r$ and r are the fundamental solutions of biharmonic operator given in (2.156). In the implementation of the MFS, however, both parts of (3.16) are needed in order to have a complete solution. Hence both the fundamental solutions of the Laplace and the biharmonic operators need to be included, that is, $\{\ln r,\, r^2 \ln r\}$ for 2D, and $\{1/r,\, r\}$ for 3D.

3.1.4 *Cauchy–Navier operator*

For the elasticity operator in (2.78), we can seek the T-complete set from its representation by potentials, similar to the biharmonic case. For 2D problems, a complete representation of the displacement vector $\boldsymbol{u}$ was given by Muskhelishvili [662] in complex variable form as

$$u_x + \mathrm{i}u_y = \frac{1}{2G}\left[\kappa\,\psi(z) - z\,\overline{\psi'(z)} - \overline{\chi'(z)}\right], \tag{3.22}$$

where $\psi(z)$ and $\chi(z)$ are complex analytic functions, the overline denotes the conjugate, the prime indicates differentiation, and the elastic constant κ is

$$\kappa = 3 - 4\nu, \quad \text{for plane strain,} \tag{3.23a}$$

$$= \frac{3 - \nu}{1 + \nu}, \quad \text{for plane stress.} \tag{3.23b}$$

The T-complete set generated from (3.22) has been used to solve 2D elasticity problems by the TCM [204, 265, 420, 488]. The generation of such a set can be tedious, and the result is limited to 2D problems; hence we shall seek alternatives.

Based on the Helmholtz theorem, the displacement vector can be split into a scalar and a vector potential as

$$\boldsymbol{u} = \nabla\phi + \nabla \times \boldsymbol{\psi}, \tag{3.24}$$

and we also impose the constraint

$$\nabla \cdot \boldsymbol{\psi} = \boldsymbol{0}. \tag{3.25}$$

If we require

$$\nabla^4 \phi = 0, \tag{3.26}$$

$$\nabla^4 \boldsymbol{\psi} = \boldsymbol{0}, \tag{3.27}$$

then (3.24) is a complete solution of the Cauchy–Navier equations (2.78) [645, 762, 811]. We can then assemble T-complete sets for $\boldsymbol{u}$ based on the sets in Section 3.1.3.

Alternatively, we can employ the Galerkin vector, and express the displacements as

$$\boldsymbol{u} = \nabla^2 \boldsymbol{w} - \frac{1}{2(1 - \nu)} \nabla(\nabla \cdot \boldsymbol{w}), \tag{3.28}$$

with

$$\nabla^4 \boldsymbol{w} = \boldsymbol{0}. \tag{3.29}$$

This representation has one fewer potential than the Helmholtz decomposition. The Helmholtz representation, however, has a constraint (3.25). Similarly, the T-complete sets for the biharmonic equation can be used.

In another approach, we may utilize the Papkovich–Neuber functions. In that case, we write

$$u = B - \frac{1}{4(1 - v)} \nabla \left(r \cdot B + \beta\right), \tag{3.30}$$

in which

$$\nabla^2 \beta = 0, \tag{3.31}$$

$$\nabla^2 B = 0. \tag{3.32}$$

This representation includes four functions in 3D, which for a convex domain can be reduced to three. In general, however, all four functions are needed to have a complete representation. These expressions allow us to use the T-complete sets of the Laplace, instead of the biharmonic, operator.

In the above representations, we observe that the number of functions does not necessarily match the number of variables (displacements), which is associated with the number of available BCs at a given point. The lack of a convenient one-to-one match can be resolved by least squares; that is, more collocation nodes and conditions can be selected, than the number of basis functions. The solution is then found in the least squares sense. Such an approach has been used by Wang *et al.* [893] to solve 3D elasticity problems utilizing the Papkovich–Neuber functions.

3.1.5 *Stokes flow*

Here we also briefly examine Stokes flow problems, defined by (2.107) and (2.108). As mentioned in Section 2.9, an analogy exists between Stokes flow and elasticity with an incompressible material, with $v = 1/2$. We can modify the above results to obtain the Stokes flow case. For the Helmholtz representation, (3.24) becomes [841]

$$v = \nabla \times \psi, \tag{3.33}$$

because the velocity field is divergence free. Equations (3.28) and (3.30) then become

$$v = \nabla^2 w - \nabla(\nabla \cdot w), \tag{3.34}$$

and

$$v = B - \frac{1}{2} \nabla \left(r \cdot B + \beta\right). \tag{3.35}$$

3.1.6 *Variable coefficients*

Similar to the cases of variable coefficient fundamental solutions presented in Sections 2.15 and 2.16, we can construct general solutions for certain special classes of variable coefficients. Although the types of variable coefficients that allow such a construction are limited, we can consider them functions that can be used to fit a data set. If the fit is performed in a local region in which the variation is mild, a good fit can be achieved.

Quasi-Harmonic Equation: The variable coefficient elliptic equation is given by (2.347). Following the method of construction in Section 2.15.1, we shall assume that a class of general solutions ψ_i satisfying (2.347) can be expressed as

$$\psi_i(x) = D^{-\frac{1}{2}}(x)\,\psi_i^*(x), \tag{3.36}$$

in which the $\psi_i^*(x)$ are the T-complete solutions of the Laplace equation introduced in Section 3.1.1. Substituting (3.36) into (2.347) and following a similar procedure as in (2.358), we obtain

$$\nabla \cdot \left[D \nabla \left(D^{-\frac{1}{2}}\psi_i^* \right) \right] = D^{\frac{1}{2}} \nabla^2 \psi_i^* - \psi_i^* \nabla^2 D^{\frac{1}{2}}. \tag{3.37}$$

We may then conclude that, if $D^{\frac{1}{2}}$ is a harmonic function,

$$\nabla^2 D^{\frac{1}{2}}(x) = 0, \tag{3.38}$$

then $\psi_i(x)$ is the solution of the quasi-harmonic equation

$$\nabla \cdot [D(x) \nabla \psi_i(x)] = 0. \tag{3.39}$$

Certain forms of $D(x)$ that can be used for data fitting, such as (2.364) and (2.365), have been discussed in Section 2.15.1.

Anisotropic Quasi-Harmonic Equation: The general solution of the anisotropic equation

$$\left[D_{ij}(x)\,\psi_{,j}(x) \right]_{,i} = 0, \tag{3.40}$$

for a special class of D_{ij} can be constructed in a similar fashion as in Section 2.16.1 for the fundamental solution.

3.2 Polynomials

In the above section, we sought approximation bases that are general solutions of homogeneous governing equations. In a numerical solution, if the governing equation is indeed homogeneous, either the fundamental solutions or the general solutions can be employed. In that case, only the BCs need to be satisfied, resulting in a smaller discrete solution system. This type of approximation is both accurate and efficient. On the other hand, if the governing equation is inhomogeneous, these basis functions are not sufficient to solve the problem. They need to be augmented, or replaced, by other approximation bases. There exist many such families of functions, such as monomials, Chebyshev polynomials, Fourier series, wavelets, and RBFs. In this section, we discuss the simplest among them, the polynomials.

3.2.1 *General polynomials*

The family of monomials, $\{1, x, y, x^2, xy, y^2, \ldots\}$ for 2D, and similarly for 3D, can be assembled to form polynomials. It is well known that fitting a function of one variable using a high degree polynomial in 1D can be problematic, due to the potential of having real roots within the fitted interval. When such roots exist, the approximation must pass through the zero value each time a root is encountered, causing the curve to oscillate. This phenomenon is most prominent near the edges of the fitting interval, and is known as the Runge phenomenon.

■ **Example 3.1**

To give an illustration, we make an attempt to fit the Runge function:

$$f(x) = \frac{1}{1 + 25x^2},$$

between $-1 \leq x \leq 1$ using a simple polynomial of degree n.

We can place $n + 1$ equally spaced nodes x_i between -1 and 1 to sample the function, and require the polynomial to exactly fit the data at these nodes. The interpolation formula is easily obtained using Lagrange polynomials. The result of using $n = 20$ is shown as the curve marked "polynomial" in Fig. 3.1. The approximation is obviously unacceptable. On the other hand, if we fit the function

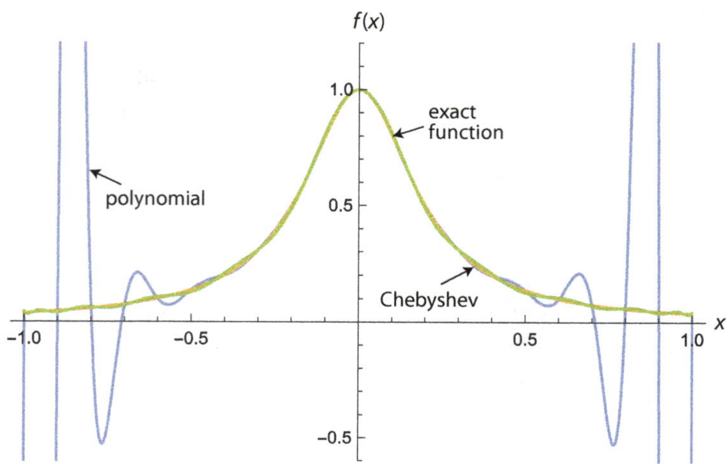

Figure 3.1. Fitting the Runge function by a simple polynomial at equally spaced nodes, and by Chebyshev polynomials, using 21 terms.

by Chebyshev polynomials (see Section 3.2.2) using terms from T_0 up to T_{20}, with data sampled at $n + 1$ roots of T_{n+1}, an excellent fit is observed in Fig. 3.1. So not all polynomials are the same.

Another way to dampen the oscillatory nature of polynomials is to use a least squares fit. With a polynomial of degree n, with $n + 1$ degrees of freedom, we can select m sampling points for the fit, with $m > n + 1$. The overdetermined system can then be solved by least squares. Figure 3.2 depicts the error of the least squares fit using $n = 20$ and $m = 61$, which is compared with the Chebyshev polynomial fit using 21 terms. We observe that the least squares fit can dampen the oscillations and deliver a performance similar to that of the Chebyshev polynomials, with a maximum error less than 2%.

In a multi-dimensional fit by a general polynomial, we encounter a different type of problem — the interpolation matrix is not positive definite, meaning that the matrix is not guaranteed to be invertible, and the result can be disastrous [37]. We can observe this by a simple example.

■ Example 3.2

In a 2D field, we select four data points located at the corners of a square, with coordinates $(0, 0)$, $(0, a)$, (a, a), and $(0, a)$. The monomial basis $\{1, x, y, x^2\}$ is used for the fit.

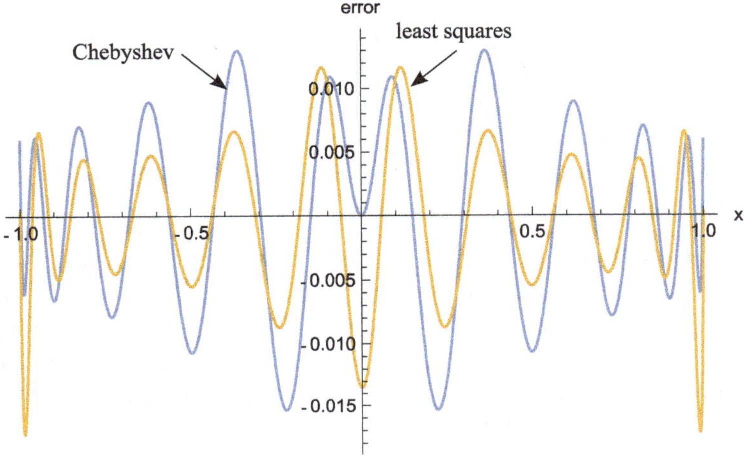

Figure 3.2. Comparison of error for fitting the Runge function using Chebyshev polynomials and the least squares fit.

It is easy to show that the resultant interpolation matrix is

$$[A] = \begin{bmatrix} 1 & 0 & 0 & 0 \\ 1 & a & 0 & a^2 \\ 1 & a & a & a^2 \\ 1 & 0 & a & 0 \end{bmatrix},$$

which is singular whatever the value of a. This is obvious because the second and the fourth column differ only by a multiplication factor a, due to the basis functions x and x^2. If we change the basis to $\{1, x, y, xy\}$, so it is unbiased in the coordinates, we obtain

$$[A] = \begin{bmatrix} 1 & 0 & 0 & 0 \\ 1 & a & 0 & 0 \\ 1 & a & a & a^2 \\ 1 & 0 & a & 0 \end{bmatrix}.$$

This matrix is no longer singular, with a determinant $|A| = -a^4$. This seems to fix the problem. However, if we use $(a/2, a/2)$ as the center to rotate the square, and keep the original sequence of node number, the determinant will change continuously. When it is rotated by $\pi/2$, we have the same appearance as the original configuration, but the node location has changed; that is, node 1 takes the position 2, and 2 moves to 3, etc. This new matrix is equivalent to three row change operations of the original matrix, and has the determinant $|A| = a^4$. As the determinant changes continuously, this indicates that there is an orientation where $|A| = 0$, which occurs at $\pi/4$. So these bases are also not safe.

It is in fact easy to prove that these interpolation matrices are never positive definite, whatever the configuration and basis selected. Given any interpolation matrix, we can switch two node numbers, say 1 to 2 and 2 to 1. This amounts to a row change, and the determinant changes sign. If a nodal configuration has symmetry, and can be rotated to the same pattern, there exist certain orientations for which the matrix is singular.

We should mention that monomial bases are indeed used in many numerical methods for local interpolation. In the FEM, the issue is resolved by mapping each element onto a fixed geometry in a local coordinate system, in which shape functions are analytically derived. In moving least squares based methods, a least squares system can always be solved, so it is also not an issue.

3.2.2 *Chebyshev polynomials*

In Chebyshev polynomial interpolation, the positions of the collocation nodes are prescribed, so it is free from the problems mentioned above. In fact, the approximation possesses spectral convergence when the polynomial degree is raised. It has been extensively utilized to solve PDEs [103, 119]. Our interest here is to use it to approximate a multi-dimensional function $f(x)$ given as the inhomogeneous part of a PDE. Its particular solution will then be sought.

Spectral Method: The Chebyshev polynomials of the first kind are defined by the recurrence relation

$$T_0(\xi) = 1, \tag{3.41}$$

$$T_1(\xi) = \xi, \tag{3.42}$$

$$T_{n+1}(\xi) = 2\xi\, T_n(\xi) - T_{n-1}(\xi), \quad n \geq 1. \tag{3.43}$$

Equivalently, in the trigonometric representation, we have

$$T_n(\cos\theta) = \cos(n\theta). \tag{3.44}$$

The polynomial of Nth degree has N roots in the interval $[-1, 1]$, given by

$$\xi_i = \cos\left(\frac{2i+1}{2N}\pi\right), \quad i = 0, 1, \ldots, N-1. \tag{3.45}$$

If these roots are used as the data point locations, we may write an approximation of a function $f(\xi)$ as

$$\hat{f}(\xi) = \sum_{k=0}^{N-1} a_k T_k(\xi), \quad -1 \le \xi \le 1, \tag{3.46}$$

where

$$a_k = \frac{2c_k}{N} \sum_{i=0}^{N-1} c_i f(\xi_i) T_k(\xi_i)$$

$$= \frac{2c_k}{N} \sum_{i=0}^{N-1} c_i f\left[\cos\left(\frac{2i+1}{2N}\pi\right)\right] \cos\left[\frac{k(2i+1)}{2N}\pi\right],$$

$$k = 0, 1, \ldots, N-1, \tag{3.47}$$

and

$$c_0 = \frac{1}{2}; \quad c_i = 1, \quad \text{for } i = 1, 2, \ldots, N-1. \tag{3.48}$$

In the above, we have used (3.44) to replace $T_k(\xi_i)$.

The effort involved in computing the coefficients a_k is of $\mathcal{O}(N^2)$. This is evident from the N-term summation in (3.47), and that there are N coefficients to compute. The effort can be reduced by utilizing the FFT. For this, we introduce the alternative expressions of (3.46) and (3.47). For (3.46), we can express f at the sampling points ξ_i as

$$f(\xi_i) = \sum_{k=0}^{N-1} a_k \cos\left[\frac{k(2i+1)}{2N}\pi\right],\tag{3.49}$$

in which we have used the relations (3.44) and (3.45). In (3.47), we have already demonstrated that

$$a_k = \frac{2c_k}{N} \sum_{i=0}^{N-1} c_i\, f(\xi_i) \cos\left[\frac{k(2i+1)}{2N}\pi\right].\tag{3.50}$$

The above pair of relations is exactly the discrete cosine transform and its inverse. Hence given the data $f(\xi_i)$, a_k can be evaluated by the FFT algorithm [103, 732], which involves $\mathcal{O}(N\log_2 N)$ operations, and is much more efficient than the direct method.

The approximation (3.46) has a spectral (exponential) error convergence rate with respect to the degree of the polynomial N (the number of nodes), as [103]

$$\varepsilon \sim O\left[(1/N)^N\right],\tag{3.51}$$

where ε is the magnitude of error. The above rate is much faster than the algebraic rate of a typical interpolation method

$$\varepsilon \sim O\left[(1/N)^k\right], \quad k = 1, 2, \ldots\tag{3.52}$$

with k a fixed number. The roots of the approximation (3.46) are all located outside of the interval $[-1, 1]$; hence within the interpolation interval, the approximation is free from the oscillation malaise observed in Fig. 3.1. We notice that the coefficients a_k in (3.46) are given explicitly in (3.47); therefore no linear system solution is required to find the coefficients. The approximation is very close to the best possible approximation by a given polynomial of certain degree, known as the minimax polynomial [732]. Hence it is recommended that, whenever possible, Chebyshev interpolation should be used, instead of a Lagrange polynomial using an arbitrary set of equally or unequally spaced data points.

In the standard form (3.46), the interpolation is performed in the interval $[-1, 1]$. For a general interpolation between $[a, b]$, a coordinate transformation is needed. For $-1 \leq \xi \leq 1$ and $a \leq x \leq b$, a linear mapping gives

$$x = \frac{b - a}{2}\xi + \frac{a + b}{2}, \tag{3.53}$$

or

$$\xi = \frac{2x - a - b}{b - a}. \tag{3.54}$$

Equation (3.46) then becomes

$$\hat{f}(x) = \sum_{k=0}^{N-1} a_k T_k\left(\frac{2x - a - b}{b - a}\right), \quad a \leq x \leq b, \tag{3.55}$$

and (3.47)

$$a_k = \frac{2c_k}{N} \sum_{i=0}^{N-1} c_i f\left[\frac{b - a}{2}\cos\left(\frac{2i + 1}{2N}\pi\right) + \frac{a + b}{2}\right] \cos\left[\frac{k(2i + 1)}{2N}\pi\right]. \tag{3.56}$$

Pseudospectral Method: We note that the approximation (3.55) is delimited by $[a, b]$, but it does not use the data at the end points, $f(a)$ and $f(b)$ in the interpolation. If the approximation is used in the solution of a BVP, it is necessary that the BCs be included at the data points. Therefore, the data sampling is modified to use the $N + 1$ extrema of the polynomial $T_N(\xi)$,

$$\xi_i = \cos\left(\frac{i\pi}{N}\right), \quad i = 0, 1, \ldots, N, \tag{3.57}$$

instead of the N roots defined in (3.45). This modified scheme is called the "pseudospectral" method. It is of interest to note that the points in (3.57) are the same as the Gauss-Lobatto abscissas used in the Gaussian quadrature method. From this point on, we shall use the pseudospectral method, instead of the spectral method.

In the pseudospectral method, the approximation of a function $f(x)$ is given by

$$\hat{f}(x) = \sum_{k=0}^{N} a_k T_k\left(\frac{2x - a - b}{b - a}\right), \quad a \le x \le b, \tag{3.58}$$

in which

$$a_k = \frac{2c_k}{N} \sum_{i=0}^{N} c_i f(x_i) \cos\left(\frac{ik\pi}{N}\right), \quad k = 0, 1, \ldots, N, \tag{3.59}$$

and

$$c_0 = c_N = \frac{1}{2}; \quad c_i = 1, \quad \text{for } i = 1, 2, \ldots, N - 1. \tag{3.60}$$

In (3.59), the function f is sampled at

$$x_i = \frac{b - a}{2} \cos\left(\frac{i\pi}{N}\right) + \frac{a + b}{2}. \tag{3.61}$$

Multi-Dimensional Chebyshev Polynomial Interpolation: The above discussion on Chebyshev approximation was for 1D problems. For multi-dimensional problems, we can utilize the product of these polynomials. For example, given a function of three spatial variables, $f(x, y, z)$, defined within $x_a \le x \le x_b$, $y_a \le y \le y_b$, and $z_a \le z \le z_b$, we can approximate the function using the pseudospectral interpolation with Chebyshev polynomials up to degrees (L, M, N), respectively, in the (x, y, z) directions, as

$$\hat{f}(x) = \sum_{i=0}^{M}\sum_{j=0}^{N}\sum_{k=0}^{L} a_{ijk}\, T_i(\xi)\, T_j(\eta)\, T_k(\zeta), \tag{3.62}$$

in which $-1 \le \xi, \eta, \zeta \le 1$, and

$$\xi = \frac{2x - x_a - x_b}{x_b - x_a}, \quad \eta = \frac{2y - y_a - y_b}{y_b - y_a}, \quad \zeta = \frac{2z - z_a - z_b}{z_b - z_a}, \tag{3.63}$$

are the transformed coordinates. The coefficients a_{ijk} are given by

$$a_{ijk} = \frac{8c_i d_j e_k}{MNL} \sum_{m=0}^{M} \sum_{n=0}^{N} \sum_{\ell=0}^{L} c_m d_n e_\ell \, f(x_m, y_n, z_\ell)$$

$$\times \cos\left(\frac{im\pi}{M}\right) \cos\left(\frac{jn\pi}{N}\right) \cos\left(\frac{k\ell\pi}{L}\right). \tag{3.64}$$

In the above equation,

$$c_0 = d_0 = e_0 = c_M = d_N = e_L = \frac{1}{2}; \quad \text{otherwise } c_i = d_i = e_i = 1; \tag{3.65}$$

and

$$x_m = \frac{x_b - x_a}{2} \cos\left(\frac{m\pi}{M}\right) + \frac{x_a + x_b}{2}, \tag{3.66}$$

$$y_n = \frac{y_b - y_a}{2} \cos\left(\frac{n\pi}{N}\right) + \frac{y_a + y_b}{2}, \tag{3.67}$$

$$z_\ell = \frac{z_b - z_a}{2} \cos\left(\frac{\ell\pi}{L}\right) + \frac{z_a + z_b}{2}. \tag{3.68}$$

The 2D version is given by

$$\hat{f}(x) = \sum_{i=0}^{M} \sum_{j=0}^{N} a_{ij} \, T_i(\xi) \, T_j(\eta), \tag{3.69}$$

and

$$a_{ij} = \frac{4c_i d_j}{MN} \sum_{m=0}^{M} \sum_{n=0}^{N} c_m d_n \, f(x_m, y_n) \cos\left(\frac{im\pi}{M}\right) \cos\left(\frac{jn\pi}{N}\right). \tag{3.70}$$

To demonstrate the efficiency of Chebyshev polynomials in approximating a function, we shall present in Example 3.3 the interpolation of a 2D function. The approximation errors are calculated at a dense set of points and plotted in the interpolated domain.

Definition of Approximation Error: Before presenting the computing example, we shall first define a few error measures. First, the relative error of an approximation is defined as

$$\varepsilon_r(x) = \frac{\hat{f}(x) - f(x)}{\max f(x) - \min f(x)}, \quad x \in \Omega, \tag{3.71}$$

in which $f(x)$ is the exact function, $\hat{f}(x)$ is the approximation, and Ω is the domain of interpolation. To give a condensed presentation, we may define a maximum relative error as

$$\varepsilon_{\max} = \frac{\|\hat{f}(x) - f(x)\|_\infty}{\max f(x) - \min f(x)}, \quad x \in \Omega, \tag{3.72}$$

in which $\|\cdot\|_\infty$ is the L-infinity norm, meaning taking the maximum absolute value of the argument. To get a sense of averaged error, we introduce the root mean square (RMS) relative error as

$$\varepsilon_{\mathrm{rms}} = \frac{1}{\max f(x) - \min f(x)} \sqrt{\frac{1}{N_e} \sum_{i=1}^{N_e} \left[\hat{f}(x_i) - f(x_i)\right]^2}, \tag{3.73}$$

where x_i, $i = 1, \ldots, N_e$, are a set of sampling points in Ω, which are at least 5 times denser than the collocation points in each spatial dimension, in the following examples.

■ Example 3.3

To test the accuracy of Chebyshev polynomial interpolation, we use it to approximate the function [198]

$$\begin{aligned}
f(x, y) = &-\frac{751\pi^2}{144} \sin\frac{\pi x}{6} \sin\frac{7\pi x}{4} \sin\frac{3\pi y}{4} \sin\frac{5\pi y}{4} \\
&+ \frac{7\pi^2}{12} \cos\frac{\pi x}{6} \cos\frac{7\pi x}{4} \sin\frac{3\pi y}{4} \sin\frac{5\pi y}{4} \\
&+ \frac{15\pi^2}{8} \sin\frac{\pi x}{6} \sin\frac{7\pi x}{4} \cos\frac{3\pi y}{4} \cos\frac{5\pi y}{4},
\end{aligned}$$

defined in $[0, 1] \times [0, 1]$, which is plotted in Fig. 3.3. We notice that the function has a few peaks and valleys in the given range.

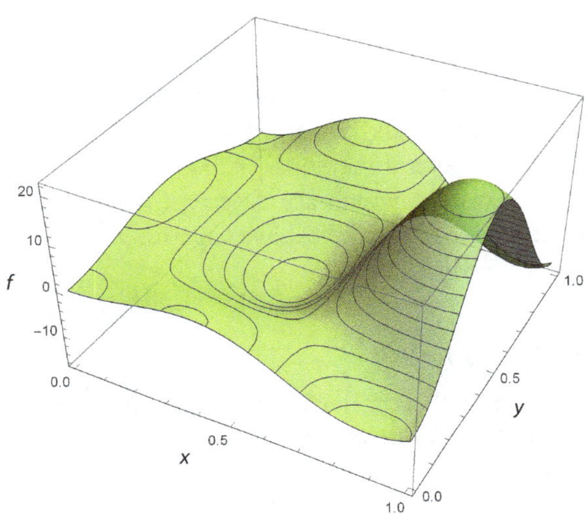

Figure 3.3. The function interpolated in Examples 3.3 and 3.4.

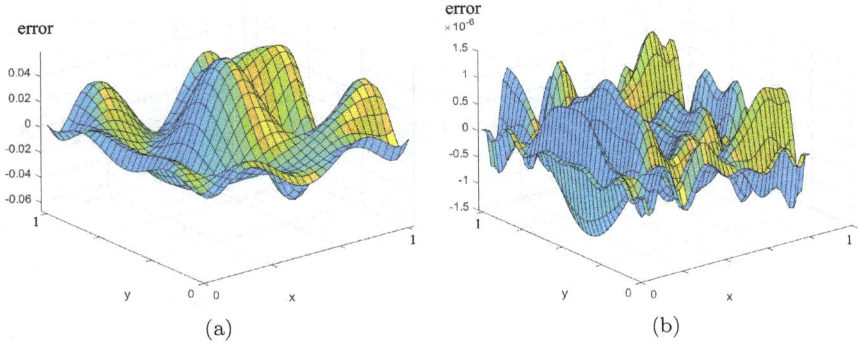

Figure 3.4. Relative error ε_r of Chebyshev interpolation. (a) $M = N = 5$, and (b) $M = N = 10$.

Equations (3.69) and (3.70) are used for the approximation. In Fig. 3.4, we plot the relative error ε_r corresponding to the $M = N = 5$ and $M = N = 10$ cases, respectively, on a dense set of points. We note in Fig. 3.4(a) that Chebyshev polynomials up to the fifth degree give a reasonable approximation with the $\varepsilon_{\max}$ of $\mathcal{O}(10^{-2})$. Figure 3.4(b), on the other hand, shows that when the polynomial degree is doubled, the $\varepsilon_{\max}$ drops four orders of magnitude to $\mathcal{O}(10^{-6})$, demonstrating the power of exponential convergence.

3.3 Radial Basis Functions

In the same spirit as the MFS, we seek basis functions that can be deployed on a scattered set of points, and are not restricted to a regular pattern. This rules out some of the spectral methods, such as Fourier series, Chebyshev polynomials, and wavelets, which are based on fixed grids. For scattered point methods, we also seek bases that have superior convergence properties. This leads us to the RBFs.

3.3.1 *Interpolation of functions*

An RBF is a function that is dependent only on the distance between two points, x and x'. The distance r is usually the Euclidean distance, $r = \|x - x'\|$, though some other measures can be used. In many applications, there is a long history of using RBFs, such as r or $1/r$, for the interpolation of scattered data in space. For example, r was used in topography modeling [370], as well as in the early versions of the dual reciprocity boundary element method (DRBEM) [106, 192, 713], and $1/r^2$ was widely adopted in the interpolation of precipitation data [207]. In 1979 and 1982, Franke [313, 314] performed a comparison of accuracy of interpolation using more than 30 methods. The Hardy multiquadric function [368, 369], an RBF, emerged as one of the most accurate and efficient of all functions tested.

In the RBF collocation method, it is typical to select the set of collocation points to coincide with the RBF centers. By the theorem of Schoenberg [783, 784] and Micchelli [641], the resulting interpolation matrix is positive definite or conditionally positive definite for a class of RBFs, as long as no two centers occupy the same location. Madych and Nelson in 1992 [605, 606] demonstrated the exponential convergence properties of a class of RBFs that included the multiquadric and the Gaussian functions. Since that time, RBFs have become highly popular basis functions used in many fields, including artificial neural networks and machine learning [111, 170, 221], data interpolation and image reconstruction [120, 363], and solving PDEs, which we will discuss in the next section. Its theoretical foundation and applications have been covered in a few books [110, 177, 309, 913].

Types of RBF: In the following, we show a few of the more frequently used RBFs:

- Pseudo-polynomial (1D, 2D) and polyharmonic spline (3D)

$$\psi = r^{2n-1}, \quad n \in \mathbb{N}. \tag{3.74}$$

- Polyharmonic spline (1D, 2D)

$$\psi = r^{2n} \ln r, \quad n \in \mathbb{N}. \tag{3.75}$$

- Generalized multiquadric (MQ) (1D, 2D, 3D)

$$\psi = (r^2 + c^2)^{\beta}, \quad c, \beta > 0, \ \beta \notin \mathbb{N}. \tag{3.76}$$

- Generalized inverse multiquadric (IMQ) (1D, 2D, 3D)

$$\psi = (r^2 + c^2)^{-\beta}, \quad c, \beta > 0. \tag{3.77}$$

- Gaussian (GA) (1D, 2D, 3D)

$$\psi = e^{-r^2/c^2}, \quad c > 0. \tag{3.78}$$

Positive Definiteness: In Table 3.1, we summarize the positive definiteness and convergence properties of these functions [201]. The positive definiteness is needed for a reliable linear system solution, because such matrices are guaranteed to be nonsingular, and have a global minimum. The former property means that the system matrix is always invertible, and the latter is needed for the convergence of a gradient based iterative solution algorithm. Not all the RBFs in Table 3.1 are positive definite, and some are only conditionally positive definite. These functions need to be augmented

Table 3.1. Order of conditional positive definiteness (m) and error estimate (ε) for various RBFs, in which $a > 0$, h is the fill distance, and $\lceil \cdot \rceil$ is the ceiling function.

RBF		m	ε
Polynomial	r^{2n-1}	n	$h^{n-(1/2)}$
Spline	$r^{2n} \ln r$	$n+1$	h^{2n}
MQ	$(r^2 + c^2)^{\beta}$	$\lceil \beta \rceil$	$\exp(-ah)$
IMQ	$(r^2 + c^2)^{-\beta}$	0	$\exp(-ah)$
GA	$\exp(-r^2/c^2)$	0	$\exp(-a \ln h/h)$

by some low degree monomials to make them fully positive definite. In Table 3.1, we show the order m of their conditional positiveness. An order m means that the representation needs to be augmented by monomial terms of degree up to $m - 1$. For example, the IMQ and GA with $m = 0$ are fully positive definite. The multiquadric $\sqrt{r^2 + c^2}$ is of the order $m = 1$, and needs to be supplemented by a constant term. The thin plate spline $r^2 \ln r$ with $n = 1$ has $m = 2$; hence the terms $\{1, x, y\}$ are needed. This means that in an approximation by the thin plate spline, the representation should be

$$\hat{f}(\boldsymbol{x}) = \sum_{i=1}^{N} c_i \, r_i^2 \ln r_i + c_{N+1} + c_{N+2} x + c_{N+3} \, y, \qquad (3.79)$$

where $r_i = \|\boldsymbol{x} - \boldsymbol{x}_i\|$, and $c_i, \ i = 1, \ldots, N + 3$, are coefficients to be determined. The above equations are augmented by the equations of moment

$$\sum_{i=1}^{N} c_i = 0, \qquad (3.80)$$

$$\sum_{i=1}^{N} c_i \, x_i = 0, \qquad (3.81)$$

$$\sum_{i=1}^{N} c_i \, y_i = 0, \qquad (3.82)$$

to form a complete solution system.

Convergence and Shape Parameter: The last column of Table 3.1 gives the error estimate ε [201, 913], where h is the fill distance, defined as

$$h = \max \min \|\boldsymbol{x} - \boldsymbol{x}_i\|, \quad \boldsymbol{x} \in \Omega, \ \boldsymbol{x}_i \in X, \qquad (3.83)$$

in which Ω is the interpolated domain, and X is the set of RBF centers contained in it. In practical terms, we may describe h as the maximum size disk and ball, respectively for 2D and 3D, that is entirely contained in Ω and can be fitted into the space among the centers without touching them.

We observe in Table 3.1 that as $h \to 0$, the error decreases in an algebraic manner for the polynomial and spline cases. The simple basis, r, which is sometimes used in geostatistics to interpolate field data, has a low convergence rate of $h^{1/2}$. For the MQ, IMQ, and GA, in which a defined in ε is a positive constant, the convergence rates are exponential. That is, as h becomes smaller, the error decreases at an increasing rate. These RBFs with spectral convergence are generally preferred.

We further observe that each of these RBFs has a scaling constant c, called the shape parameter. The shape of the functions becomes flat when c increases. It has been observed that the error can exponentially decrease with increasing c, without mesh refinement [198, 403, 404]. Such behavior was supported by the error estimate of Madych and Nelson [605, 606]. There has been a substantial effort to seek the optimal value of the shape parameter with a given mesh size [201, 298, 755, 898].

■ Example 3.4

To demonstrate the efficiency of RBFs to approximate a function, we shall test the same function shown in Example 3.3, which was fitted by Chebyshev polynomials.

Using the IMQ as the basis function, the approximation is given by

$$\hat{f}(x, y) = \sum_{j=1}^{N} c_j \frac{1}{\sqrt{r_j^2 + c^2}},$$

where $r_j = \sqrt{(x - x_j)^2 + (y - y_j)^2}$, and $(x_j, y_j) \in \Omega$ are a set of N points where the RBF centers are located. Although the RBF approximation allows the use of a set of scattered nodes, to make a comparison, we choose the same Chebyshev pseudospectral nodes for the centers. By choosing the same set of the centers as collocation points, we obtain a system of N linear equations in terms of c_j, as

$$f(x_i, y_i) = \sum_{j=1}^{N} c_j \frac{1}{\sqrt{r_{ij}^2 + c^2}}, \quad i = 1, \ldots, N,$$

where $r_{ij} = \sqrt{(x_i - x_j)^2 + (y_i - y_j)^2}$. The above equations can be solved for the undetermined coefficients c_j, and then the approximation is defined.

We interpolate the function using two resolutions, $N = 6 \times 6$ and 11×11, equivalent to the Chebyshev polynomials cases, and use

$c = 1.5$ and 1.3 respectively. The relative errors ε_r of the two cases are plotted in Figs. 3.5(a) and 3.5(b). We observe that the $\varepsilon_{\max}$ are of $\mathcal{O}(10^{-3})$ and $\mathcal{O}(10^{-6})$, respectively, which are comparable to the Chebyshev cases in Example 3.3. These results also demonstrate that the IMQ possesses the exponential convergence property, as indicated in Table 3.1.

We should mention that the IMQ and other RBFs that contain a shape parameter c are sensitive to its value. The results shown in Fig. 3.5 are based on the optimal selection found by trial and error. If we use other values, for example $c = 0.7$ and 1.0 for the 11 × 11 case, the $\varepsilon_{\max}$ are of $\mathcal{O}(10^{-4})$ and $\mathcal{O}(10^{-5})$, respectively, which are not as good as the optimal case, but still excellent.

It is also of interest to report that for the 11 × 11 case, the optimal value $c = 1.3$ caused a very large condition number of $\mathcal{O}(10^{18})$, much larger than the recommended range for matrix stability. The phenomenon of stable matrix computation with condition number exceeding $\mathcal{O}(10^{16})$ in double precision has been widely reported in collocation methods. An explanation was offered by the concept of an "effective condition number" [209]. In short, the traditional condition number is a worst case scenario estimate, which does not take into consideration the RHS of the matrix equation. The effective condition number, which can be orders of magnitude smaller than the traditional one, is a better estimate [559]. Further discussions on the optimal selection of shape parameter, condition number, and extended

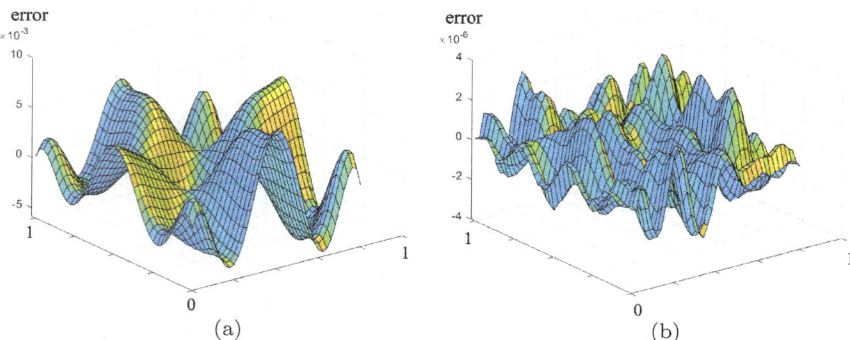

Figure 3.5. Relative error ε_r of IMQ interpolation. (a) $N = 6 \times 6$, $c = 1.5$; and (b) $N = 11 \times 11$, $c = 1.3$.

precision computations for the RBF collocation method can be found in Cheng [201].

Compactly Supported RBFs: The above-introduced RBFs are defined in the full space; that is, their influence radius reaches infinity. As a consequence, their interpolation matrices are full. A full matrix is generally solved by elimination methods, which involve $\mathcal{O}(N^3)$ operations, where N is the number of degrees of freedom of the linear system. In a multi-dimensional problem, N can increase rapidly in an effort to refine the discretization. In addition, the matrix condition number of the more accurate RBFs, such as the multiquadric and the Gaussian, can be intolerably large [201], which can cause stability problems in the system solution. One way to overcome the high solution cost and ill-conditioning issues is to use RBFs that have a finite influence radius, beyond which they are set to zero, giving zero entries to the matrix. The sparseness of the matrix can be controlled by the size of the influence radius. These are known as the compactly supported RBFs (CSRBFs).

In the following, we shall introduce the CSRBFs proposed by Wendland [911, 912]. These functions are truncated polynomials of various degrees, and can be generated by the following formulas [296, 297]:

$$\psi_{\ell,0} = (1 - \bar{r})_+^\ell, \tag{3.84}$$

$$\psi_{\ell,1} = (1 - \bar{r})_+^{\ell+1} [(\ell + 1)\bar{r} + 1], \tag{3.85}$$

$$\psi_{\ell,2} = (1 - \bar{r})_+^{\ell+2} [(\ell^2 + 4\ell + 3)\bar{r}^2 + (3\ell + 6)\bar{r} + 3], \tag{3.86}$$

$$\psi_{\ell,3} = (1 - \bar{r})_+^{\ell+3} [(\ell^3 + 9\ell^2 + 23\ell + 15)\bar{r}^3$$
$$+ (6\ell^2 + 36\ell + 45)\bar{r}^2 + (15\ell + 45)\bar{r} + 15], \tag{3.87}$$

in which $\bar{r} = r/a$, with a the user chosen influence radius, and the $+$ sign subscript of $(\cdot)_+$ means that the term is set to zero when its argument is greater than 1. These functions $\psi_{\ell,\kappa}$, $\kappa \in \mathbb{N}$, have smoothness $C^{2\kappa}$ around $r = 0$, and $C^{2\kappa + \lfloor d/2 \rfloor}$ at $r = a$. The subscript ℓ defines the dimension d that a function can be used for, with

$$\ell = \left\lfloor \frac{d}{2} \right\rfloor + \kappa + 1, \tag{3.88}$$

where $\lfloor \cdot \rfloor$ is the floor function.

■ **Example 3.5**

Find the CSRBFs in 2D and 3D that have C^2 and C^4 smoothness.

This means that $\lfloor d/2 \rfloor = 1$, $\kappa = 1, 2$, and $\ell = 3, 4$. We have, respectively,

$$\psi_{3,1} = (1 - \bar{r})_+^4 (4\bar{r} + 1),$$

$$\psi_{4,2} = (1 - \bar{r})_+^6 \left(35\bar{r}^2 + 18\bar{r} + 3\right).$$

3.3.2 Solving PDEs

As the RBFs are efficient interpolants, they can be employed to solve PDEs. Kansa in 1990 [440, 441] utilized the MQ in a collocation method to solve computational fluid dynamics problems. The RBFCM has become highly popular, solving problems such as diffusion [773], convective-diffusive solid–liquid phase change [876], fluid flow and conjugate heat transfer [261], laminated and functionally graded composite plates [300, 675], transient electromagnetics [513], time fractional nonlinear Schrödinger equation [653], and high dimensional Hamilton–Jacobi equations [122], to cite but a few. RBFs have also been used as test and trial functions in numerical methods such as the MLPG [42, 806], the local radial point interpolation method (LRPIM) [579, 580, 897], the RBF-DQ [801, 803], the RBF-FD [760, 920], among others. In the following, we shall give a brief description of the RBFCM.

Consider the BVP for the Poisson equation in 3D

$$\nabla^2 \phi(x) = f(x), \quad x \in \Omega, \tag{3.89}$$

subject to the BCs

$$\phi(x) = g(x), \quad x \in \partial\Omega_D, \tag{3.90}$$

$$\frac{\partial \phi}{\partial n}(x) = h(x), \quad x \in \partial\Omega_N. \tag{3.91}$$

We can approximate the solution as a summation of RBFs in the following form:

$$\hat{\phi}(x) = \sum_{j=1}^{N} c_j \psi(r_j), \tag{3.92}$$

where ψ is an RBF, such as the spline function, or the MQ, the c_j are undetermined coefficients, $r_j = \|x - x_j\|$, and x_j are the N selected centers for the deployment of RBFs. Of the N nodes, N_1 are located inside the domain Ω, N_2 on $\partial\Omega_D$, and N_3 on $\partial\Omega_N$, with $N = N_1 + N_2 + N_3$.

By applying the Laplacian operator to the approximation (3.92), we obtain

$$\nabla^2 \left(\sum_{j=1}^{N} c_j \psi(r_j) \right) = \sum_{j=1}^{N} c_j \nabla^2 \psi(r_j) = \sum_{j=1}^{N} c_j L(r_j). \tag{3.93}$$

In the above, for a given type of RBF, the term $L(r_j) = \nabla^2 \psi(r_j)$ can be explicitly obtained. For example, given a MQ, we find

$$L(r_j) = \nabla^2 (r_j^2 + c^2)^{1/2} = \frac{r_j^2 + 2c^2}{(r_j^2 + c^2)^{3/2}}. \tag{3.94}$$

We hence can perform the collocation of the governing equation at N_1 interior points as

$$\sum_{j=1}^{N} c_j L(r_{ij}) = f(x_i), \quad i = 1, \ldots, N_1, \quad x_i \in \Omega, \tag{3.95}$$

where $r_{ij} = \|x_i - x_j\|$.

On the Dirichlet boundary, we shall enforce the BC as

$$\sum_{j=1}^{N} c_j \psi(r_{ij}) = g(x_i), \quad i = N_1 + 1, \ldots, N_1 + N_2, \tag{3.96}$$

On the Neumann boundary, we need to take the normal derivative

$$\psi_n(x, x_j) = \frac{\partial \psi(r_j)}{\partial n(x)}$$

$$= \frac{1}{\sqrt{r_j^2 + c^2}} \left[(x - x_j)n_x + (y - y_j)n_y + (z - z_j)n_z \right], \quad (3.97)$$

where n_x, n_y, n_z are the directional cosines of the unit outward normal $n(x)$. Collocating for the BC, we obtain

$$\sum_{j=1}^{N} c_j \psi_n(x_i, x_j) = h(x_i), \quad i = N_1 + N_2 + 1, \ldots, N. \qquad (3.98)$$

Equations (3.95), (3.96), and (3.98) are N linear equations that can be solved for the N unknowns c_j, and the approximate solution (3.92) is then fully defined.

Chapter 4

Particular Solutions

As the fundamental solution satisfies only a homogeneous governing equation, for an inhomogeneous equation, we need to resolve its RHS into a particular solution. In Chapter 3, we presented the basis functions that can be used for the approximation of the RHS functions. In this chapter, we seek their particular solutions corresponding to the various partial differential operators.

4.1 Method of Particular Solutions

Seeking a particular solution for the given RHS of a linear differential equation to facilitate the solution of a BVP has a long history. Once a particular solution is found, it can be subtracted from the full solution. The remainder can then be solved by utilizing the general solutions of the resulting homogeneous equation.

For example, given a linear second order PDE of the form

$$\mathcal{L}\phi(x) = f(x), \quad x \in \Omega, \tag{4.1}$$

subject to the BCs

$$\phi(x) = g_1(x), \quad x \in \partial\Omega_D, \tag{4.2}$$

$$\mathcal{B}\phi(x) = g_2(x), \quad x \in \partial\Omega_N, \tag{4.3}$$

where $\mathcal{B}$ is a generalized normal derivative, and $\partial\Omega_D \cup \partial\Omega_N = \partial\Omega$, $\partial\Omega_D \cap \partial\Omega_N = \emptyset$, $\partial\Omega_D \neq \emptyset$, with $\partial\Omega$ the domain boundary, we may decompose the solution ϕ into two parts, a homogeneous solution part ϕ_h, and a particular solution part ϕ_p, as

$$\phi = \phi_h + \phi_p. \tag{4.4}$$

If we require ϕ_p to be a solution satisfying

$$\mathcal{L}\phi_p(x) = f(x), \tag{4.5}$$

without being constrained by the BCs (4.2) and (4.3), then we can define a new problem as

$$\mathcal{L}\phi_h(x) = 0, \tag{4.6}$$

subjected to the modified BCs

$$\phi_h(x) = g_1(x) - \phi_p(x), \qquad x \in \partial\Omega_D, \tag{4.7}$$

$$\mathcal{B}\phi_h(x) = g_2(x) - \mathcal{B}\phi_p(x), \quad x \in \partial\Omega_N. \tag{4.8}$$

Our knowledge about the general solutions of the operator $\mathcal{L}$ can now be exploited to find the homogeneous solution ϕ_h. Adding back the particular solution ϕ_p, we obtain the full solution ϕ.

The above procedure, known as the MPS, is not a useful practice for methods that do not take advantage of the general solutions as a solution tool. The solution of a BVP involves two steps, and there is no special advantage. On the other hand, for methods that use general solutions, which include fundamental solutions, such as the MFS, TCM, and BEM, this procedure is needed for solving inhomogeneous equations.

Using general solutions as basis functions has the advantage that the functions automatically satisfy the PDE; hence the solution rapidly converges. If the inhomogeneous RHS can be efficiently approximated, the combined MPS-MFS becomes an attractive methodology. The MPS-MFS will be presented and extensively discussed in Chapter 8.

Atkinson's Method: For practical problems solved by numerical methods, analytical expressions of particular solutions are generally not available. Atkinson [36] suggested to obtain the particular solution numerically as a Newton potential by performing an area or volume integral, for example,

$$\phi_p(x) = \frac{1}{2\pi} \int_{\tilde{\Omega}} f(x') \ln \|x - x'\| \, dx', \quad x \in \tilde{\Omega}, \tag{4.9}$$

where $\tilde{\Omega}$ is a larger region that incorporates the solution domain Ω. The above equation was written for the 2D Poisson equation, and uses the fundamental solution of the Laplacian operator. Similar equations can be written for other partial differential operators. The function $f(x)$ was interpolated using Fourier series, polynomials, or other basis functions to facilitate the integration [334, 726].

Dual Reciprocity BEM: In the BEM community, the DRBEM [106, 107, 713] was developed to handle the RHS functions, in order to avoid volume integration. It was shown by Cheng *et al.* [192] that the method is equivalent to an MPS. We shall give a brief description as follows.

Given the governing equation (4.1), the traditional BEM utilizes the integral equation (see Section 1.3.1)

$$c\phi(x) = \int_{\partial\Omega} \left[\phi(x')\mathfrak{B}^*G(x, x') - G(x, x')\mathfrak{B}\phi(x') \right] dx'$$

$$+ \int_{\Omega} G(x, x')f(x') \, dx'. \tag{4.10}$$

In the above, $\mathfrak{B}$ is the generalized normal derivative, the asterisk superscripts denote the adjoint operator, and G is the fundamental solution. We note that the second line of the above equation is a domain integral. To avoid the domain integral, we may write the integral equation for the homogeneous solution ϕ_h as

$$c(\phi - \phi_p) = \int_{\partial\Omega} \left[(\phi - \phi_p)\mathfrak{B}^*G - G\mathfrak{B}(\phi - \phi_p) \right] dx'. \tag{4.11}$$

To find ϕ_p, we may approximate $f(x)$ by the summation of a set of basis functions with undetermined coefficients as

$$\hat{f}(x) = \sum_{j=1}^{M} d_j \psi_j(x), \tag{4.12}$$

or

$$\hat{f}(x) = \sum_{j=1}^{M} d_j \psi(x, x_j). \tag{4.13}$$

In the first expression (4.12), the ψ_j represent different basis functions in a family, such as the monomials and Chebyshev polynomials discussed in Section 3.2; and in the second expression (4.13), ψ is the same function placed at different locations x_j, such as the RBFs discussed in Section 3.3. For RBFs, we can replace $\psi(x, x_j)$ by $\psi(r_j)$, where $r_j = \|x - x_j\|$.

In the original DRBEM, the function r was used as the basis function [712, 713], which has a poor convergence rate (see Table 3.1). Later, monomials and trigonometric functions [192], thin-plate splines and multiquadrics [336], and compactly supported RBFs [137, 196] were introduced. Using these interpolants, the undetermined coefficients d_j can be determined by solving the system resulting from collocation at a set of points in the domain. Once the coefficients are found, we may construct the approximate particular solution as

$$\hat{\phi}_p(x) = \sum_{j=1}^{M} d_j \varphi^P(x, x_j), \tag{4.14}$$

for (4.13), and similarly for (4.12). In the above, φ^P is a particular solution when using ψ as the RHS:

$$\mathcal{L}\varphi^P(x, x_j) = \psi(x, x_j). \tag{4.15}$$

The ensuing sections of this chapter are dedicated to deriving the analytical expressions of the φ^P corresponding to various basis functions.

Returning to (4.11), we may substitute the approximate particular solution (4.14) into it, and obtain [192]

$$c\phi(x) = \int_{\partial\Omega} \left[\phi(x')\mathfrak{B}^*G(x, x') - G(x, x')\mathfrak{B}\phi(x') \right] dx'$$

$$+ \sum_{j=1}^{M} d_j \left\{ c\varphi^P(x, x_j) - \int_{\partial\Omega} \left[\varphi^P(x', x_j)\mathfrak{B}^*G(x, x') \right. \right.$$

$$\left. \left. - G(x, x')\mathfrak{B}\varphi^P(x', x_j) \right] dx' \right\}. \tag{4.16}$$

We note that the above equation based on the MPS is identical to the DRBEM equation derived using a reciprocal integral equation relation. The integral in the first line of (4.16) contains boundary unknowns, either ϕ or its normal derivative, depending on the type of BCs, and needs to be discretized. The integral involving the particular solution contains all known quantities, and the integration is performed on the boundary only, and not in the domain. We hence have resolved the cumbersome domain integration issue.

MPS-MFS: For the MFS, Golberg and Chen [337] adopted a similar idea and approximated the RHS by polynomials and RBFs, as (4.12) and (4.13). Collocation was applied to solve for the undetermined coefficients to define the approximation, as described in Sections 3.2 and 3.3. The corresponding particular solutions for the various basis functions and partial differential operators were derived in many subsequent efforts, which will be reviewed in the following sections. The approximate particular solution is then given by (4.14). The knowledge of the particular solution allows us to define a new homogeneous problem given by (4.6)–(4.8). The problem is then solved by the MFS using an approximation

$$\hat{\phi}_h(\boldsymbol{x}) = \sum_{j=1}^{N} c_j\, G(\boldsymbol{x}, \boldsymbol{x}'_j), \tag{4.17}$$

where G is the fundamental solution, and $\boldsymbol{x}'_j$ are the source locations. The full solution is then assembled based on (4.4). This method is called the MPS-MFS.

We observe that the MPS-MFS bears some similarity with the RBFCM described in Section 3.3.2. They both use RBFs for collocation, which requires nodes in the domain. The RBFCM solves the problem in one step; whereas the MPS-MFS does it in two steps — finding a particular solution first, and solving the homogeneous problem next (though there is a one-step version of the MPS-MFS; see Section 8.4). This means that the RBFCM is the more straightforward method. The advantage of the MPS-MFS, however, could be its accuracy. We may consider this: in both approaches, the solution (and its error) is constructed from two parts, the homogeneous solution with the BCs, and the particular solution corresponding to the RHS. In the homogeneous part, the RBFCM has a residual error in the domain (as the basis functions do not exactly satisfy the PDE), while the MFS does not. Thus, the MFS can solve the homogeneous part more accurately. For the inhomogeneous part, the RBFCM directly approximates the solution, passing it through the partial differential operator, and collocates to the RHS function. The error committed in the approximation is generally enhanced by the differentiation. The MPS, on the other hand, approximates the RHS. The conversion to the particular solution is exact, without additional error unless there are derivative BCs, but these are of lower order than the differential operator. Hence the MPS-MFS is usually the more accurate method.

Further discussion of the MPS-MFS and its numerical implementation are presented in Chapter 8.

4.2 Monomials

We shall seek a particular solution φ^p satisfying the equation

$$\mathcal{L}\varphi^p = P(x), \tag{4.18}$$

where $\mathcal{L}$ is a linear partial differential operator, and $P(x)$ is a polynomial function approximating the RHS. As all polynomials are constructed from monomials, we shall first seek the particular solutions of the monomials. Janssen and Lambert [410] and Matthys *et al.* [632] presented a method for the systematic construction of recurrence relations of particular solutions of monomials for linear and coupled linear PDEs. The collection of particular solutions presented below, however, was largely derived heuristically.

4.2.1 *Laplacian operator*

For the Laplacian operator in 2D with a monomial as its RHS,

$$\nabla^2 \varphi_{mn}^L = x^m y^n, \quad m, n \in \mathbb{N}, \tag{4.19}$$

the particular solution has been given as [192]

$$\varphi_{mn}^L(x, y) = \sum_{i=1}^{\lfloor \frac{n}{2} \rfloor + 1} (-1)^{i+1} \frac{m!\, n!\, x^{m+2i}\, y^{n-2i+2}}{(m + 2i)!\, (n - 2i + 2)!}, \quad \text{for } m \geq n,$$

$$= \sum_{i=1}^{\lfloor \frac{m}{2} \rfloor + 1} (-1)^{i+1} \frac{m!\, n!\, x^{m-2i+2}\, y^{n+2i}}{(m - 2i + 2)!\, (n + 2i)!}, \quad \text{for } m < n, \tag{4.20}$$

where $\lfloor \cdot \rfloor$ is the floor function (the greatest integer less than or equal to the argument). In the above, we used two expressions to shorten the number of summation terms, and both expressions work in the full range. Alternatively, we can express the above in a recurrence relation:

$$\varphi_{mn}^{L}(x, y) = \sum_{i=1}^{\lfloor \frac{n}{2} \rfloor + 1} a_i \, x^{m+2i} \, y^{n-2i+2}, \tag{4.21}$$

where

$$a_1 = \frac{1}{(m+2)(m+1)}, \tag{4.22}$$

$$a_{i+1} = -\frac{(n-2i+2)(n-2i+1)}{(m+2i+2)(m+2i+1)} a_i. \tag{4.23}$$

The 3D particular solution for

$$P = x^\ell y^m z^n, \quad \text{for } \ell, m, n \in \mathbb{N}, \tag{4.24}$$

is given by Tsai [851] as

$$\varphi_{\ell mn}^{L}(x, y, z) = \sum_{i=0}^{\lfloor \frac{m}{2} \rfloor} \sum_{j=0}^{\lfloor \frac{n}{2} \rfloor} \frac{(-1)^{i+j} \, \ell! \, m! \, n! \, (i+j)! \, x^{\ell+2i+2j+2} y^{m-2i} z^{n-2j}}{i! \, j! \, (\ell+2i+2j+2)! \, (m-2i)! \, (n-2j)!}, \tag{4.25}$$

or, in recursive form [197]

$$\varphi_{\ell mn}^{L}(x, y, z) = \sum_{i=1}^{\lfloor \frac{m}{2} \rfloor + \lfloor \frac{n}{2} \rfloor + 1} \sum_{j=\max\{1, i-\lfloor \frac{n}{2} \rfloor\}}^{\min\{1, \lfloor \frac{m}{2} \rfloor + 1\}} a_{ij} \, x^{\ell+2i} y^{m-2j+2} z^{n-2i+2j}, \tag{4.26}$$

where

$$a_{ij} = -\frac{1}{(\ell+2i)(\ell+2i-1)} \Big[(m-2j+3)(m-2j+4) a_{(i-1)(j-1)}$$
$$+ (n-2i+2j+1)(n-2i+2j+2) a_{(i-1)j} \Big], \tag{4.27}$$

which is initialized by

$$a_{i0} = a_{0j} = 0, \quad a_{11} = \frac{1}{(\ell+1)(\ell+2)}. \tag{4.28}$$

The particular solutions in (4.25) and (4.26) have the fewest terms if $\ell \geq m \geq n$. Otherwise, the roles of x, y, z can be interchanged to obtain the shortest expressions.

■ Example 4.1

Generate the particular solution for the term

$$P(\boldsymbol{x}) = x^7 y^5 z^2.$$

Using (4.25), we easily find

$$\varphi_{752}^L(x, y, z) = -\frac{x^{15}y}{360360} + \frac{x^{13}y^3}{15444} + \frac{x^{13}yz^2}{10296} - \frac{x^{11}y^5}{3960}$$
$$- \frac{x^{11}y^3z^2}{396} + \frac{x^9 y^5 z^2}{72}.$$

4.2.2 *Helmholtz-type operators*

We seek particular solutions of Helmholtz-type equations with a monomial RHS

$$\nabla^2 \varphi^P \pm \lambda \varphi^P = P(\boldsymbol{x}), \tag{4.29}$$

where $\lambda > 0$. For 2D problems, $P = x^m y^n$, and the particular solutions are [340, 659]

$$\varphi_{mn}^H(x, y) = \sum_{i=0}^{\lfloor \frac{m}{2} \rfloor} \sum_{j=0}^{\lfloor \frac{n}{2} \rfloor} \frac{(-1)^{i+j}(i+j)!\,m!\,n!\,x^{m-2i}y^{n-2j}}{\lambda^{i+j+1}\,i!\,j!\,(m-2i)!\,(n-2j)!}, \tag{4.30}$$

$$\varphi_{mn}^M(x, y) = \sum_{i=0}^{\lfloor \frac{m}{2} \rfloor} \sum_{j=0}^{\lfloor \frac{n}{2} \rfloor} \frac{-(i+j)!\,m!\,n!\,x^{m-2i}y^{n-2j}}{\lambda^{i+j+1}\,i!\,j!\,(m-2i)!\,(n-2j)!}, \tag{4.31}$$

respectively, for the Helmholtz and modified Helmholtz operator. For 3D, we have $P = x^\ell y^m z^k$, and the particular solutions are

$$\varphi_{\ell mn}^H(x, y, z) = \sum_{i=0}^{\lfloor \frac{\ell}{2} \rfloor} \sum_{j=0}^{\lfloor \frac{m}{2} \rfloor} \sum_{k=0}^{\lfloor \frac{n}{2} \rfloor} \frac{(-1)^{i+j+k}(i+j+k)!\,\ell!\,m!\,n!\,x^{\ell-2i}y^{m-2j}z^{n-2k}}{\lambda^{i+j+k+1}\,i!\,j!\,k!\,(\ell-2i)!\,(m-2j)!\,(n-2k)!},$$

$$\tag{4.32}$$

$$\varphi_{\ell mn}^{M}(x, y, z) = \sum_{i=0}^{\lfloor \frac{\ell}{2} \rfloor} \sum_{j=0}^{\lfloor \frac{m}{2} \rfloor} \sum_{k=0}^{\lfloor \frac{n}{2} \rfloor} \frac{-(i+j+k)!\,\ell!\,m!\,n!\,x^{\ell-2i}\,y^{m-2j}\,z^{n-2k}}{\lambda^{i+j+k+1}\,i!\,j!\,k!\,(\ell-2i)!\,(m-2j)!\,(n-2k)!}.$$

$$(4.33)$$

■ Example 4.2

Generate the particular solution of the modified Helmholtz equation with the RHS

$$P(x) = x^5 y^3 z^2.$$

Using (4.33), we obtain

$$\begin{aligned}
\varphi_{532}^{M}(x, y, z) = &-\frac{x^5 y^3 z^2}{\lambda} - \frac{2x^5 y^3}{\lambda^2} - \frac{6x^5 yz^2}{\lambda^2} - \frac{20x^3 y^3 z^2}{\lambda^2} \\
&- \frac{24x^5 y}{\lambda^3} - \frac{80x^3 y^3}{\lambda^3} - \frac{240x^3 yz^2}{\lambda^3} - \frac{120xy^3 z^2}{\lambda^3} \\
&- \frac{1440x^3 y}{\lambda^4} - \frac{720xy^3}{\lambda^4} - \frac{2160xyz^2}{\lambda^4} - \frac{17280xy}{\lambda^5}.
\end{aligned}$$

For the particular solution of the Helmholtz equation, we can replace λ by $-\lambda$ in the above.

4.2.3 *General second-order PDEs*

For the general second order PDE in 2D with constant coefficients we have

$$\mathcal{L}\varphi_{mn}^{p} = x^m y^n, \qquad (4.34)$$

where

$$\mathcal{L} \equiv a_1 \frac{\partial^2}{\partial x^2} + a_2 \frac{\partial^2}{\partial x \partial y} + a_3 \frac{\partial^2}{\partial y^2} + a_4 \frac{\partial}{\partial x} + a_5 \frac{\partial}{\partial y} + a_6. \qquad (4.35)$$

Dangal *et al.* [233] gave the following particular solution:

$$\varphi_{mn}^{p}(x, y) = \frac{1}{a_6} \sum_{k=0}^{m+n} \left(\frac{-1}{a_6}\right)^{k} \mathcal{L}^{k}\{x^m y^n\}, \qquad (4.36)$$

where $\mathcal{L}^{k}$ means nesting the operator k times, that is, $\mathcal{L}^{k} = \mathcal{L}\{\mathcal{L}\{\cdots\}\}$. The expressions are generally very long, and can be challenging to even computer algebra for higher degree monomials.

■ Example 4.3

Find the particular solution for the RHS $x^2 y^2$.

Based on (4.36), we find

$$
\begin{aligned}
\varphi_{22}^P(x, y) = \big(& 24a_4^2 a_5^2 - 12a_3 a_4^2 a_6 - 24a_2 a_4 a_5 a_6 - 12a_1 a_5^2 a_6 + 8a_1 a_3 a_6^2 \\
& + 4a_2^2 a_6^2 - 12a_4 a_5^2 a_6 x + 8a_3 a_4 a_6^2 x + 8a_2 a_5 a_6^2 x + 2a_5^2 a_6^2 x^2 \\
& - 2a_3 a_6^3 x^2 - 12a_4^2 a_5 a_6 y + 8a_2 a_4 a_6^2 y + 8a_1 a_5 a_6^2 y \\
& + 8a_4 a_5 a_6^2 xy - 4a_2 a_6^3 xy - 2a_5 a_6^3 x^2 y + 2a_4^2 a_6^2 y^2 - 2a_1 a_6^3 y^2 \\
& - 2a_4 a_6^3 xy^2 + a_6^4 x^2 y^2 \big) / a_6^5.
\end{aligned}
$$

4.2.4 *Polyharmonic operators*

For the polyharmonic operator of degree k

$$
\Delta^k \varphi_k^P = P(\boldsymbol{x}) \tag{4.37}
$$

with the monomial RHS $P = x^m y^n$ in 2D, the particular solution is found to be [854]

$$
\varphi_{kmn}^P(x, y) = \frac{m!\, n!}{(k-1)!} \sum_{i=0}^{\lfloor \frac{n}{2} \rfloor} \frac{(-1)^i (k+i-1)!\, x^{2k+m+2i}\, y^{n-2i}}{i!\,(2k+m+2i)!\,(n-2i)!}, \tag{4.38}
$$

and with $P = x^\ell y^m z^n$ in 3D, it is

$$
\begin{aligned}
\varphi_{k\ell mn}^P(x, y, z) = {} & \frac{\ell!\, m!\, n!}{(k-1)!} \\
& \times \sum_{i=0}^{\lfloor \frac{m}{2} \rfloor} \sum_{j=0}^{\lfloor \frac{n}{2} \rfloor} \frac{(-1)^{i+j} (k+i+j-1)!\, x^{2k+\ell+2i+2j}\, y^{m-2i}\, z^{n-2j}}{i!\, j!\,(2k+\ell+2i+2j)!\,(m-2i)!\,(n-2j)!}.
\end{aligned} \tag{4.39}
$$

■ Example 4.4

Find the particular solution of (4.37) with $k = 2$ (biharmonic operator), and

$$
P(\boldsymbol{x}) = x^7 y^5 z^2.
$$

Equation (4.39) gives

$$\varphi^P_{2752}(x, y, z) = -\frac{x^{17}y}{24504480} + \frac{x^{15}y^3}{1081080} + \frac{x^{15}yz^2}{720720} - \frac{x^{13}y^5}{308880}$$

$$-\frac{x^{13}y^3z^2}{30888} + \frac{x^{11}y^5z^2}{7920}.$$

4.2.5 *Polymetaharmonic operators*

For the product of the modified Helmholtz operator of degree k

$$(\Delta - \lambda)^k \varphi^M_k = P(x) \tag{4.40}$$

with the monomial RHS $P = x^m y^n$ for 2D, the particular solution is [854]

$$\varphi^M_{kmn}(x, y) = \frac{(-1)^k m! n!}{(k-1)!} \sum_{i=0}^{\lfloor \frac{m}{2} \rfloor} \sum_{j=0}^{\lfloor \frac{n}{2} \rfloor} \frac{(k+i+j-1)! x^{m-2i} y^{n-2j}}{\lambda^{k+i+j} i! j! (m-2i)! (n-2j)!}, \tag{4.41}$$

and with $P = x^\ell y^m z^n$ for 3D, it is

$$\varphi^M_{k\ell mn}(x, y, z) = \frac{(-1)^k \ell! m! n!}{(k-1)!}$$

$$\times \sum_{i=0}^{\lfloor \frac{\ell}{2} \rfloor} \sum_{j=0}^{\lfloor \frac{m}{2} \rfloor} \sum_{k=0}^{\lfloor \frac{n}{2} \rfloor} \frac{-(i+j+k)! x^{\ell-2i} y^{m-2j} z^{n-2k}}{\lambda^{i+j+k+1} i! j! k! (\ell-2i)! (m-2j)! (n-2k)!}. \tag{4.42}$$

For the product of the Helmholtz equation, we replace λ by $-\lambda$ in the above expressions and obtain φ^H_k.

4.2.6 *Cauchy–Navier operator*

For the Cauchy–Navier equations (2.78), the RHS comes from a body force $-F$. The body force is a non-contact force generally exerted by the gradient of a physical field, such as the gravitational or electrical field. Otherwise, it can come from an analogous force field, such as the temperature field and the generated apparent stress. We shall discuss some of these body forces here [197].

Gravitational Force: For a gravitational force in the negative z-direction, we may express

$$F_z = -\rho g, \quad F_x = F_y = 0, \tag{4.43}$$

where ρ is the mass density, and g the gravitational acceleration. The displacement particular solutions are found as

$$u_x^p(\mathbf{x}) = \frac{\rho g(1 - 2v)}{2G(1 + v)} xz, \tag{4.44}$$

$$u_y^p(\mathbf{x}) = \frac{\rho g(1 - 2v)}{2G(1 + v)} yz, \tag{4.45}$$

$$u_z^p(\mathbf{x}) = \frac{\rho g(1 - 2v)}{4G(1 + v)} (z^2 - x^2 - y^2). \tag{4.46}$$

The 2D solution can be similarly constructed.

Centrifugal Force: For a body rotating about the z-axis, a centrifugal force results, which is proportional to the distance away from the axis, and points away from it. The body forces are

$$F_x = \rho \omega^2 x, \quad F_y = \rho \omega^2 y, \quad F_z = 0, \tag{4.47}$$

where ω is the angular velocity. The particular solutions are

$$u_x^p(\mathbf{x}) = -\frac{\rho \omega^2 (1 - 2v)}{16G(1 - v)} (x^2 + y^2)x, \tag{4.48}$$

$$u_y^p(\mathbf{x}) = -\frac{\rho \omega^2 (1 - 2v)}{16G(1 - v)} (x^2 + y^2)y, \tag{4.49}$$

$$u_z^p(\mathbf{x}) = 0. \tag{4.50}$$

Thermal Stress: The equation of uncoupled thermoelasticity (theory of thermal stresses) [96] is

$$G\nabla^2 \mathbf{u} + \frac{G}{1 - 2v} \nabla \nabla \cdot \mathbf{u} = \alpha_T \nabla T, \tag{4.51}$$

where T is a given temperature field, and α_T a thermoelastic constant. If we decompose the displacement vector following the Helmholtz theorem (2.116), it is obvious that we only need to retain the scalar potential, due to the gradient form of the RHS term. We may thus write

$$\boldsymbol{u}_p = \frac{\alpha_T(1-2v)}{2G(1-v)}\nabla\varphi^L. \tag{4.52}$$

Substituting the above into (4.51), and removing a gradient operator, we obtain

$$\nabla^2\varphi^L = T. \tag{4.53}$$

If we approximate the temperature field using a summation of monomials, $x^m y^n$ or $x^\ell y^m z^n$, a particular solution φ^L is given by (4.20) and (4.25), respectively, for 2D and 3D problems. The particular solution for the displacements $\boldsymbol{u}_p$ follows from (4.52).

■ **Example 4.5**

Generate the displacement particular solution for the monomial approximation term for temperature

$$T = x^5 y^3 z^2.$$

First, from (4.25), we obtain

$$\varphi^L_{532}(x, y, z) = \frac{x^{11}y}{13860} - \frac{x^9 y^3}{1512} - \frac{1}{504}x^9 yz^2 + \frac{1}{42}x^7 y^3 z^2.$$

Differentiating, we find

$$u^p_x(x, y, z) = \frac{\alpha_T(1-2v)}{2G(1-v)}\left(\frac{x^{10}y}{1260} - \frac{x^8 y^3}{168} - \frac{1}{56}x^8 yz^2 + \frac{1}{6}x^6 y^3 z^2\right),$$

$$u^p_y(x, y, z) = \frac{\alpha_T(1-2v)}{2G(1-v)}\left(\frac{x^{11}}{13860} - \frac{x^9 y^2}{504} - \frac{x^9 z^2}{504} + \frac{1}{14}x^7 y^2 z^2\right),$$

$$u^p_z(x, y, z) = \frac{\alpha_T(1-2v)}{2G(1-v)}\left(\frac{1}{21}x^7 y^3 z - \frac{1}{252}x^9 yz\right).$$

4.2.7 *Other operators*

Particular solutions for other partial differential operators with a monomial RHS have been derived by Anderson *et al.* [29], which include the quasi-harmonic equation, wave equation, advection equation, isotropic and anisotropic elasticity, elastodynamics, Stokes flow, linearized Navier–Stokes equations, and Maxwell equations.

4.3 Homogeneous Polynomials

When the monomial terms are added to the approximation to form a polynomial, it makes sense to add them in groups, complete up to a certain degree. The homogeneous polynomials $P_n(x)$, which contain all terms of monomials of degree n, serve the purpose. It is helpful to directly find the particular solutions of the homogeneous polynomials.

In 2D, the homogeneous polynomial of degree n is given by

$$P_n = \sum_{i=0}^{n} d_i\, x^{n-i} y^i, \tag{4.54}$$

where the d_i are the coefficients determined by the approximation of a function; and we seek its particular solution. For the Laplacian operator, it was shown [138] that a particular solution of a homogenous polynomial of degree n is a homogeneous polynomial of degree $n+2$. We find

$$\varphi_n^{L}(x, y) = \sum_{i=0}^{n} a_i\, x^{n-i+2} y^i, \tag{4.55}$$

where

$$a_i = \sum_{j=0}^{\lfloor \frac{n-i}{2} \rfloor} \frac{(-1)^j (i + 2j)!(n - i - 2j)!}{i!(n - i + 2)!} d_{i+2j}. \tag{4.56}$$

For 3D, given a homogeneous polynomial

$$P_n = \sum_{i=0}^{n} \sum_{j=0}^{n-i} d_{ij}\, x^{n-i-j} y^j z^i, \tag{4.57}$$

which contains $(n+1)(n+2)/2$ terms, its particular solution for the Laplacian operator is [340]

$$\varphi_n^L(x, y, z) = \sum_{i=0}^{n+2} \sum_{j=0}^{n-i+2} a_{ij}\, x^{n-i-j+2} y^j z^i, \tag{4.58}$$

in which the a_{ij} are defined by the following recurrence relation:

$$a_{ij} = \frac{d_{ij} - (j+2)(j+1)a_{i(j+2)} - (i+2)(i+1)a_{(i+2)j}}{(n-i-j+2)(n-i-j+1)},$$

$$\text{for } i = n, \ldots, 0, \ \ j = n - i, \ldots, 0, \tag{4.59}$$

which is initialized by

$$a_{i(n-i+2)} = 0, \quad \text{for } i = 0, \ldots, n+2;$$

$$a_{i(n-i+1)} = 0, \quad \text{for } i = 0, \ldots, n+1. \tag{4.60}$$

■ Example 4.6

Given the homogeneous polynomial of degree 3

$$P_3(x, y, z) = d_{00}x^3 + d_{01}x^2y + d_{02}xy^2 + d_{03}y^3 + d_{10}x^2z + d_{11}xyz$$

$$+ d_{12}y^2z + d_{20}xz^2 + d_{21}yz^2 + d_{30}z^3,$$

find its particular solution.

Utilizing (4.59), we obtain

$$\varphi_3^L(x, y, z) = \frac{1}{60}(3d_{00} - d_{02} - d_{20})x^5 + \frac{1}{12}(d_{01} - 3d_{03} - d_{21})x^4y$$

$$+ \frac{1}{12}(d_{10} - d_{12} - 3d_{30})x^4z + \frac{1}{6}d_{02}x^3y^2 + \frac{1}{6}d_{11}x^3yz$$

$$+ \frac{1}{6}d_{20}x^3z^2 + \frac{1}{2}d_{03}x^2y^3 + \frac{1}{2}d_{12}x^2y^2z + \frac{1}{2}d_{21}x^2yz^2$$

$$+ \frac{1}{2}d_{30}x^2z^3.$$

4.4 Chebyshev Polynomials

As discussed in Section 3.2.2, one of the most accurate and stable interpolations by polynomials is the Chebyshev approximation. When it is used to approximate the inhomogeneous part of a PDE, we seek its particular solution.

4.4.1 *Laplacian operator*

Suppose we are given the approximation (3.62), sampling $(L + 1) \times (M + 1) \times (N + 1)$ data points at the extrema of T_L, T_M, T_N. When expanded, it contains a large number of terms, which can be collected into $(L \times M \times N) + 1$ monomials (including the constant), or $L + M + N$ groups of homogeneous polynomials. Explicit particular solutions are available from the recursive formulas presented above. The bookkeeping and sorting of these terms can be tedious, and is usually assisted by computer algebra software, such as Mathematica® [918]. Reutskiy and Chen [750] bypassed such a process by further approximating Chebyshev polynomials using trigonometric basis functions, whose particular solutions are easier to find. Karageorghis and Kyza [449] established the relations between the coefficients of the function and its particular solution as a set of linear equations, which are then solved by matrix inversion. Tsai and Hsu [851, 857] converted the Chebyshev polynomials into monomials, and presented the explicit expression for the particular solution. We shall present a particular solution below following Tsai and Hsu's algorithm [857].

Assume that we can find the particular solution for a single term Chebyshev polynomial $T_i(\xi)T_j(\eta)T_k(\zeta)$ in (3.62) as φ_{ijk}^L, then a particular solution for the full RHS is

$$\varphi_{LMN}^L(\boldsymbol{x}) = \sum_{i=0}^{L}\sum_{j=0}^{M}\sum_{k=0}^{N} a_{ijk}\,\varphi_{ijk}^L. \tag{4.61}$$

We shall find φ_{ijk}^L in the following.

Mason and Handscomb [628] gave the following expansion of a Chebyshev polynomial into monomials:

$$T_n(x) = \sum_{i=0}^{\lfloor \frac{n}{2} \rfloor} c_i^{(n)} x^{n-2i}, \tag{4.62}$$

where

$$c_i^{(n)} = (-1)^i \, 2^{n-2i-1} \frac{n(n-i-1)!}{i!\,(n-2i)!}.$$
(4.63)

We can then express the product as

$$T_i(\xi)T_j(\eta)T_k(\zeta) = \sum_{p=0}^{\lfloor \frac{i}{2} \rfloor} \sum_{q=0}^{\lfloor \frac{j}{2} \rfloor} \sum_{r=0}^{\lfloor \frac{k}{2} \rfloor} c_p^{(i)} c_q^{(j)} c_r^{(k)} \xi^{i-2p} \eta^{j-2q} \zeta^{k-2r}.$$
(4.64)

In (4.25), we have found the particular solution for the monomial $x^\ell y^m z^n$, and we seek to make use of it. Tsai and Hsu [857] converted (4.64) into the (x, y, z) variables using the definition in (3.63) and the binomial expansion. Instead, we shall directly deal with the variables (ξ, η, ζ). We note that we can perform a variable transformation of the Laplacian operator from (x, y, z) to $(\bar{\xi}, \bar{\eta}, \bar{\zeta})$, as

$$\nabla^2_{xyz} = \frac{\partial^2}{\partial x^2} + \frac{\partial^2}{\partial y^2} + \frac{\partial^2}{\partial z^2} = \frac{4}{\ell_x^2}\frac{\partial^2}{\partial \xi^2} + \frac{4}{\ell_y^2}\frac{\partial^2}{\partial \eta^2} + \frac{4}{\ell_z^2}\frac{\partial^2}{\partial \zeta^2}$$

$$= \frac{\partial^2}{\partial \bar{\xi}^2} + \frac{\partial^2}{\partial \bar{\eta}^2} + \frac{\partial^2}{\partial \bar{\zeta}^2} = \nabla^2_{\bar{\xi}\bar{\eta}\bar{\zeta}},$$
(4.65)

where

$$\ell_x = x_b - x_a \quad \ell_y = y_b - y_a \quad \ell_z = z_b - z_a,$$
(4.66)

and

$$\bar{\xi} = \frac{\ell_x \xi}{2}, \quad \bar{\eta} = \frac{\ell_y \eta}{2}, \quad \bar{\zeta} = \frac{\ell_z \zeta}{2}.$$
(4.67)

Using (4.24) and (4.25), we can find a particular solution for

$$\xi^{i-2p} \eta^{j-2q} \zeta^{k-2r} = \frac{2^{i+j+k-2p-2q-2r}}{\ell_x^{i-2p} \ell_y^{j-2q} \ell_z^{k-2r}} \bar{\xi}^{i-2p} \bar{\eta}^{j-2q} \bar{\zeta}^{k-2r},$$
(4.68)

as

$$\bar{\varphi}_{ijk}^L$$

$$= \sum_{\alpha=0}^{\lfloor \frac{j}{2}-q \rfloor} \sum_{\beta=0}^{\lfloor \frac{k}{2}-r \rfloor} \frac{(-1)^{\alpha+\beta} (i-2p)!\,(j-2q)!\,(k-2r)!\,(\alpha+\beta)!}{4\alpha!\,\beta!\,(i-2p+2\alpha+2\beta+2)!\,(j-2q-2\alpha)!\,(k-2r-2\beta)!}$$

$$\times \ell_x^{2\alpha+2\beta+2} \ell_y^{-2\alpha} \ell_z^{-2\beta} \xi^{i-2p+2\alpha+2\beta+2} \eta^{j-2q-2\alpha} \zeta^{k-2r-2\beta}.$$
(4.69)

We then obtain the φ_{ijk}^L needed in (4.61) as

$$\varphi_{ijk}^L = \sum_{p=0}^{\lfloor \frac{i}{2} \rfloor} \sum_{q=0}^{\lfloor \frac{j}{2} \rfloor} \sum_{r=0}^{\lfloor \frac{k}{2} \rfloor} c_p^{(i)} c_q^{(j)} c_r^{(k)} \bar{\varphi}_{ijk}^L. \tag{4.70}$$

The particular solution for $f(x)$ approximated by Chebyshev polynomials is then explicitly given, without requiring the sorting and assembling of terms.

■ Example 4.7

Find a particular solution for the single Chebyshev polynomial term

$$\nabla^2 \varphi_{531}^L(x) = T_5(\xi) T_3(\eta) T_1(\zeta).$$

Before presenting the result, we should mention that as (4.61), (4.69), and (4.70) are to be numerically evaluated, there is no need to give the explicit expression of the particular solution. Here we present the result to get an idea of it, which also allows us to check the validity of the formulas. Expanding from (4.69) and (4.70), we find

$$\varphi_{531}^L = -\frac{\ell_x^2 \xi^3 \eta \zeta}{504 \ell_y^2} \left(16 \ell_x^2 \xi^6 - 72 \ell_x^2 \xi^4 + 126 \ell_x^2 \xi^2 - 192 \ell_y^2 \xi^4 \eta^2 \right.$$
$$\left. + 504 \ell_y^2 \xi^2 \eta^2 - 420 \ell_y^2 \eta^2 + 144 \ell_y^2 \xi^4 - 378 \ell_y^2 \xi^2 + 315 \ell_y^2 \right).$$

We can utilize (4.65) in either the (x, y, z) or (ξ, η, ζ) variables to show that it indeed produces the intended RHS.

4.4.2 *Helmholtz-type and other operators*

The particular solution for the Helmholtz-type and other operators with a Chebyshev polynomial as the RHS can be similarly obtained using the techniques developed for the Laplacian operator. That is, one can analytically expand the Chebyshev polynomials and sort them into monomials. As the particular solutions for monomials have been derived for the various partial differential operators in Section 4.2, those for Chebyshev polynomials can be assembled [141, 340]. The technique of a two-stage process by additional approximation using trigonometric functions can also be followed [750, 752]. Alternatively, the relations among the Chebyshev terms of the approximated function and its particular solution can be set up in matrix forms and solved [449]. The direct approach of establishing the explicit formula [259, 851, 857], as shown in the above for the Laplacian operator, can also be pursued. A method to accelerate the evaluation of the particular solutions has also been developed [516].

4.5 Polyharmonic Splines

The use of RBFs instead of monomials and polynomials to approximate a function has several advantages. First, the RBFCM allows for scattered node collocation that can adjust to an arbitrary geometry, and for uneven distribution to cope with local rapid variations. Chebyshev polynomials, on the other hand, require a domain that is rectangular or box shaped, and the data must be supplied at fixed locations. If the field to be interpolated is known only at a set of discrete data points, as is often the case in practical applications, the Chebyshev polynomial method cannot be utilized. In contrast, monomials can be collocated at arbitrary data points. However, the approximation may be plagued by oscillations and solvability issues. The RBF interpolation is guaranteed to be positive definite for a class of bases; hence the resulting linear system is always solvable. Some RBFs, such as the MQ and GA, have exponential convergence properties as is the case for Chebyshev polynomials. In this and the following sections, we seek the particular solutions for a few RBFs that are highly efficient and popular in applications.

The first RBFs to investigate are the polyharmonic splines, given as

$$\psi_n(r) = r^{2n} \ln r, \quad \text{(2D)}, \tag{4.71a}$$

$$= r^{2n-1}, \quad \text{(3D)}, \tag{4.71b}$$

where $n \in \mathbb{N}$. We observe from (2.167)–(2.173) that these are the fundamental solutions of polyharmonic operators. They are named "splines" because when used for interpolation, the lowest order ones, $r^2 \ln r$ for 2D and r for 3D, have the same smoothness property as the cubic spline used in 1D interpolation. In other words, the interpolations are smooth in the first derivatives and continuous in the second derivatives. The higher order splines, $n \geq 2$, have higher smoothness properties. However, as they lack the full smoothness, their convergence is of algebraic, rather than exponential, order, as indicated in Table 3.1. These bases are popular because they do not require the selection of a shape parameter, as needed in the MQ and GA RBFs.

4.5.1 *Laplacian operator*

In 2D, we seek particular solutions of

$$\nabla^2 \varphi_n^L = r^{2n} \ln r. \tag{4.72}$$

They can be obtained by simple integration, as

$$\varphi_n^L(r) = \frac{r^{2n+2}}{4(n+1)^2} \left(\ln r - \frac{1}{n+1} \right), \tag{4.73}$$

and the first few expressions are

$$\varphi_1^L(r) = \frac{r^4}{16} \left(\ln r - \frac{1}{2} \right), \tag{4.74}$$

$$\varphi_2^L(r) = \frac{r^6}{36} \left(\ln r - \frac{1}{3} \right), \tag{4.75}$$

$$\varphi_3^L(r) = \frac{r^8}{64} \left(\ln r - \frac{1}{4} \right). \tag{4.76}$$

The above particular solutions satisfy $\lim_{r \to 0} \varphi_n^L(r) = 0$.

The 3D particular solutions with the RHS r^{2n-1} in (4.72) are

$$\varphi_n^L(r) = \frac{r^{2n+1}}{(2n+1)(2n+2)}, \tag{4.77}$$

and the first few expressions are

$$\varphi_1^L(r) = \frac{r^3}{12}, \tag{4.78}$$

$$\varphi_2^L(r) = \frac{r^5}{30}. \tag{4.79}$$

Hence, if we are given a Poisson equation

$$\nabla^2 \phi(\boldsymbol{x}) = f(\boldsymbol{x}), \tag{4.80}$$

we can approximate the RHS by a summation of RBFs,

$$\hat{f}(\boldsymbol{x}) = \sum_{j=1}^{M} d_j \psi_n(r_j), \tag{4.81}$$

where $r_j = \|\boldsymbol{x} - \boldsymbol{x}_j\|$, with $\boldsymbol{x}_j$ the locations of the collocation nodes. By collocating with the function, we can solve for the coefficients d_j, as described in Section 3.3. Since φ_n^L is the particular solution corresponding to ψ_n, a particular solution of (4.80) is

$$\hat{\phi}_p(\boldsymbol{x}) = \sum_{j=1}^{M} d_j \varphi_n^L(r_j). \tag{4.82}$$

■ Example 4.8

For a demonstration, we shall employ the function $f(\boldsymbol{x})$ introduced in Example 3.3, plotted as Fig. 3.3, for the RHS of the Poisson equation (4.80). That function has a particular solution of the form

$$\phi_p(x, y) = \sin\frac{\pi x}{6} \sin\frac{7\pi x}{4} \sin\frac{3\pi y}{4} \sin\frac{5\pi y}{4},$$

which is plotted in Fig. 4.1(a).

As a first step, we approximate the function $f(\boldsymbol{x})$ using an 11×11 collocation grid as in Example 3.4, except that polyharmonic splines,

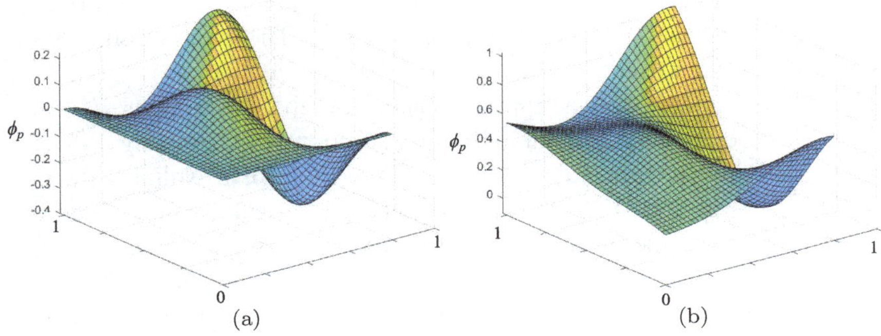

Figure 4.1. Plots of: (a) An exact particular solution of the Poisson equation with the RHS function given in Example 3.3; and (b) an exact particular solution of the approximated RHS using the thin-plate spline.

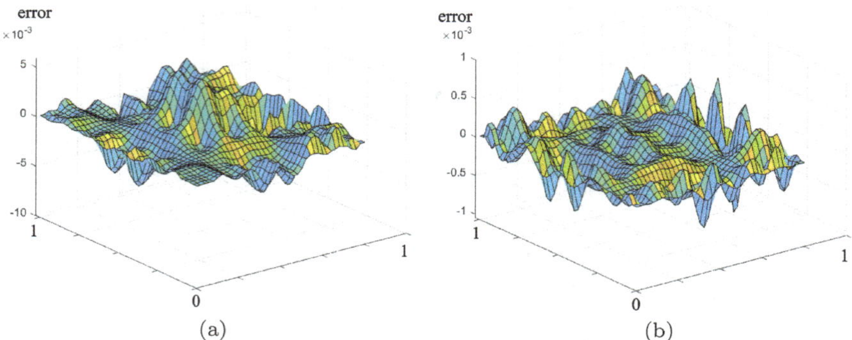

Figure 4.2. Relative error ε_r of polyharmonic spline interpolation, with $N = 11 \times 11$. (a) Thin-plate spline $r^2 \ln r$; and (b) polyharmonic spline $r^4 \ln r$.

instead of the IMQ, are used as the basis function. Two spline functions are used and compared: the thin-plate spline $r^2 \ln r$, and the second order polyharmonic spline $r^4 \ln r$. For simplicity, we omitted the augmented polynomial terms, such as those in (3.79), in both cases. Those terms are needed to guarantee the positive definiteness of the matrix, but do not appear to enhance the accuracy.

The relative error ε_r defined in (3.71) is plotted in Fig. 4.2(a) for the thin-plate spline case. We observe that the $\varepsilon_{\max}$ defined in (3.72) is of $\mathcal{O}(10^{-2})$, which is inferior to the IMQ case shown in Fig. 3.5(b). The reason is that the thin-plate spline lacks the higher order differentiability, as compared to the infinitely differentiable MQ and GA. If we raise the differentiability by using $r^4 \ln r$ instead, the error improves as shown in Fig. 4.2(b), with $\varepsilon_{\max}$ of $\mathcal{O}(10^{-3})$. These error behaviors are consistent with those reported in Table 3.1.

Once the function is approximated, and the coefficients d_j are determined, the approximate particular solution is found using (4.73) and (4.83). The particular solution for the thin-plate spline case is plotted in Fig. 4.1(b). We notice that although this solution bears some resemblance to the original generating function shown in Fig. 4.1(a), it is not an approximation of that function. The reason is that, as we stated earlier, particular solutions are not unique. This particular solution exactly produces the approximated RHS.

4.5.2 *Helmholtz-type operators*

Particular solutions for the Helmholtz and modified Helmholtz equations with polyharmonic splines as the RHS have been presented in [135, 195, 339, 658]. We shall provide a quick derivation below.

Helmholtz Operator: For the Helmholtz operator, we seek the particular solutions of

$$(\Delta + \lambda)\varphi_n^H = r^{2n}\ln r, \quad (2D), \tag{4.83a}$$

$$= r^{2n-1}, \quad (3D). \tag{4.83b}$$

We note from (2.164) and (2.171) that the RHSs of the above equations can be expressed in terms of the fundamental solutions G_{n+1}^P of the polyharmonic operators given in Section 2.11. For the 2D case, we obtain

$$(\Delta + \lambda)\varphi_n^H = 2^{2n+1}\pi\,(n!)^2 G_{n+1}^P + r^{2n}H_n, \tag{4.84}$$

where H_n is the harmonic number. The particular solution has two parts. For the first term on the RHS, we can use the annihilator method by applying the operator Δ^{n+1} to both sides of the equation to obtain

$$\Delta^{n+1}(\Delta + \lambda)\varphi_n^{H1} = 2^{2n+1}\pi\,(n!)^2\delta(\boldsymbol{x}, \boldsymbol{x}'). \tag{4.85}$$

From (2.228), the particular solution φ_n^{H1} can be expressed in terms of the fundamental solution G_{n+1}^X defined in (2.237). The second part φ_n^{H2} corresponding to the harmonic number H_n can be obtained easily. Combining the two parts, we obtain the sought for particular solution as [195]

$$\varphi_n^H(r) = \frac{(-1)^{n+1}4^n\,(n!)^2}{\lambda^{n+1}}\frac{\pi}{2}\mathrm{Y}_0\big(\sqrt{\lambda}\,r\big)$$

$$+ (n!)^2\sum_{i=1}^{n+1}\frac{(-1)^{n-i+1}4^{n-i+1}}{[(i-1)!]^2\lambda^{n-i+2}}r^{2i-2}\left(\ln r + H_n - H_{i-1}\right). \tag{4.86}$$

The first few terms are

$$\varphi_1^H(r) = \frac{4}{\lambda^2}\left[\frac{\pi}{2}Y_0\left(\sqrt{\lambda}\,r\right) - \ln r\right] + \frac{r^2\ln r}{\lambda} - \frac{4}{\lambda^2}, \tag{4.87}$$

$$\varphi_2^H(r) = -\frac{64}{\lambda^3}\left[\frac{\pi}{2}Y_0\left(\sqrt{\lambda}\,r\right) - \ln r\right] + \frac{r^2\ln r}{\lambda}\left(-\frac{16}{\lambda} + r^2\right)$$
$$+ \frac{8}{\lambda^2}\left(\frac{12}{\lambda} - r^2\right), \tag{4.88}$$

$$\varphi_3^H(r) = \frac{2304}{\lambda^4}\left[\frac{\pi}{2}Y_0\left(\sqrt{\lambda}\,r\right) - \ln r\right] + \frac{r^2\ln r}{\lambda}\left(\frac{576}{\lambda^2} - \frac{36r^2}{\lambda} + r^4\right)$$
$$- \frac{12}{\lambda^2}\left(\frac{352}{\lambda^2} - \frac{40r^2}{\lambda} + r^4\right). \tag{4.89}$$

At $r = 0$, we need to take the limit, and (4.86) becomes

$$\varphi_n^H(0) = \frac{(-1)^{n+1}4^n(n!)^2}{\lambda^{n+1}}\left(\gamma + \ln\frac{\sqrt{\lambda}}{2} - H_n\right), \tag{4.90}$$

where $\gamma = 0.577216\ldots$ is the Euler constant. The first few terms are

$$\varphi_1^H(0) = \frac{4}{\lambda^2}\left(\gamma + \ln\frac{\sqrt{\lambda}}{2} - 1\right), \tag{4.91}$$

$$\varphi_2^H(0) = -\frac{64}{\lambda^3}\left(\gamma + \ln\frac{\sqrt{\lambda}}{2} - \frac{3}{2}\right), \tag{4.92}$$

$$\varphi_3^H(0) = \frac{2304}{\lambda^4}\left(\gamma + \ln\frac{\sqrt{\lambda}}{2} - \frac{11}{6}\right). \tag{4.93}$$

The 3D case is simpler. Applying the annihilator method, (4.83b) becomes

$$\Delta^{n+1}(\Delta + \lambda)\varphi_n^H = -4\pi(2n)!\delta(x, x'), \tag{4.94}$$

and based on (2.228) and (2.238), we find

$$\varphi_n^H(r) = -4\pi(2n)!\,G_{n+1}^X$$
$$= (2n)!\left[\frac{(-1)^{n+1}}{\lambda^{n+1}}\frac{\cos(\sqrt{\lambda}r)}{r} + \sum_{i=1}^{n+1}\frac{(-1)^{n-i+1}\,r^{2i-3}}{(2i-2)!\,\lambda^{n-i+2}}\right]. \tag{4.95}$$

The first few terms are

$$\varphi_1^H(r) = \frac{2}{\lambda^2 r}\big[\cos\left(\sqrt{\lambda}\,r\right) - 1\big] + \frac{r}{\lambda}, \tag{4.96}$$

$$\varphi_2^H(r) = -\frac{24}{\lambda^3 r}\big[\cos\left(\sqrt{\lambda}\,r\right) - 1\big] - \frac{12r}{\lambda^2} + \frac{r^3}{\lambda}, \tag{4.97}$$

$$\varphi_3^H(r) = \frac{720}{\lambda^4 r}\big[\cos\left(\sqrt{\lambda}\,r\right) - 1\big] + \frac{360r}{\lambda^3} - \frac{30r^3}{\lambda^2} + \frac{r^5}{\lambda}. \tag{4.98}$$

The limit is given by $\lim_{r\to 0}\varphi_n^H(r) = 0$.

Modified Helmholtz Operator: The next case involves the modified Helmholtz operator:

$$(\Delta - \lambda)\varphi_n^M = r^{2n}\ln r, \quad (2\text{D}), \tag{4.99a}$$

$$= r^{2n-1}, \quad (3\text{D}). \tag{4.99b}$$

Following the same procedure as in the Helmholtz operator case, we find for 2D,

$$\varphi_n^M(r) = -4^n(n!)^2\frac{\mathrm{K}_0(\sqrt{\lambda}\,r)}{\lambda^{n+1}} - (n!)^2\sum_{i=1}^{n+1}\frac{4^{n-i+1}}{[(i-1)!]^2\,\lambda^{n-i+2}}$$

$$\times\, r^{2i-2}\,(\ln r + H_n - H_{i-1}), \tag{4.100}$$

with

$$\varphi_n^M(0) = \frac{4^n(n!)^2}{\lambda^{n+1}}\left(\gamma + \ln\frac{\sqrt{\lambda}}{2} - H_n\right). \tag{4.101}$$

The first few terms are

$$\varphi_1^M(r) = -\frac{4}{\lambda^2}\big[\mathrm{K}_0(\sqrt{\lambda}\,r) + \ln r\big] - \frac{r^2\ln r}{\lambda} - \frac{4}{\lambda^2}, \tag{4.102}$$

$$\varphi_2^M(r) = -\frac{64}{\lambda^3}\big[\mathrm{K}_0(\sqrt{\lambda}\,r) + \ln r\big] - \frac{r^2\ln r}{\lambda}\left(\frac{16}{\lambda} + r^2\right)$$

$$-\frac{8}{\lambda^2}\left(\frac{12}{\lambda} + r^2\right), \tag{4.103}$$

$$\varphi_3^M(r) = -\frac{2304}{\lambda^4}\left[K_0(\sqrt{\lambda}\,r) + \ln r\right] - \frac{r^2 \ln r}{\lambda}\left(\frac{576}{\lambda^2} + \frac{36r^2}{\lambda} + r^4\right)$$

$$-\frac{12}{\lambda^2}\left(\frac{352}{\lambda^2} + \frac{40r^2}{\lambda} + r^4\right). \tag{4.104}$$

For the limiting cases, $r \to 0$, we notice that (4.90) and (4.101) differ only by a $(-1)^{n+1}$ factor; hence we may refer to the results (4.91)– (4.93).

For 3D, we have

$$\varphi_n^M(r) = (2n)!\left[\frac{1}{\lambda^{n+1}}\frac{e^{-\sqrt{\lambda}\,r}}{r} - \sum_{i=1}^{n+1}\frac{1}{(2i-2)!}\frac{r^{2i-3}}{\lambda^{n-i+2}}\right], \tag{4.105}$$

which in the limit $r \to 0$, reduces to

$$\lim_{r \to 0}\varphi_n^M(r) = -\frac{(2n)!}{\lambda^{n+(1/2)}}, \tag{4.106}$$

and the first few terms are

$$\varphi_1^M(r) = \frac{2}{\lambda^2 r}(e^{-\sqrt{\lambda}\,r} - 1) - \frac{r}{\lambda}, \tag{4.107}$$

$$\varphi_2^M(r) = \frac{24}{\lambda^3 r}(e^{-\sqrt{\lambda}\,r} - 1) - \frac{12r}{\lambda^2} - \frac{r^3}{\lambda}, \tag{4.108}$$

$$\varphi_3^M(r) = \frac{720}{\lambda^4 r}(e^{-\sqrt{\lambda}\,r} - 1) - \frac{360r}{\lambda^3} - \frac{30r^3}{\lambda^2} - \frac{r^5}{\lambda}. \tag{4.109}$$

4.5.3 *Polyharmonic operators*

For polyharmonic operators of degree m, we seek particular solutions of the following:

$$\Delta^m \varphi_{mn}^P = r^{2n}\ln r, \quad (2D), \tag{4.110a}$$

$$= r^{2n-1}, \quad (3D). \tag{4.110b}$$

The solutions are given as [195]

$$\varphi_{mn}^P(r) = \frac{(n!)^2}{4^m[(m+n)!]^2}r^{2(m+n)}\left[\ln r - H_{m+n} + H_n\right], \quad (2D), \tag{4.111a}$$

$$= \frac{(2n)!}{(2m+2n)!}r^{2m+2n-1}, \quad (3D), \tag{4.111b}$$

with $\lim_{r \to 0} \varphi^P_{mn}(r) = 0$. For the biharmonic operator ($m = 2$) in 2D, we have for the first few terms:

$$\varphi^P_{21}(r) = \frac{r^6}{576} \left(\ln r - \frac{5}{6} \right), \tag{4.112}$$

$$\varphi^P_{22}(r) = \frac{r^8}{2304} \left(\ln r - \frac{7}{12} \right), \tag{4.113}$$

$$\varphi^P_{23}(r) = \frac{r^{10}}{6400} \left(\ln r - \frac{9}{20} \right). \tag{4.114}$$

4.5.4 *Polymetaharmonic operators*

For the product of Helmholtz-type operators of degree m, we seek the particular solutions of the following

$$(\Delta \varphi^P_{mn} \pm \lambda)^m = r^{2n} \ln r, \quad \text{(2D)}, \tag{4.115a}$$

$$= r^{2n-1}, \quad \text{(3D)}. \tag{4.115b}$$

The particular solutions for the bi-metaharmonic case, $m = 2$, were derived in [660], and the general polymetaharmonic case in [854].

Product of Helmholtz Operators: The particular solutions in 2D are given by

$$\varphi^H_{mn}(r) = n! \pi \sum_{i=0}^{m-1} \frac{(-1)^{n-1} 2^{2n-i-1} a_i}{\lambda^{m+n-(i/2)}} r^i Y_i \left(\sqrt{\lambda} r \right)$$

$$+ \frac{(n!)^2}{(m-1)!} \sum_{i=1}^{n+1} \frac{(-1)^{n-i+1} b_i}{\lambda^{m+n-i+1}} r^{2i-2} (\ln r + H_n - H_{i-1}), \tag{4.116}$$

where

$$a_i = \frac{(m + n - i - 1)!}{i!(m - i - 1)!}, \tag{4.117}$$

$$b_i = \frac{4^{n-i+1}(m + n - i)!}{(n - i + 1)![(i - 1)!]^2}. \tag{4.118}$$

For the bi-Helmholtz operator ($m = 2$), we find the first few terms as

$$\varphi_{21}^{H}(r) = \frac{1}{\lambda^3}\left[4\pi\, Y_0(\sqrt{\lambda}\,r) + \pi\sqrt{\lambda}\,r\,Y_1(\sqrt{\lambda}\,r) + (\lambda r^2 - 8)\ln r - 8\right], \tag{4.119}$$

$$\varphi_{22}^{H}(r) = \frac{-1}{\lambda^4}\left[96\pi\, Y_0(\sqrt{\lambda}\,r) + 16\pi\sqrt{\lambda}\,r\,Y_1(\sqrt{\lambda}\,r)\right.$$
$$\left. - (\lambda r^2 - 24)(\lambda r^2 - 8)\ln r + 16(\lambda r^2 - 18)\right]. \tag{4.120}$$

The 3D case is given by

$$\varphi_{mn}^{H}(r) = \frac{(2n)!}{n!\,r}\sum_{i=0}^{m-1}\frac{(-1)^{n-i-1}a_i}{\lambda^{m+n-i}}\frac{\partial^i \cos\sqrt{\lambda}\,r}{\partial\lambda^i}$$
$$- \frac{(2n)!}{(m-1)!}\sum_{i=1}^{n+1}\frac{(-1)^{n-i}c_i}{\lambda^{m+n-i+1}}r^{2i-3}, \tag{4.121}$$

where a_i is defined in (4.117) and

$$c_i = \frac{(m+n-i)!}{(n-i+1)!(2i-2)!}. \tag{4.122}$$

The first few terms for the 3D bi-Helmholtz operator are

$$\varphi_{21}^{H}(r) = \frac{1}{\lambda^3 r}\left[4\cos\sqrt{\lambda}\,r + \sqrt{\lambda}\,r\sin\sqrt{\lambda}\,r + \lambda r^2 - 4\right], \tag{4.123}$$

$$\varphi_{22}^{H}(r) = \frac{-1}{\lambda^4 r}\left[72\cos\sqrt{\lambda}\,r + 12\sqrt{\lambda}\,r\sin\sqrt{\lambda}\,r - \lambda^2 r^4 + 24\lambda r^2 - 72\right]. \tag{4.124}$$

Product of Modified Helmholtz Operators: The particular solutions in 2D are

$$\varphi_{mn}^{M}(r) = (-1)^m n!\sum_{i=0}^{m-1}\frac{2^{2n-i}a_i}{\lambda^{m+n-(i/2)}}r^i K_i(\sqrt{\lambda}\,r)$$

$$+ \frac{(-1)^m(n!)^2}{(m-1)!}\sum_{i=1}^{n+1}\frac{b_i}{\lambda^{m+n-i+1}}r^{2i-2}\,(\ln r + H_n - H_{i-1}). \tag{4.125}$$

For the bi-modified-Helmholtz operator, the first two solutions are

$$\varphi_{21}^{M}(r) = \frac{1}{\lambda^3}\big[8\,\mathrm{K}_0(\sqrt{\lambda}\,r) + 2\sqrt{\lambda}\,r\,\mathrm{K}_1(\sqrt{\lambda}\,r) + (\lambda r^2 + 8)\ln r + 8\big],$$

$$(4.126)$$

$$\varphi_{22}^{M}(r) = \frac{1}{\lambda^4}\big[192\,\mathrm{K}_0(\sqrt{\lambda}\,r) + 32\sqrt{\lambda}\,r\,\mathrm{K}_1(\sqrt{\lambda}\,r)$$

$$+ (\lambda r^2 + 24)(\lambda r^2 + 8)\ln r + 16(\lambda r^2 + 18)\big].$$

$$(4.127)$$

The 3D case is given by

$$\varphi_{mn}^{M}(r) = \frac{(2n)!}{n!\,r}\sum_{i=0}^{m-1}\frac{(-1)^{m-i-1}a_i}{\lambda^{m+n-i}}\frac{\partial^i e^{-\sqrt{\lambda}\,r}}{\partial\lambda^i}$$

$$+ \frac{(-1)^m (2n)!}{(m-1)!}\sum_{i=1}^{n+1}\frac{c_i}{\lambda^{m+n-i+1}}\,r^{2i-3},\qquad (4.128)$$

and the first two solutions are

$$\varphi_{21}^{M}(r) = \frac{-1}{\lambda^3 r}\big[(\sqrt{\lambda}\,r + 4)e^{-\sqrt{\lambda}\,r} - \lambda r^2 - 4\big], \qquad (4.129)$$

$$\varphi_{22}^{M}(r) = \frac{-1}{\lambda^4 r}\big[12(\sqrt{\lambda}\,r + 6)e^{-\sqrt{\lambda}\,r} - \lambda^2 r^4 - 24\lambda r^2 - 72\big]. \quad (4.130)$$

4.5.5 *Cauchy–Navier operator*

As discussed in Section 4.2.6, for elastostatic problems, there are two types of body forces: those resulting from the gradient of a physical potential, such as the temperature field, and those that are general. We shall first examine the gradient case.

Thermal Stresses: Consider the theory of thermal stresses with the governing equation (4.51). We can approximate the temperature field as

$$\hat{T}(\boldsymbol{x}) = \sum_{j=1}^{N} d_j\,\psi_n(r_j), \qquad (4.131)$$

in which ψ_n represents the polyharmonic spline functions given by (4.71). Following a similar derivation as in Section 4.2.6, we obtain

$$\hat{u}_p = \frac{\alpha_T(1-2v)}{2G(1-v)} \sum_{j=1}^{N} d_j \nabla \varphi_n^L,$$ (4.132)

where φ_n^L is a particular solution satisfying

$$\nabla^2 \varphi_n^L(r) = \psi_n(r),$$ (4.133)

and is given by (4.73) and (4.77)

General Body Force: For a body force that has three independent components, $\{F_x, F_y, F_z\}$, we make the individual substitution of $F_x = r^{2n} \ln r$, $F_y = r^{2n} \ln r$, etc., in the following equation

$$Gu_{ik,jj}^p - \frac{G}{1-2v} u_{jk,ji}^p = -\delta_{ik} r^{2n} \ln r,$$ (4.134)

in which the second subscript of u_{ik}^p indicates the force direction. The particular solution in 2D is

$$u_{ik}^p(\boldsymbol{x}) = \frac{r^{2(n+1)}}{32G(1-v)(n+1)^3(n+2)^2} \left\{ -2(n+1)(2n+3)\frac{x_i x_k}{r^2} \right.$$

$$+ 2(n+1)(n+2) \ln r \left[2(n+1)\frac{x_i x_k}{r^2} - \delta_{ik}(4n+7-4v(n+2)) \right]$$

$$\left. + \delta_{ik} \left[8n^2 + 29n + 27 - 8v(n+2)^2 \right] \right\}, \quad i,k = 1,2,$$ (4.135)

and in 3D

$$u_{ik}^p(\boldsymbol{x}) = \frac{r^{2n+1}}{8G(1-v)(n+1)(n+2)(2n+1)}$$

$$\times \left\{ (2n+1)\frac{x_i x_k}{r^2} - \delta_{ik}[4n+7-4v(n+2)] \right\}, \quad i,k = 1,2,3.$$ (4.136)

4.6 Multiquadric RBF

As we have demonstrated in Section 3.3, MQ are among the most efficient basis functions for scattered point interpolation. It is therefore important to find their corresponding particular solutions.

4.6.1 *Laplacian operator*

For MQ, we seek the particular solution of

$$\nabla^2 \varphi^L = \sqrt{r^2 + c^2},\tag{4.137}$$

in 2D and 3D. By successive integration, we find [941]

$$\varphi^L(r) = \frac{r^2 + 4c^2}{9}\sqrt{r^2 + c^2} - \frac{c^3}{3}\ln\left(\sqrt{r^2 + c^2} + c\right),\qquad \text{(2D)},\tag{4.138a}$$

$$= \frac{2r^2 + 5c^2}{24}\sqrt{r^2 + c^2} + \frac{c^4}{8r}\left[\ln\left(r + \sqrt{r^2 + c^2}\right) - \ln c\right],\ \text{(3D)}.\tag{4.138b}$$

In the limit as $r \to 0$, the above become

$$\lim_{r \to 0}\varphi^L(r) = -\frac{c^3}{3}\left(\ln 2c - \frac{4}{3}\right),\quad \text{(2D)},\tag{4.139a}$$

$$= \frac{c^3}{3},\qquad \text{(3D)}.\tag{4.139b}$$

For IMQ, we have

$$\nabla^2 \varphi^L = \frac{1}{\sqrt{r^2 + c^2}},\tag{4.140}$$

and we find [941]

$$\varphi^L(r) = \sqrt{r^2 + c^2} - c\ln\left(\sqrt{r^2 + c^2} + c\right),\qquad \text{(2D)},\tag{4.141a}$$

$$= \frac{\sqrt{r^2 + c^2}}{2} + \frac{c^2}{2r}\ln\left(\frac{r + \sqrt{r^2 + c^2}}{c}\right),\quad \text{(3D)},\tag{4.141b}$$

with the limits

$$\lim_{r \to 0} \varphi^L(r) = c - c \ln 2c, \quad \text{(2D)}, \qquad (4.142\text{a})$$

$$= c, \qquad \text{(3D)}. \qquad (4.142\text{b})$$

4.6.2 Biharmonic operator

The particular solutions for polyharmonic operators with a MQ RHS have been derived by Tsai [856, 858, 859]. The general formulas for the particular solutions are rather complex; hence only the biharmonic case is presented here. For

$$\Delta^2 \varphi^B = \sqrt{r^2 + c^2}, \qquad (4.143)$$

the particular solutions are

$$\varphi^B(r) = \frac{1}{900} \left(4r^4 + 48c^2 r^2 - 61c^4\right) \sqrt{r^2 + c^2} - \frac{c^3}{900} \left(25r^2 - 61c^2\right)$$

$$- \frac{c^3}{60} \left(5r^2 - 2c^2\right) \ln\left(\frac{c + \sqrt{r^2 + c^2}}{2c}\right), \qquad \text{(2D)}, \qquad (4.144\text{a})$$

$$= \frac{1}{1440} \left(4r^4 + 28c^2 r^2 - 81c^4\right) \sqrt{r^2 + c^2} - \frac{c^3}{90} \left(5r^2 - 6c^2\right)$$

$$+ \frac{c^4}{96} \left(6r^2 - c^2\right) \frac{1}{r} \sinh^{-1} \frac{r}{c}, \qquad \text{(3D)}. \qquad (4.144\text{b})$$

We notice that $\lim_{r \to 0} \varphi^B(r) = 0$ for both 2D and 3D cases.

4.6.3 Cauchy–Navier operator

For the case where the RHS is the gradient of a field, such as the temperature field in (4.51), we can follow the same procedure as in Section 4.5.5, using the MQ and IMQ function ψ instead of the polyharmonic spline functions ψ_n in (4.131). The particular solution u_p is given by (4.132) with φ_n^L replaced by φ^L, in which φ^L is defined in (4.138) for the MQ and (4.141) for the IMQ function.

4.7 Gaussian RBF

The Gaussian RBF is another highly efficient interpolant that possesses exponential convergence properties. Its particular solution has been investigated in [515] for the Laplacian and biharmonic operators.

4.7.1 *Laplacian operator*

For the Laplacian operator, we solve

$$\nabla^2 \varphi^L = e^{-r^2/c^2}, \tag{4.145}$$

and the solutions are

$$\varphi^L(r) = -\frac{c^2}{4}\left[2\ln r + \mathrm{Ei}\left(-\frac{r^2}{c^2}\right)\right], \quad (2\mathrm{D}), \tag{4.146a}$$

$$= -\frac{\sqrt{\pi}\,c^3}{4r}\,\mathrm{erf}\left(\frac{r}{c}\right), \quad (3\mathrm{D}), \tag{4.146b}$$

in which Ei is the exponential integral, and erf the error function. The limits are

$$\lim_{r\to 0}\varphi^L(r) = -\frac{c^2}{4}(\gamma - 2\ln c), \quad (2\mathrm{D}), \tag{4.147a}$$

$$= -\frac{c^2}{2}, \quad (3\mathrm{D}). \tag{4.147b}$$

4.7.2 *Biharmonic operator*

For

$$\Delta^2 \varphi^B = e^{-r^2/c^2}, \tag{4.148}$$

we find

$$\varphi^B(r) = -\frac{c^2}{16}\left[-2r^2 + 2(r^2 - c^2)\ln r + c^2 e^{-r^2/c^2} + (r^2 + c^2)\mathrm{Ei}\left(-\frac{r^2}{c^2}\right)\right],$$
$$(2\mathrm{D}), \quad (4.149\mathrm{a})$$

$$= -\frac{\sqrt{\pi}\,c^3}{4}\left[\mathrm{erf}\left(\frac{r}{c}\right)\left(\frac{r}{2} + \frac{c^2}{4r}\right) + \frac{c}{2\sqrt{\pi}}e^{-r^2/c^2}\right], \quad (3\mathrm{D}), \quad (4.149\mathrm{b})$$

and the limits are

$$\lim_{r \to 0} \varphi^B(r) = -\frac{c^4}{16}(1 + \gamma - 2\ln c), \quad (2D), \tag{4.150a}$$

$$= -\frac{c^4}{4}, \qquad (3D). \tag{4.150b}$$

Chapter 5

Solving Partial Differential Equations

In Chapters 2 and 3, we introduced various basis functions, such as fundamental solutions, general solutions, monomials and polynomials, and RBFs, that can be used to approximate and interpolate a function. In this chapter, we shall discuss their use for solving PDEs.

5.1   Mesh versus Meshless Methods

The earlier numerical methods use structured data points for discretization. For example, the FDM uses a rectilinear grid for the purpose of taking the Taylor series expansions. Spectral methods like the Fourier, Chebyshev polynomial, and wavelet methods, also require orthogonal grids with specified spacing, such as the roots of Chebyshev polynomials. The FEM, on the other hand, can fit irregular geometries using triangular elements of various sizes; and has therefore gained popularity in engineering and industry applications. The FEM mesh, however, is also structured in the sense that connectivity data are required to keep track of the shared nodes among elements. Such bookkeeping requirements can be challenging in the initial generation of a mesh, given a complex geometry, and remeshing is needed if the geometry evolves with the simulation.

The concern about the mesh generation has prompted the FEM community to abolish the elements. We may quote Belytschko *et al.* [74] as follows:

> *In recent years, it has become clear that in linear analysis, mesh generation is a far more time-consuming and expensive task than*

181

> *the assembly and solution of the finite element equations. Therefore, it has become expedient to explore methods which may be somewhat more expensive from the viewpoint of computer time but involve less time in the preparation of data. In this respect, it might be mentioned that even with powerful mesh generators, three-dimensional meshing is still an extremely burdensome task and that the conversion of solid models to finite element data is time-consuming and often introduces numerous ambiguities.*

For the next three decades, the movement has led to the development of dozens of "element-free", "meshfree", or "meshless" methods. We shall give a brief review in the following.

In the FEM community, the traditional non-overlapping elements, which have only C^0 continuity, have been discarded, and replaced by groups of nodes forming "covers" that overlap each other. Within a cover, the basis function is no longer limited to low degree polynomials. More efficient bases, such as RBFs, can be employed. The moving least squares method is often used for the interpolation. By choosing the compactly supported, but smooth, basis and weighting functions, higher order continuity and better accuracy can be achieved.

A variety of methods, such as the EFG [74, 602], diffuse element method (DEM) [499, 667], h-p clouds method [272, 273], MLPG [38, 41], partition of unity finite element method (PUFEM) [324, 636], generalized finite element method (GFEM) [274, 818], extended finite element method (XFEM) [651, 652], finite cloud method (FCM) [15, 16], LRPIM [578, 897], and smoothed finite element method (SFEM) [582, 583], have been developed. For the finite volume method (FVM) [283, 529], there are also meshless versions, such as the meshless finite volume method (MFVM) [43, 288, 654].

The BEM community followed suit, and spawned meshless versions like the boundary node method (BNM) [657, 962], local boundary integral equation (LBIE) method [39, 971], boundary cloud method (BCM) [532, 533], boundary point interpolation method (BPIM) [362, 581], boundary element-free method (BEFM) [566, 717], boundary face method (BFM) [737, 963], and Galerkin boundary node method (GBNM) [543, 545], etc.

The FDM has loosened its rectilinear grids requirement, and used star-shaped stencils to perform Taylor series expansion, and developed the generalized finite difference (GFD) method [76, 719]. The differential quadrature (DQ) [72, 73] and generalized differential quadrature (GDQ) [82, 800] methods are similar to the FDM in that orthogonal grids are used for taking derivatives. These methods, however, use global, rather than local, approximation, and quadrature rules are used to perform integration. Scattered node versions have been developed such as the RBF-DQ, with a global [802] and a local [801, 803] version.

In the above reviewed methods, efforts were made to convert the mesh based methodologies to meshless ones. On the other hand, there are methods that are originally based on scattered point collocation, such as the RBFCM [440, 441], MFS [285, 286], and TCM [486, 550]; hence no effort is required to undo the burden of a mesh. These methods use global interpolation covering the entire domain or boundary, as compared to the local interpolations of the FEM, FDM, FVM, etc. To avoid the difficulty of solving a full matrix, their local versions, such as the LRBFCM [261, 773], RBF-FD [308, 920], and LMFS [291, 358], have been developed.

For meshless methods, we should also mention the particle methods, of which the most prominent is the smoothed particle hydrodynamics method (SPH) [331, 603]. In these methods, the continua, or non-continua, are modeled as a set of particles that interact with each other, and are free to move. The macroscopic physical properties are taken as a certain moving local average to "smooth" out the discrete particle representations. Improved methods and variants include the reproducing kernel particle method (RKPM) [590, 591], finite point method (FPM) [689, 690], particle-in-cell (PIC) method [104, 819], material point method (MPM) [60, 820], etc.

The particle methods are the most flexible in modeling deforming and moving boundary problems, and also materials that are fragmented or discrete. As the number of particles is generally not very large, they do not actually represent the true physical material points. For modeling a simple continuum problem, these methods are not accurate; hence not efficient. They can, however, model certain complex problems and capture the qualitative physical phenomena that other methods cannot.

Our interest in this chapter is to solve a BVP for a PDE over an irregular domain using a set of scattered points, without mesh, element, or structure. We observe from the above survey, that different traditional methods have found different ways to unburden themselves from the mesh issue. In that process, they may inherit their original strengths, as well as weaknesses. For example, the FEM inspired methods typically use the weak, instead of the strong, formulation, and the FDM derived methods adhere to the series expansion. New issues may emerge as how to perform quadrature over a group of points without a fixed structure, match the BCs, or derive the finite difference formula, etc.

Alternatively, we may unburden ourselves from the inherited constraints, and start to think fresh. Given a set of scattered points, how do we approximate a function using them? For the basis functions, should we use the traditional polynomials, the highly accurate RBFs, or the fundamental and general solutions? Should we apply the strong or weak formulation to satisfy the governing equations and BCs? What are the tradeoffs of local versus global approximation? These decisions should be evaluated based on the accuracy, efficiency, and stability of the method. In the following sections, we shall discuss some basic issues arising when solving PDEs, not to offer a simple or single answer to the above questions, but to offer alternatives when a specific type of problem is encountered.

5.2 Interpolation of Scattered Data

Many methods have been put forward for the interpolation of spatial data in a multi-dimensional space. These methods may be developed for different purposes. For example, the data of a certain physical phenomenon are observed at a limited number of locations, due to practical constraints. With the belief that the phenomenon is continuous or correlated in space, the "missing data" are estimated using observations in the proximity. A popular technique is the inverse distance weighting method, which assigns larger weights on nearby than far away data. Other methods take into consideration the data measurement uncertainty and its spatial correlation to give a statistically best estimate. The kriging algorithm is one of the most used geostatistical methods [225, 252, 498, 629].

There are also methods aimed at approximating known mathematical functions using a finite set of sampled data. These methods must have convergence properties, namely, as the density of data increases, the approximation should approach the true function. The convergence may be defined as the diminishing pointwise error, or its integration over the domain. These lead to the many approximation theories; see for example Powell [730]. In this book, we are interested in yet another different purpose — solving PDEs. In such applications, the "data" used in the interpolation are in fact unknowns to be determined. So the approximation is at first symbolic. The requirements that the approximation satisfies the governing equation and BCs in a certain sense then set up the conditions for their solution.

In this section, we examine a few of the methods that are relevant to the present purpose. Particularly, we focus on the interpolation of data that are scattered in space, and not fixed on grids, for the flexibility of solving irregular domain BVPs.

5.2.1 *Interpolation using basis functions*

Consider the interpolation of a function $f(x)$ in a given domain Ω. As a continuous function has infinite degrees of freedom, it is necessary to approximate it using a finite degree system in order to handle it by numerical means. The missing information is filled using interpolation. The interpolated function can be known or unknown. Specifically, if the function is the solution of a PDE, it is not known, and the interpolation is performed symbolically. We may classify the approximation methods into two types: those explicitly using function values in the formula, and those not. For convenience, we shall call them the explicit and the implicit method, respectively, and discuss them separately in this and the next section.

So far, in Chapters 2–4 we have approximated functions by the implicit method, in the following form

$$\hat{f}(x) = \sum_{j=1}^{N} c_j \psi_j(x), \tag{5.1}$$

in which the ψ_j are basis functions, and the c_j are coefficients to be determined. The above equation may be expressed in matrix form as

$$\hat{f}(x) = \boldsymbol{\psi}^T(x)\boldsymbol{c}, \tag{5.2}$$

where $\boldsymbol{\psi}(x) = \lfloor \psi_1, \ldots, \psi_N \rfloor^T$ and $\boldsymbol{c} = \lfloor c_1, \ldots, c_N \rfloor^T$ are column matrices. The basis functions $\psi_j(x)$ can be trigonometric functions, polynomials, RBFs, general solutions or fundamental solutions, etc. They can be a subset of an infinite and complete family, $\psi_j(x)$, $j \in \mathbb{N}$, truncated at N terms; or a single function, $\psi(x, x_j)$, deployed at N centers x_j. We notice that the discrete function values do not explicitly appear in the above representation; hence we shall call (5.1) an implicit formula.

To ensure that the formula is a reasonable approximation of a function, we shall require the approximation to take the exact value of the function at a set of nodes. Assuming that such values are given at a set of points x_i, for $i = 1, 2, \ldots, N$, as $f(x_i) = f_i$, we can enforce such conditions:

$$\sum_{j=1}^{N} c_j \psi_j(x_i) = f_i, \quad \text{for } i = 1, \ldots, N. \tag{5.3}$$

In matrix form, the above can be expressed as

$$\mathbf{\Psi} c = f, \tag{5.4}$$

in which $\mathbf{\Psi} = [\psi_{ij}]$ is a square matrix with elements $\psi_{ij} = \psi_j(x_i)$, and $f = \lfloor f_1, \ldots, f_N \rfloor^T$ is a vector. As the ψ_j are known functions, they can be evaluated at the points x_i. Equation (5.3) then becomes a system of N linear equations, which can be solved for the c_j as

$$c = \mathbf{\Psi}^{-1} f. \tag{5.5}$$

With c known, the approximated function at any point $x \in \Omega$ can be evaluated using (5.1).

The above procedure of requiring the approximation to satisfy exact conditions at a set of points is known as the collocation method, or the "strong formulation". There are other methods to seek convergence of the approximation. For example, we can multiply the error by a certain weight, integrate it over a domain, and seek to minimize it with respect to the basis function of the approximation. This particular method is known as the "weak formulation". We shall discuss these procedures in Section 5.5.

For the approximation to be acceptable, a denseness property is needed. That is, we need a proof that when the number of terms N in (5.1) is continuously increased, the error will decrease toward zero. The rate of such a convergence can be strongly dependent on the smoothness of the approximation. Particularly, for the solution of a PDE, the function needs to be differentiable up to the order of the PDE. It is obvious from (5.1) that the smoothness of the approximation is the same as the smoothness of the selected basis functions. Hence, the smoother the basis function, the more rapid the error convergence. We can observe the performance of interpolation using basis functions of different smoothness in Examples 3.3 and 4.8.

5.2.2 *Interpolation using nodal values*

In another way of approximation, we utilize the function values $f(x_j) \equiv f_j$ at a set of nodes x_j to form an interpolation formula as

$$\hat{f}(x) = \sum_{j=1}^{N} \Phi_j(x) f(x_j), \tag{5.6}$$

in which Φ_j are generating functions (or weights) associated with nodes x_j. In matrix form, we write

$$\hat{f}(x) = \Phi^T(x)\,f, \qquad (5.7)$$

where $\Phi(x) = \lfloor \Phi_1, \ldots, \Phi_N \rfloor^T$ is a column matrix. In other words, we approximate the function f at a location x as the weighted sum of all available data $f(x_j)$. To avoid spurious results, it is highly desirable that the weights at any given point x be summed to unity, which is called the partition of unity, discussed in the ensuing section. As discrete function values are used in the above representation, we shall call (5.6) an explicit formula, in contrast to the implicit formula (5.1).

There are many ways in which the weights Φ_j can be constructed. For example, if the data represent physical measurements, the weights are generally made to be inversely proportional to the distance, as it is believed that the farther the data points, the less relevant they are. Alternatively, we can perform a least squares fit to generate the weights. In that case, the exact data values are generally not reproduced. In selecting the weights, we should recall the comment in Section 5.2.1 that the smoothness of the approximation is dictated by its interpolants, and in the current case, by the weights. Therefore, if the weights lack sufficient smoothness, the approximation is generally not very accurate.

We note that it is possible to transform the implicit formula (5.1) into the explicit one (5.6) by the following procedure. Substituting (5.5) into (5.2), we obtain

$$\hat{f}(x) = \psi^T(x)\,\Psi^{-1}f. \qquad (5.8)$$

It is obvious that the above formula is just (5.7) if we define

$$\Phi^T(x) = \psi^T(x)\,\Psi^{-1}. \qquad (5.9)$$

We note that the weight function $\Phi(x)$ so constructed preserves the smoothness of the original interpolants, and there is no loss of accuracy in the conversion. In fact, if the explicit formula were originally constructed using the same set of basis functions as the implicit one, then these two approximations would be exactly the same.

In the above discussion we assumed that the approximation used all the data points for a global interpolation. It is in fact common that (5.6) uses only data in the vicinity of a point to perform a local interpolation. The formula can include a fixed number of nearest nodes, or use all nodes that fall within a radius of influence. In the latter case, this can be accomplished using weights that are compactly supported. In other words, the weights are non-zero only within a predefined radius, such that the influence of the farther nodes is automatically eliminated. We, however, need to be careful about the property of the weights. If the weights are continuous functions that are arbitrarily truncated at the radius of influence, they lack the smoothness needed for accuracy. The CSRBFs discussed in Section 3.3.1 were particularly designed to have smoothness up to a certain degree at a given radius, and can be good candidates for the weights. We shall further discuss the pros and cons of global versus local interpolation in Section 5.3.

5.2.3 *Partition of unity*

For a consistent interpolation, the weights in (5.6) need to possess certain properties. In the partition of unity (PU) method [54], the weights $\Phi_j(x)$ are required to take the value of 1 when $x = x_j$, and 0 when $x = x_i$ for $i \neq j$. This guarantees that the data values are exactly reproduced at the nodes. Also, at any location, the weights should sum to unity. The above statements can be expressed as

$$\Phi_j(x_i) = \delta_{ij}, \tag{5.10}$$

$$\sum_{j=1}^{N} \Phi_j(x) = 1. \tag{5.11}$$

We shall discuss a few PU cases below.

Shepard Function: For physical applications, a popular interpolation method is the inverse distance weighting algorithm for estimating missing data. For example, terrain elevations, hydrological measurements, and pollutant concentration are generally sampled at a sparse set of locations. At a location where there is no data, an estimate is obtained by referring to measurements at nearby sites. Rather than using a single nearest value, a weighted average is taken from a number of locations. The nearer data should have the larger weights, and the farther ones the smaller weights. Hence the weights are made to be inversely proportional to the distance, raised to a power p. A consistent formula was given by Shepard [793]:

$$\hat{f}(x) = \frac{\sum_{j=1}^{N} w_j(x) f(x_j)}{\sum_{i=1}^{N} w_i(x)}, \quad \text{for } x \neq x_j, \tag{5.12a}$$

$$= f(x_j), \quad \text{for } x = x_j, \tag{5.12b}$$

in which

$$w_j(x) = \frac{1}{r_j^p}, \tag{5.13}$$

are weights associated with data points x_j, and $r_j = \|x - x_j\|$ are the distances to the data points. The power p is typically selected in the range $1 \leq p \leq 3$. Comparing (5.12) with (5.6), we can identify the generating function

$$\Phi_j(x) = \frac{w_j(x)}{\sum_{i=1}^{N} w_i(x)}. \tag{5.14}$$

As $x \to x_j$, the weight (5.13) becomes singular, such that all other weights can be neglected, leading to $\Phi_j(x_j) \to 1$. Equation (5.14) hence satisfies the PU properties (5.10) and (5.11).

■ Example 5.1

In a 1D example, visualize the Shepard generating functions. Within the interval $[0, 1]$, data are given at 6 equally spaced points,

$$x_j = \{0, 0.2, 0.4, 0.6, 0.8, 1.0\}.$$

Find the generating functions corresponding to the inverse distance squared weight $w_j(x) = 1/(x - x_j)^2$.

First, we notice that data on these nodes are not needed for the determination of the generating functions. The six functions corresponding to x_j, $j = 1, \ldots 6$, are obtained from (5.14). Figure 5.1 shows the first three functions, Φ_1, Φ_2 and Φ_3. The remaining three are symmetrical about $x = 0.5$; hence are not plotted. We observe that these functions rapidly decay with the distance, and possess the PU properties. That is, they have the value 1 when $i = j$ and 0 when $i \neq j$, and at any $0 \leq x \leq 1$, the three functions add to 1.

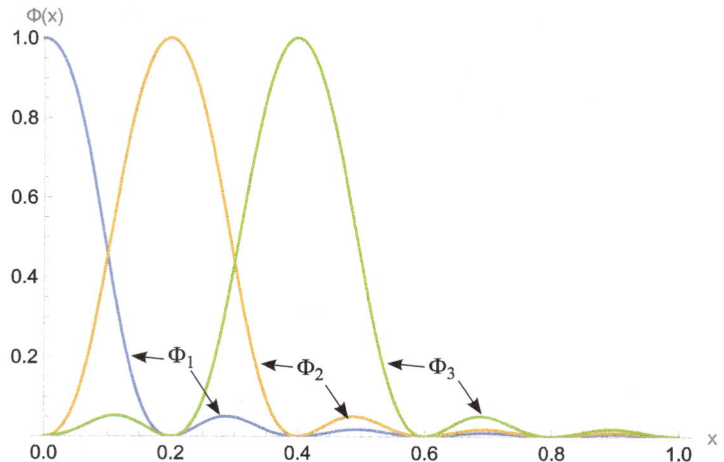

Figure 5.1. Plot of Shepard generating functions Φ_1, Φ_2 and Φ_3.

The Shepard function (5.12) is a popular formula for interpreting missing data given a sparse set of physical measurements. Particularly, it is an explicit formula without the need of solving a system of equations. For approximating mathematical functions, however, it has the undesirable property that for the cases $p > 1$, its partial derivatives are zero at every data point, creating flat surfaces around them [786]. As demonstrated in Franke's survey articles [313, 314], as well as in the example below, its accuracy is poor. Furthermore, it certainly should not be used to calculate the derivatives of the approximated function, due to the many flat regions of the approximation. There are many modifications of Shepard's original formula [342, 439, 634, 786], and most use least squares to smooth out the flatness. The least squares fits are discussed in Section 5.2.4.

■ Example 5.2

Approximate the function $f(x, y)$ defined in Example 3.3 using Shepard's formula (5.12).

Following the previous examples, we lay an 11×11 grid over the domain $[0, 1] \times [0, 1]$, where the data are sampled. The resultant fit is plotted as Fig. 5.2(a), which can be compared with the exact function shown in Fig. 3.3, The relative error ε_r is presented in Fig. 5.2(b), which can be as large as 30% and is therefore unacceptable.

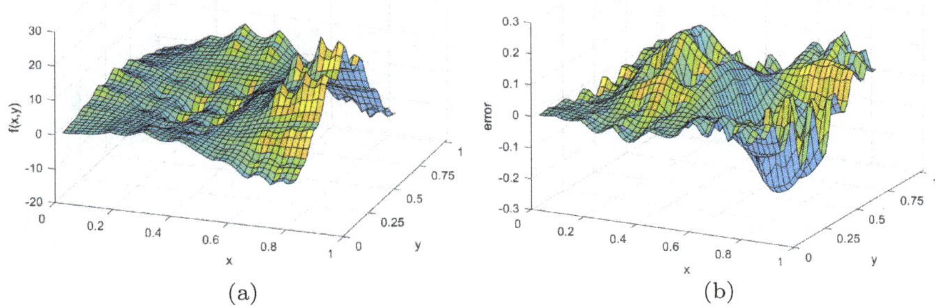

Figure 5.2. Approximating the function in Example 3.3 using Shepard's formula with an 11×11 grid: (a) Approximation of the function f; and (b) relative error ε_r of the approximation.

FEM Shape Functions: In the classical FEM, polynomial bases are used to construct the PU relations within a local element Ω_i, delineated by a certain number of nodes, which also define the degrees of freedom of the interpolation. We may rewrite (5.6) in the following form:

$$\hat{f}(\boldsymbol{x}) = \sum_{j=1}^{k} N_j(\boldsymbol{x}) f(\boldsymbol{x}_j), \tag{5.15}$$

where k is the number of nodes on the element, and the N_j play the same role as the generating functions Φ_j, and are called shape functions. The shape functions have the PU properties, and can be derived in analytical forms for different types of elements — triangular, quadrilateral, tetrahedron, linear, quadratic, etc. We shall give a brief illustration for the simplest element in the example below.

■ **Example 5.3**

Given a triangular element defined by the three nodes (x_1, y_1), (x_2, y_2), and (x_3, y_3) as vertices (see Fig. 5.3), find a proper interpolation of the function $f(x, y)$ within the triangle using its nodal values.

Due to the small number of degrees of freedom, $k = 3$, we are limited to linear interpolation. We hence define

$$N_j(x, y) = a_j x + b_j y + c_j, \quad \text{for } j = 1, 2, 3.$$

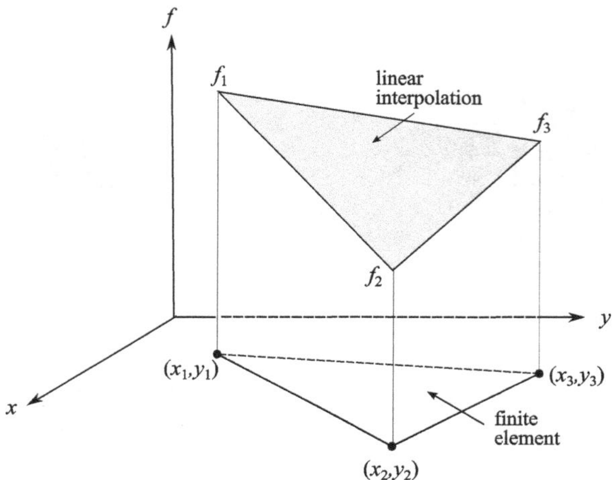

Figure 5.3. Linear interpolation of a function in a triangular finite element.

Substituting the above into (5.15) shows that $f(x, y)$ is approximated as a linear function:

$$\hat{f}(x, y) = ax + by + c.$$

Using (5.10), we can establish the nine equations

$$N_1(x_1, y_1) = a_1 x_1 + b_1 y_1 + c_1 = 1,$$

$$N_1(x_2, y_2) = a_1 x_2 + b_1 y_2 + c_1 = 0,$$

$$\vdots$$

$$N_3(x_3, y_3) = a_3 x_3 + b_3 y_3 + c_3 = 1,$$

which allow us to solve for the 9 coefficients $\{a_j, b_j, c_j\}$, $j = 1, 2, 3$, as

$$a_1 = \frac{y_2 - y_3}{2A}, \quad a_2 = \frac{y_3 - y_1}{2A}, \quad a_3 = \frac{y_1 - y_2}{2A},$$

$$b_1 = \frac{x_3 - x_2}{2A}, \quad b_2 = \frac{x_1 - x_3}{2A}, \quad b_3 = \frac{x_2 - x_1}{2A},$$

$$c_1 = \frac{x_2 y_3 - x_3 y_2}{2A}, \quad c_2 = \frac{x_3 y_1 - x_1 y_3}{2A}, \quad c_3 = \frac{x_1 y_2 - x_2 y_1}{2A},$$

where

$$A = \frac{1}{2} \begin{vmatrix} x_1 & y_1 & 1 \\ x_2 & y_2 & 1 \\ x_3 & y_3 & 1 \end{vmatrix}$$

is the area of the triangle. We hence have found the shape functions N_j. By substitution, we can prove that the condition

$$\sum_{j=1}^{3} N_j(x, y) = 1.$$

is satisfied.

As commented by Babuška and Melenk [54], the classical FEM uses non-overlapping patches $\{\Omega_i\}$ to cover the entire domain Ω. Low degree polynomials are used to construct the PU shape functions. These lead to schemes that are low in accuracy and slow in convergence. It was suggested that more efficient basis functions, such as RBFs, can be used to build shape functions on overlapping patches to cover the domain. These ideas have led to a whole new class of meshless methods. A brief discussion will be given in Section 5.3.

5.2.4 *Least squares approximation*

For the approximation given by (5.1), we can form a uniquely solvable system by using a family of basis functions having the same number of degrees of freedom as that of data points. Sometimes, however, for stability reasons, it is desirable to use a lower degree of freedom system than the available data for fitting. It is obvious that the exact satisfaction of the conditions at the data points is no longer possible; hence a new interpretation of error is needed to make the system solvable.

Least Squares Approximation: In a least squares (LS) approximation, monomials up to degree k are used as the basis functions $\psi_j(x)$ to represent the lower degree of freedom system. Written in matrix form, we express

$$\boldsymbol{\psi}(x) = \lfloor 1, x, y, x^2, xy, y^2, \ldots \rfloor^T, \tag{5.16}$$

for 2D, and similarly for 3D. The number of elements of the column vector $\boldsymbol{\psi}$ is given by

$$M = \frac{(d+k)!}{d!\,k!},\tag{5.17}$$

where d is the number of spatial dimensions, and k is the highest degree of the polynomial. We shall require $M \leq N$, with N the number of available data points. We then write the approximation of $f(\boldsymbol{x})$ as a polynomial

$$\hat{f}(\boldsymbol{x}) = p(\boldsymbol{x}) = \boldsymbol{\psi}^T(\boldsymbol{x})\boldsymbol{c},\tag{5.18}$$

where $\boldsymbol{c} = \lfloor c_1, \ldots, c_M \rfloor^T$. A square error functional can be defined as

$$E(\boldsymbol{c}) = \|p(\boldsymbol{x}_i) - f_i\|^2 = \sum_{i=1}^{N} \left[\boldsymbol{\psi}^T(\boldsymbol{x}_i)\boldsymbol{c} - f_i\right]^2,\tag{5.19}$$

where $f_i = f(\boldsymbol{x}_i)$ are given data. We seek to minimize the square error with respect to the undetermined coefficients c_j by setting the gradient to zero,

$$\nabla E = \boldsymbol{0},\tag{5.20}$$

where $\nabla = \lfloor \partial/\partial c_1, \ldots, \partial/\partial c_M \rfloor^T$. This results in the following set of M linear equations in c_j,

$$\sum_{i=1}^{N} \psi_j(\boldsymbol{x}_i)\left[\boldsymbol{\psi}^T(\boldsymbol{x}_i)\boldsymbol{c} - f_i\right] = 0, \quad \text{for } j = 1, \ldots, M,\tag{5.21}$$

or, in matrix form

$$\boldsymbol{A}\boldsymbol{c} - \boldsymbol{B}\boldsymbol{f} = \boldsymbol{0},\tag{5.22}$$

in which

$$\boldsymbol{A} = \boldsymbol{\Psi}^T\boldsymbol{\Psi},\tag{5.23}$$

$$\boldsymbol{B} = \boldsymbol{\Psi}^T,\tag{5.24}$$

where $\boldsymbol{\Psi} = [\psi_{ij}]$ is an $N \times M$ matrix with $\psi_{ij} = \psi_j(\boldsymbol{x}_i)$, and $\boldsymbol{A}$ is $M \times M$, called the moment matrix. The above system can be solved for $\boldsymbol{c}$ as

$$\boldsymbol{c} = \boldsymbol{A}^{-1}\boldsymbol{B}\boldsymbol{f}.\tag{5.25}$$

Substituting the above into (5.18), we obtain (5.7), with the generating function $\boldsymbol{\Phi}$ given by

$$\boldsymbol{\Phi}^T(\boldsymbol{x}) = \boldsymbol{\psi}^T(\boldsymbol{x}) \boldsymbol{A}^{-1} \boldsymbol{B}. \tag{5.26}$$

Weighted Least Squares Approximation: The LS approximation gives a single formula that is valid for all $\boldsymbol{x} \in \Omega$. We may interpret it as, at a given location $\boldsymbol{x}$, all data points $f(\boldsymbol{x}_i)$, near or far, contribute equally to the interpolation. In a better approximation, we may consider that the nearby data have more relevance to the interpolation. Similar to the inverse distance weighting method, we may put weights that diminish with the distance from the data point. This leads to the weighted least squares (WLS) method.

We may modify the error functional in (5.19) to

$$E(\boldsymbol{c}, \boldsymbol{x}) = \sum_{i=1}^{N} w(\boldsymbol{x}, \boldsymbol{x}_i) \left[\boldsymbol{\psi}(\boldsymbol{x}_i) \boldsymbol{c}(\boldsymbol{x}) - f_i \right]^2 . \tag{5.27}$$

We note in the above that the weight w is a function of $\boldsymbol{x}$, and so is $\boldsymbol{c}$. The weight is typically made to be a function of the distance, $w = w(r_i)$, with $r_i = \|\boldsymbol{x} - \boldsymbol{x}_i\|$. There are many possible forms for w, and a popular one is the Gaussian function

$$w(r_i) = e^{-r_i^2/h^2}, \tag{5.28}$$

where h is a scaling factor that controls the rate of decay of the function. The minimization process is similar to that of the LS case, and we obtain

$$\boldsymbol{c}(\boldsymbol{x}) = \boldsymbol{A}^{-1}(\boldsymbol{x}) \boldsymbol{B}(\boldsymbol{x}) \boldsymbol{f}, \tag{5.29}$$

where

$$\boldsymbol{A}(\boldsymbol{x}) = \boldsymbol{\Psi}^T \boldsymbol{W}(\boldsymbol{x}) \boldsymbol{\Psi}, \tag{5.30}$$

$$\boldsymbol{B}(\boldsymbol{x}) = \boldsymbol{\Psi}^T \boldsymbol{W}(\boldsymbol{x}), \tag{5.31}$$

in which $\boldsymbol{W}$ is an $N \times N$ diagonal matrix with diagonal elements $w_i = w(r_i)$. Substituting (5.29) into (5.18), we recover the interpolation formula (5.7), with the generating function

$$\boldsymbol{\Phi}^T(\boldsymbol{x}) = \boldsymbol{\psi}^T(\boldsymbol{x}) \boldsymbol{A}^{-1}(\boldsymbol{x}) \boldsymbol{B}(\boldsymbol{x}). \tag{5.32}$$

Comparing the above with (5.26), we notice the difference in the definitions of $\boldsymbol{A}$ and $\boldsymbol{B}$.

We may investigate the special cases of the WLS approximation. First, rather than using a smooth weight as (5.28), we may use singular weights, such as the inverse distance function given in (5.13) for the Shepard method. The singularity at a data point forces all other weights to drop out, and the exact data value is recovered from the interpolation. Second, we can consider a degenerate case by using a single term in the basis function (5.16), as $\psi = \{1\}$, for the least squares fit. This leads to $\Psi = \lfloor 1, 1, 1, \ldots \rfloor^T$. It is easy to prove that the resultant generating functions are the same as (5.14). If singular weights are used, we arrive at exactly the Shepard formula (5.12). Hence we may view the WLS as a generalization of the Shepard method [517].

5.2.5 *Moving least squares approximation*

The LS and WLS are rather inefficient ways of approximating mathematical functions. To obtain better accuracy, a large number of data points are sampled, and the polynomial degree needs to be raised accordingly. In the WLS, the evaluation of the generating function (shape function) $\Phi(x)$ requires the inversion of an $M \times M$ moment matrix $A(x)$, in which M is the number of the monomial basis functions used for the interpolation. As A is a function of x, it needs to be generated and inverted each time a location is referred; thus the amount of work can be considerable.

As a remedy, Lancaster and Salkauskas [517] proposed the moving least squares (MLS) method. Rather than using all the data points, it uses a small number of points in the neighborhood to perform LS fits. Each of the local fits is called a cover, and they can be patched together to form a global interpolation. Due to its increased efficiency, the MLS and its variants have played an important role in the development of meshless methods for solving PDEs [75]. Methods that have adopted such an interpolation strategy include the DEM [667], EFG [74, 602], MLPG [38, 42], PIC [819], *h-p* clouds method [272, 273], FPM [689, 691], least squares collocation method (LSCM) [302, 899, 965], MPM [60, 253, 402], and LMFS [356–358, 740, 741].

There are a number of key issues in the implementation of the MLS. We shall give a brief discussion on some of them in the following.

Local Interpolation: For the interpolation at a given location, only a small number of nearby nodes are utilized in the formula. Two strategies are generally used for the selection of data points. One is based on a fixed

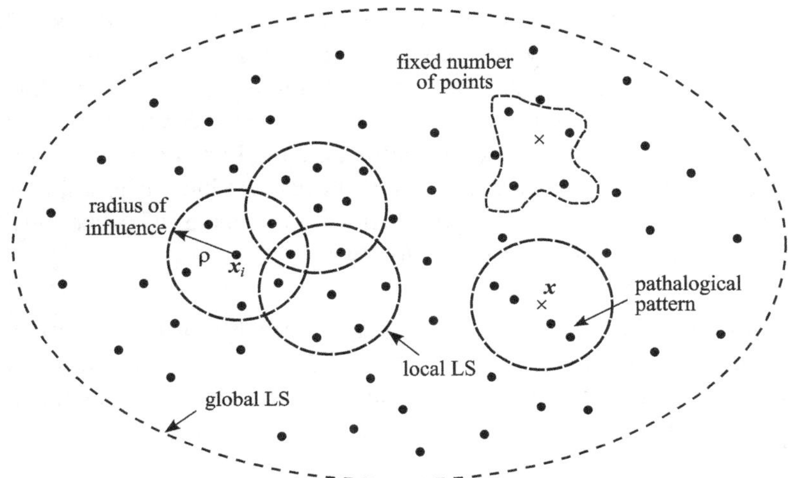

Figure 5.4. Local covers of moving least squares.

number, and the other on a fixed radius of influence ρ (see Fig. 5.4). For the former, the number of nodes is generally small, and it is possible to derive explicit formulas of the shape functions Φ [109]. For the latter, numerical matrix inversion and multiplication are performed.

The selection of nodes may not be automatic, and sometimes interventions are required. For example, data should be randomly distributed in all quadrants, and not all gathered in one. For the influence radius, it needs to include a minimum number of nodes for the LS up to a certain polynomial degree. Also, as demonstrated in Section 3.2.1, the interpolation matrix based on monomial bases can be ill conditioned. Certain pathological patterns, such as collinear placement, positioning on four corners of a square, and on a circle, should be avoided (see Fig. 5.4). The influence zone need not be a circle or sphere, and it can be elliptical or rectangular. In the weak formulation of solving PDEs, the approximated function needs to be integrated, and the quadrature rules are easier to apply for rectangular or box regions.

In the MLS, as an interpolated point x continuously moves across the domain, data nodes can join in or drop out of the influence zone. To have a continuous interpolation, the shape functions should naturally drop to zero at their edges, and not be truncated. For the approximation to be differentiable, derivatives of the shape functions should also be zero at the radius of influence of the circle/sphere. As mentioned above, the smoothness of the interpolation is defined by the smoothness of the shape functions.

For the solution of PDEs, a finite number of interpolations are performed on zones (covers) anchored at a set of nodes. These covers overlap each other, and together should cover the entire domain. For a point falling on the overlapped region of several covers, its interpolated value can be ambiguous; hence it is desirable that the shape functions possess the PU properties. We shall further discuss these issues as follows.

Basis Functions: The traditional LS approximation uses monomials as basis functions, as shown in (5.16). As discussed in Chapter 3, monomials are in fact not efficient interpolants, and more efficient basis functions can be used. For example, we may use RBFs, and write (5.16) as

$$\boldsymbol{\psi}(\boldsymbol{x}) = \lfloor \psi_1, \psi_2, \ldots, \psi_M \rfloor^T, \tag{5.33}$$

in which $\psi_j = \psi(r_j)$, with $\psi(r)$ one of the RBFs given in Section 3.3. For a given set of N local nodes, M of them can be selected as the centers of the RBFs. The minimization procedure follows the same lines as that in Section 5.2.4, and the same matrix equations (5.26) and (5.32) for the shape functions $\boldsymbol{\Phi}$ are obtained. The RBF interpolation matrices are always invertible, and free from the potential ill-conditioning of monomial matrices.

Furthermore, if there is some knowledge about the local behavior of the function, such as it being a solution of a homogeneous and (locally) constant coefficient PDE, then the general solutions presented in Section 3.1 can be employed. For example, as pointed out by Babuška and Melenk [54], if the governing equation is the 2D Laplace equation, then the use of $2p + 1$ harmonic polynomials (see (3.3a)) complete up to degree p is equivalent to the use of $(p + 2)(p + 1)/2$ monomials up to the same degree. By the same token, the fundamental solutions can be utilized as basis functions, as long as they are distributed outside of the influence zone. This scheme has already been implemented in the LMFS [356–358].

Weighting Functions: As commented above, the shape function $\boldsymbol{\Phi}$, which is contributed by both the basis and the weighting functions, needs to have compact support and smoothness properties. The basis functions, such as monomials and RBFs, can be smooth to a certain degree or infinitely smooth;

but are not compactly supported. Hence the burden of being compactly supported falls on the weighting functions. In the following, we present a few of the commonly used weighting functions.

Spline Functions: A commonly used weight is the cubic spline function, given as

$$w(s) = (1 - 3s^2 + 2s^3)[1 - \mathrm{H}(s - 1)], \quad s \geq 0, \qquad (5.34)$$

in which H is the Heaviside unit step function, $s = r_i/\rho$, $r_i = \|\boldsymbol{x} - \boldsymbol{x}_i\|$, and ρ is the radius of influence. The function is plotted in Fig. 5.5. It is obvious that $w(0) = 1$ and $w(1) = 0$. Its first derivative vanishes at $s = 1$ as well; so the function has C^1 continuity.

Compactly Supported RBFs: The CSRBFs of Wendland [911], presented in Section 3.3, provide a range of choices of smoothness. For example, choosing $d = 2$ for 2D, and $k = 1$, (3.85) gives

$$w(s) = (1 - s)_+^4 (4s + 1). \qquad (5.35)$$

As observed in Fig. 5.5, the function is much flatter around $s = 1$ than the spline, as it has up to $w'''(1) = 0$. However, it only has C^2 continuity at $s = 0$.

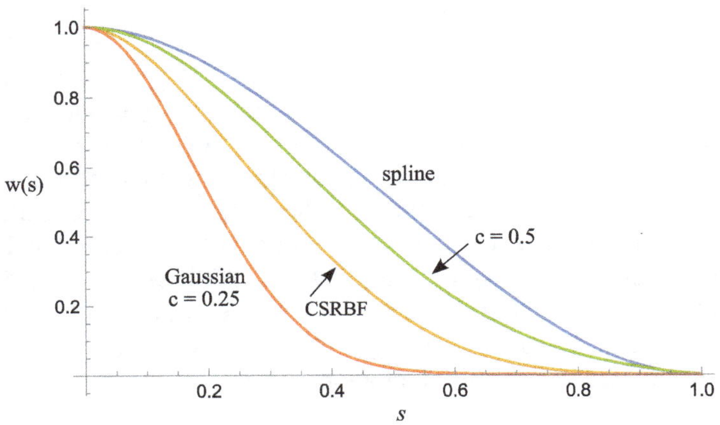

Figure 5.5. Various weighting functions.

Gaussian Function: A weight function used in EFG [74] is a normalized Gaussian curve

$$w(s) = \frac{\exp(-(s/c)^{2k}) - \exp(-(1/c)^{2k})}{1 - \exp(-(1/c)^{2k})}[1 - H(s-1)], \quad s \geq 0,$$

(5.36)

such that $w(1) = 0$. The function contains two free parameters, c and k, which can be adjusted to manipulate its shape. Figure 5.5 presents two such curves obtained using $k = 1$, and $c = 0.5$ and 0.25. Theoretically, these functions have C^0 continuity only. However, we observe that for a smaller c value, the curve can be made flat around $s = 1$. Although the derivatives there are not exactly zero, they are extremely small.

Derivatives: In the interpolation of known functions, and the solution of PDEs, the derivatives of the function are often needed. Particularly, when solving PDEs by the strong formulation, the derivatives of the function up to the order of the governing equation are needed. For the weak formulation, the smoothness requirement can be lowered. Generally, the derivatives are obtained by differentiating the approximated function.

Take, for example, the approximation given by (5.7). We can take the kth partial derivative of f with respect to x as

$$\frac{\partial^k \hat{f}}{\partial x^k} = \frac{\partial^k \mathbf{\Phi}^T}{\partial x^k} \mathbf{f}.$$

(5.37)

We notice in the above that as $\mathbf{f}$ are given data, only the shape functions need to be differentiated. The shape functions $\mathbf{\Phi}(x)$, however, are made of the product of the basis functions $\boldsymbol{\psi}^T(x)$ and the weighting functions $\mathbf{W}(x)$, as indicated in (5.30)–(5.32). Although these matrices can be analytically differentiated, this involves some work, particularly for second and higher derivatives. In methods like the EFG [74], MLPG [38], and LSCM [827], such differentiations are duly carried out. In the DEM [667], however, a "diffuse" derivative was introduced, such that the weighting functions are treated as constants in the differentiation, that is,

$$\frac{\partial^k \hat{f}}{\partial x^k} \approx \frac{\partial^k \boldsymbol{\psi}^T}{\partial x^k} \mathbf{A}^{-1} \mathbf{B}.$$

(5.38)

The above approximation simplifies the numerical implementation considerably. One may regard the above equation as a crude approximation of the derivative by neglecting the contribution from $W(x)$. Mirzaei *et al.* [648], on the other hand, sought to perform the least squares approximation directly on the derivatives, rather than on the function. They found that minimization led to the equivalent of diffuse derivatives. This was applied to the MLPG with better results in [649]. Hence the diffuse derivatives can be the simpler and more accurate alternatives.

In another scenario, CSRBFs can be used as the basis functions, which can provide the needed compact support as well as the smoothness. Hence the weighting function may no longer be needed. The formulation simplifies and the derivatives can be calculated without the complication of the weights.

Interpolating MLS: As mentioned above, the LS and WLS do not actually interpolate the data; that is, at the nodes, the data values are not reproduced. There are occasions when it is desirable that the original data be reproduced. Particularly, in the solution of PDEs, there is a need to match the nodal values with the BCs. The non-interpolating approximation may cause ambiguity or loss of accuracy in those situations.

Lancaster and Salkauskas [517] demonstrated that the WLS can interpolate data, if the weights are made singular at the data points. For the interpolating MLS (IMLS), we may take one of the compactly supported weights $w(s)$, such as those given in (5.34)–(5.36), and multiply it by an inverse distance weight to obtain a modified weight as

$$\tilde{w}(s) = \frac{w(s)}{s^p}, \tag{5.39}$$

in which $p \geq 1$. Another formula is given by

$$\tilde{w}(s) = \frac{w(s)}{1 - w(s)}. \tag{5.40}$$

The resultant shape function Φ then has the PU properties; that is

$$\Phi_i(x_j) = \delta_{ij}, \tag{5.41}$$

and

$$\sum_{i=1}^{N} \Phi_i(x) = 1. \tag{5.42}$$

Furthermore, its gradient has the partition of nullity property [540, 672]

$$\sum_{i=1}^{N} \nabla \Phi_i(x) = 0. \tag{5.43}$$

■ Example 5.4

For a demonstration of the IMLS, we examine a 1D example. In the interval $[0, 1]$, we are given six equally spaced data points, $x_i = \{0, 0.2, 0.4, 0.6, 0.8, 1.0\}$. The basis functions used are $\boldsymbol{\psi} = \{1, x, x^2\}$, which give a quadratic polynomial fit. For the weighting function, we select the compactly supported cubic spline (5.34), multiplied by the inverse distance with power $p = 2$, to obtain

$$w(s) = \frac{1 - 3s^2 + 2s^3}{s^2}[1 - \mathrm{H}(s - 1)], \quad s \geq 0.$$

On each node x_i, the weight $w_i(x, x_i) = w(|x - x_i|/\rho)$ is assigned. The radius of influence is selected as $\rho = 0.7$, such that a minimum of 4 nodes will be included in the LS approximation.

Using the formulas (5.30)–(5.32), we evaluate the six shape functions Φ_i as functions of x, and present them in Fig. 5.6(a). As the six functions are symmetrical about the midpoint $x = 0.5$, only the first three, $\{\Phi_1, \Phi_2, \Phi_3\}$, are plotted. We can make the following observations. First, these functions satisfy (5.41); that is, Φ_i takes the value 1 at x_i, and 0 at x_j when $j \neq i$. The dashed line on top shows the sum of the six functions, although only three are plotted in the figure. We also observe the compact support property of the functions; for example, Φ_1 becomes flat and equal to 0 for $x \geq 0.7$.

For comparison, we perform the regular, non-interpolating MLS, for which the nonsingular weighting function (5.34) is used. Going through the same procedure, we present the first three shape functions

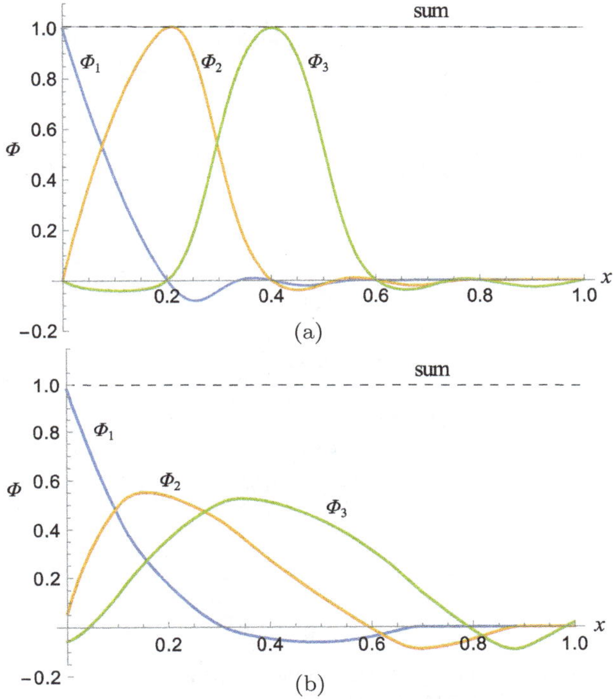

Figure 5.6. Shape functions of (a) interpolating MLS; and (b) non-interpolating MLS.

in Fig. 5.6(b). We can see that these shape functions do not possess the PU properties.

To test the performance of an actual fit, we seek to interpolate the following simple function

$$f(x) = \cos \pi x, \quad 0 \le x \le 1.$$

The exact function values are given on the six data points $f_i = f(x_i)$. The resultant fits are shown in Fig. 5.7(a) together with the exact function. We observe that the IMLS gives an excellent fit, and the exact and approximation curves nearly coincide. The non-interpolating approximation is also good, but slightly off. In Fig. 5.7(b), we plot the errors of these two fits. It seems that forcing the fit to go through the data points has the tendency of reducing the overall error.

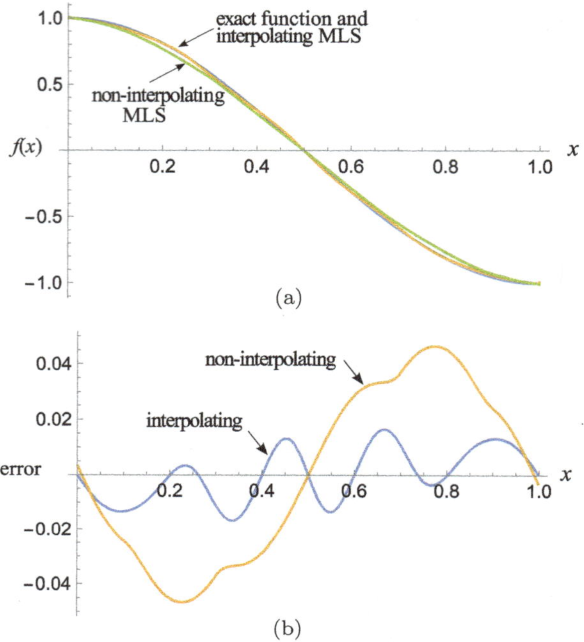

Figure 5.7. Least squares fitting of data: (a) Plot of the exact function compared to interpolating and non-interpolating MLS; and (b) error of the approximations.

5.3 Local versus Global Interpolation

As already demonstrated in the preceding section, generally there are two approaches to approximate a function $f(x)$ over a given domain Ω. We can create a single representation defined over the entire domain; or use many local approximations, which can be overlapping or non-overlapping, patched together to cover the whole domain. We have examined some examples in the above. In this section, we shall discuss the pros and cons of local versus global approximations.

As we are interested in the solutions of PDEs, the functions to be approximated are generally smooth, with the exception of certain local features, such as corners with mixed BCs, discontinuous BCs, discontinuities in the geometry, such as fractures, and singularities. In many numerical methods, these features can be separately treated to enhance the accuracy of the solution, when desirable. Meshless methods are amenable to such treatments by the use of specialized basis functions [75], or local particular solutions

[548]. In the following, we shall discuss only the approximation of continuous functions.

5.3.1 *Smoothness*

The solution of a PDE should be differentiable at least up to the order of the governing differential equation, and is likely to be smoother than that. However, non-smooth bases, such as piecewise constants, have been used in numerical methods. For example, the discontinuous FEM [63, 390], and the constant elements in the early implementation of the BEM [412], have been successfully applied to a large variety of problems. In general, though, we anticipate that the accuracy of the approximation will increase if we make the interpolants smoother.

Observing from the two methods, (5.1) utilizing basis functions, and (5.6) interpolating among nodal values, it is clear that the smoothness of the approximation is the same as the basis and the shape functions adopted. For global interpolation, the approximation can be made infinitely smooth by choosing bases such as RBFs and fundamental solutions. The smoothness of local interpolations, however, is generally limited. In the traditional FEM, non-overlapping elements are used. The differentiability of the function within an element can be increased by raising p (the polynomial degree) in the p or h-p version of the FEM [52, 53]. This, however, does not change the fact that there is a lack of inter-element differentiability. The overall approximation has only C^0 continuity across the element interface.

The situation is improved in the local meshless methods that use overlapping covers [75], in which a variety of basis functions can be used. Indeed, RBFs have been adopted in many versions of local meshless methods [42, 579, 806, 897]. However, to create the needed compact support, weighting functions that have lower order of continuity, such as CSRBFs, are used. Hence the smoothness is also limited.

■ **Example 5.5**

Given the function $f(x, y)$ defined in Example 3.3 and plotted in Fig. 3.3, we seek to interpolate it using FEM-like elements.

Figures 5.8(a) and 5.8(b) give the interpolation using quadrilateral elements with bilinear interpolation, using, respectively, 6×6 and 11×11 data points. We acknowledge that it is not fair to use a crude 6×6 grid to approximate this moderately complicated function.

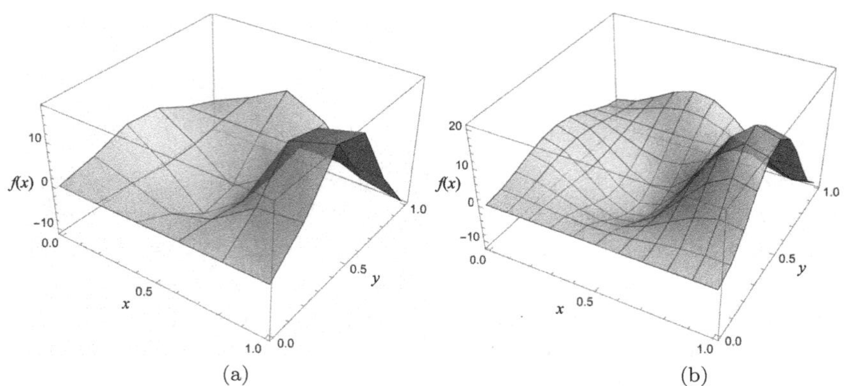

Figure 5.8. Local interpolation of the function in Example 3.3 and Fig. 3.3 by FEM quadrilateral elements. (a) 5 × 5 elements; and (b) 10 × 10 elements.

The poor representation is shown in order to provide a contrast with the global interpolation to be presented in the next example. In Fig. 5.8(b), we observe that the more refined grid yields a better visual result. The approximated function is continuous; however, it has no derivative across the edges of the elements. It has only C^0 continuity.

5.3.2 *Accuracy and efficiency*

An interpolation method is more accurate than another method if it achieves a smaller error using the same number of data points. A method is more efficient if it delivers the same accuracy, yet consumes less CPU time. An accurate approximation is not necessarily efficient, and vice versa. One may choose the method depending on the need.

Accuracy: In the example below, we give a demonstration of the superior accuracy of a global approximation using highly smooth basis functions.

■ Example 5.6

We seek to approximate the same function as in Example 5.5 by performing a global interpolation using the IMQ basis.

This has been done in Example 3.4 using 6 × 6 and 11 × 11 grids, with errors plotted in Fig. 3.5. For a visual comparison with the FEM-type interpolation, we plot the IMQ approximated function using a 6 × 6 grid in Fig. 5.9(a), which can be compared to the exact function

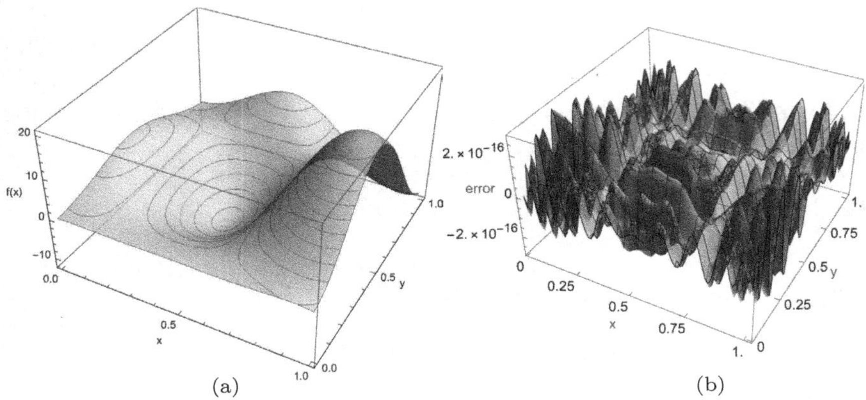

Figure 5.9. Global interpolation of the function in Example 3.3 and Fig. 3.3 by IMQ. (a) Plot of the approximated function based on a 6×6 grid and $c = 1.5$; and (b) Plot of the relative error ε_r based on a 21×21 grid and $c = 3$.

in Fig. 3.3, and the FEM-type approximation in Fig. 5.8(a). We observe that the global interpolation is smooth, and visually indistinguishable from the exact function, despite the fact that only a sparse data set is used in the fit. The infinite smoothness of the basis function enables it to capture the slopes and curvatures to produce a fit that is less than 1.5% in relative error.

As illustrated in Fig. 3.5(b), an even better fit can be achieved by halving the fill distance among the nodes, using an 11×11 grid. We find a nearly 4 orders of magnitude improvement of accuracy. To further demonstrate the power of the global interpolation, we half the fill distance again, using a 21×21 grid. Selecting a shape parameter $c = 3$, we run into the computational difficulty of an extremely high condition number. Using a special technique to overcome this difficulty, which will be explained in Section 5.4, the relative error drops by another 10 orders of magnitude, with an $\varepsilon_{\max}$ of $\mathcal{O}(10^{-16})$, as shown in Fig. 5.9(b). This shows the power of global interpolation.

We observe in the above example the power of the IMQ (and similarly the MQ and GA) to approximate a function in a global interpolation. Such power can be further enhanced if the chosen basis functions belong to the same class as the function to be interpolated. For example, solutions of the Laplace equation belong to a class of harmonic functions, and those of the Helmholtz equation belong to the metaharmonic functions. If the basis functions are chosen from the same class, such as the general and fundamental solutions of the PDE, then the function can be highly efficiently approximated. This in essence is the concept of the MFS and TCM, which will be discussed in detail in Chapter 6. In the following example, we give a quick demonstration of such approximation power.

■ Example 5.7

Solve a 2D exterior domain BVP governed by the Helmholtz equation. The boundary is given by the parametric equation

$$x = r(\theta)\cos\theta, \quad y = r(\theta)\sin\theta, \quad 0 \le \theta < 2\pi,$$

where

$$r(\theta) = \sqrt{(a+b)^2 + 1 - 2(a+b)\cos(a\theta/b)},$$

with $a = 3$ and $b = 1$. The curve is bounded between $3 \le r \le 5$. Its shape is shown as the extruded cylinder in Fig. 5.10. A Dirichlet BC is prescribed based on the function

$$\phi = J_3(kr)\sin(3\theta) + J_4(kr)\cos(4\theta),$$

where J_n is the Bessel function of the first kind of order n, k is the wave number, and for the present example, $k = 2$ is selected. We note

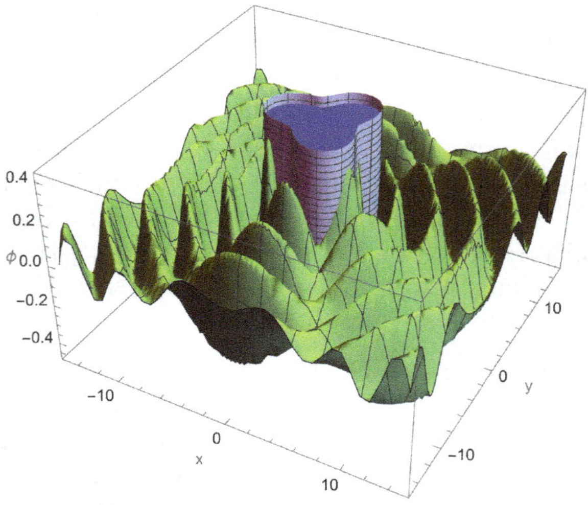

Figure 5.10. Exact solution of an exterior domain Helmholtz problem.

that the above function satisfies the Helmholtz equation, hence is the exact solution of the BVP.

For a visualization, we plot the solution in Fig. 5.10 in the near field. We observe a highly fluctuating behavior; hence it is a challenging function to approximate. To solve the problem, we assume that the solution is approximated by a summation of such basis functions

$$\hat{\phi}(x) = \sum_{j=1}^{N} c_j \, \mathrm{J}_0(k r_j),$$

where $r_j = \|x - x'_j\|$. We observe that $\mathrm{J}_0(kr)$ is a general solution of the Helmholtz equation. Following the MFS procedure, we place the x'_j on a circle of radius 0.5, centered at the origin, with equal spacing. For the result presented below, we use 20 terms in the summation, i.e. $N = 20$. To fit the BC, the same number of collocation points is selected on the boundary at equal spacing in θ. In Fig. 5.11 (a), we plot the BC for $0 \le \theta \le 2\pi$ and the 20 collocated values. It is clear that the sampled values can hardly represent the fluctuating nature of the function. Nevertheless, we obtain a highly accurate solution.

To demonstrate the accuracy, we calculated the error at a distance from the boundary, on a circle with radius $r = 10$. The relative error ε_r

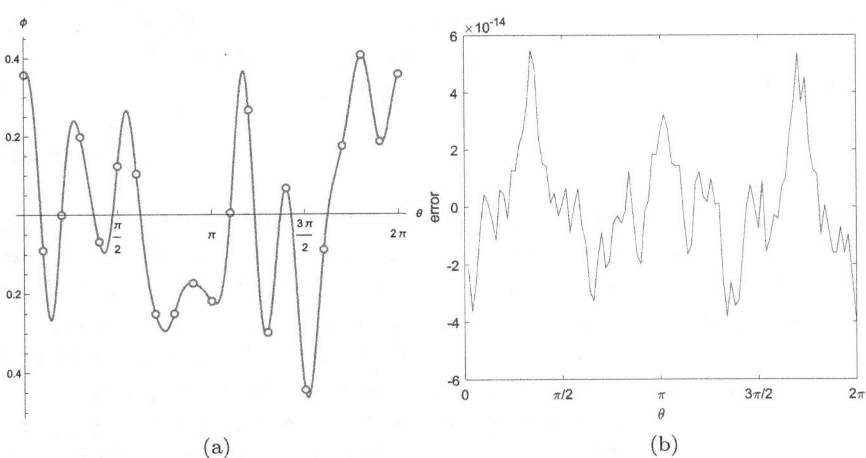

(a)   (b)

Figure 5.11. (a) Plot of the boundary condition as the solid line and the 20 collocated values in symbols $\circ$, and (b) relative error ε_r at $r = 10$, with $N = 20$.

is presented in Fig. 5.11 (b). We observe extremely high accuracy of the approximation, with the $\varepsilon_{\max}$ about 6×10^{-14}, using merely 20 terms. At a large distance, say $r = 100$, we find the accuracy undiminished. This is possible only because the basis function belongs to the same PDE solution space as the specific BVP.

Efficiency: Next, we shall address the issue of efficiency; that is, the comparison of computational cost to produce the same level of accuracy. The answer can be complicated as there are different parts in the computational cost, such as the effort to assemble a linear system, and the cost of manipulating and solving it. Here we address only the effort of solving a linear system, and only in very general terms.

For simplicity, let us consider the geometry of a square (2D) or a box (3D), which is divided in each direction by n nodes. The number of unknowns is n^d, where d is the dimension of the domain. In principle, the matrix for the linear system is of size $n^d \times n^d$, though for a sparse matrix, the null entries need not be stored. The cost (operation counts) of solving a linear system is dependent on the size, sparsity, and bandwidth of the matrix.

For a global interpolation, the matrix is full and it is usually solved by Gaussian elimination, which involves $\mathcal{O}(n^{3d})$ operations. For local interpolations, such as the FEM, the matrix is sparse. The effort of solving such a matrix is proportional to $b^2 n^d$, where b is the semi-bandwidth of the matrix. As the bandwidth is proportional to n for 2D, and n^2 for 3D, the operation counts are of $\mathcal{O}(n^4)$ and $\mathcal{O}(n^7)$, respectively [84]. We therefore write the CPU time t_{cpu} needed for the matrix solution as

$$t_{\mathrm{cpu}} \sim \mathcal{O}(n^k), \tag{5.44}$$

where k takes the value of 6 (2D) and 9 (3D) for global interpolation, and 4 (2D) and 7 (3D) for local interpolation. We note that the ratio n^2 in CPU time between the global and the local method is very large. However, for a comparison of efficiency, we must also address the accuracy of solution.

Next we address the issue of approximation error. The error is dependent on the fill distance h defined in (3.83), which can be interpreted as the radius of the largest possible empty disk/sphere that can be fitted among the scattered data points. Or, we may refer to it as the mesh size. For the FEM, the most popular quadratic elements lead to an error estimate of $\mathcal{O}(h^2)$. The use of cubic elements in an h–p adaptive scheme [247, 684] can achieve an

$\mathcal{O}(h^3)$ convergence; but the complexity of forming the elements and matrix considerably increases. We can write, in general, the error convergence rate of a local method as

$$\varepsilon \sim \mathcal{O}(h^{\ell}), \tag{5.45}$$

where $\ell = 2, 3$, etc. For a given domain in d dimensions, the mesh size is inversely proportional to the number of nodes in one direction; hence we may write

$$h \sim \mathcal{O}(n^{-1}). \tag{5.46}$$

Combining this with (5.45), we obtain

$$\varepsilon \sim \mathcal{O}(n^{-\ell}). \tag{5.47}$$

The error for global methods belongs to a different category, characterized by exponential convergence. For example, the RBF collocation method using MQ has an error convergence rate of [913]

$$\varepsilon \sim \mathcal{O}(e^{-a/h}), \tag{5.48}$$

where a is a positive constant. In the above equation, we did not factor in the additional convergence power of an optimal shape factor c in an h-c scheme [198], which can change the exponent from $-a/h$ to $-a/h^{3/2}$. We also note that the GA RBF has an even faster convergence rate of $\mathcal{O}(h^{1/h})$. Using (5.46) in (5.48), we can express

$$\varepsilon \sim \mathcal{O}(e^{-n}), \tag{5.49}$$

where we have neglected the factor a.

Now, eliminating n between (5.44) and (5.47), we obtain a relation between the CPU time and the error as

$$t_{\text{cpu}} \sim \mathcal{O}(\varepsilon^{-k/\ell}), \tag{5.50}$$

for a local method. Similarly for a global method, elimination between (5.44) and (5.49) yields

$$t_{\text{cpu}} \sim \mathcal{O}((-\ln \varepsilon)^k). \tag{5.51}$$

For a comparison, we plot in Figs. 5.12(a) and 5.12(b) these two estimates in log–log scale, respectively, for 2D and 3D problems. We note that in

    An Introduction to the Method of Fundamental Solutions

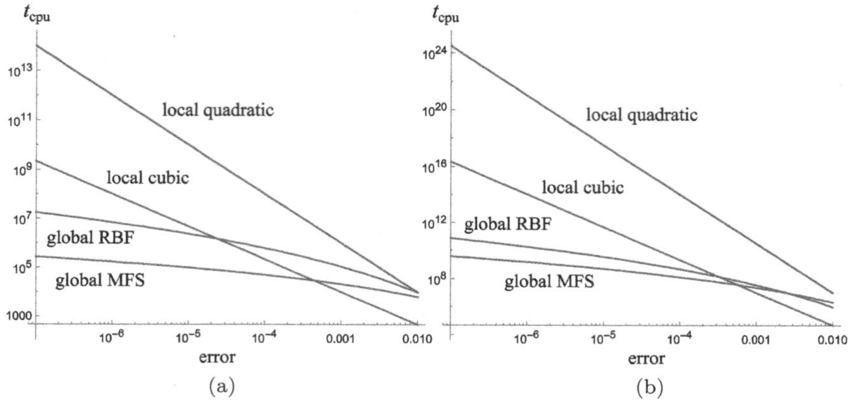

Figure 5.12. Plot of CPU time versus approximation error for (a) 2D problems, with $k = 6$ for global and $k = 4$ for local approximation, and (b) 3D problems, with $k = 9$ for global and $k = 7$ for local approximation. In both cases, $\ell = 2$ for quadratic, and $\ell = 3$ for cubic approximation.

arriving at (5.50) and (5.51), we have ignored a number of constant factors of unknown magnitude; so the values of t_{cpu} and ε should not be taken literally. These equations only give the trend.

For 2D problems, we observe in Fig. 5.12(a) that if a quadratic approximation is used in the local approximation, we have $\ell = 2$ and $k = 4$, and $t_{\text{cpu}} \sim \mathcal{O}(\varepsilon^{-2})$. This means that to reduce the error by one order of magnitude, the required CPU time will increase by 100-fold. If the error is to be reduced by several orders of magnitude, the CPU time can become unattainable with the modern day computational capabilities. For a cubic approximation, $\ell = 3$, we may improve the situation to $t_{\text{cpu}} \sim \mathcal{O}(\varepsilon^{-4/3})$. The effort of seeking higher accuracy is even more challenging for 3D problems. Figure 5.12(b) shows that for quadratic approximation, $t_{\text{cpu}} \sim \mathcal{O}(\varepsilon^{-7/2})$, and for cubic approximation, $t_{\text{cpu}} \sim \mathcal{O}(\varepsilon^{-7/3})$. These suggest that for quadratic approximation, a reduction of error of one order of magnitude requires a 3,000-fold increase in CPU time, and for cubic approximation, 200-fold.

As shown in these log-log plots, the global RBF approximation (5.51) is not of constant slope. Not only it has smaller slope, but also the slope decreases in the direction of smaller error. So the global approximations become progressively more efficient compared to the local ones, when higher accuracy is sought. We have already demonstrated in Example 5.6 that very high accuracy can be attained using the RBFCM, which is beyond the reach of local methods.

The RBFCM is a domain discretization method. We can perform a similar analysis for the MFS for homogeneous PDEs. As the MFS nodes are distributed on a boundary-like geometry, the number of degrees of freedom for the above-mentioned 2D square and 3D box geometry is proportional to $4n$ and $6n^2$, instead of n^2 and n^3, respectively. The MFS yields a full matrix and has similar exponential convergence as (5.49). These lead to the CPU time estimates

$$t_{\text{cpu}} \sim \mathcal{O}(4^3(-\ln \varepsilon)^3), \quad (2D), \tag{5.52a}$$

$$\sim \mathcal{O}(6^3(-\ln \varepsilon)^6), \quad (3D). \tag{5.52b}$$

Once again cautioning that the above do not represent the actual CPU time, we also plot these estimates in Figs. 5.12(a) and 5.12(b) for comparison. We observe that the MFS can be even more efficient than the RBFCM.

In addressing the efficiency, we may look at another issue: the degrees of freedom N needed to achieve a given accuracy. Recalling that $N \sim \mathcal{O}(n^d)$ for domain discretization methods, we may express (5.47) and (5.49) as

$$N \sim \mathcal{O}(\varepsilon^{-d/\ell}), \tag{5.53}$$

for local approximation, and

$$N \sim \mathcal{O}((-\ln \varepsilon)^d), \tag{5.54}$$

for global approximation. For the MFS, we obtain

$$N \sim \mathcal{O}(-4\ln \varepsilon), \quad (2D), \tag{5.55a}$$

$$\sim \mathcal{O}(6(-\ln \varepsilon)^2), \quad (3D). \tag{5.55b}$$

These relations are plotted in Fig. 5.13. We can make similar observations as in Fig. 5.12. That is, for the same accuracy, the number of degrees of freedom needed for the global approximation is much smaller than that for the local one.

Is High Accuracy Needed? In making the statement that it is difficult for local methods to attain high accuracy, we should recognize that high accuracy may not be needed for problems of engineering interest. For many practical problems, errors in the range of 10^{-2} and 10^{-3} may be satisfactory.

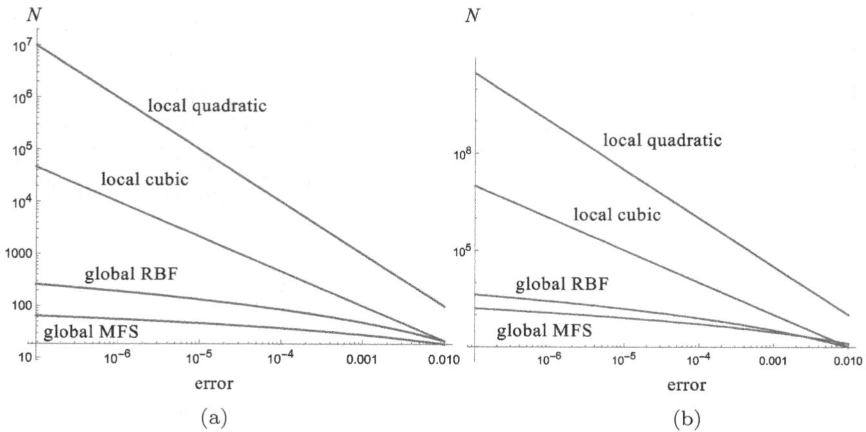

Figure 5.13. Plot of degrees of freedom N versus approximation error for (a) 2D problems; and (b) 3D problems. In both cases, $\ell = 2$ for quadratic, and $\ell = 3$ for cubic approximation.

However, for certain problems, the critical information is provided by the derivatives of the solution, such as the flux, which is obtained as the first derivative. Other applications may require even higher derivatives, such as the curvature of a surface, which is required for the evaluation of surface tension. When the interpolation of the discrete data set is differentiated once, we may lose one order of magnitude in accuracy. Higher accuracies may be needed to avoid such losses. Depending on the efficiency and accuracy needed in an application, among other factors, the user can judicially choose global or local approximation schemes. Although the MFS and RBFCM are presented here as global approximation methods, they both have their local versions, as reviewed in Section 1.7. The LMFS will be presented in Chapter 10.

5.4 Condition Number and Stability

In demonstrating the high accuracy of global approximations, we did not mention a critical issue — the high condition number associated with the fully populated interpolation matrix. When higher and higher accuracy is being pursued, the condition number grows rapidly. A high condition number can interfere with matrix operations by roundoff errors, causing instability. Hence the concern about large condition numbers can become a deterrence to the application of global approximations.

On the other hand, it has been curiously observed that in solving PDEs, and particularly when the collocation method is used, accurate solutions can be obtained even with very large condition numbers, often exceeding the perceived limit of 10^{16} for double precision computation. We shall discuss these issues in the following.

5.4.1 *Condition number*

The condition number is a measure of sensitivity of a solution system, particularly a matrix solution system. If there is a perturbation of data, either from the uncertainty of the physical input, the truncation of an approximation series, or the rounding off of computational digits, we would like to know how much the output will change. If a perturbation is much amplified in the output, the solution system is unstable.

The numerical solution of linear PDEs typically leads to a system of linear equations, expressed in matrix form as

$$A x = b, \tag{5.56}$$

where A is an $N \times N$ nonsingular matrix, and x and b are $N \times 1$ column vectors. If there exist small perturbations in A and b, expressed as ΔA and Δb, we would like to know their effect on x, expressed as Δx. We can write the above statement as follows:

$$(A + \Delta A)(x + \Delta x) = b + \Delta b. \tag{5.57}$$

For our purposes, we shall consider (5.56) an exact expression, with each element given in infinite precision, and (5.57) the actual expression in computation that contains roundoff errors ΔA and Δb, due to the finite precision of the computing tool. When the linear system is solved by techniques such as Gauss elimination, or QR factorization, we would like to know its effect on Δx. More specifically, we seek the following bound statement:

$$\frac{\|\Delta x\|}{\|x\|} \leq \kappa_A \frac{\|\Delta A\|}{\|A\|} + \kappa_b \frac{\|\Delta b\|}{\|b\|}, \tag{5.58}$$

where the proportionality constants κ_A and κ_b are the condition numbers of A and b, respectively. It is obvious that κ_A and κ_b magnify the roundoff errors. In particular, when the condition numbers approach 10^{16}, the inverse of machine precision for double precision, the solution system breaks down, as x will not have a single digit that is reliable. In fact, long before that limit is reached, the desirable accuracy is lost. As a rule of thumb, we may consider that we seek the reliability of the solution up to the nth significant digit. If the condition number is of the order 10^k, then the condition $n + k < 16$ is necessary. In fact, it is desirable to leave a gap of, say, 2–3 orders of magnitude; hence we may adopt a criterion for the condition number as

$$\kappa < 10^k, \quad \text{with } 16 - n - k > 2 \text{ or } 3, \tag{5.59}$$

for computation stability.

Most numerical analysis textbooks [37, 386, 520] derive the condition number by assuming that the perturbation comes form b. By dropping the ΔA term from (5.57), and multiplying both sides by A^{-1}, we obtain

$$\Delta x = A^{-1} \Delta b, \tag{5.60}$$

and (5.58) reduces to

$$\frac{\|\Delta x\|}{\|x\|} \leq \kappa_b \frac{\|\Delta b\|}{\|b\|}. \tag{5.61}$$

We seek a practical way to estimate the condition number κ_b. To do so we perform the singular value decomposition of A to obtain

$$A = U \Sigma V^T, \tag{5.62}$$

where U and V are unitary matrices, and Σ is a diagonal matrix with non-negative real diagonal elements. The singular values are contained in Σ as

$$\sigma = \text{diag} \, \Sigma, \tag{5.63}$$

in which σ is a column matrix with elements arranged in a descending sequence, $\sigma_1 \geq \sigma_2 \geq \cdots \geq \sigma_N$. We may express the ratio of the errors in (5.61) as

$$\frac{\|\Delta x\|}{\|x\|} \bigg/ \frac{\|\Delta b\|}{\|b\|} = \frac{\|A^{-1}\Delta b\|}{\|x\|} \bigg/ \frac{\|\Delta b\|}{\|Ax\|} = \frac{\|A^{-1}\Delta b\|}{\|\Delta b\|} \cdot \frac{\|Ax\|}{\|x\|}$$

$$\leq \frac{\|A^{-1}\| \cdot \|\Delta b\|}{\|\Delta b\|} \cdot \frac{\|A\| \cdot \|x\|}{\|x\|}$$

$$= \|A^{-1}\| \cdot \|A\| = \frac{\sigma_1}{\sigma_N}, \tag{5.64}$$

in which we note that $\|A\| = \sigma_1$ and $\|A^{-1}\| = \sigma_N^{-1}$. Comparing the above with (5.61), we can define the condition number as the ratio of the largest to the smallest singular value:

$$\kappa_b = \frac{\sigma_1}{\sigma_N}. \tag{5.65}$$

Other numerical analysis textbooks [341, 816, 915] considered the effect of ΔA. In that case, we drop Δb in (5.57) and utilize (5.56) to obtain

$$A \, \Delta x = \Delta A \, (x + \Delta x). \tag{5.66}$$

Multiplying both sides by A^{-1}, and taking the norm, we find

$$\|\Delta x\| = \|A^{-1}\Delta A \, (x + \Delta x)\| \leq \|A^{-1}\| \cdot \|\Delta A\| \cdot \|x + \Delta x\|. \tag{5.67}$$

We can define a condition number

$$\frac{\|\Delta x\|}{\|x + \Delta x\|} \leq \kappa_A \frac{\|\Delta A\|}{\|A\|}, \tag{5.68}$$

where κ_A is the same as κ_b in (5.65). As $x \gg \Delta x$, we can combine (5.61) and (5.68) to obtain [341]

$$\frac{\|\Delta x\|}{\|x\|} \leq \kappa \left(\frac{\|\Delta A\|}{\|A\|} + \frac{\|\Delta b\|}{\|b\|} \right), \tag{5.69}$$

in which

$$\kappa = \frac{\sigma_1}{\sigma_N} \tag{5.70}$$

shall be called the traditional condition number. The question is, is κ a good stability measure for matrices derived from a global approximation? The answer is that it is not, as we shall demonstrate in the following example.

■ Example 5.8

We once again examine the function defined in Example 3.3, and interpolated using the IMQ in Examples 3.4 and 5.6. Here we shall observe the relation between the condition number and the error of the approximation.

Table 5.1. Condition number κ, effective condition number κ_{eff}, maximum relative error $\varepsilon_{\max}$, and RMS relative error ε_{rms}, for interpolating the function f in Example 3.3, using the IMQ with various shape parameters c, based on the double precision and high precision computation.

c	κ	κ_{eff}	Double precision		High precision	
			$\varepsilon_{\max}$	ε_{rms}	$\varepsilon_{\max}$	ε_{rms}
0.2	3.9(13)	4.1(7)	1.3(−4)	3.1(−5)	1.4(−4)	3.1(−5)
0.4	1.9(19)	1.3(8)	1.1(−5)	2.7(−6)	2.8(−6)	1.2(−5)
0.6	4.0(19)	3.9(6)	7.0(−7)	1.5(−7)	3.5(−7)	7.9(−8)
0.8	2.2(19)	3.9(5)	2.4(−6)	8.5(−7)	6.1(−9)	1.6(−7)
1.0	4.8(19)	2.1(6)	**2.6(−7)**	**7.7(−8)**	4.6(−10)	1.2(−10)
1.5	2.4(20)	4.6(4)	3.6(−6)	9.0(−6)	9.9(−13)	3.2(−12)
2	8.5(19)	1.3(3)	2.2(−4)	6.6(−5)	4.0(−14)	1.4(−14)
3	—	—	—	—	2.9(−16)	8.8(−17)
5	—	—	—	—	2.4(−18)	7.8(−19)
7	—	—	—	—	**1.6(−18)**	**5.7(−19)**
10	—	—	—	—	1.1(−15)	4.5(−16)

Note: Best results are highlighted in boldface. Numbers in parentheses () indicate the power of 10.

We interpolate the function f by fixing the collocation grid to 21×21, and raising the value of the shape parameter c to make the IMQ increasingly flat. As the basis function becomes flat, it is less sensitive to the distances r among the nodes. Consequently, the elements of the matrix become nearly identical to each other. Naturally, the condition number increases. Table 5.1 shows that for $c = 0.2$, the maximum relative error, $\varepsilon_{\max}$, and the RMS relative error, ε_{rms} (see (3.72) and (3.73) for definitions), in the "double precision" columns are of $\mathcal{O}(10^{-4})$ and $\mathcal{O}(10^{-5})$, respectively; and the condition number κ is of $\mathcal{O}(10^{13})$. Continuing to increase c to 1.0, we obtain the best results with $\varepsilon_{\max}$ and ε_{rms} of $\mathcal{O}(10^{-7})$ and $\mathcal{O}(10^{-8})$. Further increasing c to 2.0 causes the results to deteriorate. For the best result, we observe a condition number κ of $\mathcal{O}(10^{19})$.

The interesting fact in the above example is that highly accurate results are obtained with κ much greater than 10^{16}. As mentioned above, in such cases, not a single digit is supposed to be accurate. This phenomenon was in fact observed in solving ordinary differential equations (ODEs) [239], and in the MFS [203, 554], TCM [492, 974], integral equations [209, 210], as well as the RBFCM [198, 201, 403]. These observations prompted the re-examination of the condition number, leading to the effective condition number [124, 210, 559, 753], as discussed in the section below.

5.4.2 *Effective condition number*

First, we may observe a curiosity in the derivation of κ_b. Although κ_b is a measure of sensitivity of $\|\Delta x\|/\|x\|$ caused by $\|\Delta b\|/\|b\|$, (5.65) shows that it is dependent on A only, and not b. This suggests that (5.64) may not be a sharp bound. Hence we give it a re-derivation by taking a different path, as follows:

$$\frac{\|\Delta x\|}{\|x\|} \Big/ \frac{\|\Delta b\|}{\|b\|} = \frac{\|A^{-1}\Delta b\|}{\|x\|} \Big/ \frac{\|\Delta b\|}{\|b\|} = \frac{\|A^{-1}\Delta b\|}{\|\Delta b\|} \cdot \frac{\|b\|}{\|x\|}$$

$$\leq \frac{\|A^{-1}\| \cdot \|\Delta b\|}{\|\Delta b\|} \cdot \frac{\|b\|}{\|x\|} = \frac{1}{\sigma_N} \frac{\|b\|}{\|x\|}. \tag{5.71}$$

Based on the above, we can redefine κ_b as an effective condition number defined by

$$\kappa_{\text{eff}} = \frac{1}{\sigma_N} \frac{\|b\|}{\|x\|}. \tag{5.72}$$

By virtue of the inequality $\|A\| \cdot \|x\| \geq \|b\|$, it is easy to show from (5.64) and (5.71) that

$$\kappa_{\text{eff}} \leq \kappa, \tag{5.73}$$

with κ is defined in (5.70). Hence κ is an overestimate of the instability caused by Δb.

The expression in (5.72) is an *a posteriori* estimate, as it requires the knowledge of x. For an *a priori* estimate, the following formula is offered [405, 559]

$$\kappa_{\text{eff}} = \frac{1}{\sigma_N} \frac{\sqrt{\beta_1^2 + \cdots + \beta_N^2}}{\sqrt{(\beta_1/\sigma_1)^2 + \cdots + (\beta_N/\sigma_N)^2}}, \tag{5.74}$$

where the β_i are the elements of the column matrix

$$\boldsymbol{\beta} = \boldsymbol{U}^T \boldsymbol{b}, \tag{5.75}$$

in which $\boldsymbol{U}$ is the unitary matrix defined in (5.62).

Although (5.72) gives a sharper upper bound for κ_b, we still need to address the effect of κ_A in (5.58), as it can overwhelm κ_{eff}. First, Banoczi *et al.* [59] argued that for an arbitrary matrix $\boldsymbol{A}$, the dominant error is caused by $\Delta\boldsymbol{A}$, and not $\Delta\boldsymbol{b}$. On the other hand, by studying the Vandermonde matrix, defined as

$$\boldsymbol{V} = \begin{bmatrix} 1 & x_1 & x_1^2 & \cdots & x_1^{N-1} \\ 1 & x_2 & x_2^2 & \cdots & x_2^{N-1} \\ \vdots & \vdots & \vdots & \ddots & \vdots \\ 1 & x_N & x_N^2 & \cdots & x_N^{N-1} \end{bmatrix}, \tag{5.76}$$

which is well known for its poor condition, Chan *et al.* [124] showed that the stability is controlled by $\Delta\boldsymbol{b}$. Hence, despite their notoriously large condition number, Vandermonde matrix systems can be solved accurately [90, 385] due to their much smaller effective condition numbers. We shall explain below that the MFS and RBFCM matrices are similar to the Vandermonde matrices. Banoczi *et al.* [59] also acknowledged that "*a fortunate right-hand side $\boldsymbol{b}$ can significantly reduce the sensitivity of a system of linear equations $\boldsymbol{Ax} = \boldsymbol{b}$.*" Christiansen and coworkers [209, 210] argued that in the approximate solution of a PDE, particularly by boundary collocation methods, we must also consider a second type of error — the data error due to the discretization of the BCs, which is much larger than the roundoff errors of $\boldsymbol{A}$ and $\boldsymbol{b}$. In that case, the effective condition number (5.72) is the relevant one. Furthermore, Li and Huang [552] demonstrated that in solving PDEs by methods such as the FDM [549], FEM [555], TCM [553], MFS and RBFCM [554], the condition

$$\kappa \frac{\|\Delta\boldsymbol{A}\|}{\|\boldsymbol{A}\|} \ll \kappa_{\text{eff}} \frac{\|\Delta\boldsymbol{b}\|}{\|\boldsymbol{b}\|}, \tag{5.77}$$

holds, again pointing to the relevance of κ_{eff} that is based on $\Delta\boldsymbol{b}$. Hence we shall re-examine Example 5.8 using the effective condition number.

■ Example 5.9

We solve the problem defined in Example 5.8 by also calculating the effective condition number κ_{eff}.

The effective condition numbers are shown in the third column of Table 5.1. First we observe that for the cases up to $c = 1.0$, κ_{eff} is of $\mathcal{O}(10^6)$. The best errors, marked in boldface, are of $\mathcal{O}(10^{-7})$ and $\mathcal{O}(10^{-8})$. Based on (5.59), we may get $16 - 6 - 2 = 8$ significant digits. This is exactly reflected in the result.

To further support the claim, we perform an arbitrary precision computation [201, 403, 404] using 100 digits. In the last two columns of Table 5.1, we show the "true" errors of the approximation, without the interference of double precision roundoff. For $c = 0.2$ and 0.4, the double and high precision errors are consistent, in the range of $\mathcal{O}(10^{-4})$ to $\mathcal{O}(10^{-6})$, meaning that there is no interference of roundoff. As c continues to increase, the true error improves to $\mathcal{O}(10^{-10})$ for $c = 1.0$. The double precision error, however, cannot progress, bound by the magnitude of κ_{eff}. To further experiment on the power of c in obtaining better solutions with a given grid, we increase it to up to 10, as shown in Table 5.1. We observe that an optimal c exists roughly at 7, at which the true approximation error is of $\mathcal{O}(10^{-18})$.

As a side note, we should mention that using double precision, κ is observed to flatten out around $\mathcal{O}(10^{19})$, while κ_{eff} fluctuates. These values may not be accurate as their evaluation relies on the smallest of the singular values, σ_N. When computed using double precision, σ_N can lose its accuracy in the large condition number range. However, κ_{eff} can still give a qualitative assessment of computational stability.

In addition to the theoretical arguments and numerical experiment on the significance of κ_{eff} presented above, we can offer another explanation. First, we notice that the Vandermonde matrix is the result of interpolating a function $f(x)$ using monomial bases $\{1, x, x^2, \ldots\}$. It was commented that although the large condition number may cause the interpolation coefficients to be inaccurately calculated, the interpolation can nevertheless be highly accurate [732]. We observe that the RBFCM solutions become more accurate when c is increased. It has been proven for univariate [65, 270] and multivariate [305, 519, 777] functions that the MQ and GA bases effectively degenerate into polynomial interpolants when $c \to \infty$. This is an indication that the RBF interpolation matrix may behave in the limit like the Vandermonde matrix, whose stability is controlled by κ_{eff}. We observe a similar behavior in the MFS computations, namely that more and more accurate results can be obtained by increasing the radius of the auxiliary circle/sphere used for the distribution of sources. In the limit, the interpolants behave like the harmonic polynomials [94, 162, 781]. These arguments support the use of κ_{eff}, instead of κ, as the predictor for the reliability of computed solutions by collocation methods.

5.4.3 *Schaback's uncertainty principle*

There is an abundance of observations in the MFS and RBFCM, as well as other numerical methods, that when we seek solutions of higher accuracy, the matrix sensitivity in terms of the condition number increases. This is also supported by many theoretical results, in which both the error estimate and the condition number are derived. Hence in all or most numerical methods, we anticipate that when the discretization is refined, the error decreases, and the condition number increases. We also observe the paradox in the MFS and RBFCM that error can be drastically reduced without refining the discretization, by simply making the interpolants flat. However, the principle is the same: the error and the condition number move in opposite directions.

Schaback [775, 776] commented on this situation as: *In particular, there is no case known where the error and the sensitivity are both reasonably small. There is a dichotomy: Either one goes for a small error and gets a bad sensitivity, or one wants a stable algorithm and has to take a comparably large error. This effect is reminiscent the Uncertainty Principle in quantum mechanics....*

We can provide some examples as follows. For the RBFCM, the error estimate for the MQ and IMQ, according to Wendland [913], has been given in (5.48). For the condition number, the following upper bounds were presented [58, 201, 665, 666]

$$\kappa \leq \frac{48D^2\sqrt{1+D^2}}{h^4} \exp\left(\frac{24}{h^2}\right), \quad \text{(2D)}, \qquad (5.78\text{a})$$

$$\leq \frac{68D^3\sqrt{1+D^2}}{h^5} \exp\left(\frac{32}{h^2}\right), \quad \text{(3D)}, \qquad (5.78\text{b})$$

in which h is the fill distance (see (3.83)), and $D = \max \|x_i - x_j\|$ is the diameter of the data set (or loosely, the size of the domain). It is obvious that as the error decreases exponentially when h is made smaller, the condition number increases at an even faster rate.

In the above and below, we should clarify that the error ε presented here is the theoretical error, without being affected by the roundoff error of finite precision matrix operations. Also, the condition number κ derived in the literature was based on the perturbation of A, and is the traditional condition number defined in (5.70).

The error estimate in (5.48) did not include the effect of the shape parameter c. Madych [605] gave the following error estimate for a class of RBFs that include MQ, IMQ, and GA, as

$$\varepsilon \sim \mathcal{O}(e^{bc^2} e^{-ac/h}), \tag{5.79}$$

in which a and b are positive constants. We may compare the above with (5.48). Equation (5.79) shows that for a sufficiently small h, the error decreases exponentially with increasing c, until a minimum is reached [201]. On the other hand, the condition number for the MQ has been reported as [58, 110, 201]

$$\kappa \leq C\sqrt{D^2 + c^2}\, \frac{D^d}{h^{d+1}} \exp\left(\frac{8dc}{h}\right), \tag{5.80}$$

where C is a constant, and d is the number of dimensions. It is clear that the condition number increases exponentially with c.

We can observe similar behavior for the MFS. From various studies [94, 208, 478, 490], the error estimate can be given as

$$\varepsilon \sim \mathcal{O}\left(\frac{\rho}{R}\right)^{N/2}, \tag{5.81}$$

where ρ is the size of the domain, R is the radius of the auxiliary boundary circle, and N is the number of sources. As $\rho/R < 1$, the error decreases exponentially with increasing N, and also with increasing R. For the condition number, an estimate is given as

$$\kappa \sim \mathcal{O}\left(\frac{N}{2}\left(\frac{\rho}{R}\right)^{-N/2} \ln R\right), \tag{5.82}$$

and we then find

$$\kappa\varepsilon \sim \mathcal{O}\left(\frac{N}{2} \ln R\right). \tag{5.83}$$

That is, the condition number increases more rapidly than the error decreases, when we increase the number of sources, or the radius of the auxiliary circle.

The rapid increase of the condition number has been a great concern for collocation methods that produce a full matrix. The reason is explained by the "uncertainty principle". Such a principle can be interpreted two ways. A pessimistic interpretation is that, in global collocation methods, the effort of seeking higher accuracy is thwarted by the increasing condition number. But we may offer an opposite view: a higher condition number means higher accuracy. When a high condition number is encountered, it means that we have reached an accuracy that is not achieved by other methods. Hence a high condition number is not to be feared. We should also be aware that a large traditional condition number does not prevent accurate results from being obtained, though a large effective condition number will. When the roundoff error starts to interfere with the theoretical error, high precision computation can be considered.

We may also take another consideration. In many engineering applications, high precision may not be needed, due to the practical nature of the problems. When the effective condition number has reached a moderate level, it may mean that a desirable accuracy has been reached. A further refinement at the risk of a much greater condition number may not be necessary. Ultimately, it is the efficiency discussed in Section 5.3.2 that rules. For a given desirable error, the global approximation using an efficient basis function may deliver the accuracy with a coarse discretization, which does not cause a high condition number, and requires less CPU time.

5.4.4 *High precision computation*

A digital computer performs arithmetic operations using a finite number of digits. In 1985, IEEE standardized the binary floating point format for single precision as 8 significant digits using 32 bits, and double precision as 16 digits using 64 bits. These standards have been universally implemented in hardware as well as software. The 2008 IEEE standard revision further introduced a quadruple precision format with 32 digits, which has a number of software implementations, but is rarely adopted in hardware [387].

We have observed in Examples 5.8 and 5.9 that finite precision caused roundoff in matrix operations, which can interfere with the theoretical accuracy of the approximation. It can also cause problems in the evaluation of asymptotic series with terms of increasing magnitude and alternating signs, in the numerical quadrature of highly fluctuating integrals, and other challenging problems [100]. Increased precision can alleviate these difficulties.

The arbitrary precision capability exists in computer algebra software, such as Mathematica® [918] and Maple™ [612]. The popular numeric computation tool MATLAB® [631] has incorporated Maple™ in its Symbolic Math Toolbox. Variable precision arithmetic can be carried out using the vpa function. On its floating point part, MATLAB® does not officially support precision beyond double precision. There is, however, a third party toolbox *Advanpix* [5] for multi-precision computing. There are also many subroutine libraries and software that support arbitrary precision arithmetic for programming languages such as C, C^{++}, Java, Python, etc. [914].

5.5 Solving PDEs — Strong versus Weak Form

In Section 5.2, we demonstrated how to interpolate/approximate a function using a set of discrete data. We also commented that solving a PDE follows a similar process. The difference is that, while in the function case, the approximation is required to satisfy a set of function values either exactly, or in an average sense as in least squares; in the solution of a PDE, it is the governing equation and BCs that need to be satisfied.

To satisfy the governing equation, the approximation formula, such as (5.1) or (5.6), is substituted into the PDE operator, with the differentiations carried out. The resultant is set to equal the RHS, either exactly on a set of points, or in some average sense over a region. For the BCs, either the function itself or its derivative(s) is given. These conditions can be exactly or approximately enforced on a set of points on the boundary. The resultant is a set of linear equations to be solved. We shall give a description of these processes in the following sections.

5.5.1 *Strong form — Collocation method*

Given a second order linear PDE in spatial variables

$$\mathcal{L}\phi(x) = f(x), \quad x \in \Omega, \tag{5.84}$$

subject to the BCs

$$\phi(x) = g(x), \quad x \in \partial\Omega_D, \tag{5.85}$$

$$\mathcal{B}\phi(x) = h(x), \quad x \in \partial\Omega_N, \tag{5.86}$$

where $\partial\Omega_D \cup \partial\Omega_N = \partial\Omega$, with $\partial\Omega$ the boundary of the domain Ω, and $\mathfrak{B}$ is a linear partial differential operator of one order or more lower than $\mathfrak{L}$. Other conditions, such as $\partial\Omega_D \cap \partial\Omega_N = \emptyset$ and $\partial\Omega_D \neq \emptyset$, may be needed to ensure the existence and uniqueness of the solution. We shall present two specific solution methods below.

RBFCM: As discussed in Section 3.3.2, we may approximate ϕ by

$$\hat{\phi}(x) = \sum_{j=1}^{N} c_j \psi(x, x_j), \tag{5.87}$$

where $\psi(x, x_j) = \psi(r_j)$, with $r_j = \|x - x_j\|$, are RBFs. Substituting the above into the partial differential operator, and noticing that $\mathfrak{L}$ applies only to the basis functions, and not the c_j, we obtain

$$\mathfrak{L}\hat{\phi}(x) = \sum_{j=1}^{N} c_j L(x, x_j), \tag{5.88}$$

in which

$$L(x, x_j) = \mathfrak{L}\psi(x, x_j). \tag{5.89}$$

If we select N_1 scattered points x_i, $i = 1, \ldots, N_1$, in the domain, we may require (5.88) to equal the RHS of (5.84) at these nodes, to obtain

$$\sum_{j=1}^{N} c_j L(x_i, x_j) = f(x_i), \quad i = 1, \ldots, N_1. \tag{5.90}$$

Similarly, we may select N_2 nodes on $\partial\Omega_D$ and N_3 nodes on $\partial\Omega_N$, and enforce the BCs as

$$\sum_{j=1}^{N} c_j \psi(x_i, x_j) = g(x_i), \quad i = N_1 + 1, \ldots, N_1 + N_2, \tag{5.91}$$

and

$$\sum_{j=1}^{N} c_j B(x_i, x_j) = h(x_i), \quad i = N_1 + N_2 + 1, \ldots, N, \tag{5.92}$$

in which $B(x_i, x_j) = \mathfrak{B}\psi(x_i, x_j)$, and $N = N_1 + N_2 + N_3$.

As an example, we may take $\mathfrak{L} \equiv \nabla^2$, and ψ the IMQ defined in (3.77) with $\beta = 1/2$. We then obtain

$$L(\boldsymbol{x}_i, \boldsymbol{x}_j) = L(r_{ij}) = \frac{r_{ij}^2 - 2c^2}{(r_{ij}^2 + c^2)^{5/2}}, \quad (2D), \tag{5.93a}$$

$$= \frac{3c^2}{(r_{ij}^2 + c^2)^{5/2}}, \quad (3D), \tag{5.93b}$$

with $r_{ij} = \|\boldsymbol{r}_i - \boldsymbol{r}_j\|$. We also find $\mathfrak{B} \equiv \partial/\partial n$, and

$$B(\boldsymbol{x}_i, \boldsymbol{x}_j) = -\frac{\boldsymbol{r}_{ij} \cdot \boldsymbol{n}_i}{(r_{ij}^2 + c^2)^{3/2}}, \tag{5.94}$$

where $\boldsymbol{r}_{ij} = \boldsymbol{x}_i - \boldsymbol{x}_j$, and $\boldsymbol{n}_i = \boldsymbol{n}(\boldsymbol{x}_i)$ is the outward normal of $\partial\Omega_N$.

We recognize that (5.90)–(5.92) form a set of N linear equations with the c_j, $j = 1, \ldots, N$, as unknowns. The system can be solved for the c_j, and the approximate solution is given by (5.87). The above solution procedure that enforces the governing equation and BCs on a set of points is called the collocation method. The convergence of the solution is guaranteed by taking more and more points in the domain and the boundary, until they become totally dense, that is, $N \to \infty$. This method of solution is called the strong form, in contrast to the weak form, to be presented in the next section.

MFS: We may also adopt the above solution procedure for the MFS. Given a homogeneous PDE

$$\mathfrak{L}\phi(\boldsymbol{x}) = 0, \quad \boldsymbol{x} \in \Omega, \tag{5.95}$$

with the BCs (5.85) and (5.86), we may approximate the solution as

$$\hat{\phi}(\boldsymbol{x}) = \sum_{j=1}^{N} c_j G(\boldsymbol{x}, \boldsymbol{x}_j'), \quad \boldsymbol{x} \in \overline{\Omega}, \ \boldsymbol{x}_j' \in \partial\Omega' \in \Omega_e, \tag{5.96}$$

where $G(\boldsymbol{x}, \boldsymbol{x}_j')$ is the fundamental solution of $\mathfrak{L}$, with its singularity located at $\boldsymbol{x}_j'$, $\overline{\Omega} = \Omega \cup \partial\Omega$, $\partial\Omega'$ is an auxiliary boundary, and Ω_e is the exterior of Ω. We note that the above approximation automatically satisfies the governing equation (5.95); hence no collocation within the domain is necessary.

To satisfy the BCs, we select N_1 nodes on $\partial\Omega_D$ and N_2 nodes on $\partial\Omega_N$, with $N_1 + N_2 = N$. On each node x_i, we require

$$\sum_{j=1}^{N} c_j\, G(x_i, x'_j) = g(x_i), \quad x_i \in \partial\Omega_D, \quad i = 1, \ldots, N_1, \tag{5.97}$$

$$\sum_{j=1}^{N} c_j\, B(x_i, x'_j) = h(x_i), \quad x_i \in \partial\Omega_N, \quad i = N_1 + 1, \ldots, N, \tag{5.98}$$

in which $B(x_i, x'_j) = \mathcal{B}G(x_i, x'_j)$, and $N = N_1 + N_2$. Equations (5.97) and (5.98) form a linear system to solve for the undetermined coefficients c_j.

In the case of the Laplace equation in 2D, with the fundamental solution given as (2.23a), we may write (5.97) and (5.98) as

$$\sum_{j=1}^{N} c_j \ln r_{ij} = g(x_i), \tag{5.99}$$

$$\sum_{j=1}^{N} c_j \frac{x_{ij}\, n_i^x + y_{ij}\, n_i^y}{r_{ij}^2} = h(x_i), \tag{5.100}$$

in which $r_{ij} = \|x_i - x'_j\|$, $x_{ij} = x_i - x'_j$, $y_{ij} = y_i - y'_j$, and n_i^x and n_i^y are the components of the outward normal n of $\partial\Omega$ at x_i. We notice that in the above we have dropped the constant factor in front of the fundamental solution (2.23a), as it is immaterial to the numerical solution. After solving for the c_j, the approximate solution is then

$$\hat{\phi}(x) = \sum_{j=1}^{N} c_j \ln r_j, \tag{5.101}$$

with $r_j = \|x - x'_j\|$.

Similarly for 3D, corresponding to (5.97) and (5.98), we have

$$\sum_{j=1}^{N} c_j \frac{1}{r_{ij}} = g(x_i), \tag{5.102}$$

$$\sum_{j=1}^{N} c_j \frac{-x_{ij}\, n_i^x - y_{ij}\, n_i^y - z_{ij}\, n_i^z}{r_{ij}^3} = h(x_i), \tag{5.103}$$

and the approximate solution is

$$\hat{\phi}(\boldsymbol{x}) = \sum_{j=1}^{N} c_j \frac{1}{r_j}. \qquad (5.104)$$

In Examples 1.1 and 1.2 we implemented the above procedure to solve an interior domain and an exterior domain problem, respectively, with the MATLAB® codes provided. Further applications of the MFS are explored in Chapter 6.

5.5.2 *Weak form — Method of weighted residuals*

A weak form solution of a PDE is an approximation that satisfies the equation and its associated BCs not pointwise, like the collocation method, but in an average sense. The idea, which can be traced to Ritz [318, 756], is to integrate the differential equation and the BCs, weighted by some "test functions", over the domain and boundary, respectively. Integration by parts is often performed to lower the order of the differential equation, and therefore the original smoothness requirement of the solution. Hence such a solution method is called a "weak form".

Finite Element Method: The are several ways that the FEM can be formulated [64, 978]. One is based on the physical law of the principle of minimum potential energy. It applies to physical problems obeying such a law. For example, the static response of an elastic solid subjected to a force load is governed by the work-energy principle. Another way is through the Rayleigh–Ritz method, utilizing the calculus of variations. This approach requires the knowledge of a functional, whose variation corresponds to the governing PDE and its existence is limited to the class of Euler–Lagrange differential equations. A third method is rooted in the method of weighted residuals (MWRs). By allowing the use of different weights, it is the most flexible; and it can incorporate the above two methods. It applies to a wide range of PDEs and is the most popular of the formulations. Our discussion below is focused on the MWR and its weak form. However, as the focus of the book is on the strong form, i.e. the collocation method, the discussion is conducted on a philosophical level, without the details of implementation.

Consider the BVP formulated as (5.84)–(5.86). We may generalize the problem and rewrite those equations as

$$\mathcal{A}u(x) = 0, \quad x \in \Omega, \tag{5.105}$$

$$\mathcal{B}u(x) = 0, \quad x \in \partial\Omega, \tag{5.106}$$

In the above, the dependent variable u is a vector $u = \lfloor u_1, u_2, \ldots, u_d \rfloor^T$ with d the number of variables (or spatial dimensions), $\mathcal{A}$ is a new partial differential operator that incorporates the RHS as $\mathcal{A} = \mathcal{L} - f$, which is generalized into an array $\mathcal{A} = \{\mathcal{A}_1, \mathcal{A}_2, \ldots, \mathcal{A}_d\}$ to accommodate the multiple variables. These equations are coupled because $\mathcal{A}_1 = \mathcal{A}_1(u_1, \ldots, u_d)$, etc. Similarly, we define $\mathcal{B} = \{\mathcal{B}_1, \ldots, \mathcal{B}_d\}$ as the boundary operators that incorporate (5.85) and (5.86) with their RHSs, which are called the essential and the natural BCs, respectively, in the FEM literature.

Next, we define a functional, which is the weighted integration of the above two equations over the domain and the boundary, respectively, as

$$\mathcal{F}(u) = \int_\Omega w^T \mathcal{A}u \, dx + \int_{\partial\Omega} v^T \mathcal{B}u \, dx = 0, \tag{5.107}$$

in which w and v are weighting functions. These vector products give $w^T \mathcal{A} = w_1 \mathcal{A}_1 + \cdots + w_d \mathcal{A}_d$, etc.; hence $\mathcal{F}$ is a scalar. We notice that the functional is identically zero for any weighting functions, by virtue of (5.105) and (5.106).

In a numerical solution, we may approximate u using a set of basis functions or discrete values of u, as discussed in Sections 5.2.1 and 5.2.2. Following the origin of the FEM, it is more appealing for engineers to use discrete values $u_j = u(x_j)$ in an explicit approximation than an implicit one. Similar to (5.6), we can express

$$\hat{u}(x) = \sum_{j=1}^{N} N_j(x) u_j = (N^T \tilde{u})^T, \tag{5.108}$$

in which N_j is the shape function associated with x_j, $N = \lfloor N_1, \ldots, N_N \rfloor^T$, and $\tilde{u} = [u_{ji}]$ where i indicates the vector component, $j = 1, \ldots N$, and $i = 1, \ldots, d$. The shape functions need to have the PU properties, as discussed in Section 5.2.3. If we substitute $\hat{u}$ for u in (5.105) and (5.106), the RHSs are no longer zero. They produce errors, called residuals, of the approximation. Obviously (5.107) can no longer be satisfied for *any* choice of the weighting functions. However, we may choose *some* weighting functions and enforce such a condition. As the approximation (5.108) has N degrees of freedom, we may choose N weighting functions $w_i = \{w_1, \ldots, w_N\}$ and $v_i = \{v_1, \ldots, v_N\}$, and require

$$\int_\Omega w_i^T \mathcal{A}\{(N^T \tilde{u})^T\}\, dx + \int_{\partial\Omega} v_i^T \mathcal{B}\{(N^T \tilde{u})^T\}\, dx = 0, \quad i = 1, 2, \ldots, N.$$
(5.109)

In principle, the system of N equations allows us to solve for the N unknowns u_j.

The above process, in mathematical terms, is to construct integrals of the inner product of the residuals and the weight functions, and set them to zero. At this point, (5.109) is quite general, and can incorporate all three approaches mentioned above. For example, in solving elasticity problems, we may assign $\mathcal{A}$ as the equilibrium equations expressed in terms of stress variables, which are tied to the displacements through constitutive equations. The virtual displacement vector δu is used as the weight w. By performing integration by parts, which transfers the divergence operator from the equilibrium equations to the virtual displacements, we obtain products of stresses with virtual strains, i.e., the virtual strain energy, as the integrand. The integral is then minimized with respect to virtual displacements, in accordance with the minimum potential energy principle. The second approach, the Rayleigh–Ritz variational method, requires the knowledge of a corresponding Euler–Lagrange equation to the governing equation as the integrand of a functional $\mathcal{F}$. Its variation with respect to u then produces the desirable governing equation as $\mathcal{A}$, with δu as w. The WRM approach as presented above offers a wide choice of weighting functions, which may not have any physical significance. As the weighting functions can be arbitrary, it can be viewed as a generalization of the above two approaches. In fact, when the Dirac delta function is chosen as the weight, the WRM becomes the collocation method. A discussion of the equivalence of the various methods can be found in Zienkiewicz *et al.* [978].

We observe that in the integral equation (5.109), the integrations are performed over the entire domain Ω. Hence, up to this point, the WRM can be viewed as a global approximation method. It is, however, generally not feasible to implement the WRM as a global method. First, we consider the node based approximation (5.108). As mentioned earlier, it is not easy to construct the shape functions analytically for a large number of scattered nodes, with the exception of utilizing the Shepard formula (5.12) and its weights (5.14). The accuracy of this interpolation, however, is poor, as demonstrated in Example 5.2. Another issue is the integration over an irregular domain. Efficient numerical integration uses quadrature rules, which are developed for simple geometries, such as rectangles, triangles, etc. The global geometry is not amenable to these treatments.

To overcome these difficulties, the domain Ω is subdivided into many regular shaped "finite elements", Ω_e. The minimization of weighted residuals shown in (5.109) is performed on each Ω_e, which are then patched together to form a global system. The conversion from the global to local approximation also gives rise to a few issues. One is that to have global C^0 continuity, nodes shared by multiple elements need to be referred to the same degrees of freedom. This requires the construction of connectivity data, which could be a costly process, as commented in Section 5.1. To have higher global continuity, such as C^1 and C^2, local interpolants must have continuous first and second derivatives across the interfaces shared with neighboring elements. Such elements are difficult to construct [34, 235, 711], and are rarely used. Hence the standard FEM uses C^0 elements, whose first derivatives have finite jumps across element interfaces, and their second derivatives are unbounded. However, the global convergence of integral equations allows only finite jumps in the integrand. This means that the PDE operator should contain only first derivatives. This requires the lowering of order of $\mathcal{A}$, which is accomplished by integration by parts. Assuming that $\mathcal{A}$ in (5.109) is a second order operator, such a process yields

$$\int_\Omega \mathcal{C}^T\{w_i\}\mathcal{D}\{(N^T\tilde{u})^T\}\,dx + \int_{\partial\Omega} \mathcal{E}^T\{v_i\}\mathcal{F}\{(N^T\tilde{u})^T\}\,dx = 0,$$

$$i = 1, 2, \ldots, N, \tag{5.110}$$

in which $\mathcal{C}$, $\mathcal{D}$, $\mathcal{E}$, and $\mathcal{F}$ are first order PDE operators. The above equation that changes from the original higher order PDE to a lower order one, in order to accommodate the C^0 continuity of the local approximation, is known as the weak form.

The MWR can allow a wide choice of shape and weighting functions, such as polynomials, RBFs, general solutions, and the Dirac delta function, as long as the resulting integrands are integrable within the elements, and have global convergence properties. In the standard FEM, polynomials of linear, quadratic, and up to cubic degrees are used to construct the PU shape functions. Furthermore, the Galerkin formulation, which uses the shape functions as the weighting functions in order to create symmetric solution matrices, is adopted. The use of polynomials for the shape and weighting functions also facilitates the adoption of a Gaussian quadrature, which can integrate the polynomials exactly. The details of such implementations for various PDEs can be found in the many FEM textbooks.

Element-Free and Meshless Methods: As reviewed in Section 5.1, in order to be free from the constraints of the traditional FEM, many meshless methods have been developed. Particularly, the MLPG adopts the Petrov–Galerkin formulation, which allows a wider selection of test and trial functions, and different methods of approximation, such as the MLS approximation [42, 806]. In most versions of the method, the weak form was used for the satisfaction of the governing equation.

Boundary Methods: The weak form can be applied to the domain as well as the boundary type numerical methods. In boundary methods, the minimization of weighted residuals is applied to BCs only, as the differential equation is automatically satisfied. Examples of such implementations include the Galerkin BEM [98, 608, 664], the Petrov–Galerkin meshless integral equation method [41], the Galerkin MFS [80], and the weak form based Trefftz methods [550, 556].

Comparison of Strong and Weak Form Solutions: In the following example, we shall solve a simple problem to gain some insights into the local versus global approximation, and the weak versus strong formulation.

■ **Example 5.10**

Solve the following 1D steady state heat conduction equation

$$\frac{d^2 T(x)}{dx^2} = -Q(x), \quad 0 \le x \le 1,$$

with the BCs

$$T(0) = T(1) = 0.$$

In the above, $Q(x)$ is a distributed heat source, and is given by

$$Q(x) = 4\pi^2 \sin 2\pi x,$$

for the current problem. The exact solution of the problem is

$$T(x) = \sin 2\pi x,$$

which is a full period sine curve within $x = [0, 1]$.

The problem is first solved using the global approximation. We discretize the domain using seven equally spaced nodes, $x_i = \{0, 1/6, 1/3, \ldots, 1\}$. The solution is approximated by a polynomial

$$\hat{T}(x) = a_1 + a_2 x + a_3 x^2 + a_4 x^3 + a_5 x^4 + a_6 x^5 + a^7 x^6.$$

The governing equation is satisfied by setting

$$\left. \frac{d^2 \hat{T}(x)}{dx^2} \right|_{x=x_i} = -Q(x_i), \quad i = 2, \ldots, 6,$$

with the BCs

$$\hat{T}(x_1) = \hat{T}(x_7) = 0.$$

These equations form a linear system to be solved for the undetermined coefficients a_i. In Fig. 5.14 we plot the relative error of the approximation as the global strong form curve. We observe that $\varepsilon_{\max}$ is about 0.06.

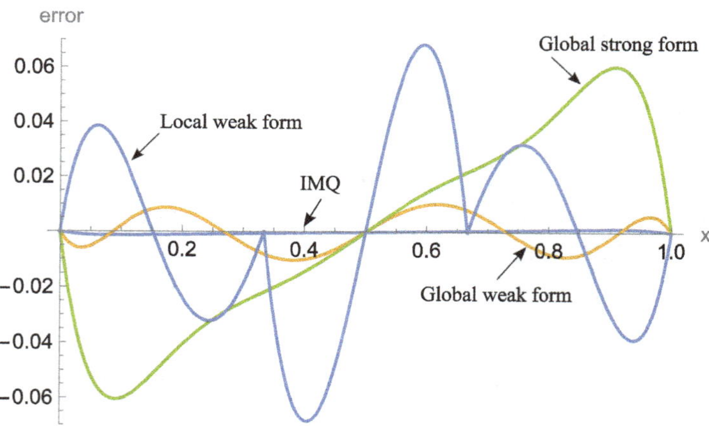

Figure 5.14. Comparison of relative error from solving an ODE using global or local approximation, and weak or strong form.

Next, we solve the same problem by using the global weak form. Again, the monomial basis functions

$$\boldsymbol{\psi} = \{1, x, x^2, \ldots, x^6\},$$

are used to approximate the solution. The global shape functions can be constructed following (5.9). To do so, we first assemble $\boldsymbol{\Psi}$ by substituting in turn x_i for x in $\boldsymbol{\psi}$ to form the rows of the matrix. Using its inverse in (5.9) produces the global shape functions $\boldsymbol{N} = \lfloor N_1, N_2, \ldots, N_7 \rfloor$, which are plotted in Fig. 5.15. These shape functions have the PU properties.

We notice in the figure that these shape functions are very different from those obtained from the inverse distance weight, such as the Shepard functions shown in Fig. 5.1. For the interpolation of physical data, we may consider that all data give positive contributions to an interpolated value, and data at a large distance is of less importance. For the current shape functions, we observe that the contribution can be positive or negative, and there is no obvious decay with distance.

With the shape functions, the approximation is given by

$$\hat{T}(x) = \boldsymbol{N}^T(x)\,\boldsymbol{T},$$

in which $\boldsymbol{T} = \lfloor T_1, \ldots, T_7 \rfloor^T$, and $T_1 = T(x_1)$, etc. Using the Galerkin method in (5.110), we obtain the matrix system

$$\boldsymbol{K}\,\boldsymbol{T} = \boldsymbol{b},$$

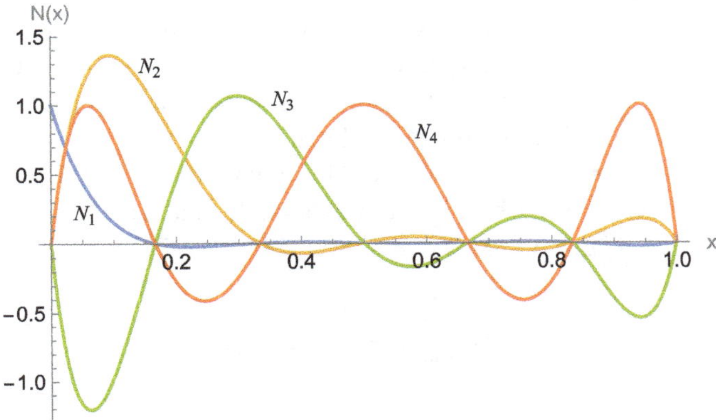

Figure 5.15. Plot of global shape functions N_1–N_4. N_5–N_7 are symmetrical about $x = 0.5$, and not plotted.

in which K is the "stiffness matrix", with elements

$$K_{ij} = \int_0^1 \frac{dN_i}{dx} \frac{dN_j}{dx} \, dx,$$

and b is given by

$$b_i = \int_0^1 N_i(x) Q(x) \, dx.$$

In view of the BCs, $\hat{T}_1 = \hat{T}_7 = 0$, we are in fact solving for $\{\hat{T}_2, \ldots, \hat{T}_6\}$, utilizing $\{N_2, \ldots, N_6\}$. Solving the matrix system, we plot the resultant interpolation in Fig. 5.14 as the global weak form curve. We observe that $\varepsilon_{\max}$ of the weak form is about 0.01, six-fold better than the strong form. Hence, by minimizing the error overall, and not focussing on discrete nodes, the weak form can produce the more accurate results with the same degrees of freedom. However, we should mention that the above procedure involved the solution of a matrix system twice, once to construct the shape functions, and the second time to solve for the discrete nodal values; while the strong form method requires only one matrix system solution.

The current problem is in 1D. As discussed above, the global weak form is not feasible in multi-dimensional geometries. Therefore we shall solve the problem in the local weak form. We may group the six segments into three quadratic elements. The basis functions are limited to $\{1, x, x^2\}$. The problem is solved and the approximate solution plotted as the local weak form curve. We observe that $\varepsilon_{\max}$ is about 0.07, which is comparable with the global strong form. Hence the weak form does not seem to have an advantage. We further notice that the local weak form does not have a continuous first derivative, and its derivative needs to be re-interpolated.

Finally, we realize that the global approximation is not limited to the use of polynomial bases. We may switch to an RBF, and particularly the IMQ, and write

$$\hat{T}(x) = \sum_{j=1}^{7} a_j \frac{1}{\sqrt{(x - x_j)^2 + c^2}}.$$

Following a similar collocation procedure as in the polynomial case, and using an optimal shape parameter $c = 1.8$, we produce a solution with $\varepsilon_{\max}$ about 0.002. The approximation is plotted as the IMQ curve in Fig. 5.14, which is no doubt much better than all the other solutions.

5.6 Summary

In this chapter, we have briefly examined several different approaches for solving PDEs, and discussed their advantages and disadvantages. In the following, we shall give a summary.

Mesh versus Meshless: In the development of the FEM, the variational principle and weight residual approaches led to integral equations. The need for numerical integration necessitated the division of the domain into finite elements with a local approximation. To ensure continuity across elements, the connectivity data was built to keep track of the node-element relations. In the FDM, the use of series expansions led to a rectilinear grid arrangement. In recent decades, many meshless methods were introduced to be free of these awkward requirements. The meshless developments allowed for the flexibility to use more efficient basis functions, such as RBFs, different methods for approximation, such as the MLS, and particle based methods. With each new method, new issues may arise, which need to be addressed.

We may envision the building of a numerical method this way. Given a PDE, a set of BCs, and a domain geometry, what are the best basis functions for interpolating the solution? Should we perform a global interpolation over the entire domain, or subdivide it into overlapping or non-overlapping local covers? What principle should we apply to fulfill the governing equation and BCs, namely, the strong or weak formulation, or least squares? How do we judge a better or best method, by its accuracy, efficiency, stability, ease of programming, and/or adaptability to different geometries and types of problems? We may not need to start from a traditional method to improve it, and may start fresh by answering the above questions.

Basis Functions: Polynomials seem to be the most intuitive of the basis functions, and have been widely used. The resulting interpolation matrix, however, is not guaranteed to be invertible. Also, without a proper placement of the nodes, the interpolation can be oscillatory. Polynomials in special forms, such as Chebyshev polynomials, are stable and have spectral convergence properties. Fourier and other orthogonal series have similar convergence, hence are highly efficient. These approximations, however, require collocation at fixed locations, and are not suitable for application in problems with multi-dimensional irregular geometries.

RBFs use scattered nodes for interpolation, and can adjust to arbitrary domains. The matrices are guaranteed to be nonsingular, barring overlapping nodes. Scalable functions such as the MQ, IMQ, and GA, exhibit exponential convergence properties not only by increasing the density of nodes, but also by changing their shapes (increasing flatness). Additionally, if the basis functions satisfy the PDE to be solved either globally or locally, such as the general and fundamental solutions, they are the most efficient interpolants.

Approximation Methods: The approximation of a function can be performed implicitly by summing up a set of basis functions with undetermined coefficients, or explicitly by interpolating among discrete function values. The latter approach is more appealing to engineers, and is how the FEM and FDM were developed.

Explicit interpolation like the inverse distance weighting are physically motivated with the assumption that nearby data bear stronger relation than faraway ones. Their ability to approximate mathematical functions is generally poor. For an accurate approximation, the interpolating weights, also known as shape functions, should be constructed from a set of efficient basis functions. If the same basis functions are used, the two approximation methods are the same, as one can be converted to the other. Given a large number of nodes, the shape functions need to be numerically constructed, losing the appeal of an explicit representation. Hence, an explicit formula is usually used to interpolate a small number of local nodes, in which analytical shape functions can be derived. For global approximations, an implicit formula is generally used.

Accuracy: A local, element-based approximation only has C^0 smoothness, and is not efficient to approximate a continuous function. It is not the method of choice if high accuracy is needed. Meshless methods that use overlapping local covers can achieve higher smoothness with properly selected basis and weighting functions. The global approximation allows for the use of infinitely smooth basis functions to achieve exponential convergence.

If the solution is not infinitely smooth, for example, containing discontinuities in its derivatives, such as across the interface of material zones, or near discontinuous BCs, a continuous approximation can cause local inaccuracies. These situations can be remedied by dividing the domain into zones and applying interface conditions, or by approximating the local condition using local nonsingular or singular particular solutions.

Efficiency: Global approximations, though highly accurate, produce full systems that can be costly to solve, when the size is large. Matrices from local approximations are sparse, and the corresponding systems can be efficiently solved using gradient search algorithms without the tedious elimination process. A comparison of efficiency of numerical algorithms should not be based on accuracy or CPU time alone. It should be based on the CPU time required to achieve a target accuracy using whatever the necessary discretization. In general, when high accuracy is desired, the global methods are more efficient.

Condition Number: In many collocation methods, including the RBFCM, MFS, and TCM, it was often observed that excellent results were obtained when the traditional condition number indicated that they should not. The condition number can often exceed 10^{16} yet without interfering with the accuracy of the solution. It has been demonstrated that a more relevant measure of the solution stability is the effective condition number. Schaback's uncertainty principle implies that the higher the condition number, the higher the accuracy, until the instability sets in. Hence a high condition number can be a blessing. In fact, the magnitude of the effective condition number has been used as an error indicator to guide the search of optimal shape parameter of the RBF and source locations for better accuracy in the MFS [147, 271, 680].

Strong versus Weak Form: The strong form works by requiring that the governing equation and BCs be satisfied on a set of points. In between the points, it has no control, and the error can fluctuate from one sign to the other. The convergence is achieved by making the points denser and denser. Its concept and numerical implementation are straightforward.

The weak form is based on the concept of weighted residuals, which integrates the weighted error over a domain, and seeks its overall minimization. In a global application, this process tends to distribute the error over the domain, hence lowering the maximum error. It, however, requires extra work, particularly because of the integration. In fact, the need of integration can limit its advantage in accuracy. In the traditional FEM, to facilitate the integration, the domain is subdivided into elements. As the elements have only C^0 continuity across them, the order of the PDE is lowered to the first derivatives. Low degree polynomials are used for local interpolation. These steps transform the original method into a less accurate and slowly converging algorithm, not to mention the burden of managing the elements. The new generation of meshless methods can alleviate some of the above issues.

We may assess the merits of the strong and weak form as follows. When they are applicable to the same region, the weak form involves more work, but can reduce the interpolation error. The execution of the weak form, however, generally limits it to smaller regions, which are patched together to cover the entire domain. The patching can erode the accuracy of the weighted residual approach, particularly in the traditional FEM. In addition, the pursuit of an easy and accurate integration algorithm may affect the selection of basis functions. In contrast, the strong form has no integration requirement and can adjust to regions of any size and shape. It is also flexible in adopting different basis functions and special solutions. However, for complex geometries, the discretization and the matrix size can become unwieldy in a global implementation, and a local scheme also needs to be developed. Based on the above insights, the choice between a strong and a weak formulation should be carefully assessed in the selection of a numerical method.

Chapter 6

The Method of Fundamental Solutions

We have presented in Chapter 5 a broad discussion of numerical methods. Among them, the MFS is a special method, and not one for solving all types of PDEs. However, when it is applicable, it can be the most accurate and efficient method. Based on the types of problems, the numerical modeler should make a judicious choice of the method to use. With such an understanding, in this chapter we focus on the MFS and its variants. We shall discuss issues such as solvability, uniqueness, stability, error convergence, and other issues and remedies.

6.1 Singular and Nonsingular Basis Functions

The MFS utilizes fundamental solutions as basis functions. Strictly speaking, a fundamental solution is a solution to a PDE with the Dirac delta function as its RHS. For the purpose of the numerical method, however, we shall extend the definition to incorporate any particular solution of a PDE with an inhomogeneous RHS, which is nonzero only at a point, along a line segment, or within a small area or volume. In other words, we may take a fundamental solution and differentiate it, or we may integrate it along a line, over a finite area, or volume, and still call it a fundamental solution. For the integrated types, we may call them the distributed singularities. The resultant basis functions may be hypersingular, singular, or not.

We may recognize another type of solutions of a PDE, the general solutions, which are solutions of homogeneous PDEs. Both fundamental solutions and general solutions can be used as basis functions to approximate

241

the solution of a PDE. In the latter case, the numerical method is the TCM, which we shall discuss in Section 6.2.

The numerical implementation of these two types of basis functions has some fundamental differences. One is that the inhomogeneous part of a fundamental solution, a singular point, an integrated line, etc., must be placed outside the solution domain. The general solution procedure has no such issue. The second is that fundamental solutions are generally of radial nature, possessing a "center". By anchoring the center of a single fundamental solution at different locations, each one becomes an independent basis function. An approximate solution of a PDE can thus be written as follows:

$$\hat{\phi}(x) = \sum_{j=1}^{N} c_j G(x, x_j'). \tag{6.1}$$

We note in the above that G is a single fundamental solution. By placing its center at various locations x_j', it forms a complete interpolation set.

In contrast, general solutions do not have such a center. It requires a family of different functions to form a complete set. Although each function has an origin, its translation or rotation produces a linear combination of other functions within the same family, and not a new basis. Hence the approximation formula is given as

$$\hat{\phi}(x) = \sum_{j=1}^{N} c_j \psi_j(x), \tag{6.2}$$

in which each ψ_j is a different function from the family of functions. In this section, we shall discuss fundamental solutions and their variants. The methodology for general solutions will be presented in Section 6.2.

6.1.1 *Potential problems*

In Section 5.5.1 and Examples 1.1 and 1.2 , we have demonstrated how to solve a BVP for the Laplace equation using the MFS based on the source fundamental solutions (2.23), together with MATLAB® codes. Rather than repeating that methodology, here we shall present a variant of that approach.

In Sections 1.3.2 and 1.5, we mentioned the physical interpretation of an electric dipole as the result of pushing a positive and a negative charge infinitely close together, with their strengths increasing proportionally to the inverse distance between them. In mathematical terms, this is equivalent to taking the spatial derivative of the source solution, which results in a solution of higher singularity. When the dipole solution is used in an integral equation such as (1.30), it forms the basis of the double layer integral equation method, as compared to the simple layer method based on (1.33). Similarly, we can extend the concept to the MFS, using dipoles instead of sources for basis functions.

We can obtain a dipole solution in the x_i direction by differentiating (2.23), which gives

$$\frac{\partial G(x, x')}{\partial x_i} = \frac{1}{2\pi} \frac{x_i - x_i'}{r^2}, \quad \text{(2D)}, \tag{6.3a}$$

$$= \frac{1}{4\pi} \frac{x_i - x_i'}{r^3}, \quad \text{(3D)}, \tag{6.3b}$$

where $r = \|x - x'\|$. As the dipole is a vector, when it is distributed, we need to choose a direction. It is customary to align it with the normal of the auxiliary boundary $\partial \Omega'$, though not necessarily so. If the normal is chosen, the approximate solution can then be written as

$$\hat{\phi}(x) = \sum_{j=1}^{N} c_j \frac{\partial G(x, x_j')}{\partial n(x_j')}, \quad x \in \overline{\Omega}, \; x_j' \in \partial \Omega'. \tag{6.4}$$

Similar to the source case, (5.97) and (5.98), we can collocate for the Dirichlet and Neumann BCs as

$$\sum_{j=1}^{N} c_j \frac{\partial G(x_i, x_j')}{\partial n(x_j')} = g_1(x_i), \quad x_i \in \partial \Omega_D, \; i = 1, \ldots, N_1, \tag{6.5}$$

$$\sum_{j=1}^{N} c_j \frac{\partial^2 G(x_i, x_j')}{\partial n(x_i) \partial n(x_j')} = g_2(x_i), \quad x_i \in \partial \Omega_N, \; i = N_1 + 1, \ldots, N,$$

$$\tag{6.6}$$

where $\partial \Omega_D \cup \partial \Omega_N = \partial \Omega$. The above differentiations are easier to perform if we transform the coordinates to a local system centered at the dipole with the η-axis pointing to the dipole direction (see Fig. 6.1); and we obtain for

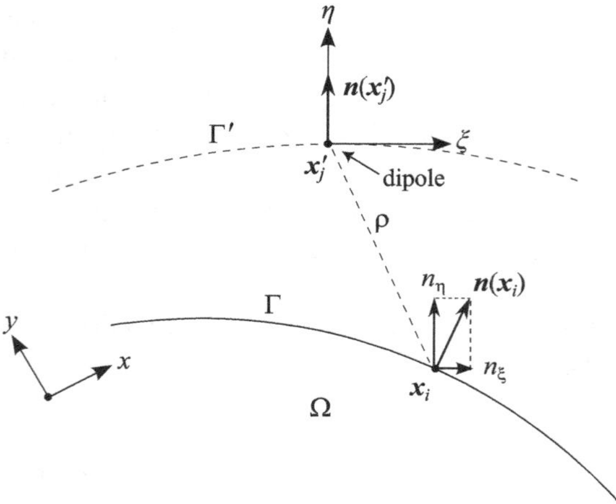

Figure 6.1.   Local coordinate system centered at a dipole.

the 2D case

$$\frac{\partial G(x_i, x_j')}{\partial n(x_j')} = \frac{\eta}{\rho^2}, \tag{6.7}$$

$$\frac{\partial^2 G(x_i, x_j')}{\partial n(x_i)\,\partial n(x_j')} = -\frac{2\xi\eta}{\rho^4}\,n_\xi + \frac{\rho^2 - 2\eta^2}{\rho^4}\,n_\eta, \tag{6.8}$$

where ρ is the distance between x_i and x_j', and n_ξ and n_η are the projections of $n(x_i)$ in these directions. The set of collocation equations, (6.5) and (6.6), then form a linear system for the determination of the coefficients c_j.

In general, there is no special advantage in using dipoles, instead of sources, in an approximation. In fact, the accuracy can be slightly inferior. However, the logarithmic function contained in 2D fundamental solutions can be ill-behaved. An issue with the $\ln r$ term is that it is unbounded at infinity, which can be a problem in solving exterior domain problems. The dipole function decays at infinity, thus automatically satisfying the far-field condition. Another issue is that the interpolation matrix based on the logarithmic function is not positive definite; hence for a given type of geometry, there can exist degenerate scales such that the numerical solution is nonunique. As will be demonstrated in Section 6.4, using dipoles instead of sources can circumvent the issue. Also, in Section 6.5, it is shown that the use of dipoles for the Helmholtz equation can assist in the elimination of spurious frequencies. Yet another use of the dipole or dipole-type solution is in problems with degenerate geometry, such as an infinitely thin impermeable barrier in a potential flow, or a fracture in a solid. Dipoles can be used together with sources to solve the duality of potential at the same location. These issues will be discussed in later sections.

■ Example 6.1

In the following, we solve the 2D ideal fluid flow problem of a uniform flow passing an ellipse. The ellipse is centered at the origin, with semi-major and semi-minor axis, a and b, respectively. The far-field uniform flow is of velocity U and comes at an angle β with respect to the x-axis.

The exact solution of the problem can be derived by the Joukowski transform. The complex potential is

$$W(z) = \phi + \mathrm{i}\,\psi = -U\left[\zeta\,e^{-\mathrm{i}\beta} + \frac{(a+b)^2}{4}\zeta^{-1}e^{\mathrm{i}\beta}\right],$$

in which ϕ is the velocity potential, and ψ the stream function. The mapping function that maps a circle onto an ellipse is

$$\zeta = \frac{1}{2}\left[z + \sqrt{r_1 r_2}\,e^{\mathrm{i}(\theta_1+\theta_2)/2}\right].$$

In the above,

$$r_1 = |z - c|,$$
$$r_2 = |z + c|,$$
$$\theta_1 = \arg(z - c),$$
$$\theta_2 = \arg(z + c),$$

where $c = \sqrt{a^2 - b^2}$, and $(\pm c, 0)$ are the foci of the ellipse, r_1 and r_2 are radial distances measured from the foci, and θ_1 and θ_2 are their respective angles. In Fig. 6.2, we plot the streamline pattern for the case of $U = 1$, $a = 1$, $b = 0.4$, and $\beta = 30°$.

To solve the BVP for the Laplace equation, we seek the exterior domain solution that vanishes at infinity. Hence we subtract the

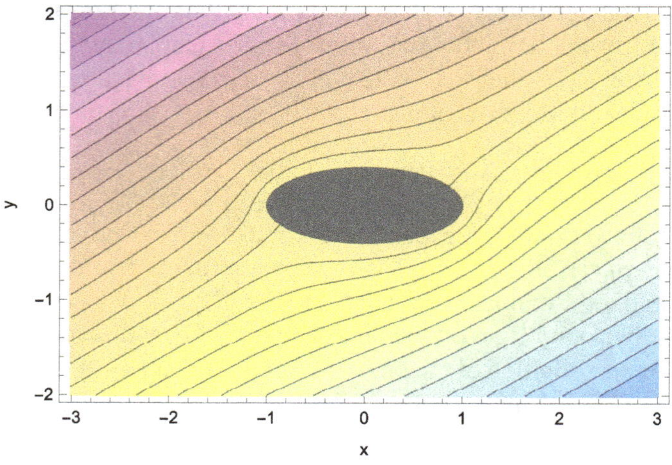

Figure 6.2. Streamlines of a uniform flow at $30°$ angle passing an ellipse.

far-field uniform flow from the potential ϕ, and solve for

$$\phi' = \phi + \Re\{Uze^{-i\beta}\} = \phi + U(x\cos\beta + y\sin\beta),$$

which satisfies the condition

$$\phi' \to 0 \quad \text{as} \quad r \to \infty.$$

For the BC, the no flux condition $\partial\phi/\partial n = 0$ on the ellipse is modified to

$$\frac{\partial\phi'}{\partial n} = U\left(n_x\cos\beta + n_y\sin\beta\right),$$

in which

$$n_x = \frac{b^2 x}{\sqrt{b^4 x^2 + a^4 y^2}},$$

$$n_y = \frac{a^2 y}{\sqrt{b^4 x^2 + a^4 y^2}},$$

are the components of the outward normal to the ellipse. Hence we are solving an all-Neumann BC problem. As this is an exterior domain problem, it has a unique solution.

We examine the case of an ellipse with an aspect ratio of 0.4. The problem is solved by distributing either sources or dipoles on an interior ellipse with a semi-major axis of 0.9 and a semi-minor axis of 0.01. The thin ellipse nearly coincides with the x-axis. The reason for such a pattern is that the analytical solution shows that the major axis of the ellipse is a branch cut across which there is a jump in potential. The sources/dipoles and collocation points are distributed at equal arc length. These nodes are slightly rotated to avoid symmetry about the axes. The dipoles are distributed with their directions tangential to the auxiliary contour.

Using 200 nodes, we find that $\varepsilon_{\max}$ for the source case is 1.5×10^{-5} and for the dipole 3×10^{-5}. The exact and numerical solutions of the potential ϕ' around the ellipse surface are plotted in Fig. 6.3. We note that in the source case, we did not impose an auxiliary condition to ensure that all source strengths sum to zero. It did not seem to pose an issue. In fact, the source solution is slightly better than the dipole

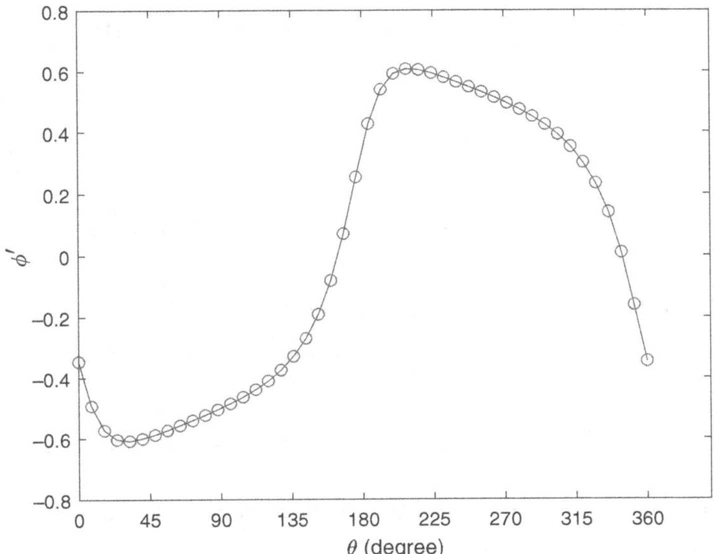

Figure 6.3. Plot of potential ϕ' for flow around an ellipse with aspect ratio $b/a = 0.4$ and flow angle $\beta = 30°$. The Exact solution is shown as the solid line, and the numerical solution in symbols ($\circ$).

one. The above results show that the MFS can utilize different types of singularities.

6.1.2 *Elastostatics*

The elasticity fundamental solution presented in Section 2.8.1 has the physical meaning of a force acting on a point of the medium. It can be differentiated to form higher order singular solutions. As the force fundamental solution is a second rank tensor, there are a variety of ways the higher order solutions can be assembled, as demonstrated in Sections. 2.8.2–2.8.4. Of particular interest is the displacement discontinuity solution in Section 2.8.4.

First, we shall examine the MFS based on the point force solution [77, 246, 729]. On a point x'_j outside of the domain Ω, we can apply forces pointing to the x_k direction, where $k = 1, 2$, or $1, 2, 3$, respectively for 2D and 3D problems, to generate the displacement fields $u^F_{\ell k}$, where the subscript ℓ denotes the vector component. These expressions are given in (2.80). If we distribute these forces at N such locations, the displacement field becomes, for the 2D case,

$$\hat{u}_1(x) = \sum_{j=1}^{N} \left[c_{1j} u^F_{11}(x, x'_j) + c_{2j} u^F_{12}(x, x'_j) \right], \tag{6.9}$$

$$\hat{u}_2(x) = \sum_{j=1}^{N} \left[c_{1j} u^F_{21}(x, x'_j) + c_{2j} u^F_{22}(x, x'_j) \right]. \tag{6.10}$$

In the above, the c_{1j} and c_{2j} are undetermined coefficients. In a condensed form, we may write

$$\hat{u}_\ell(x) = \sum_{j=1}^{N} \sum_{k=1}^{D} c_{kj} u^F_{\ell k}(x, x'_j), \tag{6.11}$$

in which $\ell = 1, 2$, or $1, 2, 3$, and $D = 2$ or 3, respectively for 2D and 3D cases.

On the boundary $\partial\Omega_u$, where the displacement vector is prescribed, we can collocate for the BCs as

$$\bar{u}_\ell(x_i) = \sum_{j=1}^{N} \sum_{k=1}^{D} c_{kj} u^F_{\ell k}(x_i, x'_j), \quad x_i \in \partial\Omega_u, \ i = 1, \ldots N_1, \tag{6.12}$$

where $\bar{u}_\ell$ is the boundary value. On $\partial\Omega_t$, where the traction vector is prescribed as $\bar{t}_\ell$, we can collocate for

$$\bar{t}_\ell(\boldsymbol{x}_i) = \sum_{j=1}^{N}\sum_{k=1}^{D} c_{kj}\, t_{\ell k}^{F}(\boldsymbol{x}_i, \boldsymbol{x}'_j), \quad \boldsymbol{x}_i \in \partial\Omega_t,\ i = N_1 + 1, \ldots N, \quad (6.13)$$

in which

$$t_{\ell k}^{F}(\boldsymbol{x}_i, \boldsymbol{x}'_j) = \sum_{m=1}^{D} \sigma_{\ell m k}^{F}(\boldsymbol{x}_i, \boldsymbol{x}'_j)\, n_m(\boldsymbol{x}_i), \quad (6.14)$$

with $\sigma_{\ell m k}^{F}$ given in (2.89), and n_m being the components of the outward normal. Equations (6.13) and (6.14) yield $D \times N$ equations for the $D \times N$ unknowns, c_{kj}.

Similar to the dipole case for potential problems, we can also utilize higher order singular solutions, such as the double force, etc., in the above approximation formulas. Among these solutions, the displacement discontinuity has received the most attention. Not only can it be used as a fundamental solution to solve general BVPs [229], but it also has the physical appeal of mimicking the displacement jump of a fracture. The method, known as the displacement discontinuity method (DDM), has been applied to solving fracture problems in elastic [93, 228, 497, 924] as well as thermoelastic and poroelastic [250, 328, 546, 872] media. In the following, we shall give a brief discussion.

We observe that the displacement discontinuity solution u_{ijk}^{d}, given in (2.106), is a third rank tensor, in which the first index i indicates the vector component of $\boldsymbol{u}$. The second index j denotes the surface orientation of a fracture defined by its normal, and the third index k represents the opening mode for a jump normal to the fracture surface when $j = k$, and shear modes when $j \neq k$. For example, in 2D, to create a fracture surface located at the x-axis with constant opening, we first seek the displacement field u_y due to a displacement opening in the y-direction, which is given by (2.106a) as

$$u_{222}^{d} = -\frac{1}{4\pi(1-\nu)}\frac{1}{r}\left[(1-2\nu)\frac{y}{r^2} + 2\frac{y^3}{r^4}\right]. \quad (6.15)$$

If we replace the x contained in the above by $x - x'$, and integrate the expression with respect to x' from -1 to 1, we can obtain a mode 1 constant

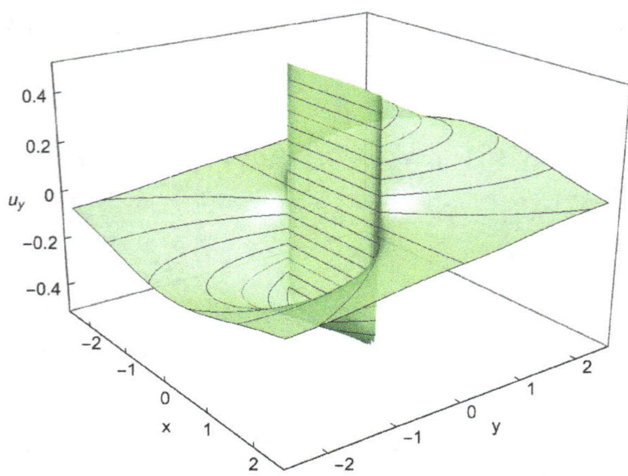

Figure 6.4. The displacement field u_y for a fracture located at the x-axis and between $-1 \leq x \leq 1$, which opens in the y-direction with unit width. The Poisson ratio is $\nu = 0.25$.

fracture opening between $-1 \leq x \leq 1$ as

$$
u_{222}^{d\ell} = \frac{1}{4\pi (1 - \nu)} \left[\frac{2y(1 - x^2 + y^2)}{x^4 - 2x^2(1 - y^2) + (1 + y^2)^2} \right.
$$
$$
\left. + 2(1 - \nu) \left(\tan^{-1} \frac{1 - x}{y} + \tan^{-1} \frac{1 + x}{y} \right) \right], \qquad (6.16)
$$

which is plotted in Fig. 6.4 for the case $\nu = 0.25$. Similarly, we can obtain $u_{122}^{d\ell}$, as well as the stresses. The mode 2 (shear mode) displacements can be created by integrating u_{121}^{d} and u_{221}^{d} along the segment. In the DDM, these segments can be stacked along the fracture trajectory with undetermined magnitude, and one may collocate for the stresses on the fracture surface, given as the BC, to solve a fracture problem.

6.1.3 *Elastodynamics*

Elastodynamic problems in the frequency domain can be treated in the same way as the elastostatic case. By employing the force fundamental solutions u_{ik}^{F} in (2.137) and (2.138), and their stress σ_{ijk}^{F} and traction t_{ik}^{F} expressions, the MFS follows (6.11)–(6.13). Here, however, we shall present an alternative method using the spherical wave fundamental solutions presented in Section 2.10.2.

Spherical Wave Fundamental Solution: As commented in Section 2.10.2, the force fundamental solutions can take some effort to derive, particularly for coupled wave phenomena. The elastodynamic equations can be decomposed into several scalar wave equations (or Helmholtz equations in the frequency domain), whose fundamental solutions are readily available. These solutions do not have a direct physical meaning, such as the force solutions, but they are fundamental solutions just the same. In Section 2.10.2 we have presented spherical wave solutions for elastodynamics. Other spherical wave solutions have been derived for poroelastodynamics [598], and double-porosity dual-permeability poroelastodynamics [599].

For the elastodynamic spherical wave MFS, we may approximate the displacement solution as

$$\hat{u}_\ell(\boldsymbol{x}) = \sum_{j=1}^{N} \sum_{k=0}^{D} c_{kj} u_{\ell k}^*(\boldsymbol{x}, \boldsymbol{x}_j'), \tag{6.17}$$

in which $\ell = 1, 2$, or $1, 2, 3$, and $D = 2$ or 3, respectively for 2D and 3D cases. We note in particular that the index k starts from 0 and not 1, because there is an extra wave component based on the Helmholtz decomposition. On the boundary $\partial\Omega_u$, we can collocate for the BCs as

$$\bar{u}_\ell(\boldsymbol{x}_i) = \sum_{j=1}^{N} \sum_{k=0}^{D} c_{kj} u_{\ell k}^*(\boldsymbol{x}_i, \boldsymbol{x}_j'), \quad \boldsymbol{x}_i \in \partial\Omega_u, \ i = 1, \ldots N_1. \tag{6.18}$$

On $\partial\Omega_t$, we collocate for

$$\bar{t}_\ell(\boldsymbol{x}_i) = \sum_{j=1}^{N} \sum_{k=0}^{D} c_{kj} t_{\ell k}^*(\boldsymbol{x}_i, \boldsymbol{x}_j'), \quad \boldsymbol{x}_i \in \partial\Omega_t, \ i = N_1 + 1, \ldots N. \tag{6.19}$$

In (6.17), we note that on each source point $\boldsymbol{x}_j'$ we distribute $D + 1$ fundamental solutions, with undetermined intensities c_{kj}, as k goes from 0 to D. So there are $(D + 1) \times N$ coefficients to be determined. Equations (6.18) and (6.19), however, only provide $D \times N$ equations, and are not sufficient to solve the system. The missing equations are supplied by the divergence condition (2.123). We hence introduce the following constraint equations:

$$\sum_{j=1}^{N} \sum_{k=1}^{D} c_{kj} \frac{\partial \psi_{kk}^*(\boldsymbol{x}_i, \boldsymbol{x}_j')}{\partial x_k} = 0, \quad \boldsymbol{x}_i \in \partial\Omega, \ i = 1, \ldots, N, \tag{6.20}$$

in which

$$\frac{\partial \psi_{kk}^{*}(r)}{\partial x_k} = -\frac{ik_p x_k}{4r} H_1^{(1)}(k_p r), \qquad (2D), \qquad (6.21a)$$

$$= \frac{x_k}{4\pi r^3}(-1 + ik_p r)e^{ik_p r}, \quad (3D). \qquad (6.21b)$$

A complete solution system is then formed. Numerical implementations utilizing this type of methodology can be found in [598, 599] for poroelastic materials.

6.1.4 *Non-singular bases*

In the MFS, the distribution of singular fundamental solutions should be located outside the domain, or else the governing equation is not satisfied in the domain. In a typical numerical implementation, an auxiliary boundary is introduced to guide the distribution of the singularities. This step leaves open the uncertainty of choosing a proper auxiliary boundary. In order to eliminate the uncertainty, there have been efforts to use nonsingular basis functions, or to desingularize singular solutions, so that they can be placed on the boundary. In particular, it is desirable to let the source nodes coincide with the collocation nodes to simplify the data preparation process.

One technique follows the lead of the BIEM, as discussed in Sections 1.3 and 1.5, and places the singular sources on the boundary. When the source and collocation nodes coincide, the diagonal terms of the matrix become singular. The singularities are regularized by using a subtracting and adding back technique [774, 946, 950] similar to that practiced in the BEM [108, 230, 592], or evaluated using an inverse interpolation technique [176, 352, 571].

Another technique is to use basis functions that are nonsingular, which by definition are general solutions. The general solution must take the form of an RBF, such that following (6.1), the same function can be distributed at various locations to form the approximation basis. Although their existence is rare, there are indeed general solutions that are RBFs. For the Helmholtz and modified Helmholtz equations, we find in (3.7) the general solutions $J_0(kr)$ for 2D and $j_0(kr)$ for 3D, and in (3.8) $I_0(kr)$ for 2D and $i_0^{(1)}(kr)$ for 3D, which serve such purposes. These nonsingular basis functions have been used in the BKM [173, 174, 392] to solve Helmholtz-type as well as other governing equations. We shall test these basis functions in the following examples.

Modified Helmholtz Equation: First, we examine the modified Helmholtz equation.

■ Example 6.2

A 2D BVP for the modified Helmholtz equation is defined as follows. The solution domain Ω is the interior of a Cassini curve:

$$r(\theta) = \left(\cos 4\theta + \sqrt{\frac{18}{5} - \sin^2 4\theta} \right)^{1/3}, \quad \text{for } 0 \le \theta < 2\pi.$$

The shape of the domain can be observed as the projected area on the x-y plane in Fig. 6.5. A Dirichlet BC is prescribed:

$$\phi(\boldsymbol{x}) = \sin cx [a \cosh(\sqrt{k^2 + c^2}\, y) + b \sinh(\sqrt{k^2 + c^2}\, y)], \quad \boldsymbol{x} \in \partial\Omega.$$

In the above, k is the wave number of the modified Helmholtz equation, and the parameter values are: $k = 2$, $c = 2$, $a = 1$, and $b = 0.3$. We note that the right hand side of the above equation is a metaharmonic

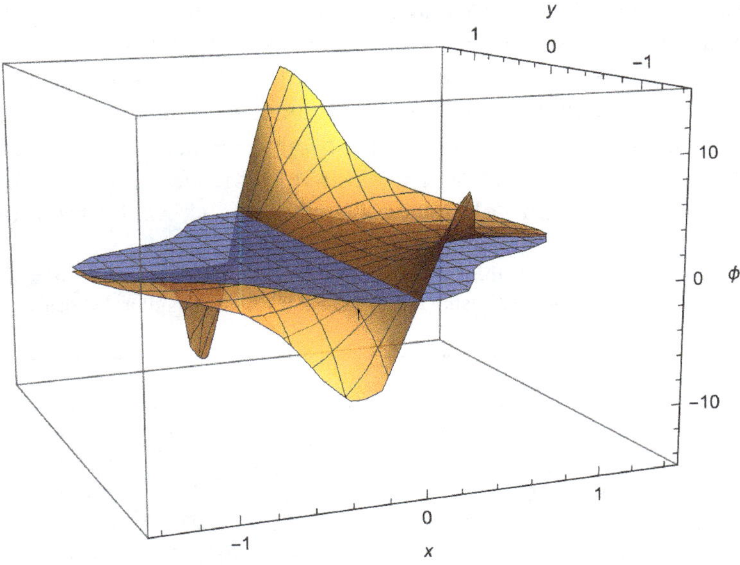

Figure 6.5. Plot of the exact solution ϕ with the domain defined by a Cassini curve, which is projected on the x–y plane.

function; i.e. it satisfies the modified Helmholtz equation. Hence the exact solution of the problem is the same function, which is plotted in Fig. 6.5.

To solve the BVP, we approximate the solution by the linear combination of nonsingular basis functions

$$\hat{\phi}(x) = \sum_{j=1}^{N} c_j I_0(k r_j).$$

Twenty points are distributed on the Cassini curve with equal spacing in θ, and used both as centers and collocation points. The error of approximation is evaluated at a dense set of points in the domain. The errors $\varepsilon_{\max}$ and $\varepsilon_{\mathrm{rms}}$, as defined in (3.72) and (3.73), are shown in the first row of Table 6.1. We observe good accuracy in representing this function using relatively few terms in the approximation.

Although the nonsingular function is not intended for the purpose, we also solve the problem by placing the sources on an auxiliary contour. The contour is defined by dilating the boundary curve using a new radial distance $r'(\theta) = \eta r(\theta)$ in the parametric equation of the Cassini curve, where η is a dilation factor. For the first row of the table, we have $\eta = 1$. Using $\eta = 1.1$, the second row of Table 6.1 shows that the error is the same. We further realize that the nonsingular nature of the basis function allows its placement not only on and outside the boundary, but also within it. By using $\eta = 0.9$, we observe the same error. Hence the error is insensitive to the size of the auxiliary boundary. In the last column of the table, we show the traditional

Table 6.1. Error of the solution of the modified Helmholtz equation with metaharmonic BC using the nonsingular (N) and singular (S) basis functions.

S/N	N	η	$\varepsilon_{\max}$	$\varepsilon_{\mathrm{rms}}$	κ
N	20	1	1.2×10^{-3}	1.3×10^{-4}	5.0×10^{12}
N	20	1.1	1.2×10^{-3}	1.3×10^{-4}	2.3×10^{12}
N	20	0.9	1.2×10^{-3}	1.3×10^{-4}	1.2×10^{13}
N	40	1	5.4×10^{-4}	7.9×10^{-5}	2.6×10^{17}
N	80	1	3.3×10^{-4}	3.9×10^{-5}	1.5×10^{18}
S	20	3	2.6×10^{-3}	2.9×10^{-4}	1.9×10^{4}
S	40	3	4.0×10^{-9}	2.1×10^{-10}	2.2×10^{10}
S	80	3	4.1×10^{-13}	2.0×10^{-14}	1.9×10^{19}

condition numbers κ, which are rather large. If we refine the mesh to 40 and 80 nodes, the error is only minimally reduced, and we do not observe the rapid convergence of the MFS.

We may compare the performance of the nonsingular bases with the singular ones. The singular function $K_0(kr)$ cannot be placed on the boundary, and should be located at a distance. Using $\eta = 3$ and $N = 20$, Table 6.1 shows that the accuracy is slightly inferior, but the condition number is much smaller. When N is doubled and doubled again, we observe the exponential convergence that is anticipated of the MFS.

◼ Example 6.3

In this example, we again solve the 2D modified Helmholtz equation in the interior domain defined in Example 6.2. However, the BC is changed to the following:

$$\phi(x) = e^x \cos y, \quad x \in \partial\Omega.$$

The significance of this BC is that it does not satisfy the governing equation, and we do not know the exact solution of the problem.

The problem is solved in the same way as the above example. To check for the error, we compute the approximate solution on a dense set of points on the boundary, and compare it with the exact BC. Based on the maximum principle for the modified Helmholtz equation, which we will discuss in Section 6.7.1, the maximum error is located on the boundary; hence we have a way to find the maximum error of the approximation.

As explained in Section 6.7, problems with nonmetaharmonic BCs are more difficult to solve. In fact, the nonsingular basis performs poorly. Using $N = 20$ and $\eta = 1$, the maximum error is shown in Table 6.2. Further refinement leads to an unstable solution due to the rapidly increasing condition number. For the singular basis, better results are found when the auxiliary boundary is close to the true boundary. Table 6.2 shows that as N increases, a better approximation is obtained. By doubling the nodes, we do see the convergence of solution, though not at the exponential rate observed in Table 6.1 for the metaharmonic BC case. Figure 6.6 depicts the approximate solution.

Table 6.2. Error of the solution of the modified Helmholtz equation with nonmetaharmonic BC using the nonsingular (N) and singular (S) basis function.

S/N	N	η	ε_{max}	κ
N	20	1	3.7×10^{-2}	5.0×10^{12}
N	40	1	4.7×10^{-1}	2.6×10^{17}
S	20	1.3	1.4×10^{-2}	1.6×10^{1}
S	40	1.2	4.7×10^{-3}	1.8×10^{2}
S	80	1.1	1.1×10^{-3}	1.2×10^{3}
S	160	1.05	4.6×10^{-4}	1.3×10^{4}

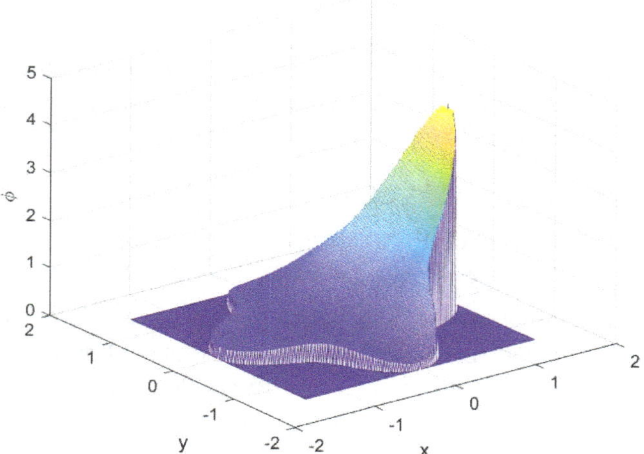

Figure 6.6. Plot of the approximate solution ϕ for the nonmetaharmonic BC problem.

Helmholtz Equation: Next, we examine the Helmholtz equation.

■ Example 6.4

In this example we solve the Helmholtz equation in the interior domain defined in Example 6.2. The BC is

$$\phi(x) = \sin cx \left[a \cos \left(\sqrt{k^2 - c^2}\, y \right) + b \sin \left(\sqrt{k^2 - c^2}\, y \right) \right], \quad x \in \partial\Omega,$$

with $k = 4$, $c = 3$, $a = 1$, and $b = 2$. The above function is a metaharmonic function that satisfies the Helmholtz equation; hence it is the exact solution of the problem. The function is plotted in Fig. 6.7.

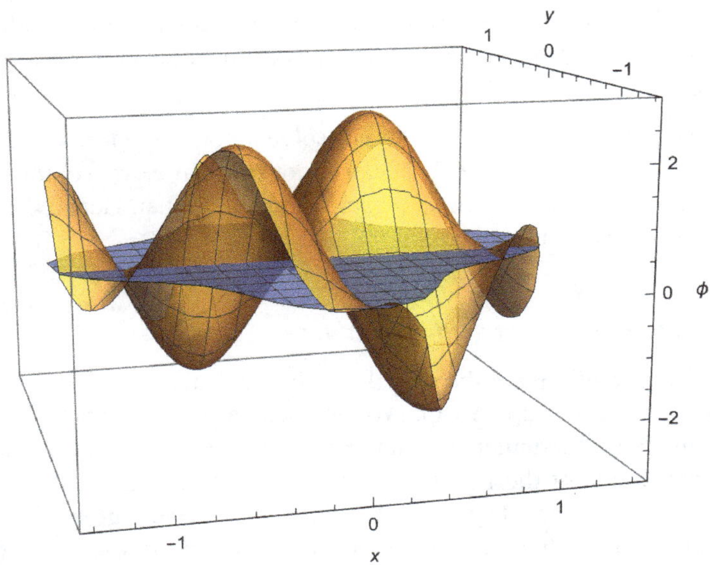

Figure 6.7. Plot of the exact solution ϕ for the Helmholtz equation and metaharmonic BC case.

Table 6.3. Error of the solution of the Helmholtz equation with metaharmonic BC using the nonsingular (N) and singular (S) basis functions.

S/N	N	η	$\varepsilon_{\max}$	$\varepsilon_{\mathrm{rms}}$	κ
N	20	1	1.6×10^{-3}	2.9×10^{-4}	4.5×10^{5}
N	40	1	3.2×10^{-8}	4.5×10^{-9}	6.1×10^{16}
N	80	1	1.6×10^{-7}	5.4×10^{-8}	5.9×10^{17}
S	20	3	6.0×10^{-2}	$9,0 \times 10^{-3}$	1.1×10^{4}
S	40	3	5.6×10^{-9}	3.0×10^{-10}	5.2×10^{9}
S	80	3	1.1×10^{-13}	5.9×10^{-15}	1.6×10^{17}

The problem is solved using both the nonsingular basis $J_0(kr)$ and the singular basis $Y_0(kr)$, and the results are presented in Table 6.3. We observe similar behavior as in the modified Helmholtz equation case in Table 6.1. The nonsingular basis cases, however, have smaller condition numbers compared to the modified Helmholtz equation case; hence exponential convergence is observed when N is doubled from 20 to 40. Further refinement to $N = 80$ does not improve the result, likely due to the interference of the exceedingly large condition number. The singular basis demonstrates exponential convergence, as expected.

For the Helmholtz equation, we do not solve a nonmetaharmonic BC case corresponding to Example 6.3, because not only an exact solution is not available, but also the maximum principle does not exist. Hence we cannot determine the maximum overall error.

6.1.5 *Distributed fundamental solutions*

Another way to eliminate the singularity in the fundamental solution is to integrate it over a line, an area, or a volume, with a given distribution density. Using integrated fundamental solutions has a long history of application. For example, one of the first implementations of the BEM used constant elements, which is the integration of a source over a line segment [411, 412, 823]. For elasticity, the DDM integrates a displacement discontinuity over an element to simulate a crack, as well as to solve BVPs [228, 229]. We also mentioned in Section 1.4 the panel method, which integrates vortices over surface panels to simulate subsonic and supersonic flows [382–384].

In the integral equation method, the boundary is subdivided into line or surface elements, and the integration is performed over them. In the MFS, there is no requirement that the distributed bases should cover the entire surface; hence it has more flexibility and is a meshless method.

The simplest of the distributed fundamental solutions is the integration of $\ln r$ over a line. Using local coordinates (ξ, η), we may place the line segment on the ξ-axis between $-a \le \xi \le a$. Integrating with the intensity $1/a$, the result is

$$G_\ell(\xi, \eta) = \frac{1}{2a} \left\{ -4a + 2\eta \left(\tan^{-1} \frac{a - \xi}{\eta} + \tan^{-1} \frac{a + \xi}{\eta} \right) \right.$$

$$\left. + (a - \xi) \ln \left[(a - \xi)^2 + \eta^2 \right] + (a + \xi) \ln \left[(a + \xi)^2 + \eta^2 \right] \right\}.$$

$$(6.22)$$

We may place the center of the segment on a set of boundary points x'_j, and align it to be tangential to the boundary. The approximation formula is then

$$\hat{\phi}(x) = \sum_{j=1}^{N} c_j G_\ell(x, x'_j), \quad x'_j \in \partial\Omega. \tag{6.23}$$

We however observe a problem with the function (6.22). Although the function is continuous along the ξ-axis, when $\xi = \pm a$ are crossed, its normal derivative $\partial G_\ell / \partial \eta$ is not. When the segment is placed on a straight line boundary, the evaluation of the normal derivative at or near the end points will be affected.

As a remedy, we may use line sources with distribution densities dropping to zero at the edges. For example, we can use a triangle shaped source density $(1 - |\xi|/a)/a$ between $-a \leq \xi \leq a$, and obtain

$$
\begin{aligned}
G_\ell(\xi, \eta) = \frac{1}{4a^2} \Bigg\{ &-6a^2 - 8\xi\eta \tan^{-1} \frac{\xi}{\eta} + 4\eta(a - \xi) \tan^{-1} \frac{a - \xi}{\eta} \\
&+ 4\eta(a + \xi) \tan^{-1} \frac{a + \xi}{\eta} + [(a - \xi)^2 - \eta^2] \ln[(a - \xi)^2 + \eta^2] \\
&+ [(a + \xi)^2 - \eta^2] \ln[(a + \xi)^2 + \eta^2)] \\
&- 2(\xi^2 - \eta^2) \ln(\xi^2 + \eta^2) \Bigg\}.
\end{aligned}
\tag{6.24}
$$

The normal derivative $\partial G_\ell / \partial \eta$ is then continuous along the ξ-axis. The performance of these distributed bases has been tested in [203].

Another issue with the distributed bases is that they may invade the solution domain, particularly if the object is an area or a volume. For example, disk shaped bases have been used in 2D potential [485, 594, 718], Stokes flow [222], and elasticity [584, 585] problems; and spheres have been used in 3D [223, 587]. If the center is placed on the boundary, half of the object will encroach the domain. The homogeneous governing equation is violated in the invaded parts, and the accuracy is affected [203]. The same issue can exist for line segments in 2D as well, if a part of the boundary is concave. In those circumstances, the size of the object should be kept as small as possible.

■ Example 6.5

Create a distributed force fundamental solution over a line segment for application in 2D elasticity.

To address the issue of discontinuity of normal derivatives at the end points of a constant distribution, we shall use a triangular distribution that vanishes at these points, the same as that used in (6.24).

For an easier derivation, we avoid the integration of force fundamental solutions given in Section 2.8.1. Instead, we utilize the Galerkin vector potential $\boldsymbol{w}$ introduced in Section 3.1.4. The vector $\boldsymbol{w}$ satisfies the biharmonic equation as shown in (3.29), and its relation with the displacement vector is given by (3.28).

To derive fundamental solutions, we introduce three separate Dirac delta functions, one for each vector component of $\boldsymbol{w}$, to the right-hand side of (3.29), and define

$$\nabla^4 w_{ik}^* = \delta_{ik}\,\delta(\boldsymbol{x}, \boldsymbol{x}').$$

Based on (2.156a), we have

$$W_{ik}^* = \delta_{ik}\, r^2 \ln r,$$

in which we have omitted the constant factor. Integrating the above along a line segment with the triangular distribution, we obtain

$$
\begin{aligned}
W_{ik}^{\ell} = \frac{-\delta_{ik}}{144a^2}\Bigg\{ &\, 2a^2(7a^2 + 42\xi^2 + 78\eta^2) + 96\eta^3 \\
&\times \left[2\xi \tan^{-1}\frac{\xi}{\eta} - (a-\xi)\tan^{-1}\frac{a-\xi}{\eta} - (a+\xi)\tan^{-1}\frac{a+\xi}{\eta} \right] \\
&- 6\left[(a-\xi)^4 + 6(a-\xi)^2\eta^2 - 3\eta^4\right]\ln[(a-\xi)^2 + \eta^2] \\
&- 6\left[(a+\xi)^4 + 6(a+\xi)^2\eta^2 - 3\eta^4\right]\ln[(a+\xi)^2 + \eta^2] \\
&+ 12(\xi^4 + 6\xi^2\eta^2 - 3\eta^4)\ln(\xi^2 + \eta^2) \Bigg\}.
\end{aligned}
$$

Substituting the above into the displacement expression, we obtain

$$
\begin{aligned}
u_{xx}^{\ell} = \frac{-1}{4a^2(1-v)}\Bigg\{ &\, 8a^2(1-v) + 16(1-v)\eta \\
&\times \left[2\xi \tan^{-1}\frac{\xi}{\eta} - (a-\xi)\tan^{-1}\frac{a-\xi}{\eta} - (a+\xi)\tan^{-1}\frac{a+\xi}{\eta} \right] \\
&- \left[(3-4v)(a-\xi)^2 - (5-4v)\eta^2\right]\ln[(a-\xi)^2 + \eta^2] \\
&- \left[(3-4v)(a+\xi)^2 - (5-4v)\eta^2\right]\ln[(a+\xi)^2 + \eta^2] \\
&+ 2\left[(3-4v)\xi^2 - (5-4v)\eta^2\right]\ln(\xi^2 + \eta^2) \Bigg\},
\end{aligned}
$$

$$u_{xy}^{\ell} = u_{yx}^{\ell}$$

$$= \frac{\eta}{2a^2(1-v)} \left\{ 2\eta \left[2\tan^{-1}\frac{\xi}{\eta} + \tan^{-1}\frac{a-\xi}{\eta} - \tan^{-1}\frac{a+\xi}{\eta} \right] \right.$$

$$+ (a-\xi)\ln[(a-\xi)^2 + \eta^2] - (a+\xi)\ln[(a+\xi)^2 + \eta^2]$$

$$\left. + 2\xi\ln(\xi^2 + \eta^2) \right\},$$

$$u_{yy}^{\ell} = \frac{-1}{4a^2(1-v)} \left\{ 4a^2(1-v) + 8(1-2v)\eta \right.$$

$$\times \left[2\xi\tan^{-1}\frac{\xi}{\eta} - (a-\xi)\tan^{-1}\frac{a-\xi}{\eta} - (a+\xi)\tan^{-1}\frac{a+\xi}{\eta} \right]$$

$$- \left[(3-4v)(a-\xi)^2 - (1-4v)\eta^2 \right]\ln[(a-\xi)^2 + \eta^2]$$

$$- \left[(3-4v)(a+\xi)^2 - (1-4v)\eta^2 \right]\ln[(a+\xi)^2 + \eta^2]$$

$$\left. + 2\left[(3-4v)\xi^2 - (1-4v)\eta^2 \right]\ln(\xi^2 + \eta^2) \right\}.$$

We note that the above solutions are not distributed force solutions, but they are fundamental solutions just the same, and can be used in a formula like (6.11) to solve BVPs for the Cauchy–Navier equations.

6.2 Trefftz Collocation Method

The original Trefftz method (TM) was proposed by Erich Trefftz [842, 843] as an improvement to the Ritz method [756]. The Ritz method used polynomial basis functions and the calculus of variations as a means of solving PDEs. As discussed in Section 5.5.2, it is one of the foundations of the FEM. Trefftz's idea was to use general solutions of the PDE, such as harmonic polynomials for the Laplace equation, rather than general polynomials, as basis functions. The solution methodology was based on the weak form, as is the case with the Ritz method [200, 203].

The TM has been applied as a boundary method for solving PDEs [204, 420, 486, 973, 974]. The idea has also been adopted in the FEM as the Trefftz element (T-element) [421, 422, 736, 977]. There are also variants like the Trefftz-Herrera method [379–381], and other implementations taking advantage of Trefftz functions [574, 575, 723, 873]. Our interest lies in the TCM; that is, the strong form collocation [487, 526]. It has been demonstrated that among the different implementations of the TM through the weak and strong forms, the TCM performed the best in accuracy and efficiency [550, 556]. In addition, the TCM is a meshless method and easier to implement.

The TCM methodology is very similar to that of the MFS. In fact, we may classify the TCM as a subset of the MFS, and vice versa. These two methods differ from each other by the two approximation equations, (6.1) and (6.2), and the basis functions they use. We also note the following degeneracy of the MFS [94, 162, 163]: for the Laplace and biharmonic equations, when the radius of the auxiliary boundary approaches infinity, the interpolation matrix degenerates into that of the harmonic polynomials, and the MFS and the TCM become equivalent.

As an example of the TCM, we may write an approximate solution of the 2D Laplace equation as follows:

$$\begin{aligned}
\hat{\phi}(\boldsymbol{x}) = {} & c_1 + c_2 x + c_3 y + c_4 xy + c_5(x^2 - y^2) \\
& + c_6(x^3 - 3xy^2) + c_7(y^3 - 3x^2 y) \\
& + c_8(x^4 - 6x^2 y^2 + y^4) + c_9(4x^3 y - 4xy^3) + \cdots
\end{aligned} \qquad (6.25)$$

in which the basis functions are taken from (3.3a). As the above terms are all distinct, a certain algorithm is needed to generate them in the computer program. For this purpose, either the polar forms in (3.1), or a certain iterative algorithm [791], can be used. Equation (6.25) can be employed in a similar collocation procedure as (5.97) and (5.98) for the BCs, and form a set of linear equations to solve for the undetermined coefficients c_j. In Section 6.6.4 we shall compare the performance of the TCM and MFS in an example.

6.3   Invertibility of MFS Matrices

By the theorems of Schoenberg [783, 784] and Micchelli [641], for a class of RBFs, the interpolation matrices are positive definite, hence always invertible, as long as the collocation points and centers coincide, and no two centers occupy the same location. Although fundamental solutions are also RBFs, their singularity prevents the centers from being placed on collocation nodes, and the above theorems do not apply. Schaback [779] appears to be the first to warn that the lack of positive definiteness can make the MFS matrix singular. Alves [22] demonstrated that the collinear (2D) and coplanar (3D) placement of source nodes lead to non-invertible matrices. Further demonstration of nodal arrangements that lead to singular matrices were given in [203, 470]. In the following, we shall give an example.

■ Example 6.6

Solve a potential problem in 2D. The domain is the square $[0, 1] \times [0, 1]$. A Dirichlet BC is prescribed based on the function

$$\phi = e^x \cos y,$$

which is also the exact solution of the problem.

We solve the problem using the standard MFS with 20 equally spaced collocation points over the boundary as shown in Fig. 6.8. The fundamental solutions are placed on an auxiliary circle of radius $R = 2$, centered at the center of the square. The sources are located on the circle every $18°$, starting from $\theta = 0°$. The matrix solver, however, reports that the matrix is singular. This is a demonstration that not all MFS collocation matrices are invertible.

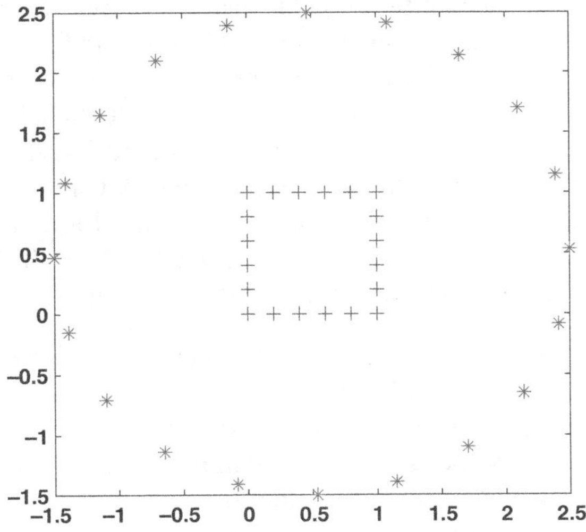

Figure 6.8.   Source ($*$) and collocation ($+$) points for solving the BVP.

Table 6.4. Maximum relative error and RMS relative error, with $R = 2$ and $N = 20$. $\Delta\theta$ is the degree of rotation to break the symmetry.

$\Delta\theta$	R	$\varepsilon_{\max}$	$\varepsilon_{\mathrm{rms}}$
$0°$	2	matrix is singular	
$1°$	2	7.2×10^{-9}	9.3×10^{-10}
$9°$	2	9.5×10^{-9}	7.7×10^{-10}
$0.1°$	2	2.0×10^{-8}	3.5×10^{-9}

To circumvent the issue, we rotate all sources by $1°$ to break the symmetry. The matrix now can be inverted, and the approximation error is evaluated on a dense mesh covering the domain and boundary. Table 6.4 depicts the maximum relative error $\varepsilon_{\max}$ and the RMS relative error $\varepsilon_{\mathrm{rms}}$. We observe a highly accurate solution, with errors of $\mathcal{O}(10^{-8})$ and $\mathcal{O}(10^{-9})$. To confirm the solution, sources are rotated $9°$. We find essentially the same accuracy. To test the range of the instability near the singular matrix, we perturb sources by a mere $0.1°$ from their symmetrical positions. Table 6.4 depicts that the accuracy is only slightly affected. Hence the solution system is stable, if one does not hit the exact matrix singularity.

To understand the cause of the singular matrix, we may solve the problem using $N = 4$, with sources located at $0°$, $90°$, $180°$, and $270°$, and the collocation nodes located at the corners of the square. It is easy to observe that among these nodes, there are only two distinct distances. The 4×4 matrix has rank 3; hence is singular. Although symmetrical node arrangements reduce the number of distinct distances, the matrix is not necessarily singular. For the above example, the matrix is singular for a symmetric node arrangement with $N = 4, 12, 20, \ldots$, but not for $N = 8, 16, 24, \ldots$ To avoid these "accidents", we recommend the following:

- Use scattered nodes without a pattern. In that case, the chances of hitting a singular matrix is virtually zero.
- If a symmetric arrangement is convenient, a slight break of symmetry, such as a rotation of the auxiliary circle in Example 6.6, or a perturbation of a few node locations, can avoid the issue.
- The problem can be formulated as an overdetermined system, that is, use more collocation nodes than sources. The least squares system is then solvable.

- Methods used to stabilize a system solution, such as the Tikhonov regularization [836], can make the singular system solvable.

6.4 Degenerate Scale

A number of the 2D fundamental solutions presented in Chapter 2 contain the logarithmic function, which has a zero at argument 1. For integral equations, there exist geometries, known as degenerate scales, for which the matrix suffers from rank deficiency and the solution is not unique [155, 156, 371, 412]. In particular, the unit circle has such an issue [720], which was referred to as the Γ-contour [412]. For example, the simple layer integral equation (1.33), which is a Fredholm integral equation of the first kind, expressed in 2D becomes

$$\phi(x) = \int_{\partial\Omega} \sigma(x') \ln r(x, x')\, dx', \quad x \in \Omega. \tag{6.26}$$

If we place x at the origin, and choose $\partial\Omega$ to be the unit circle, we then produce the result

$$\int_{\partial\Omega} \sigma(x') \ln 1\, dx' = 0. \tag{6.27}$$

The above equation is satisfied by any function $\sigma(x)$. It is obvious that the integral equation cannot be used to solve a BVP in such a geometry.

The above issue of nonuniqueness is not limited to first kind integral equations. It exists for Green's formula (1.14) in 2D as well, manifesting in the form of the matrix condition number approaching infinity [254]. This problem is also not confined to the unit circle. Based on the Riemann mapping theorem [754], an arbitrary simply connected region of the plane can be mapped conformally onto a unit disk. The theorem implies that a result proven for the unit disk can often be transferred to a more general region. The theory is also valid for mapping the exterior of any closed Jordan curve onto the interior of a unit disk [824], and any doubly connected region onto a circular annulus [825]. The above statements suggest that the malady found in the unit disk can exist in other geometries.

This phenomenon has been studied under the subject of logarithmic capacity or transfinite diameter [220, 503, 938]. The theory predicts that for a given geometric shape, there is a size for which the integral equation becomes unsolvable. The issue stems from the interaction between the zero of the logarithmic function and the eigenvalue of the geometry. Using logarithmic capacity as a measure, it was demonstrated that integral equations can admit a unit logarithmic capacity, causing the nonuniqueness of solution [254, 938].

Analytical results for degenerate scale have been found for the circle, ellipse, triangle, square, multiply connected, and exterior domain geometries for potential problems [153, 158, 160, 504]. As the logarithmic function appears in the kernel of integral equations in 2D elasticity, Stokes flow, and biharmonic problems, the degenerate scale can also find its way into these problems [155, 159, 167, 877]. For elasticity, the degenerate scale is also dependent on the Poisson ratio [155].

A number of remedies have been offered to eliminate the ill-conditioning mentioned above. The simplest one is to rescale the logarithmic function from $\ln r$ to $\ln(r/L)$, where L is a scale larger than the maximum distance between the source and collocation nodes to avoid the zero of the function [97]. Alternatively, we can reinterpret the physical dimension by using different units. For example, 1 m is the same as 100 cm, to circumvent the problem.

More rigorously, we may use Fichera's method [301], which modifies the Fredholm integral equation of the first kind (6.26) to an augmented set

$$\phi(x) = \int_{\partial\Omega} \sigma(x') \ln r(x, x')\, dx' + \sigma_o, \quad x \in \Omega, \qquad (6.28)$$

$$0 = \int_{\partial\Omega} \sigma(x')\, dx', \qquad (6.29)$$

with an additional undetermined coefficient σ_o. The above formulation removes the nonuniqueness [166]. In another remedy, we can use the double layer potential integral equation (1.30), which is a Fredholm integral equation of the second kind, expressed in 2D,

$$\phi(x) = \int_{\partial\Omega} \mu(x') \frac{\partial \ln r(x, x')}{\partial n(x')}\, dx', \quad x \in \Omega. \qquad (6.30)$$

The simultaneous use of the single and double layer integral equations, (6.26) and (6.30), known as the dual boundary integral equation method [157], can overcome the nonuniqueness issue.

As we have commented in Section 1.3.2, the MFS is closely related to integral equations. Hence, issues plaguing integral equations are likely to be also found in the MFS. By the same token, remedies for integral equations should work for the MFS. There is, however, an important difference between the two methods: the degenerate scale for the MFS exists not in the problem geometry, but in the auxiliary boundary for the distribution of sources, which will be demonstrated in the examples below.

To fix the nonuniqueness of the MFS, Bogomolny [94] suggested adding a constant to the approximation

$$\hat{\phi}(x) = c_0 + \sum_{j=1}^{N} c_j \ln r(x, x'_j), \quad x \in \overline{\Omega}, \ x'_j \in \partial\Omega'. \tag{6.31}$$

For the added degree of freedom, a constraint equation is introduced

$$\sum_{j=1}^{N} c_j = 0. \tag{6.32}$$

We observe that the above treatment is essentially the same as Fichera's for integral equations in (6.28) and (6.29).

In fact, based on the asymptotic behavior of $\ln r$ approaching infinity and the requirement of orthogonality, Bogomolny [94] in his density proof suggested to always include a constant term in (6.31) for the sake of completeness. Others [22, 285, 334, 731] have made similar recommendations, including adding low degree polynomial terms. Alves and Silvestre [17] stated the need of adding a constant term to 2D Stokes flow problems, due to the presence of the logarithmic term. Such practice should also be extended to the 2D equations of elasticity. It should be mentioned that the above issue is associated with the logarithmic term in 2D fundamental solutions. Such an issue does not exist in 3D problems, and no precaution is needed.

In another treatment, we may follow the lead of the dual integral equations (6.26) and (6.30) by the simultaneous use of sources and dipoles in the MFS. In fact, for the MFS, we only need to use the dipole equation

$$\hat{\phi}(x) = \sum_{j=1}^{N} c_j \frac{\partial \ln r(x, x'_j)}{\partial n(x'_j)}, \quad x \in \overline{\Omega}, \ x'_j \in \partial\Omega', \tag{6.33}$$

and there is no issue of nonuniqueness.

Table 6.5. Breakdown of solution with the auxiliary boundary of a circle with $R = 1$, and remedies by adding a constant term, and using dipole solution.

Fundamental solution	With constant	R	ε_{max}	ε_{rms}
source	no	1.0	breaks down	
source	yes	1.0	9.8×10^{-5}	1.2×10^{-5}
dipole	no	1.0	1.9×10^{-3}	1.8×10^{-4}
source	no	0.9	2.3×10^{-3}	2.9×10^{-4}
source	no	1.1	5.4×10^{-5}	6.6×10^{-6}

■ Example 6.7

Solve the same BVP defined in Example 6.6. The only difference is that instead of using a circle with radius $R = 2$ as the auxiliary boundary, we shall use $R = 1$.

We solve the problem by the standard MFS using 20 collocation points, with source points rotated by $1°$. With $R = 1$, the solver reports that the matrix is singular, and the problem cannot be solved. We then try the remedy defined by (6.31) and (6.32). The matrix becomes nonsingular, and errors are of $\mathcal{O}(10^{-5})$, as shown in Table 6.5. We also solve the problem using dipoles as in (6.33). Again there is no degenerate scale issue, although errors are about one order of magnitude larger.

We observe that the errors reported in Table 6.5 are much larger than those in Table 6.4. This is not caused by the erosion of accuracy by remedial measures. It is because the solution accuracy is sensitive to the size of the auxiliary boundary. To demonstrate this, we also solve the problem using $R = 0.9$ and 1.1. We observe in Table 6.5 that the errors progressively decrease as R is made larger. We shall further discuss this effect in Section 6.6.

To demonstrate that the degenerate scale for the MFS is associated with the auxiliary boundary only, we solve the following problem.

■ Example 6.8

We consider a 2D potential problem in a circular disk of radius R_o, centered at (R_o, R_o). A Dirichlet BC is prescribed using the same exact solution as in Example 6.6.

Table 6.6. Demonstration of no degenerate scale with respect to the domain geometry.

R_o	ε_{max}	ε_{rms}
0.99	5.0×10^{-6}	5.6×10^{-7}
1.00	5.7×10^{-6}	6.4×10^{-7}
1.01	6.5×10^{-6}	7.3×10^{-7}

The problem is solved with three domain sizes, $R_o = 0.99, 1.00$, and 1.01. The auxiliary boundary is a circle with radius $R = 2$. Twenty source and collocation points are uniformly distributed. The errors are reported in Table 6.6. We notice a smooth transition of errors. There is no degenerate scale issue at $R_o = 1$.

6.5 Spurious Frequency of Wave Equation

Acoustic and seismic wave propagation, reflection, and scattering problems in a homogeneous medium are most efficiently solved by the integral equation method and the MFS. For exterior domain problems, only the surface of the reflector and scatterer needs to be discretized, and the far-field radiation condition is automatically satisfied.

In solving wave problems, the matrix can become singular at certain frequencies. When this occurs, it does not necessarily mean that the solution has failed. Rather, it may represent the physical phenomenon of resonance, when the response becomes unbounded. In numerical solutions, however, there can exist "spurious frequencies", which are artifacts of the mathematical formulation and numerical discretization, and not due to the physical resonance.

Copley [218, 219] was the first to report the existence of eigenfrequencies in the integral equation solution of the Helmholtz equation. Its cause can be traced to the multiple zeros of the wavy fundamental solutions, such as the Bessel and trigonometric functions. This creates the opportunity for artificial eigenvalues in the discretized matrix, which are dependent on the kernel, problem geometry, wave number, and the type of boundary data prescribed. This issue is more serious than the degenerate scale in potential problems, because a wave problem is typically solved for a range of frequencies, which essentially sweeps through a range of sizes of a fixed geometry.

There has been a large amount of effort devoted to the understanding and removal of spurious frequencies in integral equations. We shall mention a few that are relevant to the present purposes. One technique is to use the complex fundamental solutions, as in (2.41), rather than their real parts as in (2.32). As the zeros of the real and imaginary parts never coincide, when the problem is solved in the complex variable domain, it is possible to eliminate the spurious frequencies. Tai and Shaw [826] performed such a computation and demonstrated the removal of spurious frequencies. However, it was theoretically proven and numerically demonstrated that the above statement is true only for simply connected domains [125, 154]. Spurious frequencies can still exist for multiply connected domains, created by the eigenvalues of the interior boundaries.

Another idea was contributed by Burton and Miller [115]. Similar to the dual integral equation method mentioned in Section 6.4, the Burton-Miller formulation utilized both the single and double layer potentials by taking a linear combination of them. For example, for exterior domain problems, the following integral equation was employed:

$$\phi_s(x) = \int_{\partial\Omega} \sigma(x') \left[G(x, x') + i\frac{\partial G(x, x')}{\partial n(x')} \right] dx', \quad x \in \overline{\Omega}_e, \qquad (6.34)$$

where $\overline{\Omega}_e = \Omega_e \cup \partial\Omega$, ϕ_s is the solution of the scattered wave field, $\sigma(x')$ is the source density, and Ω_e is the external domain. The solution of the above integral equation is an indirect boundary integral equation method. The Burton–Miller approach has also been adopted for the direct boundary integral equation method (BIEM) based on Green's theorem (Section 1.3.1) [28, 640, 804, 967].

Yet another method is the CHIEF (combined Helmholtz integral equation formulation) method [782, 789], which uses the Helmholtz integral equation

$$c\phi_s(x) = \int_{\partial\Omega} \left[G(x, x')\frac{\partial \phi_s(x')}{\partial n(x')} - \phi_s(x')\frac{\partial G(x, x')}{\partial n(x')} \right] dx', \quad x \in \overline{\Omega}_e,$$

$$(6.35)$$

where $c = 1$ for $x \in \Omega_e$, and $c = 1/2$ for x on a smooth part of $\partial\Omega$. In the numerical implementation, x is placed not only on $\partial\Omega$, but also at a few points in the interior domain Ω, which leads to $c = 0$, and

$$0 = \int_{\partial\Omega} \left[G(x, x')\frac{\partial \phi(x')}{\partial n(x')} - \phi(x')\frac{\partial G(x, x')}{\partial n(x')} \right] dx', \quad x \in \Omega. \qquad (6.36)$$

The above is a null-field integral equation [164, 901, 902] as discussed in Section 1.3.3, and has been used to solve acoustic and electromagnetic problems through the T-Matrix method [650, 903]. In the CHIEF method, (6.35) and (6.36) are combined and solved by the least squares algorithm. A discussion of these methods and their implementation using the indirect boundary integral equation method can be found in Lee *et al.* [524].

We now examine the implications of the above techniques in the MFS. Similar to the integral equation scheme, Sánchez–Sesma as well as others [496, 766, 767] used the complex fundamental solutions in (2.41) in formulas like

$$\hat{\phi}_s(x) = \sum_{j=1}^{N} c_j\, G(x, x_j'), \quad x \in \overline{\Omega}_e, \ x_j' \in \partial\Omega', \qquad (6.37)$$

where the coefficients c_j are complex numbers, $\partial\Omega' \in \Omega$ is an auxiliary boundary in the interior of the scatterer. Alternatively, it was suggested [61] that a form like

$$G(x, x') = \mathrm{Y}_0(kr) + \mathrm{i}\,\eta \mathrm{J}_0(kr), \quad (2\mathrm{D}), \qquad (6.38)$$

for any η with $\Re(\eta) \neq 0$, can be used. For elastic waves, the complex fundamental solution of elastodynamics was used [495], and it was reported that spurious frequencies were eliminated [286, 644].

Lee *et al.* [524], however, demonstrated that in analogy to the integral equation case, spurious frequencies could still exist, even though complex fundamental solutions were used. In a CHIEF-like strategy, they modified (6.37) to

$$\hat{\phi}_s(x) = \sum_{j=1}^{N} c_j\, G(x, x_j') + \sum_{k=1}^{N_r} d_k\, G(x, x_k'),$$

$$x \in \overline{\Omega}_e, \ x_j' \in \partial\Omega' \subseteq \Omega, \ x_k' \in \Omega, \ x_k' \notin \partial\Omega'. \qquad (6.39)$$

In the above equation, in addition to deploying x_j' on an auxiliary contour $\partial\Omega' \in \Omega$, a few extra sources are placed at $x_k' \in \Omega$, deliberately away from $\partial\Omega'$, in order to break the regular geometric shape of the auxiliary contour. The overdetermined system is solved by least squares.

Also, Chen [151] and Zhang *et al.* [964] followed the Burton–Miller scheme for integral equations, and combined sources and dipoles to give the following formula

$$\hat{\phi}_s(x) = \sum_{j=1}^{N} c_j \left[\frac{\partial G(x, x_j')}{\partial n(x_j')} \pm \mathrm{i}\eta\, G(x, x_j') \right], \quad x \in \overline{\Omega}_e, \; x_j' \in \partial\Omega' \subseteq \Omega,$$

(6.40)

in which the constant η is chosen as the wave number k. A mathematical proof was provided for the case when $\partial\Omega'$ is a circle, to show that the solution is unique. Hence (6.39) and (6.40) provide two methodologies for the MFS solving exterior wave scattering problems without the spurious frequency issues.

■ Example 6.9

We solve an exterior domain problem governed by the Helmholtz equation in 2D. The boundary is the same as that defined in Example 5.7. A Dirichlet BC is prescribed using the function

$$\phi = \frac{\mathrm{H}_n^{(1)}(kr)}{\mathrm{H}_n^{(1)}(kr_o)} \cos(n\theta).$$

As the above expression is a metaharmonic function, it is also the exact solution of the BVP. For the current solution, we choose $r_o = 1$, $n = 0, 1, 2, 3, 4$, and k is the wave number to be varied. To get some idea of the solution, in Fig. 6.9 we plot the exact solutions for the cases of $k = 2.4$, and $n = 0$ and 4, and observe their wavy nature.

To solve the BVP, we distribute sources on a circle in the interior of the boundary, centered at the origin with radius $R = 1$. Thirty equally spaced sources are distributed on the circle. The same number of collocation points is selected on the boundary with equal angular increment. The problem is solved using both the standard MFS formula (6.37) and the modified formula (6.40). As the domain is infinite, to simplify the reporting of error, only errors on the boundary are checked. Due to the decaying nature of the solution, the error is anticipated to decrease outward.

As our purpose is to demonstrate the existence of spurious frequencies, and the effectiveness of their removal, the simulation will

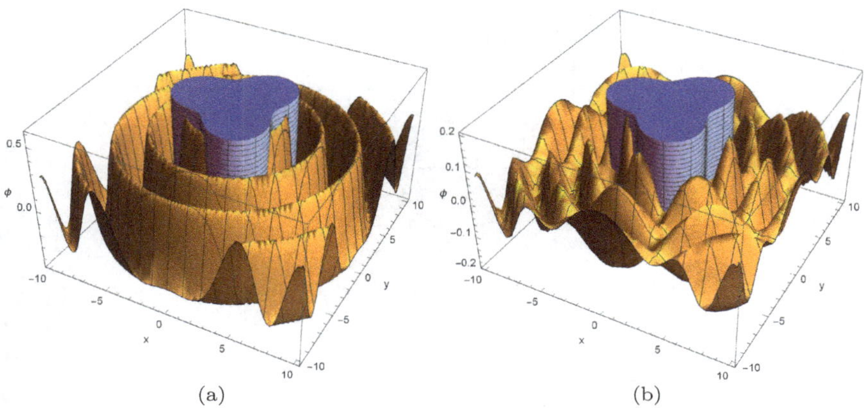

Figure 6.9. Plot of exact solution with $r_0 = 1$, $k = 2.4$, and (a) $n = 0$, (b) $n = 4$.

be performed near these frequencies. According to Lee *et al.* [524] and Zhang *et al.* [964], spurious frequencies can exist at the eigenfrequencies of the source circle, which are located at the zeros of the Bessel function of the first kind,

$$\mathrm{J}_n(kR) = 0; \quad n = 0, 1, 2, \ldots$$

The first few roots $j_{n,i}$ of the above are

$$j_{0,1} = 2.4048255557695773, \quad j_{1,1} = 3.8317059700207512,$$

$$j_{2,1} = 5.135622301840683, \quad j_{0,2} = 5.520078110286311,$$

where the first subscript corresponds to the order of the Bessel function, and the second is the sequence of roots.

We first test the case of $n = 0$ and $k = j_{0,1}$, using its full 16 digits. The result is shown as the first row of Table 6.7. We observe that the solution by the standard MFS breaks down, giving a huge response. Hence we have indeed found an artificial resonance at $j_{0,1}$. However, if we solve the problem using the modified method given by (6.40), an accurate solution is obtained. Next, we truncate the input of the wave number k to single precision, with 7 digits, and present the result in the second row. The standard method gives a maximum error of $\mathcal{O}(10^{-9})$, while the modified method gives $\mathcal{O}(10^{-14})$. The standard MFS result shows some influence of the spurious frequency if not treated. With a further truncation of k to 2.40, a mere 0.0048 away from the true root, we achieve near full accuracy with an error

Table 6.7. Maximum relative error at the boundary of the external domain Helmholtz problem, solved by the standard MFS and the modified MFS.

n	k	Standard MFS $\varepsilon_{\max}$	Modified MFS $\varepsilon_{\max}$
0	2.404825557695773	3.4×10^{1}	1.8×10^{-14}
0	2.404825	6.2×10^{-9}	1.6×10^{-14}
0	2.4	5.3×10^{-13}	2.0×10^{-14}
0	3.831705970207512	4.6×10^{-15}	2.0×10^{-14}
0	5.135622301840683	2.9×10^{-14}	5.7×10^{-14}
0	5.520078110286311	8.1×10^{0}	8.2×10^{-14}
1	2.404825557695773	6.9×10^{-15}	2.2×10^{-14}
2	2.404825557695773	3.5×10^{-15}	3.5×10^{-14}
3	2.404825557695773	3.5×10^{-15}	3.5×10^{-14}
4	2.404825557695773	4.6×10^{-13}	2.8×10^{-12}

Note: In the above, n is the Hankel function order in the exact solution, and k the wave number.

of $\mathcal{O}(10^{-13})$. These results indicate that the instability interval of the spurious frequency is extremely narrow. To find it, one more or less needs to hit it exactly with many digits of precision.

We continue to search the spurious frequencies using the 2nd, 3rd, … roots. Table 6.7 shows that the standard MFS is unstable only at $j_{0,2}$. We also test the higher order Hankel functions with $n = 1, 2, 3, 4$. No unstable behavior of the standard MFS was detected. We note that the modified MFS is always stable, free from spurious behavior.

Summary: Based on the above observations, we may summarize as follows:

- Similar to integral equations, distributing a fundamental solution that contains zeros on the boundary of a domain can create geometry-dependent eigenmodes, which manifest themselves as spurious frequencies in the numerical solution. For integral equations, the eigenmodes are associated with the problem geometry; while for the MFS, with the auxiliary boundary.
- For multiply connected domains, both internal and external auxiliary boundaries are needed, but only the internal auxiliary boundary can create the spurious frequencies.

- For the MFS, the instability interval of a spurious frequency appears to be very narrow. It may take some effort even to find it.
- Any of the integral equation inspired remedial methods, such as the CHIEF and the Burton–Miller formulation, can be used in the MFS to free it from the spurious frequency problem.

6.6   BVPs with Harmonic and Metaharmonic BCs

In discussing strategies for solving the Laplace equation by the MFS, we need to differentiate between two types of BVPs. In the first type, BCs are given by harmonic functions, and in the second, by nonharmonic functions. In the harmonic case, the function defining the BC satisfies the governing equation; hence the exact solution of the BVP is just the extension of that function. Obviously, such problems need not be solved. The reason that these problems are studied is because the exact solution allows us to analyze the accuracy and efficiency of a numerical method. Most examples studied in the literature for testing numerical methods belong to this type. BCs for real world problems, however, are dictated by the physical environment they are exposed to. It is unlikely that they happen to be harmonic functions. Similarly, we can make the same statement for solving Helmholtz-type equations with metaharmonic and nonmetaharmonic BCs, as well as other types of governing equations.

The reason that we distinguish between these two types of BCs is that they behave differently when solved by the MFS. Different solution strategies are needed. In the harmonic type, the best solution is obtained when sources are placed far away from the boundary, and the MFS is extremely efficient. A small number of terms can solve the problem to a near machine precision accuracy. In the nonharmonic type, on the other hand, sources need to stay close to the boundary. More terms are needed to achieve high accuracy, which can cause the condition number to be exceedingly large, and one must watch out for the stability of the system.

In this section, we shall investigate problems with harmonic BCs. Despite the fact that few real world problems are of this type, we shall still investigate them because most theoretical analyses were performed for this type. As exact solutions are available, the approximation error can be easily computed. The insight gained from such analyses can help the understanding of the nonharmonic BC case.

6.6.1 *Error estimate and condition number*

In Sections 1.6 and 5.3.2, we briefly demonstrated the high accuracy of the MFS, owing to its exponential error convergence; and in Section 5.4, we studied the challenge caused by a high condition number, leading to solution instability. We also discussed in Section 5.4.3 that the error and the condition number are closely linked: a smaller error inevitably leads to a larger condition number. In this section, we shall explore these issues further.

Error Estimate: For the Laplace equation, Bogomolny [94], Katsurada [477–479], among others [208, 489, 490, 542, 808], have proven density results for the MFS approximation, subject to various smoothness requirements of the boundary geometries and conditions, for interior and exterior domain problems, and for simply and multiply connected regions.

For interior domain problems, we may loosely state the error estimate obtained in various studies as follows: if the boundary is sufficiently smooth, and the solution has analytic continuation inside and outside the domain, covering the region enclosed by the auxiliary boundary, the approximation (6.31) has the following error estimate

$$\varepsilon \sim \mathcal{O}\left(\frac{\rho}{R}\right)^{N/n}, \tag{6.41}$$

where ρ describes the size of the domain, such as the minimum radius of a circle enclosing Ω, R is the radius of the auxiliary boundary enclosing Ω, with $R > \rho$, N is the number of sources, and $n = 1, 2,$ or 3, depending on the proofs by various authors. As $\rho/R < 1$, the error decreases exponentially with increasing N. The exponential convergence of (6.41) can be compared to the algebraic rate of the FEM and FDM (see Section 5.3.2)

$$\varepsilon \sim \mathcal{O}(h^{\ell}) \sim \mathcal{O}(N^{-\ell/d}), \tag{6.42}$$

in which h is the mesh size, ℓ is a power typically between 2 (quadratic convergence) and 3 (superconvergence), depending on the interpolation scheme, N is the number of nodes (degrees of freedom), and d is the spatial dimension. The ratio ℓ/d is between 1 and 3/2, hence the convergence is slow.

For exterior domain problems, the following error estimate is given

$$\varepsilon \sim \mathcal{O}\left(\frac{R}{\rho}\right)^{N}, \tag{6.43}$$

for $R < \rho$, which means that convergence can be achieve by taking $N \to \infty$, and the scheme becomes more efficient if R is made as small as possible.

Condition Number: The above presented powerful convergence property, however, is not without a check. Its performance can be limited by rapidly increasing condition numbers. Based on various studies [94, 208, 478, 490, 808], an estimate of the condition number for interior domain problems can be expressed as

$$\kappa \sim \mathcal{O}\left(\frac{N}{2} \left(\frac{R}{\rho} \right)^{N/2} \ln R \right), \tag{6.44}$$

for $R > \rho$. The condition number hence grows exponentially with N and R/ρ. For exterior domain problems, we have

$$\kappa \sim \mathcal{O}\left(\frac{\rho}{R} \right)^{N}, \tag{6.45}$$

for $R < \rho$.

In the pursuit of an ever higher accuracy, the solution system can become too sensitive to the roundoff of a finite precision computation. Such a phenomenon has been described by Schaback [775, 776] as reminiscent to the uncertainty principle of quantum mechanics—one cannot make both the error and the sensitivity (inverse of stability) small. For problems with harmonic BCs, however, solution stability is generally not an issue. We shall demonstrate in the examples below that the MFS is highly efficient for this type of problems, and near machine precision accuracy can often be obtained. This is however not the case for nonharmonic BVPs, as we shall observe in Section 6.7.

6.6.2 *Limit as $R \to \infty$*

An interesting observation about the error estimate (6.41) is the role played by the ratio R/ρ. In a computation using a fixed N, the error can decrease exponentially by increasing R. While the computational cost rapidly increases with increasing N, there is no extra cost in using a large R. Hence we should use an R as large as possible. However, we should note that although (6.41) indicates that $\varepsilon \to 0$ for either $N \to \infty$ or $R \to \infty$, the latter cannot be true. We may consider the fundamental principle that no matter how efficient the interpolants become by taking $R \to \infty$, a convergence to zero is not possible with a finite number of degrees of freedom. In fact, it has been shown that as $R \to \infty$, the fundamental solution interpolation matrix behaves like harmonic polynomials with the same number of degree of freedom [94, 162, 163, 779].

As R gets larger, the estimate (6.44) shows that the condition number grows rapidly; hence it is not practical to use too large an R. Nevertheless, it is of interest to explore its limiting behavior in this direction. As pointed out by Schaback [779], for 2D potential problems, when $R \to \infty$, the fundamental solution goes asymptotically to

$$\ln \|x - x'\| \to \ln \|x'\|, \quad \text{as } \|x'\| \to \infty, \tag{6.46}$$

in which x' is the source location, and $x \in \Omega$. Hence the interpolant becomes essentially "flat", and the elements of the interpolation matrix A are nearly identical to each other. The matrix naturally becomes singular. This effect is similar to the behavior of the MQ and GA basis functions in the RBFCM [198, 201, 307, 780], in which the flattening is accomplished by adjusting the shape parameters (see Section 3.3). In both cases, as basis functions become more flat, the condition number increases, and the error decreases, until the computational instability caused by the roundoff error sets in.

In an attempt to alleviate the singularity of the matrix, Schaback [779] suggested the expansion

$$\ln \|x - x'\| \simeq \ln \|x'\| - \frac{xx' + yy'}{\|x'\|^2} + \cdots, \quad \text{as } (x, y)/R \to 0. \tag{6.47}$$

We may continue the above expansion for more terms of harmonic polynomials in (x, y). However, as commented by Schaback [779], in that case, we might as well directly use harmonic polynomials for the interpolation, which yields the TCM. Hence we shall use only the leading term to alleviate the matrix singularity and modify the MFS approximation to

$$\hat{\phi}(x) = c_0 + \sum_{j=1}^{N} c_j \left(\ln \|x - x'_j\| - \ln \|x'_j\| \right)$$

$$= c_0 + \sum_{j=1}^{N} c_j \ln \frac{\|x - x'_j\|}{\|x'_j\|}. \tag{6.48}$$

In the above, the coefficient c_0 is added to compensate for the lack of a constant term in the expansion. We notice that each term on the right-hand side of (6.48) satisfies the Laplace equation, and this new formula amounts to the rescaling of the fundamental solution by its distance to the origin.

By the same token, for 3D potential problems, we may modify the approximation formula to

$$\hat{\phi}(\boldsymbol{x}) = \sum_{j=1}^{N} c_j \left(\frac{1}{\|\boldsymbol{x} - \boldsymbol{x}'_j\|} - \frac{1}{\|\boldsymbol{x}'_j\|} \right). \tag{6.49}$$

These two formulas, (6.48) and (6.49), are tested in the following examples against the traditional ones without removing the asymptotic values. We find that the modified formulas perform only marginally better than the traditional ones. We observe little improvement in the condition number, but a slight increase in accuracy. In cases with a large N, the modified formula can extend the flat part of the error curve to a slightly larger R before the instability sets in. In the following examples involving harmonic BCs, all reported results are based on the modified formulas.

Before leaving this subject, we should also mention that there have been efforts to show that the limit to the approximation accuracy imposed by the "uncertainty principle" is not necessarily true. For the RBFCM, it was demonstrated that the use of equivalent bases that span the same space as the flattened RBFs can create a well-conditioned system [306, 307, 921]. A similar concept was extended to the MFS in 2D to remove the ill-conditioning caused by $R \to \infty$ [33]. Such a technique, however, is not tested here.

6.6.3 *2D interior potential problem*

We shall examine the application of the MFS to a 2D Laplace BVP for its convergence behavior with respect to the number of degrees of freedom, and the placement of the sources.

■ **Example 6.10**

We examine the same mixed BVP solved in Example 1.1. The purpose here is to test the error estimate (6.41) in relation to N and R.

We first run cases by fixing N and varying R. The unit square domain can be fitted by a circle of $R = \sqrt{2}/2 \approx 0.707$. We hence start the simulation using a source circle close to the boundary with $R = 0.8$, and progressively increase it to 50. Each side of

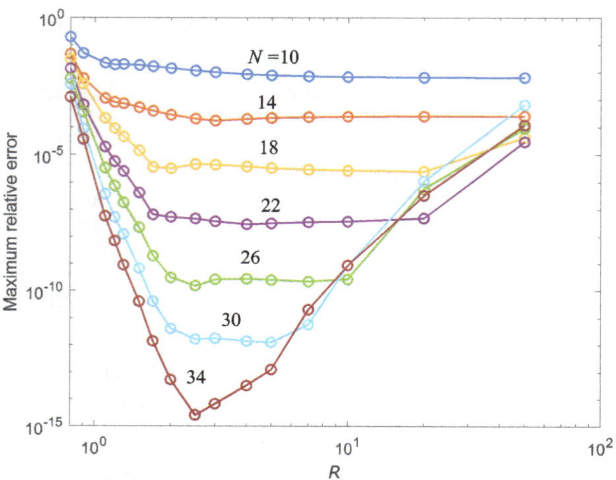

Figure 6.10. Plot of maximum relative error $\varepsilon_{\max}$ versus source circle radius R for various source number N, in log–log scale.

the square is divided into 2, 3, …, up to 8 equal parts, resulting in $N = 10, 14, \ldots, 34$. (Here we recall that there are two corner nodes with dual boundary conditions.) In Fig. 6.10, we plot the curves of $\varepsilon_{\max}$ versus R for various N in log-log scale. Based on (6.41), such curves should appear as straight lines.

We first examine the two curves corresponding to $N = 10$ and 14. At $R = 0.8$, the maximum relative errors are of $\mathcal{O}(10^{-1})$, and are unacceptable. When R is increased to 1.2, the errors drop at smaller slopes to 2% and 0.09%, respectively. These errors are already acceptable. Further increasing R, however, has minimal effect, and the curves flatten out. This is because the approximations have reached their polynomial limits.

For cases with $N = 18$ to 30, the errors drop at ever steeper slopes, as predicted by (6.41). These curves also flatten out after certain R values, reaching their approximation limits. Further increasing R causes the errors to turn upward, due to the exceedingly large condition numbers. For the last curve, $N = 34$, it seems that the solution becomes unstable before the approximation limit is reached. The optimal result is of $\mathcal{O}(10^{-15})$, near the machine precision of 10^{-16}.

Next, we examine cases of fixing R and varying N. With R fixed at 0.8, 0.9, 1.05, 1.2, 1.5, 2 and 3, N is increased from 10 to 42. Figure 6.11 shows $\varepsilon_{\max}$ versus N in semi-log scale. As predicted by

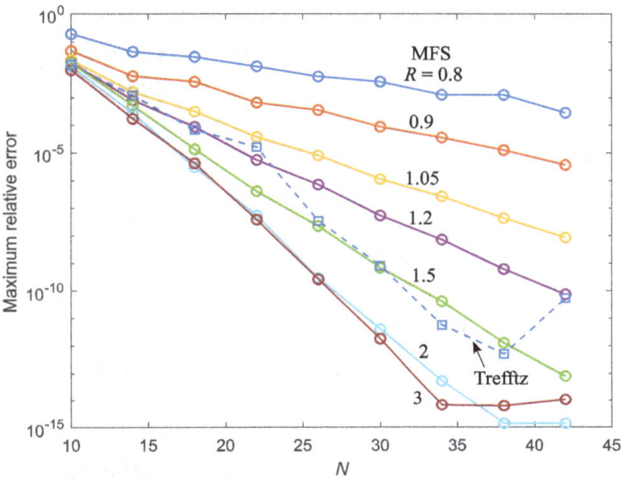

Figure 6.11. Plot of the maximum relative error $\varepsilon_{\max}$ versus the number of terms N in the approximation, for various source circle radii R of the MFS (in solid lines and symbols $\circ$), and for the TCM (in dashed line and symbols $\square$), in semi-log scale.

(6.41), these curves appear as straight lines. The slopes of these lines steepen with increasing R. Hence it is important to place the sources at a large R. However, we also observe that the effect of increasing R wears off beyond $R = 2$. The curve for $R = 3$ does not improve, and in fact breaks down when N becomes too large.

6.6.4 *Trefftz collocation method*

As commented in Section 6.2, for the MFS, as $R \to \infty$, the approximation approaches the sum of harmonic polynomials [94]. As the accuracy of the MFS improves with increasing R, this may suggest that the MFS can be no better than the TCM. Schaback [778, 781] further suggested that if the solution is a harmonic function that can be efficiently expanded into a polynomial, it is a "crime" not to solve it by harmonic polynomials, that is, by the TCM. Hence we shall test the TCM for its performance.

■ **Example 6.11**

We solve the same problem in Example 6.10, but by the TCM, instead of the MFS.

The TCM does not need an auxiliary boundary, and only collocation nodes are needed. We solve the problem using the same collocation nodes as the MFS, including the dual BC corner nodes. The maximum relative error is plotted in Fig. 6.11 together with the MFS results. We observe that the TCM has a similar exponential convergence. Its performance, however, is about two orders of magnitude inferior to the $R = 2$ case, despite the theoretical prediction. It also breaks down earlier than the MFS with a large N, due to a larger condition number. Nevertheless, the TCM, like the MFS, is highly effective in solving problems with harmonic BCs.

The above example may suggest that the TCM is a better choice than the MFS. We observe that although the MFS can deliver a better result if an optimal auxiliary boundary is found, it also can give a worse result if the radius R is not properly chosen. The TCM, on the other hand, is free from such uncertainty and gives a good result. However, such a statement can be deceiving, as the above function belongs to a "nice" category that can be expanded into a power series; that is, a polynomial. In the following example, we shall test a different function, which cannot be expanded into a polynomial.

■ Example 6.12

We solve the Laplace equation over the square domain $\Omega = (0, 1) \times (0, 1)$, with a Dirichlet BC prescribed by the function

$$\phi = \Re\{z^{3/2}\cos(z)\}$$

$$= (x^2 + y^2)^{3/4}\left[\cos\left(\frac{3}{2}\tan^{-1}\frac{y}{x}\right)\cos x \cosh y\right.$$

$$\left. + \sin\left(\frac{3}{2}\tan^{-1}\frac{y}{x}\right)\sin x \sinh y\right],$$

in which $z = x + iy$ is the complex variable, and $\Re$ denotes the real part. The above function is harmonic, and gives the exact solution of the BVP. It is, however, not analytic everywhere: at the origin $(0, 0)$, it is only once differentiable. Moreover, the function cannot be expanded into a power series.

Table 6.8. Comparison of error by the MFS and the TCM.

Method	N	R	ε_{max}	ε_{rms}	κ
MFS	40	1.1	6.7×10^{-6}	2.2×10^{-4}	3.7×10^{6}
MFS	80	1.1	9.9×10^{-7}	6.6×10^{-5}	1.0×10^{13}
MFS	160	1.1	3.3×10^{-7}	3.4×10^{-5}	1.7×10^{18}
TCM	40	—	1.5×10^{-3}	3.6×10^{-2}	3.9×10^{11}
TCM	80	—	4.1×10^{-3}	1.7×10^{-1}	1.0×10^{22}
TCM	160	—	1.0×10^{-2}	4.1×10^{-1}	8.3×10^{22}

We first solve the problem by the MFS. The sources are distributed over an auxiliary circle of $R = 1.1$. The results of using 10, 20, and 30 collocation nodes per side of the square are shown in Table 6.8. We observe that the solution is more difficult to approximate. More collocation nodes are needed and we cannot reach the ultra high accuracy of the preceding example. The maximum error occurs near the origin due to the local discontinuity. The RMS errors are generally two orders of magnitude smaller than the maximum.

We next solve the same problem by the TCM using the same collocation nodes. The results are also reported in Table 6.8. We observe that the TCM gives inferior results, and in fact becomes unstable with the increase of collocation nodes, due to the extremely high condition numbers. This example shows that the TCM, being polynomial based, has difficulty in coping with functions that are not amenable to a power series expansion.

We should comment that in solving the above problem, the maximum error is concentrated near the corner $(0, 0)$, where the solution has a discontinuity in its second derivatives. Errors elsewhere are generally 4 to 5 orders of magnitude smaller. In the above MFS simulation, sources and collocation nodes are uniformly distributed, without an effort to concentrate them near the corner. Like any other numerical method, a refinement near the steep variation can improve the accuracy. This is, however, not attempted here.

6.6.5 *3D interior potential problem*

In the following example, we solve a 3D mixed BVP with a harmonic BC to demonstrate the exponential convergence condition predicted in (6.41).

■ Example 6.13

The boundary of a 3D domain is given by the "bumpy sphere" defined by the following parametric equations:

$$x = \rho \sin\theta \cos\varphi, \quad y = \rho \sin\theta \sin\varphi, \quad z = \rho \cos\theta,$$

$$0 \le \varphi < 2\pi, \quad 0 \le \theta < \pi,$$

in which φ and θ are respectively the azimuth and polar angle, and

$$\rho(\theta, \varphi) = 1 + \frac{1}{6}\sin(7\theta)\sin(6\varphi).$$

The geometry is shown in Fig. 6.12(a). A mixed BVP is created based on the harmonic function

$$\phi = \sin x \cosh\sqrt{5}\, y \sin 2z.$$

On the left side of the boundary, $x \le 0$, a Dirichlet BC is prescribed; while on the right side, $x > 0$, a Neumann BC is applied; both are derived from the above exact solution.

Following the lead of 2D problems, the 3D problem is solved using an auxiliary sphere of radius R enclosing the domain, and the approximation formula (6.49). Sources and collocation points are uniformly

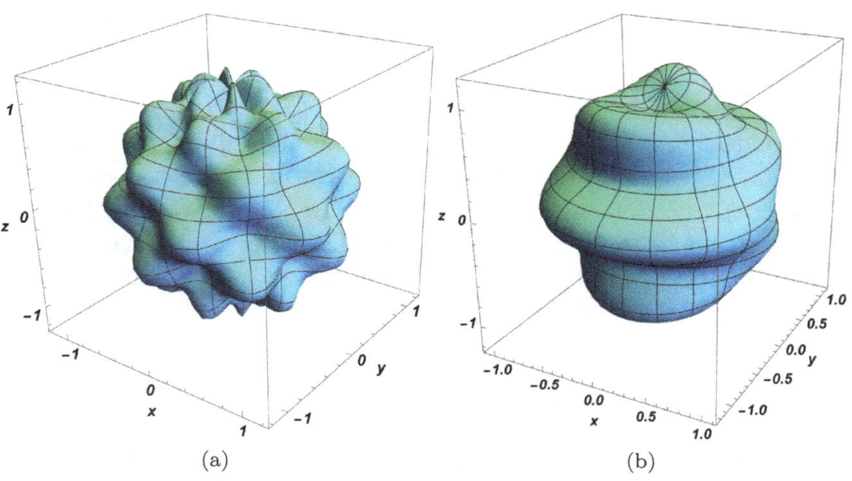

(a)   (b)

Figure 6.12. Geometry of bumpy spheres. (a) A bumpy sphere; and (b) a simplified bumpy sphere.

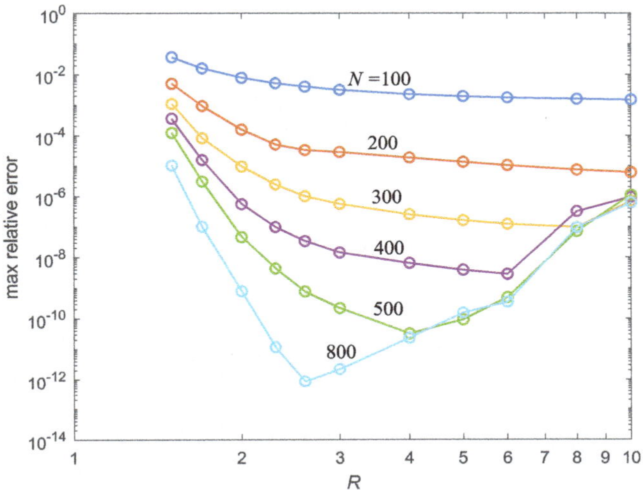

Figure 6.13. Plot of maximum relative error $\varepsilon_{\max}$ versus source sphere radius R for various source number N, in log–log scale.

distributed over the surfaces based on polar and azimuth angles. First we run cases with fixed N by varying R. Curves of the maximum relative error $\varepsilon_{\max}$ versus R are plotted in Fig. 6.13 in a log–log scale for N values ranging from 100 to 800. We observe that due to the increase in dimensionality from an essentially 1D to a 2D mesh, a larger number of sources are needed.

The simulations begin with $R = 1.5$, close to the boundary (the size of the domain is $\rho = 7/6 \approx 1.17$), and increased to 10. For smaller numbers of nodes, $N = 100, 200,$ and 300, we observe that errors decrease with increasing R, and approach asymptotic values when R becomes large. For larger N, we observe that errors drop off sharply, but become unstable before asymptotic values are reached, due to exceedingly large condition numbers. For the case $N = 800$, the drop is rather dramatic, and the best result is of $\mathcal{O}(10^{-12})$, before the instability sets in.

In Fig. 6.14, we plot the maximum relative error $\varepsilon_{\max}$ versus N for various R values in semi-log scale. We observe that errors drop in nearly straight lines, as predicted by the error estimate. The slopes of these lines increase with increasing R. However, the effectiveness of R diminishes after it reaches 2.5. Too large an R causes an early onset of instability when N is increased.

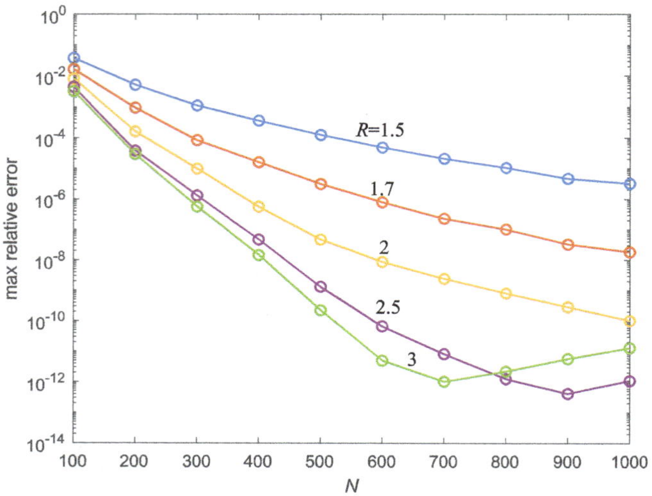

Figure 6.14. Plot of maximum relative error $\varepsilon_{\max}$ versus source number N for various R, in semi-log scale.

6.6.6 *Helmholtz-type equations*

The above examples concern the Laplace equation. Similar convergence and stability properties exist for the Helmholtz equation with metaharmonic BCs. For Helmholtz equation exterior domain problems, the following error estimate was presented [205, 685]

$$\varepsilon \sim \mathcal{O}\left(\frac{R}{\rho}\right)^N, \quad \text{for } \rho > R. \tag{6.50}$$

For interior domain problems, the opposite is true [61]

$$\varepsilon \sim \mathcal{O}\left(\frac{\rho}{R}\right)^N, \quad \text{for } R > \rho. \tag{6.51}$$

For modified Helmholtz equation interior domain problems, the following error estimate:

$$\varepsilon \sim \mathcal{O}\left(N^2 \left(\frac{\rho}{R}\right)^N\right), \quad \text{for } R > \rho, \tag{6.52}$$

and condition number estimate

$$\kappa \sim \mathcal{O}\left(N^2 \left(\frac{R}{\rho}\right)^{N/2}\right), \quad \text{for } R > \rho, \tag{6.53}$$

were given [835].

The accuracy of the MFS for solving problems with metaharmonic BCs has been demonstrated in Examples 5.7 and 6.9 for the exterior domain Helmholtz equation, and 6.2 and 6.4 for the interior domain Helmholtz and modified Helmholtz equation, respectively. In the following, we shall test the above suggested exponential convergence with respect to N and R using an interior Helmholtz BVP example.

■ Example 6.14

We shall solve the Helmholtz equation in the interior domain bounded by the Cassini curve defined in Example 6.2. A Dirichlet BC is prescribed using the metaharmonic function given in Example 6.4. To increase the complexity of the BC, the following set of parameters are used: $k = 10$, $c = 5$, $a = 1$, and $b = 2$. Figure 6.15 shows the variation of the BC on the curve as a function of θ.

The problem is solved by the MFS using as an auxiliary boundary a circle with the radius R; while N source points are uniformly distributed and slightly rotated to avoid symmetry. For a comparison, the problem is also solved using the TCM. The results for a range of N and R values are presented in Fig. 6.16.

Comparing these results to those of Fig. 6.11 of Examples 6.10 and 6.11 for the Laplace equation, we observe somewhat different behaviors. First, we observe that at a low number of sources, $N = 40$, we obtain acceptable results with $\varepsilon_{\max}$ of $\mathcal{O}(10^{-3})$. The errors, however, are insensitive to the radius R, ranging from 2 to 4 (the size of the

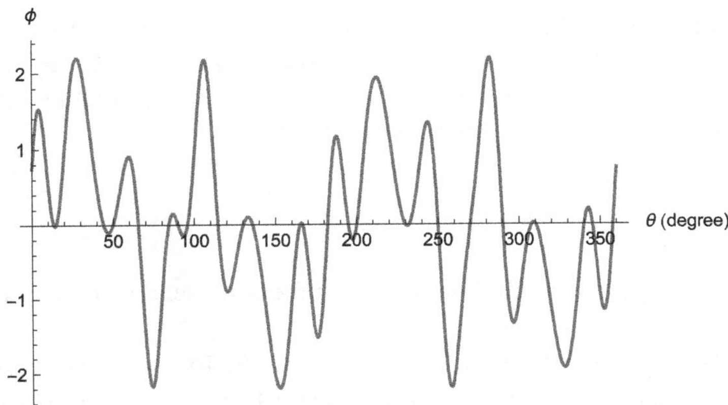

Figure 6.15. Metaharmonic BC for the Helmholtz equation on the Cassini curve as a function of θ.

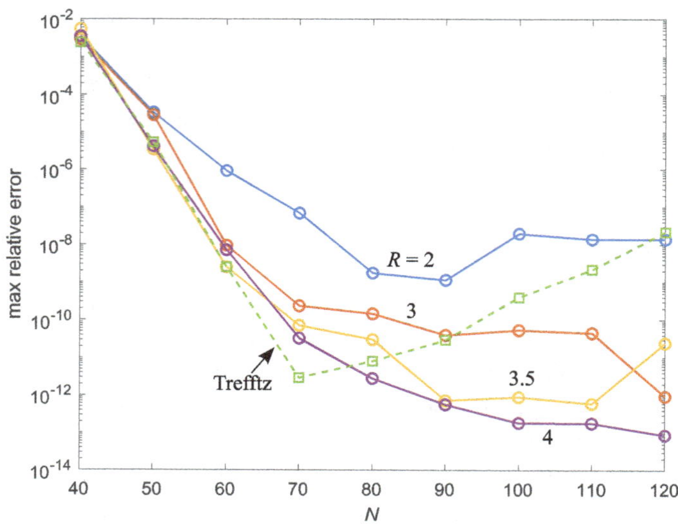

Figure 6.16. Plot of maximum relative error $\varepsilon_{\max}$ versus source number N for various R, in semi-log scale.

domain is $\rho = 1.42$). For larger N values, increasing R significantly reduces the error. Similar to the Laplace equation, the effect of R wears off when it is greater than 4; hence it is not beneficial to use too large an R. It is curious that for smaller R, say $R = 2$, the result becomes unstable beyond $N = 90$. The case of $R = 4$ appears to be stable at large N. The curve, though, can no longer sustain a straight line drop.

We also examine the results obtained by the TCM, using the set of general solutions in (3.7a). It seems that the TCM gives the best result of the MFS at large R, without the need of choosing an R. However, it becomes unstable at larger N and cannot reach the smallest error obtained by the MFS, which is of $\mathcal{O}(10^{-13})$.

6.6.7 *Summary*

We may summarize the findings based on the above examples as follows:

- For the Laplace, Helmholtz-type, and other PDEs, when the prescribed BC satisfies the governing equation, the BVP can be efficiently solved to an accuracy near the machine precision by the MFS.

- When such BCs can be expanded into power series for the Laplace equation, and general solutions for other equations, the TCM utilizing general solutions as basis functions can be the best method. It can deliver among the best results of the MFS, yet without the need of searching for an optimal auxiliary boundary.
- When the BC cannot be expanded into a summation of general solutions, the TCM becomes inefficient, and the MFS is preferred. The MFS has the ability of not only adjusting the source distances, but also using scattered distributions to concentrate near the part of the boundary where the solution is not fully smooth.

6.7 BVPs with Non-harmonic and Non-metaharmonic BCs

In Section 6.6 we have demonstrated the efficiency of the MFS for solving harmonic and metaharmonic BVPs by placing the sources far away from the solution boundary. We also commented that behaviors of the non-harmonic and nonmetaharmonic BVPs are different, and different strategies are needed. The root of such differences can be traced to the potential theory [481], which states that given a Dirichlet BC and a boundary of certain smoothness, there exist a unique interior domain solution and a unique exterior domain solution. These two solutions have C^0 continuity, because they share the same BC; but their normal derivatives generally have a jump. By deploying fundamental solutions on an auxiliary boundary enclosing the solution domain and boundary, the enclosed field is C^∞, which is not fully consistent with the sought for solution. It becomes more difficult to approximate the solution if the continuous field becomes ever larger. Hence the sources should be placed as near the boundary as possible, yet without letting the singularities affect the accuracy. The situation, however, changes if the BC is given by a harmonic function that has analytic continuation inside and outside the domain. In that case, pushing sources away to create a large continuous field becomes beneficial, which is what was observed in the preceding section.

6.7.1 *Maximum principle*

Before solving nonharmonic BVPs, we need to first address the issue of how to estimate the error of a numerical solution, when the exact solution is not available. One way to estimate the error is by checking on the residuals;

that is, the error committed against the governing equation and BC of the BVP. Methods like the FEM generally rely on estimating the residuals in the domain to perform an *a posteriori* error analysis [8, 51, 70, 482]. However, we should stress that a residual error is not the approximation error. It can give the trend of convergence, but does not give the actual magnitude of the approximation error.

For the MFS, the fundamental solution exactly satisfies the governing equation, so there is no residual error in the domain, and only the boundary error needs to be examined. The ability to check the residual of boundary values is unique to collocation methods, such as the MFS, TCM, and RBFCM, as compared to methods like the FEM and FDM. Consider a finite element adjacent to the boundary. For the element face that is on the boundary, the BC is enforced on the nodes, and the boundary value at any point on the face is interpolated using shape functions—linear, quadratic, etc. The boundary error is determined *a priori* by the discretization and the interpolation scheme, although an *a posteriori* "smoothing" of nodal values is sometimes performed [975, 976].

For collocation methods, once the unknown coefficients are determined, the approximate solution, as well as its derivatives, can be evaluated at any point in the domain and on the boundary. Hence the error can be found on the boundary. Once we obtain the boundary error, the next question is, how does it relate to the overall error? For this, we can invoke the maximum principle.

The maximum principle of a PDE, when it exists, states that the maximum of the solution occurs at the boundary [330, 733]. For the MFS, since both the exact solution and the approximate solution satisfy the PDE, their difference, that is, the error, also satisfies the PDE. The maximum principle then applies to the error, and the maximum error occurs at the boundary. Hence by checking the boundary error, we can determine the maximum error of the approximation.

It has been shown that the maximum principle exists for the following elliptic operators: [733]

$$\mathcal{L}\phi(x) \equiv a_{ij}(x)\frac{\partial^2 \phi}{\partial x_i \partial x_j} + b_i(x)\frac{\partial \phi}{\partial x_i} - c(x)\phi \geq 0,$$

$$\text{for } c(x) \geq 0, \quad i, j = 1, 2, 3. \tag{6.54}$$

In the above, a_{ij} is a symmetric positive definite tensor, ensuring that the operator is elliptic. The condition $c \geq 0$ means that the maximum principle exists for the Laplace and modified Helmholtz equation, but not the Helmholtz equation. A maximum principle is also found for the biharmonic [275, 721] and polyharmonic [722] operators, but may not for the Cauchy–Navier equations of elasticity [668, 815].

For the Helmholtz equation, as the solutions are waves that oscillate, it is natural that the maximum principle does not exist. However, for exterior domain Helmholtz problems, although the solutions are wavy, their amplitudes decay outwardly. Hence the maximum amplitude is dictated by the Dirichlet BC. As the boundary error satisfies the Helmholtz equation as well, we anticipate that the maximum error amplitude occurs on the boundary [551, 964]. Hence we have a similar way of assessing the overall error by information gathered on the boundary.

As mentioned above, most problems tested in various studies are based on harmonic BCs, because the exact solution is needed to evaluate the error. The difference in behavior between the harmonic and nonharmonic types in the MFS went largely unnoticed. Schaback [779] appears to be the first to have raised such caution in solving piecewise harmonic and nonharmonic BC problems. Drombosky *et al.* [271] investigated these problems for their effect on effective condition numbers. Only relatively recently the treatment of the nonharmonic type by the MFS was systematically investigated for the optimal placement of the auxiliary boundary [145, 367, 538, 604, 686, 715]. It was demonstrated that the sources should stay near, rather than far away, from the boundary. In the following, we shall examine the different strategies needed for solving problems of this kind.

6.7.2 *Overdetermined MFS by least squares*

Up to this point, the MFS has been applied based on the strong formulation. By selecting the same number of collocation nodes as sources, the BC can be exactly satisfied at the nodes, and the solution matrix is square. It is however possible to apply the MFS as an optimization problem by minimizing the error in a certain sense, similar to the weak formulation. This is generally done by choosing more collocation nodes than sources to create an overdetermined system. The solution matrix is then rectangular in shape. As there is no exact solution to this system, a linear LS procedure is applied to find an optimal solution. A description is given as follows.

If we choose N sources and M collocation nodes, the MFS discretization (1.49), (1.51), and (1.52) leads to a matrix system

$$\boldsymbol{A}\boldsymbol{c} = \boldsymbol{b}, \tag{6.55}$$

where A is an $M \times N$ matrix with $M > N$, and c and b are respectively $N \times 1$ and $M \times 1$ column vectors. The LS solution can be obtained through the normal equations

$$c = (A^T A)^{-1} A^T b. \tag{6.56}$$

However, a direct solution based on the above formula is generally undesirable, because the condition number of the matrix $(A^T A)$ is κ^2, where κ is the condition number of A. As A is a dense matrix and typically has a high condition number, there can be concern for instability. Instead, the system is usually solved by applying the QR factorization. In MATLAB® [631], the command `alpha = A\b` automatically detects a non-square matrix and solves it by the QR algorithm.

We should mention that one of the earliest implementations of the MFS was formulated as a nonlinear LS problem, in which not only the source strengths c_j, but also the source locations x'_j, are unknowns [630]. The system was solved as an optimization problem using for example the gradient search. A number of subsequent MFS implementations followed this approach [430, 443, 445]. However, the computational cost is generally high, and the procedure is more complex. This approach is largely out of favor for straightforward MFS problems [145]. It can still be useful for solving nonlinear, optimization, and inverse problems, in which iterations are needed.

As declared by Schaback's uncertainty principle (see Section 5.4.3), the pursuit of high accuracy leads to a high condition number. One way to overcome the instability caused by high condition numbers is to formulate the problem as an overdetermined system and solve it by LS. Other methods like the singular value decomposition (SVD) [489], damped and truncated SVD [414, 744], and Tikhonov regularization [557, 886, 939] have been used to stabilize ill-conditioned systems. However, it has been demonstrated [140, 568] that these techniques are no more reliable than the regular Gaussian elimination, and they do not improve the accuracy of the solution.

LS has been used in solving inverse and ill-posed problems by the MFS [458]. Such problems, hence the resulting solution systems, are inherently unstable. An overdetermined system, together with regularization techniques, can help stabilize the numerical solution and produce a satisfactory approximation. For these and other applications, we should recognize two types of problems and instabilities. In solving inverse and ill-posed problems, the problems themselves are inherently unstable. Without treatment, we cannot obtain stable solutions, and stabilization techniques are essential. The well-posed direct problems, on the other hand, are stable. The instability in the numerical solution is caused by the method of approximation and the finite precision of the computing machinery, which truncate the information needed for accuracy. Stabilization may allow the system be solved, but it cannot recover the accuracy information that is already lost.

In solving nonharmonic BC problems using the collocation based MFS, we face another issue—by forcing the error to go through zero at collocation nodes, the approximated BC can have large fluctuations between adjacent nodes, similar to the Runge phenomenon observed in Example 3.1. In the same example, we have also demonstrated that the use of LS can much reduce the fluctuation. In the examples below, we shall show that LS can stabilize the MFS solution and reduce the error of nonharmonic BC problems.

We should clarify that the error fluctuation also exists in harmonic BC problems solved by the collocation method. However, as demonstrated in Section 6.6, the MFS is highly efficient in approximating harmonic functions, and problems can be solved to an accuracy near the machine precision using relatively few degrees of freedom. Additional smoothing of the error is not needed. This is generally not the case for nonharmonic BC problems, and the use of LS can be essential.

The use of LS is not without a cost. We can judge the cost by the number of arithmetic operations needed to solve an $M \times N$ matrix system to produce the LS solution [248]. If we solve the matrix system using the normal equations (6.56) by the Cholesky factorization, the operation count is of the order $MN^2 + N^3/3$ by neglecting order N^2 and lower terms. On the other hand, if we employ the QR factorization, the count is $2MN^2 - 2N^3/3$, and if we solve it by the SVD, the count is $4MN^2 - 4N^3/3$. For example, if we select $M = 2N$, the respective counts are $7N^3/3$, $10N^3/3$, and $20N^3/3$. For a square matrix $N \times N$, the above counts reduce to $4N^3/3$, $4N^3/3$, and $8N^3/3$, which can be compared to the count $2N^3/3$ by Gaussian elimination and LU factorization. For the LS case, we notice that solving it by Cholesky factorization with the normal equations has the least cost, and by the SVD the most. As commented above, due to the potential instability associated with the normal equations, generally the QR method is used, as in the following examples.

■ Example 6.15

We shall solve BVPs for the Laplace equation by the MFS to examine the effectiveness of the LS algorithm. The solution domain is the interior bounded by the Cassini curve given in Example 6.2. We shall solve two problems defined by different BCs. For the first, a Dirichlet BC is prescribed using the following function:

$$\phi = x^2 y^3,$$

and for the second, the function

$$\phi = x^2 y^3 - \frac{x^4 y}{2} - \frac{y^5}{10},$$

is used.

These two BCs are plotted on the Cassini curve as functions of θ in Fig. 6.17. We observe that these two functions are of similar complexity. A critical difference, however, is that the first function is nonharmonic, while the second is harmonic. We shall examine the harmonic case first.

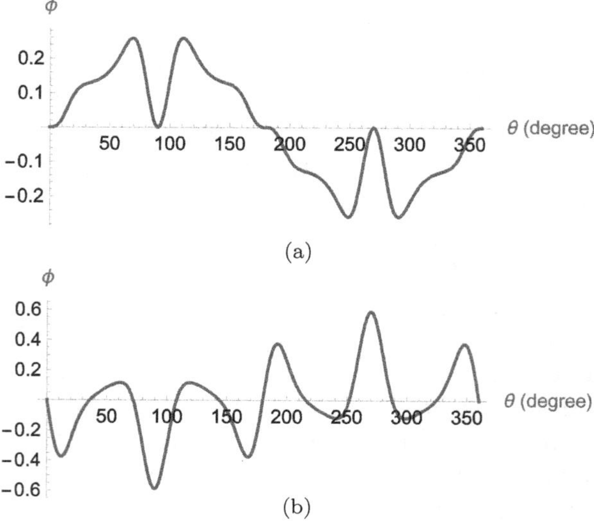

(a)

(b)

Figure 6.17. Boundary conditions on the Cassini curve as function of θ, for the (a) non-harmonic BVP, and (b) harmonic BVP.

For the harmonic BC problem, we distribute 30 sources on the contour of a circle with normalized radius $R/\rho = 2$. The same number of collocation points are placed on the boundary. These nodes are distributed with equal increment of θ. The errors are sampled on a dense set of points on the boundary. The maximum relative error ε_{max} and RMS relative error ε_{rms}, and the condition number κ, are presented in the first row of Table 6.9. We observe the high accuracy of the approximation, with a maximum relative error of $\mathcal{O}(10^{-7})$.

We next solve the problem using the same set of sources, $N = 30$, but the collocation nodes are doubled and tripled to $M = 60$ and 90. The solutions by the LS are reported in the second and third row of Table 6.9. We find that by doubling the collocation nodes, ε_{max} is cut to 1/3, and ε_{rms} to 1/2. The condition number is also reduced. Further increasing the collocation nodes to $M = 3N$ and higher, however, has minimum to no effect in reducing the error.

Based on the above discussion on operation counts, the use of an overdetermined system increases the computational cost to five-fold and eight-fold, respectively, using the QR factorization for $M = 2N$ and 3N, as compared to the case of $M = N$ using Gaussian elimination. As an alternative method for reducing the error, we may simply increase N from 30 to 32, and use $M = N$. The result is shown in row 4. We observe that it is slightly better than the overdetermined

Table 6.9. Results of the MFS solution of harmonic and nonharmonic BC problems, with and without the LS enhancement.

Harmonic (Y/N)	Auxiliary boundary	R/ρ or η	N	M	ε_{max}	ε_{rms}	κ
Y	Circle	2	30	30	1.9×10^{-7}	4.0×10^{-8}	2.9×10^{7}
Y	Circle	2	30	60	6.2×10^{-8}	2.2×10^{-8}	4.9×10^{6}
Y	Circle	2	30	90	5.9×10^{-8}	2.1×10^{-8}	4.9×10^{6}
Y	Circle	2	32	32	3.3×10^{-8}	8.4×10^{-9}	8.8×10^{7}
Y	Circle	2	50	50	1.5×10^{-12}	3.4×10^{-13}	2.0×10^{12}
N	Dilated	1.1	50	50	1.4×10^{-2}	2.7×10^{-3}	1.6×10^{2}
N	Dilated	1.1	50	100	7.5×10^{-3}	1.9×10^{-3}	1.4×10^{2}
N	Dilated	1.1	50	150	7.3×10^{-3}	1.9×10^{-3}	1.4×10^{2}
N	Dilated	1.1	85	85	1.1×10^{-3}	2.1×10^{-4}	7.0×10^{4}
N	Dilated	1.5	50	50	1.2×10^{-1}	2.7×10^{-2}	3.9×10^{7}
N	Dilated	1.5	50	100	1.6×10^{-3}	8.3×10^{-4}	7.1×10^{5}

cases. The computational cost, according to the operation counts, is only 21% higher than the $N = M = 30$ case. Furthermore, we may increase the number of nodes to $N = M = 50$, which increases the computational cost to $(50/30)^3 = 4.6$ folds, which is about the same as the $N = 30$ and $M = 60$ case. We then achieve errors in the $\mathcal{O}(10^{-12})$ range, as shown in Table 6.9. Hence, due to the superior exponential convergence rate with respect to N, it is generally not worthwhile to use an overdetermined system for harmonic BC problems.

Next we examine the nonharmonic BC case. As the sources generally need to stay close to the boundary, it is more effective to use an auxiliary boundary that is similar to the original one, rather than a circle. Hence we define a dilated boundary using

$$r'(\theta) = \eta r(\theta),$$

in the parametric equation of the Cassini curve for the distribution of sources, in which $\eta > 1$ is a dilation factor.

We shall start the simulation using $N = 50$, $\eta = 1.1$, and $M = 50$, 100, and 150. The results are presented in the first three rows of the lower part of Table 6.9. Comparing with the upper part, we observe that the accuracy for the nonharmonic cases is much inferior to that of the harmonic ones, even though more collocation points are used. Without the LS, for $N = M = 50$, the $\varepsilon_{\max}$ is 1.4×10^{-2}. Doubling M to 100 for a LS solution reduces the error by half, to 7.5×10^{-3}. Increasing M to 150 has no further effect.

In Fig. 6.18(a) we plot the relative error on the boundary between $0 \leq \theta \leq \pi/2$ for the two cases, $M = 50$ and 100. We observe that for $M = 50$ (solid line, no LS), the error passes through zero at each collocation point, but fluctuates in between. For $M = 100$ (dashed line, with LS), the error curve does not pass through zero at collocation points, but the fluctuation is somewhat dampened. This shows that the LS is effective in reducing the error. However, the LS procedure increases the CPU time. For a comparison, we run the case without LS using $N = M = 85$, which requires about the same number of operations as the $N = 50$ and $M = 100$ case. From the fourth row of the lower part of Table 6.9, we find that the result is better. Hence the LS is not necessarily the first choice in an attempt to reduce the error.

In the next test, we increase the size of the dilation factor to $\eta = 1.5$. Using $N = M = 50$, the result is also shown in Table 6.9, and plotted

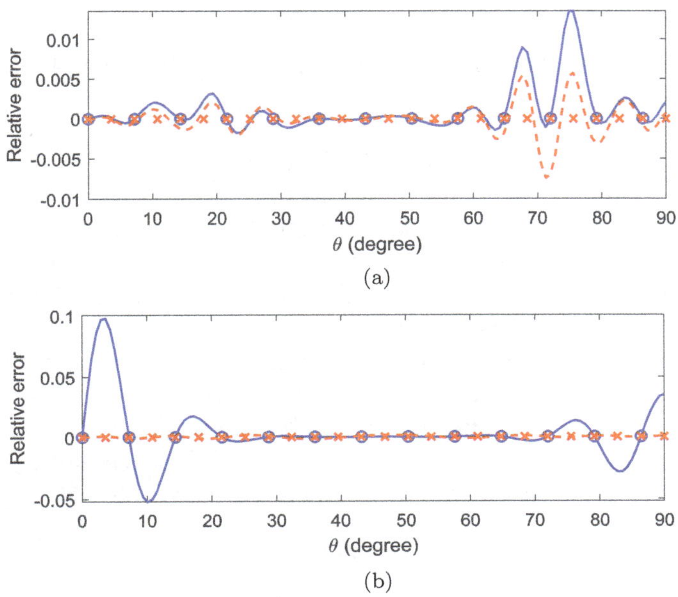

Figure 6.18. Relative error on the boundary between $0 \leq \theta \leq 90°$ for $M = 50$ (without LS) in solid line, with collocation points marked in ∘, and for $M = 100$ (with LS) in dashed line, with collocation points marked in ×, for cases (a) $\eta = 1.1$, and (b) $\eta = 1.5$.

in Fig. 6.18(b). In this case, we find very large fluctuation amplitudes near $\theta = 0, \pi/2, \pi, 3\pi/2$, where the BC is both zero and having zero slope (see Fig. 6.17(a)). The fluctuation is similar to the Runge phenomenon illustrated in Example 3.1, and the LS can dampen it. Using LS with $M = 100$, we observe a nearly two orders of magnitude drop in error, as shown in Table 6.9 and Fig. 6.18(b). In fact, the result is better than the $\eta = 1.1$ and $M = 100$ case.

From the above example, we observe that for the nonharmonic BC case, the error convergence with respect to N is not as rapid as in the harmonic case, making LS a competitive methodology for reducing the error. In addition, when the sources are moved farther away from the boundary, the traditional MFS can become unstable with large fluctuations in between collocation nodes. A simple application of LS can cure this malady. Hence LS is recommended for nonharmonic BC problems. We have also shown that when M is raised to beyond $2N$, the benefit of error reduction wears off.

6.7.3 *Error estimate and condition number*

For the case of nonharmonic BCs, Katsurada and Okamoto [480] gave the following error estimate:

$$\varepsilon \sim \mathcal{O}\left(\tau^N\right), \quad \text{for } 0 < \tau < 1. \tag{6.57}$$

The above expression predicts an exponential convergence with an increasing N. However, unlike the harmonic case given by (6.41), in which τ is replaced by ρ/R, the above formula does not predict a relation with the size of the auxiliary boundary. We may argue that even if τ is related to ρ/R, such a ratio is only slightly less than 1 because the auxiliary boundary needs to stay close to the boundary due to the lack of analytic continuation of the solution. Hence the convergence of (6.57) is much slower than (6.41). For interior and exterior domain Helmholtz and modified Helmholtz problems, only an algebraic convergence rate was proven [835, 868, 869]

$$\varepsilon \sim \mathcal{O}\left(N^{-p}\right), \tag{6.58}$$

where $p \geq 1$.

For the condition number κ, we note that it is independent of the BCs (see Section 5.4). Hence it is predicted by the same formulas as for the harmonic cases. The effective condition number, however, will change, due to its dependence on the right hand side. We shall test and observe the above theoretical predictions.

6.7.4 *2D interior potential problem*

We shall examine the application of the MFS to a 2D Laplace BVP for its convergence with respect to the number of degrees of freedom, and the position of the auxiliary boundary.

■ **Example 6.16**

We solve the same problem defined in Example 6.15, but only for the nonharmonic BC case. The focus here is on the error convergence with respect to N and η. The LS method is applied to all solutions using $M = 2N$.

Similar to Example 6.10, we run cases of fixed η with varying N, and vice versa. Figure 6.19 presents the maximum relative errors $\varepsilon_{\max}$

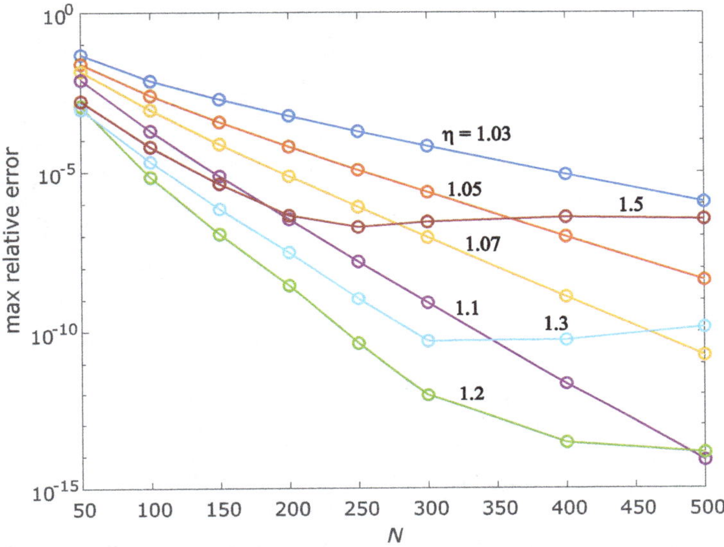

Figure 6.19. Plot of the maximum relative error $\varepsilon_{\max}$ versus the number of terms N, in semi-log scale, for various dilation factors η.

for η fixed at 1.03, 1.05, 1.07, 1.1, 1.2, 1.3, and 1.5, and N varied from 50 to 500. The error estimate (6.57) should appear as a straight line in the semi-log plot. Indeed we find it true for the cases of $\eta = 1.03$ up to 1.1. For $\eta \geq 1.2$, the solutions become unstable at large N due to exceedingly large condition numbers. The slopes of these lines are not as steep as those for the harmonic case, as demonstrated in Fig. 6.11. According to (6.57), the slope is proportional to $\log \tau < 1$. We observe that τ is dependent on $1/\eta$, similar to the harmonic case, which is dependent on ρ/R, as shown in (6.41). Since $1/\eta$ is just slightly less than 1, the convergence rate is much slower. Nevertheless, using $\eta = 1.1$, we can achieve an error of $\mathcal{O}(10^{-14})$, which is near the machine precision.

Next, we plot the results of fixed N by varying η in Fig. 6.20. We may compare the curves with those in Fig. 6.10 for the harmonic case. These curves are similar but different. In the harmonic case, as R/ρ increases, the error reaches an asymptotic value characterized by the harmonic polynomial limit, until the computation breaks down due to large condition numbers. For the nonharmonic case, there is no such asymptotic limit. Each curve seems to have an optimal η where the

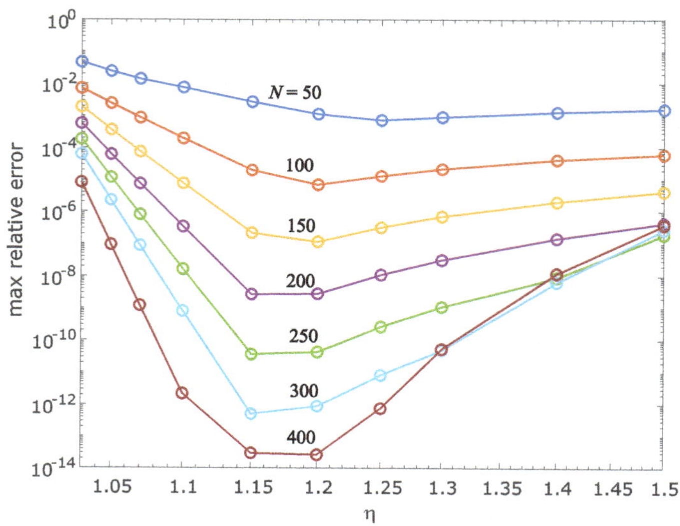

Figure 6.20. Plot of the maximum relative error $\varepsilon_{\max}$ versus the dilation factor η, in log–log scale, for various fixed N.

error is the smallest. The optimal η is not a constant for all N, but it is close to the boundary.

The above example shows that finding an optimal η within a small range near the boundary can reduce the error by orders of magnitude. Hence it is worthwhile to conduct a few trials to find such a value before increasing N for better solution accuracy. This brings the question of how to find the optimal η. One of the strengths of the MFS is that we can find the *a posteriori* error on the boundary by comparing with the exact BC. Hence maximum error curves like those in Fig. 6.20, or their approximations by taking a small number of samples, can be explicitly constructed. To find the optimal η, a 1D search is required. As only an approximate η is needed, the search should not involve more than a few iterations. Alternatively, one may use more sophisticated algorithms, such as the LOOCV presented in Section 7.1.

6.7.5 *3D interior potential problem*

In the following example, we examine 3D nonharmonic BC problems in order to confirm that the error behavior observed in 2D problems applies to 3D as well.

■ Example 6.17

We solve interior domain problems for two geometries: the bumpy sphere defined in Example 6.13, shown in Fig. 6.12(a), and a simplified geometry in Fig. 6.12(b). The simplified bumpy sphere is given by the parametric equation

$$\rho(\theta, \varphi) = 1 + \frac{1}{6} \sin(7\theta) \sin(\varphi).$$

Definitions of these parameters are given in Example 6.13. For both geometries, the Dirichlet BC is prescribed using the function

$$\phi = x^2 y^2 z^2,$$

which is nonharmonic.

The BVP involving the bumpy sphere geometry and the above nonharmonic BC has been previously investigated [145, 203, 604]. However, obtaining acceptable results was challenging. Part of the reason is the complexity of the geometry. In Fig. 6.21, we zoom in to the north pole of the sphere and plot the surface between the elevation angles $0 \le \theta \le \pi/8$. We observe multiple ridges that may need special attention in modeling. The previous attempts used a roughly uniform distribution of collocation and source nodes.

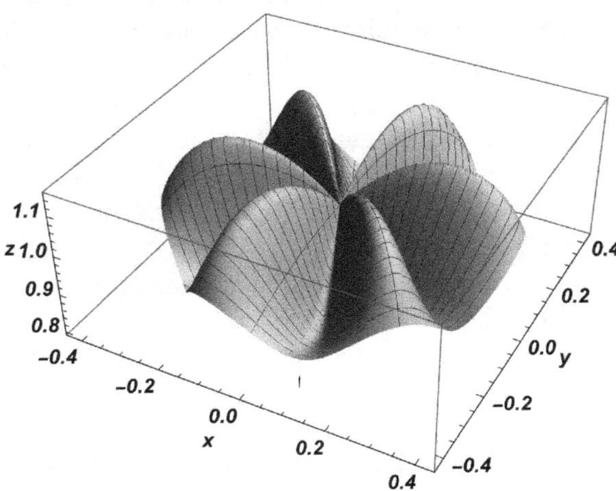

Figure 6.21. Partial plot of the bumpy sphere around the elevation angles $0 \le \theta \le \pi/8$.

We observed in examples with harmonic BCs that the complexity of geometry does not hamper our ability to obtain accurate solutions. In Example 6.13, the same bumpy sphere geometry with harmonic BC can be solved to an accuracy of $\varepsilon_{\max}$ less than $\mathcal{O}(10^{-7})$ with 300 nodes, and $\mathcal{O}(10^{-10})$ with 500 nodes. These collocation nodes can hardly give an accurate representation of the bumpy sphere geometry. The reason for obtaining so accurate solutions is that for problems using the same harmonic function as the BC, the sought for solutions are always the same, whatever the geometry. Different boundary geometries only change the locations of sampling points to recover the same function. This is not the case for nonharmonic BVPs. Using the same function as the BC, each different geometry represents a different solution. A poor representation of a local geometry gives a poor local solution. This situation is not unique to the MFS; it is true for all numerical methods, and a fine local mesh is needed.

In the current example, we devise a scheme that somewhat concentrates nodes around the two poles. For the generation of collocation nodes, we utilize a unit sphere with its z-axis divided into M increments. In previous attempts at generating boundary points, these increments were equal. Here the increments are spaced based on roots of Chebyshev polynomials, which have a gradual increase in concentration toward the two ends. At each level of z, a point is created on the surface of the sphere in the direction of φ using a step increment $\Delta\varphi$, which is an irrational number to avoid nodes lining up in the longitudinal direction. These points spiral downward to fill the space of φ. The (θ, φ) coordinates on the sphere are then substituted into the parametric equation of $\rho(\theta, \varphi)$ to generate the collocation points on the bumpy sphere surface. Figure 6.22(a) gives an illustration of collocation nodes generated using $M = 3000$ ($N = 1000$), which have a slight concentration near the poles. The source nodes are similarly generated using $\rho' = \eta\rho$.

The problem is solved using a dilated boundary with η changing from 1.03 to 1.07, and N from 1000 to 5000, with $M = 3N$. Error is sampled at $5N$ points on the surface similarly distributed as the collocation points, such that there is also a concentration near the poles. In Fig. 6.23, the maximum and RMS relative errors of the simulated results are presented. Similar to the 2D case in Section 6.7.4, we observe that for each N there exists an optimal η where the error is the smallest. The optimal distances, however, are closer to the boundary.

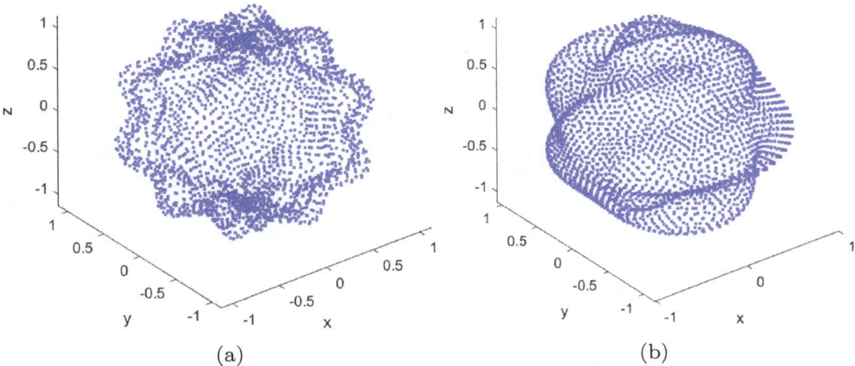

Figure 6.22. Collocation nodes for the (a) bumpy sphere with node concentration near the poles, and (b) simplified bumpy sphere with relatively uniform node distribution.

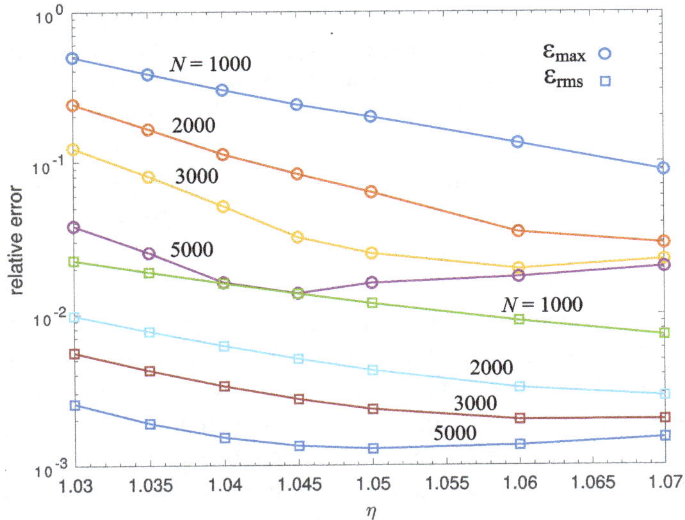

Figure 6.23. Plot of maximum relative error $\varepsilon_{\max}$ in ○, and RMS relative error $\varepsilon_{\mathrm{rms}}$ in □, versus dilation factor η for various source number N, in log–log scale, for the bumpy sphere problem.

Also, increasing N reduces the error, though not as dramatically as in the harmonic cases, because η is only slightly greater than unity. We observe that the maximum relative errors are rather large, due to a not-fine-enough representation of the pole geometry. Using $N = 5000$

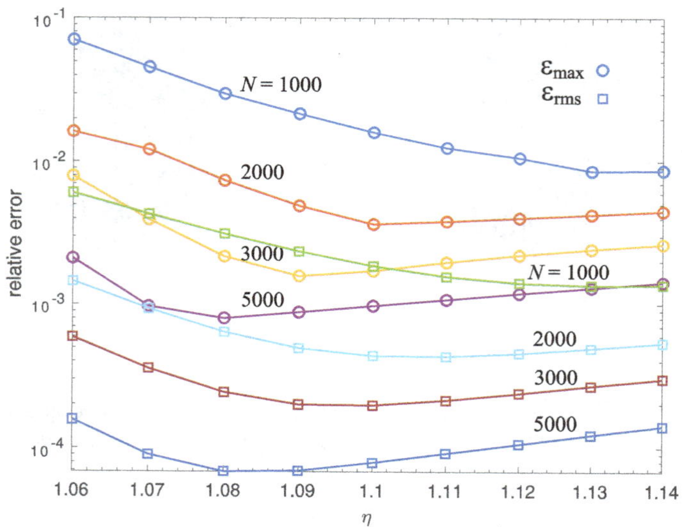

Figure 6.24. Plot of maximum relative error $\varepsilon_{\max}$ in $\circ$, and RMS relative error $\varepsilon_{\mathrm{rms}}$ in $\square$, versus dilation factor η for various source number N, in log–log scale, for the simplified bumpy sphere problem.

with $\eta = 1.045$ we are able to reduce the $\varepsilon_{\max}$ to 1.4%. The best RMS relative error is observed at $\eta = 1.05$ and $N = 5000$ as $\varepsilon_{\mathrm{rms}} = 0.13\%$.

Next, we solve the problem of the simplified bumpy sphere with η from 1.06 to 1.14, and N from 1000 to 5000. Using equal increments of z, a relatively uniform distribution of collocation nodes is created; see Fig. 6.22(b). The maximum and RMS relative errors are plotted in Fig. 6.24. We observe that using $N = 1000$, an $\varepsilon_{\max}$ less than 1% can be achieved. With $N = 5000$, the error can be reduced to 8.0×10^{-4} at $\eta = 1.08$. The RMS relative error is roughly one order of magnitude smaller, and at $N = 5000$ and $\eta = 1.08$, we find $\varepsilon_{\mathrm{rms}} = 6.8 \times 10^{-5}$. This example may explain that the relatively inferior performance of the bumpy sphere case was caused by the complex geometry at the poles.

6.7.6 *Exterior domain Helmholtz problem*

The efficiency of the MFS for solving Helmholtz-type interior and exterior domain problems with metaharmonic BCs has been demonstrated in various examples, such as 5.7, 6.2, 6.4, 6.9, and 6.14. Here we examine the application of the MFS to an exterior domain Helmholtz problem with a nonmetaharmonic BC to test its convergence behavior with respect to N and η.

■ Example 6.18

Consider the same exterior domain Helmholtz problem defined in Example 5.7, except that the wave number is set at $k = 20$, and the Dirichlet BC is replaced by the function

$$\phi = x^2 y^3,$$

which is nonmetaharmonic. The exact solution of this problem is not known.

To solve the nonmetaharmonic BC problem, a larger number of sources and collocation nodes than the metaharmonic cases are needed. In the results presented in Fig. 6.25, the number of sources N ranges from 150 to 400. For the LS procedure, $M = 2N$ is used. The complex fundamental solution $H_0^{(1)}(kr)$ is employed, rather than

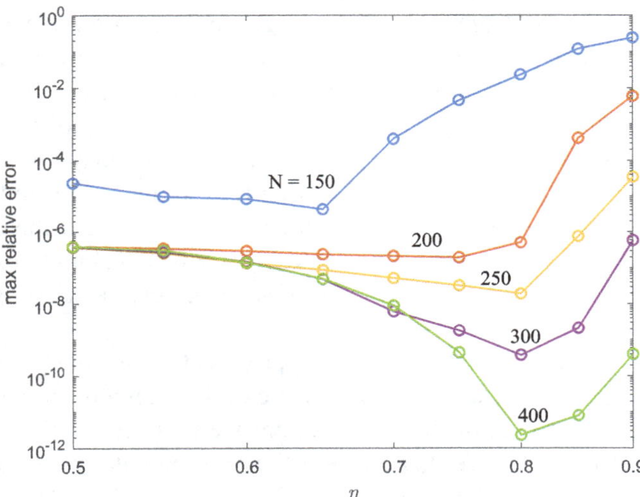

Figure 6.25. Plot of maximum relative error $\varepsilon_{\max}$ versus dilation factor η for various source number N, in log–log scale, for the nonmetaharmonic BC exterior domain Helmholtz problem.

the real valued one, to give a more stable result (see Section 6.5). The sources are distributed in the interior using an auxiliary contour that is contracted from the original boundary, with a contracting factor η ranging from 0.5 to 0.9.

As the exact solution is not known, the error is sampled at the boundary using a dense set of nodes to capture the maximum amplitude. For exterior domain problems extending to infinity, the amplitude of the wave solution decays outward. Hence the error amplitude decays as well, and the boundary maximum error serves as a good estimate of the overall maximum error.

With each given N, the problem is solved using a range of η values. Different from the metaharmonic case (6.50), the error estimate for the nonmetaharmonic case (6.58) does not provide a guideline for the location of the auxiliary boundary. However, in view of the lack of analytic extension between the interior and exterior solutions, the sources should be placed as close to the boundary as possible, but not too close to allow the rapid change near the singularities to interfere with the accuracy. This means that there is an optimal location near the boundary where the error is the smallest.

In Fig. 6.25, we plot the maximum relative error $\varepsilon_{\max}$ versus η for each N to show the trend of convergence. In the figure, we clearly observe the existence of an optimal η for each N. Also, the error rapidly decreases with increasing N, as predicted by (6.58). For $N = 400$, the optimal η is roughly 0.8, at which we find the $\varepsilon_{\max}$ of $\mathcal{O}(10^{-12})$. We also find that the shapes of the error versus η curves are benign enough to allow for a simple 1D search for the minimum. In fact, an accurate determination of the optimum is not necessary. As there exists a good *a posteriori* error estimator, a few trials of η, can help locate a good value. Further improvement of accuracy can be accomplished by increasing N.

Once the coefficients c_j of the approximation formula are determined, the solution at any point in the exterior domain can be evaluated. To get an idea of the full solution, we first plot in Fig. 6.26(a) the prescribed BC. We then present the approximate solution at the radial distance $r = 10$ for $0 \le \theta \le 2\pi$ in Fig. 6.26(b). We observe the complexity of the solution due to a relatively large wave number $k = 20$, as well as decay in amplitude as the wave moves outward.

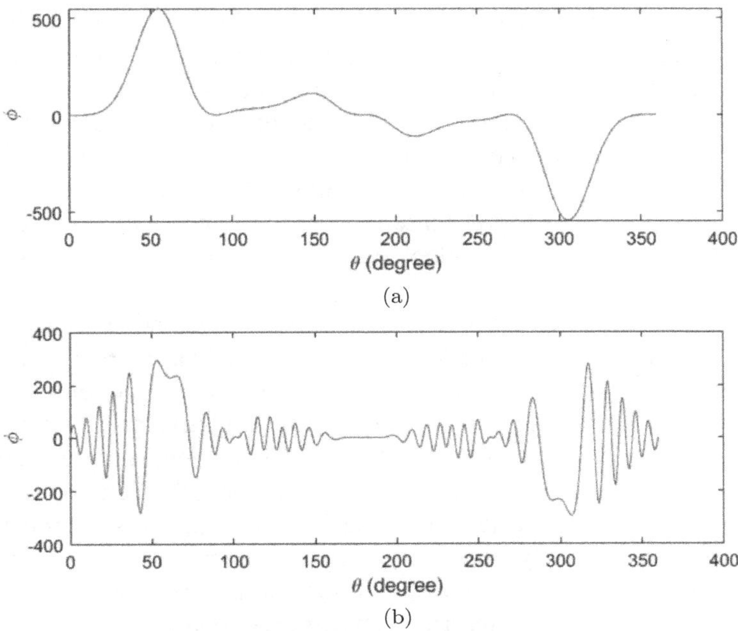

Figure 6.26. Plot of the Helmholtz potential: (a) Exact BC along the boundary; and (b) numerical solution at a distance $r = 10$.

6.7.7 *Summary*

From the above examples, we observe that the nonharmonic and non-metaharmonic BVPs behave quite differently from their harmonic and meta-harmonic counterparts, when they are approximated by the MFS. Different solution strategies are needed. In the following, we give a summary of their characteristics and recommended solution strategies.

- We may view the MFS procedure as fitting a function, that is, the BCs, in a 1D and 2D geometry, respectively for 2D and 3D BVPs, using fundamental solutions as basis functions. Once the fitting is complete, the solution is obtained. For the Laplace equation, if the BC is given by a harmonic function that is analytic, meaning that it can be expanded into a convergent power series, the fit is highly efficient. The placement of the sources should be as far away from the boundary as possible, so that the interpolants behave like harmonic polynomials, but not too far such that the condition number becomes exceedingly large, affecting stability. Generally, a small number of terms can fit the function to near machine precision accuracy. The efficiency of the fit is not hampered by the complexity of the boundary geometry. Similar statements can be made for Helmholtz-type equations with metaharmonic BCs.

- When the BC is only piecewise harmonic, there can exist discontinuities of certain order, or even singularities, at the junctures (corners). Singularities require special treatment by adding singular terms that mimic the local behavior. Lower order discontinuities can be treated by the placement of extra sources and collocation nodes near these corners, in addition to sources on the auxiliary boundary.

- When the BC is nonharmonic or nonmetaharmonic, the fitting using fundamental solutions is less efficient. A larger number of terms are needed. As the function to be fitted does not have an analytic continuation into the region outside the domain, the sources need to stay as close to the boundary as possible to minimize the extended region, yet not too close to allow the singularities and the steep variation around them to invade the domain. Between these two opposing goals, there exists an optimal location. As we can easily compute an *a posteriori* error from the boundary, and a near optimum is sufficient, the search may take but a few tries.

- For nonharmonic BCs prescribed on complex geometries, the local solution can be sensitive to the local geometry. To obtain a uniformly accurate solution, the sources and collocation nodes should concentrate near the geometry to adequately represent it.

- The collocation method can suffer from oscillations in between nodes. An LS fit can dampen the oscillations, and is recommended for nonharmonic BVPs.

6.8   Localized Method of Fundamental Solutions

The MFS uses scattered nodes distributed over a boundary type geometry, and is much more efficient and accurate than domain type numerical methods. The application of the MFS, however, requires the explicit knowledge of a fundamental solution of the given partial differential operator. In Chapter 2, fundamental solutions for a wide range of linear partial differential operators with constant coefficients have been compiled. When the coefficients become functions of space, however, fundamental solutions are available only for a few special cases. In many practical applications, these coefficients are material constants, which can change from one zone to another in composite materials, or continuously vary in functionally graded materials. To deal with these problems, a domain type method is necessary in order to capture the spatial variation.

In Chapter 5, we discussed that in many traditional domain discretization methods, such as the FEM and FDM, the modeling strategy has changed from non-overlapping "finite" covers to overlapping "cloud" covers. Using a set of scattered points for interpolation in the domain, the cloud type engages points in the neighborhood to perform local approximations. Certain basis functions are chosen, such as polynomials, RBFs, etc. Here we may ask the question: of all the basis functions, which can best approximate the solution? The answer is likely to be fundamental solutions or T-complete general solutions. These bases can locally satisfy the governing equation by using averaged material constants within the cover as constant coefficients for the PDE. We notice that the same strategy is used in the FEM [978], in which coefficients are taken as constants within each element, and vary from element to element.

The above considerations give the opportunity to apply the MFS as a local cloud method. In a small enough local region, a variable material property can be approximated as a constant, or interpolated using certain functions such as the fundamental solutions of the variable coefficient operator (see Section 2.15). As fundamental solutions have been shown to be highly efficient basis functions, a localized method of fundamental solutions (LMFS) can be an appealing meshless method for solving PDEs. A description of the methodology is given in the following.

6.8.1 *The LMFS methodology*

Referring to Fig. 6.27, we may represent the domain Ω bounded by $\partial\Omega$ using a set of scattered nodes in the domain and on the boundary, denoted as x_i, with $i = 1, \ldots, N$, and $N = N_d + N_{b_1} + N_{b_2}$, where N_d is the number of interior nodes, and N_{b_1} and N_{b_2} are, respectively, the number of nodes located on the Dirichlet and the Neumann part of the boundary. For each global node x_i, we may define a local support by drawing a circle/sphere of a certain radius $R^{(i)}$ to form a cover $\Omega^{(i)}$, which encloses m_i neighboring nodes, labelled as $x_k^{(i)}$, $k = 1, 2, \ldots, m_i$. We shall refer to x_i as the master node, and the supporting nodes $x_k^{(i)}$ as the slave nodes. We also give the master node x_i a local designation as $x_i = x_0^{(i)}$. We note that each of these local nodes corresponds to a global node.

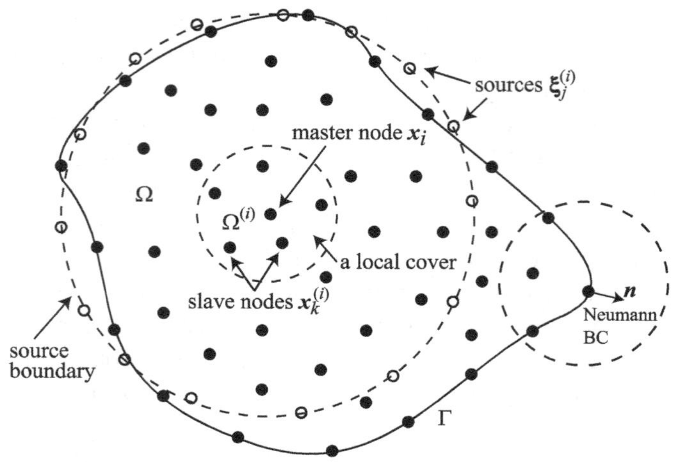

Figure 6.27. Illustration of support basis for the LMFS.

To approximate the solution within the cover, we create a circle/sphere centered at the master node, with radius $R_s^{(i)}$, and $R_s^{(i)} > R^{(i)}$, for distributing sources. These sources are located at a set of points $\xi_j^{(i)}$, with $j = 1, \ldots, n_i$. Similar to the MLS (see Section 5.2.5), we may use fewer sources (basis functions) than collocation nodes, that is, $n_i < m_i$, to form an overdetermined system, which is the strategy taken in most LMFS implementations [291, 358, 741]. However, for internal covers that do not touch the boundary, the functions to be interpolated are harmonic, and an LS procedure is not necessary. Hence here we choose $n_i = m_i$ to perform a direct collocation.

With the above, we interpolate the solution within the cover $\Omega^{(i)}$ as

$$\hat{\phi}^{(i)}(x) = \sum_{j=1}^{n_i} c_j^{(i)} G\left(x, \xi_j^{(i)}\right), \quad x \in \Omega^{(i)}, \tag{6.59}$$

in which G is the fundamental solution of the PDE, and the $c_j^{(i)}$ are undetermined coefficients. We note that the fundamental solution is generally not available for a PDE with non-constant coefficients. Here, however, we assume that coefficients can be locally approximated as constant for a small cover. This assumption is consistent with a method like the FEM. With this assumption, the fundamental solution for the local approximation is available. We shall discuss the case of locally variable coefficient in Section 6.8.2.

Equation (6.59) can be collocated at each of the slave nodes $x_k^{(i)}$ to form a set of linear equations

$$\phi_k^{(i)} := \phi\left(x_k^{(i)}\right) = \sum_{j=1}^{n_i} c_j^{(i)} G_{kj}^{(i)}, \quad k = 1, \ldots, n_i, \tag{6.60}$$

in which $G_{kj}^{(i)} = G(x_k^{(i)}, \xi_j^{(i)})$. We note in particular that the master node $x_0^{(i)}$ is not included in the collocation. Although we do not yet know $\phi_k^{(i)}$, (6.60) can be solved for $c_j^{(i)}$ as

$$c = G^{-1}\phi = H\phi, \tag{6.61}$$

in which $H = G^{-1}$ is the matrix inverse of G, and we have dropped superscripts (i) for convenience. The above equation can be written in the summation form as

$$c_j^{(i)} = \sum_{j=1}^{n_i} H_{kj}^{(i)} \phi_k^{(i)}, \quad j = 1, \ldots, n_i, \tag{6.62}$$

where the $H_{kj}^{(i)}$ are elements of H. Substituting (6.62) into (6.59), we obtain the interpolation formula for $\phi(x)$ within $\Omega^{(i)}$ based on nodal values as

$$\hat{\phi}^{(i)}(x) = \sum_{k=1}^{n_i}\sum_{j=1}^{n_i} H_{kj}^{(i)} G_j^{(i)}(x)\phi_k^{(i)} = \sum_{k=1}^{n_i} \Phi_k^{(i)}(x)\phi_k^{(i)}, \tag{6.63}$$

where $G_j^{(i)}(x) = G(x, \xi_j^{(i)})$, and

$$\Phi_k^{(i)}(x) = \sum_{j=1}^{n_i} H_{kj}^{(i)} G_j^{(i)}(x), \tag{6.64}$$

or in matrix form

$$\Phi(x) = H^T G(x), \tag{6.65}$$

are generating functions (shape functions). It can be shown that Φ has PU properties, which are demonstrated in the example below.

■ Example 6.19

We form a cover centered at the origin using $R = 1$, which contains 6 slave nodes located at $x_k = (1, 0)$, $(0.7, 0.7)$, $(-0.3, 0.8)$,

$(-0.5, 0.5)$, $(-0.8, -0.2)$, and $(0, -0.8)$. Assuming that the governing equation is the Laplace equation, find the generating functions based on fundamental solutions.

For the distribution of sources, we choose a circle of $R_s = 2$, and place 6 sources at $\boldsymbol{\xi}_j$, each $\theta = \pi/3$ apart, starting from $\theta = 0$. For the Laplace equation, the matrix elements G_{kj} are simply $\ln r_{kj}$, with $r_{kj} = \|\boldsymbol{x}_k - \boldsymbol{\xi}_j\|$. With the coordinates of $\boldsymbol{x}_k$ and $\boldsymbol{\xi}_j$ given, the matrix $\boldsymbol{G}$ can be inverted. Substituting into (6.64), we obtain 6 shape functions Φ_k, which are functions of $\boldsymbol{x}$. In Fig. 6.28, we present the first two functions, Φ_1 and Φ_2, in the range $[-1, 1] \times [-1, 1]$ as contour plots, in which the collocation points $\boldsymbol{x}_k$ are marked in dots. We observe that these functions take the value of 1, respectively at $\boldsymbol{x}_1$ and $\boldsymbol{x}_2$, and decrease to 0 toward all other nodes. At a given location $\boldsymbol{x}$, we can find the contribution from each interpolation node. For example, substituting the coordinates of the origin $(0, 0)$, we find

$$\boldsymbol{\Phi} = \{0.09934, 0.1909, -0.2373, 0.6278, -0.009357, 0.3285\}.$$

It is easy to show that these weights sum to 1.

Given an interpolation within the cover, such as (6.63) for the present case, the next step is to require it to satisfy the governing equation. This can be

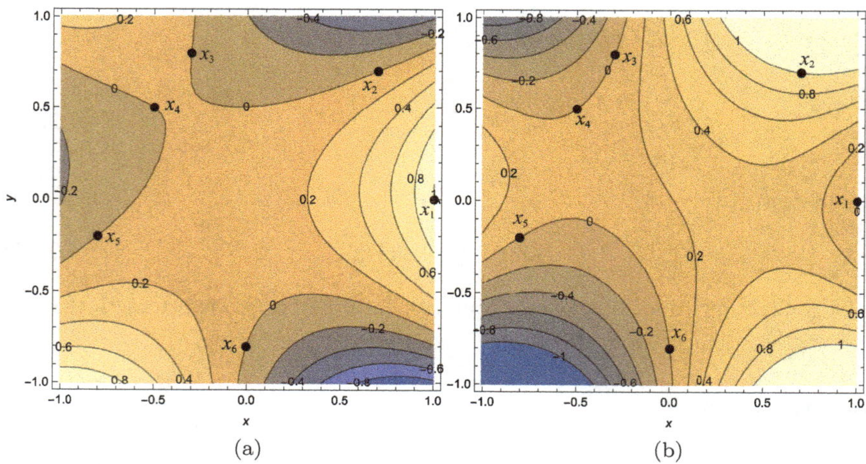

Figure 6.28.　Contour plots of shape functions: (a) Φ_1, and (b) Φ_2.

done by using a strong or weak formulation in the many MLS based methods. In the case of weak formulation, integration needs to be performed. In the present case, as each of the shape functions $\Phi_k^{(i)}(x)$ satisfies the governing equation, no such steps are needed. We can directly write a representation at the master node for each cover $\Omega^{(i)}$ centered at an interior node as

$$\phi_i := \phi(x_0^{(i)}) = \phi(x_i) = \sum_{k=1}^{n_i} \Phi_k^{(i)}(x_i)\phi_k^{(i)}, \quad \text{for } i = 1, \ldots, N_d. \quad (6.66)$$

By recalling that each local node $x_k^{(i)}$ corresponds to a global node, (6.66) are N_d linear equations that tie the global $\phi_i = \phi(x_i)$, $i = 1, \ldots, N$, together.

To complete the solution, we need to satisfy the BCs. On the Dirichlet boundary, we can simply assign the BC as

$$\phi_i = g(x_i), \quad i = N_d + 1, \ldots, N_d + N_{b_1}, \quad (6.67)$$

where $g(x)$ is the given BC. The above equations can be used to condense the system by removing these ϕ_i as unknowns, or we can consider them as N_{b_1} linear equations.

For the Neumann BC involving the normal derivative, we need to differentiate (6.63) in the direction of the boundary normal n, as

$$\frac{\partial \phi^{(i)}}{\partial n}(x) = \sum_{k=1}^{n_i} \sum_{j=1}^{n_i} \left[n(x) \cdot \nabla G_j^{(i)}(x) \right] H_{kj}^{(i)} \phi_k^{(i)} = \sum_{k=1}^{n_i} \Psi_k^{(i)}(x)\phi_k^{(i)},$$

$$(6.68)$$

where $\Psi_k^{(i)}(x)$ are shape functions for the normal derivative. We notice in the above that as the $G_j^{(i)}(x) = G(x, \xi_j^{(i)})$ are fundamental solutions, then the terms in the brackets are analytical expressions. At the Neumann boundary nodes, we obtain

$$h(x_i) = \sum_{k=1}^{n_i} \Psi_k^{(i)}(x_i)\phi_k^{(i)}, \quad i = N_1 + N_{b_1} + 1, \ldots, N, \quad (6.69)$$

where $h(x)$ is the Neumann boundary value. Combining and assembling (6.66), (6.67), and (6.69), we then have a sparse linear system of N equations in N unknowns. Further discussion of the LMFS and its implementation can be found in Chapter 10.

6.8.2 *Locally variable coefficient*

For certain PDEs, the variable coefficients cannot be treated as locally constant, as their influence can disappear entirely. For example, we take the quasi-harmonic equation resulting from steady state heat conduction or porous medium flow with a variable conductivity coefficient, as

$$\nabla \cdot [D(x)\nabla\phi(x)] = 0. \tag{6.70}$$

The discretized FEM equation may be written as

$$K\phi = f, \tag{6.71}$$

in which K is the stiffness matrix, and f contains the BCs. Based on the Galerkin formulation, the stiffness matrix for an internal element can be written as

$$K^e = \int_{\Omega_e} (\nabla N)^T D \nabla N \, d\Omega, \tag{6.72}$$

where $N = N(x)$ is the shape function. For an element without internal source, the right hand side f is zero. It is obvious that if we take D as a local constant within the element, it has no effect on the linear equation. Hence D must be modeled as a variable and interpolated using nodal values as

$$\hat{D}(x) = \sum_{i=1}^{n_e} N_i(x)D_i. \tag{6.73}$$

A similar situation is found in the LMFS. If D is approximated as a constant in a cover, the fundamental solution is independent of it, and its value is lost. Hence D must be modeled as a variable within a cover.

For the LMFS, this issue can be addressed by the use of variable coefficient fundamental solutions. In Sections 2.15 and 2.16, closed form fundamental solutions have been derived for coefficients that can be fitted by certain functions. For example, fundamental solutions for (6.70) have been presented in simple forms in Section 2.15.1 if the coefficient D can be fitted by the square of any harmonic function. Equations (2.364) and (2.365) are based on harmonic polynomials that allow the fitting using 4 and 8 data points, respectively, in 2D and 3D. These functions can be adjusted for the fitting of any number of data points. In Example 2.1, we performed a 2D fit of 4 data points with the resulting surface plotted in Fig. 2.3. With such a fit, the fundamental solution is simply given by (2.363), and the methodology developed in Section 6.8.1 can be directly applied.

Part II

Advanced Topics

Chapter 7

Solution of Elliptic BVPs

In Chapter 6, we described the MFS for solving various types of BVPs. The effectiveness of the MFS is affected by several factors such as the location/distribution of the boundary collocation and source points, the shape of the boundary, and the number of collocation and source points. Some of these issues have been addressed in Chapter 6. In this chapter, we will focus on the optimal choice of the source points location and the impact of the shape of the boundary on the accuracy of the method.

The determination of the source location for the optimal performance of the MFS is a major challenge and there are numerous numerical algorithms available in the literature attempting to address this issue. In Chapter 6, the determination of the optimal source location was carried out by the scanning of a dilation factor of a certain fixed auxiliary boundary shape, as demonstrated in Figs. 6.10, 6.13, 6.20, etc. In recent years, the leave-one-out cross validation (LOOCV) algorithm has become a popular method for selecting a good location of the MFS sources. The LOOCV was originally proposed by Rippa [755] for the identification of a suitable shape parameter in RBF approximation. The LOOCV was first applied for the identification of an appropriate location of the MFS sources in [145]. In this chapter, we will apply the LOOCV in the context of the MFS for the determination of the source location for various PDEs.

7.1   LOOCV

Consider the following BVP

$$\mathcal{L}\phi(x) = 0, \quad x \in \Omega, \tag{7.1a}$$

$$\phi(x) = f(x), \quad x \in \partial\Omega, \tag{7.1b}$$

where $\mathcal{L}$ is a second order linear elliptic differential operator. In the MFS, the approximate solution of BVP (7.1) is expressed as a linear combination of the fundamental solutions of (7.1a) as

$$\hat{\phi}(x) = \sum_{j=1}^{N} c_j \, G(x, x'_j), \quad x \in \overline{\Omega}, \tag{7.2}$$

where G is a fundamental solution of the operator $\mathcal{L}$ in (7.1a) and $\{x'_j\}_{j=1}^{N}$ are source points. Here we shall use the same notations as those in Chapter 6: the number of boundary collocation points will be denoted by M and the number of source points by N.

In the application of LOOCV to the MFS, one removes a single point, say x'_k, from the set of source points $\{x'_j\}_{j=1}^{N}$, which yields the reduced set of sources $\{x'_j{}^{[k]}\}_{j=1}^{N-1}$. Similarly, the point x_k is removed from the set of collocation points $\{x_i\}_{i=1}^{M}$, where $N \leq M$, yielding the reduced set of collocation points $\{x_i^{[k]}\}_{i=1}^{M-1}$. Collocation of the reduced MFS approximation $\hat{u}^{[k]}(x)$ at the reduced set of collocation points gives

$$\hat{\phi}^{[k]}(x_i^{[k]}) = \sum_{j=1}^{N-1} c_j^{[k]} G(x_i^{[k]}, x'_j{}^{[k]}) = f(x_i^{[k]}), \quad i = 1, 2, \ldots, M - 1.$$

$$\tag{7.3}$$

We then use the solution of the resulting system of equations to evaluate the error at the removed boundary point x_k; i.e.

$$\varepsilon_k = \phi(x_k) - \hat{\phi}^{[k]}(x_k). \tag{7.4}$$

The procedure is repeated for each of the N sources and corresponding N collocation points. A cost function can then be determined by taking the norm of the error vector $\boldsymbol{\varepsilon} = [\varepsilon_1, \varepsilon_2, \ldots, \varepsilon_N]^T$. Thus, for each choice of auxiliary boundary we can compute $\|\boldsymbol{\varepsilon}\|$ and the optimal source location can be identified by checking the minimum value of $\|\boldsymbol{\varepsilon}\|$. Note that as we still need to solve the problem for different auxiliary boundaries, the above LOOCV algorithm is still computationally rather intensive, of $\mathcal{O}(N^4)$. To overcome this, following the idea of Rippa [755], the procedure can be simplified by taking

$$\varepsilon_k = \frac{c_k}{A_{kk}^{-1}} \tag{7.5}$$

where c_k is the kth coefficient in the (full) MFS approximation (7.2) and A_{kk}^{-1} is the element (k, k), $k = 1, \ldots, N$, of the Moore–Penrose pseudoinverse of the corresponding MFS matrix A in (6.55).

The following script `Costeps.m` can be used to perform the above LOOCV algorithm when the sources are located on a circle. Note that when the number of source points equals the number of boundary collocation points ($M = N$), `pinv` can be replaced by `inv`. The following code is similar to the LOOCV code for RBFs in [298]. The fundamental solution of the Laplace equation is given in Line 4.

```
% Costeps.m
% bdpt: boundary collocation points; source: on a circle/sphere;
% ctrs: geometric centre of the domain; rhs: boundary conditions
1   function ceps = Costeps(ep,bdpt,rhs,source,ctrs)
2   source=source*ep+ctrs;
3   DM=pdist2(bdpt,source);
4   A=(1/(2*pi))*log(DM);
5   invA=pinv(A);
6   errorvector=(invA*rhs)./diag(invA);
7   ceps=norm(errorvector);
8   end
```

To find the minimum of the cost function we can apply the MATLAB® function `fminbnd` as

```
R=fminbnd(@(ep) Costeps(ep,bdpt,rhs,source,ctrs),r_min,r_max);
```

where `r_min` and `r_max` define the search interval for the optimal value of the radius R of the circle on which the sources are placed. These will be denoted by $R_{\min}$ and $R_{\max}$ in the sequel.

7.2 Laplace Equation

In this section, we consider the Laplace equation with harmonic and non-harmonic BCs as described in Sections 6.6 and 6.7.

In particular, we consider the Laplace equation in $\mathbb{R}^d$, $d = 2, 3$, subject to a Dirichlet BC

$$\Delta \phi(x) = 0, \qquad x \in \Omega, \tag{7.6a}$$

$$\phi(x) = f(x), \quad x \in \partial\Omega, \tag{7.6b}$$

where f is a known function.

7.2.1 *Numerical implementation in 2D*

From [145, 538], see also Sections 6.6 and 6.7, it is known that the source points should be placed far away from the boundary for the case of harmonic BCs, and close to the boundary for the case of nonharmonic BCs. The profiles of typical distributions of collocation and source points for harmonic and nonharmonic BCs are shown in Fig. 7.1. The issue of nonharmonic Dirichlet BCs for the Laplace equation was first raised by Schaback [779] and has since become an attractive research topic, see Chapter 6.

In [145, 538], four approaches were proposed for the distribution of boundary collocation and source points in the MFS solution of BVP (7.6). Furthermore, two efficient algorithms were put forward for the selection of the source locations. In practice, from the findings of the above studies, for harmonic BCs the source points should be placed relatively far from

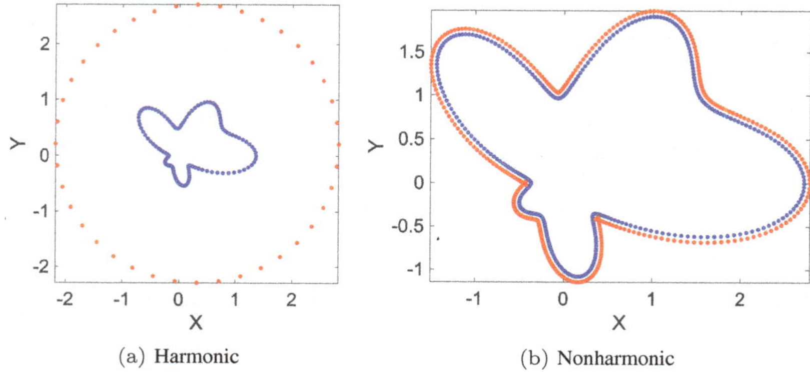

(a) Harmonic   (b) Nonharmonic

Figure 7.1. Profiles of typical distributions of collocation and source points for the cases of (a) harmonic BC (b) nonharmonic BCs.

the boundary. Moreover, Approach 1 in [145, 538] is recommended, which means the source points should be placed on a circle centered at the geometric center of the domain. In contrast, for nonharmonic BCs, the source points should be placed near the boundary. In this case, Approaches 3 and 4 in [145, 538] should be adopted. In Approach 3, the sources are placed at equal distance from the boundary, and are distributed at equal polar angles. In Approach 4, the sources are also at equal distance from the boundary, but are equally spaced. We shall call Approach 3 the nonuniform distribution, and Approach 4 the uniform distribution, and compare them with respect to the accuracy of the MFS approximation. It should be noted that the construction of an equal space distribution of sources can be rather challenging when the parametric equation of the boundary curve is not available.

Before the actual implementation, we introduce a frequently used subprogram in the upcoming examples called `DistanceMatrix.m`. It is used to compute the matrix of pairwise Euclidean distances between two sets of observation points; i.e. the collocation points $\{x_i\}_{i=1}^{M}$ and source points $\{x'_j\}_{j=1}^{N}$. The distance matrix can be represented as follows:

$$
\text{DM} = \begin{bmatrix}
\|x_1 - x'_1\| & \|x_1 - x'_2\| & \cdots & \|x_1 - x'_N\| \\
\|x_2 - x'_1\| & \|x_2 - x'_2\| & \cdots & \|x_2 - x'_N\| \\
\vdots & \vdots & \ddots & \cdots \\
\|x_M - x'_1\| & \|x_M - x'_2\| & \cdots & \|x_M - x'_N\|
\end{bmatrix}. \tag{7.7}
$$

If we let $x = [x_1, x_2, \ldots, x_M]^T$, $x' = [x'_1, x'_2, \ldots, x'_N]$, then in MATLAB®, $x - x'$ is equal to $\texttt{repmat}(x,1,\text{N}) - \texttt{repmat}(x',\text{M},1)$, which is

$$
x - x' = \begin{bmatrix} x_1 \\ x_2 \\ \vdots \\ x_M \end{bmatrix}_{M \times 1} [1, 1, \ldots, 1]_{1 \times N} - \begin{bmatrix} 1 \\ 1 \\ \vdots \\ 1 \end{bmatrix}_{M \times 1} [x'_1, x'_2, \ldots, x'_N]_{1 \times N}
$$

$$
= \begin{bmatrix}
x_1 & x_1 & \cdots & x_1 \\
x_2 & x_2 & \cdots & x_2 \\
\vdots & \vdots & \ddots & \vdots \\
x_M & x_M & \cdots & x_M
\end{bmatrix}_{M \times N} - \begin{bmatrix}
x'_1 & x'_2 & \cdots & x'_N \\
x'_1 & x'_2 & \cdots & x'_N \\
\vdots & \vdots & \ddots & \vdots \\
x'_1 & x'_2 & \cdots & x'_N
\end{bmatrix}_{M \times N}
$$

$$
= \begin{bmatrix}
x_1 - x'_1 & x_1 - x'_2 & \cdots & x_1 - x'_N \\
x_2 - x'_1 & x_2 - x'_2 & \cdots & x_2 - x'_N \\
\vdots & \vdots & \ddots & \vdots \\
x_M - x'_1 & x_M - x'_2 & \cdots & x_M - x'_N
\end{bmatrix}_{M \times N}. \tag{7.8}
$$

As shown above, by performing MATLAB® matrix operations, the distance matrix DM in (7.7) can be calculated efficiently. Let A be a two-column array with its first column containing the x-coordinates of the collocation points and the second column the y-coordinates. Similarly, let B be a two-column array containing the x and y coordinates of the source points. Thus, in 2D, the following function handle may be employed to generate DM:

```
DM = @(A,B) sqrt((A(:,1)-B(:,1)').^2+(A(:,2)-B(:,2)').^2);
```

The above distance matrix DM can also be calculated using the MATLAB® command `pdist2` from the Statistics and Machine Learning Toolbox. For convenience, we will use this built-in command in our examples. Users without access to the above toolbox can use the function handle DM shown above.

Geometric Shape: The geometric shape of the boundary plays an important role in the accuracy of the MFS approximation. For illustration, we consider the following boundary shapes which include smooth symmetric, asymmetric, and non-smooth boundaries (see Fig. 7.2). The first three shapes can be represented by a polar equation $r(\vartheta)$ with $0 \leq \vartheta < 2\pi$. The equations for their representation are given below:

$$\text{Cassini}: \quad r(\vartheta) = \sqrt[3]{\cos(4\vartheta) + \sqrt{\frac{18}{5} - \sin^2(4\vartheta)}}, \tag{7.9a}$$

$$\text{Star}: \quad r(\vartheta) = 1 + \cos^2(5\vartheta/2), \tag{7.9b}$$

$$\text{Amoeba}: \quad r(\vartheta) = e^{\sin\vartheta}\sin^2(2\vartheta) + e^{\cos\vartheta}\cos^2(2\vartheta), \tag{7.9c}$$

$$\text{L-shape}: \quad (-0.25, 0.75)^2 \setminus (0.25, 0.75)^2. \tag{7.9d}$$

In this section, for the purpose of comparison, we rescaled these domains so that they have similar size. For the Cassini domain, we shrink the size by 2/3 and the star and amoeba by 1/2. In other sections, the full sizes will be used.

Definitions of Error: Before examining numerical examples, we shall first present the various measures of error of numerical solutions. The error is defined as the difference between the approximate solution $\hat{\phi}$ and the exact solution ϕ. To obtain an approximate measure, we calculate the approximate solution $\hat{\phi}$ at a set of L test points that do not coincide with the collocation points. For BVPs that obey the maximum principle (see Section 6.7.1), such

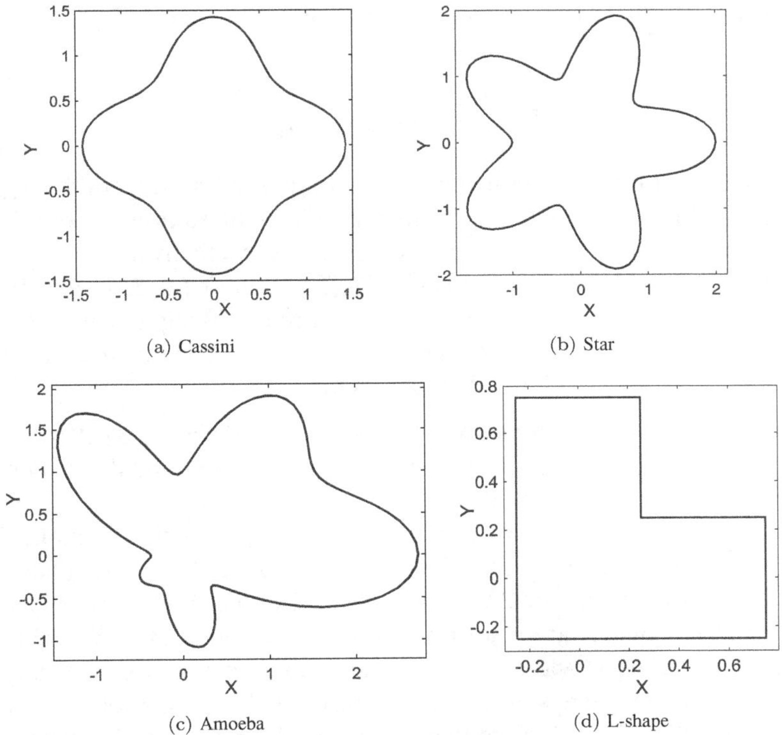

(a) Cassini

(b) Star

(c) Amoeba

(d) L-shape

Figure 7.2. Four geometric domains: (a) smooth symmetric, (b) and (c) smooth asymmetric, and (d) non-smooth L-shaped domain.

as the potential problems solved in this section, the maximum error occurs at the boundary; hence the L test points are selected on the boundary $\partial\Omega$. For BVPs that do not obey the maximum principle, such as those for the Helmholtz operator in Section 7.3, test points are taken on the boundary $\partial\Omega$ as well as in the domain Ω.

For a measure of accuracy, we may define the maximum absolute error as

$$\mathcal{E}_{\max} = \|\hat{\phi}(x) - \phi(x)\|_{\infty,\partial\Omega}, \tag{7.10}$$

and the root mean square error (RMSE) as

$$\mathcal{E}_{\mathrm{rms}} = \sqrt{\frac{1}{N_e}\sum_{i=1}^{N_e}\left[\hat{\phi}(x_i) - \phi(x_i)\right]^2}. \tag{7.11}$$

We also define the maximum relative error by normalizing $\mathcal{E}_{\max}$

$$\epsilon_{\max} = \frac{\|\hat{\phi}(x) - \phi(x)\|_{\infty,\partial\Omega}}{\|\phi(x)\|_{\infty,\partial\Omega}}. \tag{7.12}$$

We notice that the $\epsilon_{\max}$ defined above is slightly different from the $\varepsilon_{\max}$ in (3.72) used in Part I, in that they are normalized differently. Other relative errors used in Part I of the book, ε_r and $\varepsilon_{\mathrm{rms}}$, are respectively defined in (3.71) and (3.73). The above definitions (7.10)–(7.12) will be used in Part II. We also remark that for BVPs in which the maximum principle does not exist, the test points needs to be distributed over the entire domain, and the $\partial\Omega$ in the definitions (7.10)–(7.12) should be replaced by $\overline{\Omega}$.

Harmonic and Non-Harmonic BCs: As discussed above, BVPs with harmonic or nonharmonic BCs require different treatment and this is demonstrated in the programs below.

In the script `Laplace2D.m`, we solve BVP (7.6) with a harmonic BC, when Ω is the reduced size amoeba-shaped domain of Fig. 7.2(c) (Line 27). For other geometries, Line 27 can be replaced by other parametric equations in (7.9). The sources are placed on an auxiliary boundary $\partial\Omega'$ which is a circle with radius R and centered at the geometric center of the domain. Once the distance matrix in Line 13 is constructed, the interpolation matrix is computed in Line 14. Note that Lines 13–15 implement the main steps for solving the Laplace equation using the MFS.

```
   %MATLAB code: Laplace2D.m
1  function Laplace2D
2  clear;   warning off all;
3  u = @(x,y) exp(2*x).*cos(2*y); % Exact solution & BC
%  M = # of boundary points, N = # of source points
4  M=120; N=120; MT=101;   % MT= # of test points
%  [x,y]:boundary points;[xs0,ys0]:sources;[xt,yt]:test points
5  [x,y,xs0,ys0,xt,yt]=coll_pts(M,N,MT);
6  xc=(max(x)-min(x))/2+min(x); % geometric center x coordinate
7  yc=(max(y)-min(y))/2+min(y); % geometric center y coordinate
8  ut = u(xt,yt);   % exact solution on the test points
9  R=linspace((max(x)-min(x))/2+0.2,4,30);
10 error=zeros(30,1);
11 for k=1:30
12     xs=R(k)*xs0+xc;   ys=R(k)*ys0+yc; % location of sources
13     DM=pdist2([x y],[xs  ys]);
14     FS=(1/(2*pi))*log(DM); % MFS interpolation matrix
15     coef=FS\u(x,y);   % system solution
16     DMT=pdist2([xt yt],[xs ys]);
```

```
17      uh = (1/(2*pi))*log(DMT)*coef;   % approximate solution
18      error(k) = norm(uh-ut,inf);  % maximum relative error
19 end
20 [max_err,m]=min(error);
21 fprintf('R =%5.3f, Max_errr=%10.3e\n',R(m),max_err)
22 figure (1) % plot of errors vs radii of source circle
23 semilogy(R,error,'-','LineWidth',2); set(gca,'FontSize',14)
24 xlabel('\it R'); ylabel('Abs. Max. Errors');
25 end
   %%%%% Generation of boundary, source, and test points.
26 function [x,y,xs0,ys0,xt,yt]=coll_pts(M,N,MT)
27 r=@(x) (exp(sin(x)).*sin(2*x).^2+exp(cos(x)).*(cos(2*x)).^2)/2;
28 t=(1:M)'/M*2*pi;
29 [x,y]=pol2cart(t,r(t)); % boundary collocation points
30 t=(1:N)'/N*2*pi;
31 [xs0,ys0]= pol2cart(t,1); % source points on unit circle
32 t=(1:MT)'/MT*2*pi;
33 [xt,yt]=pol2cart(t,r(t)); % test points on the boundary
34 end
```

The corresponding script for the nonharmonic BC can be obtained by replacing certain lines of the code `Laplace2D.m` as shown below. Primarily, we replace R in the code `Laplace2D.m` by h which denotes the distance between the boundary and auxiliary boundary, see Approach 3 in [145, 538].

```
Line 3:   u = @(x,y) x.^2.*y.^3;
Line 6:   dx=gradient(xs0); dy=gradient(ys0);
Line 7:   xn=dy./sqrt(dx.^2+dy.^2);   yn=-dx./sqrt(dx.^2+dy.^2);
Line 9:   h=linspace(0.01,0.2,30);
Line 12:  xs=h(k)*xn+xs0;   ys=h(k)*yn+ys0;
Line 31:  [xs0,ys0]= pol2cart(t,r(t));
```

■ Example 7.1

We shall examine two cases, a case with a harmonic BC, and one with a nonharmonic BC. In the harmonic case, we chose the BC as

$$f(x, y) = \exp(2x)\cos(2y),$$

and the numbers of collocation and source points are $M = N = 120$. In the nonharmonic case, we chose

$$f(x, y) = x^2 y^3,$$

with $M = N = 400$.

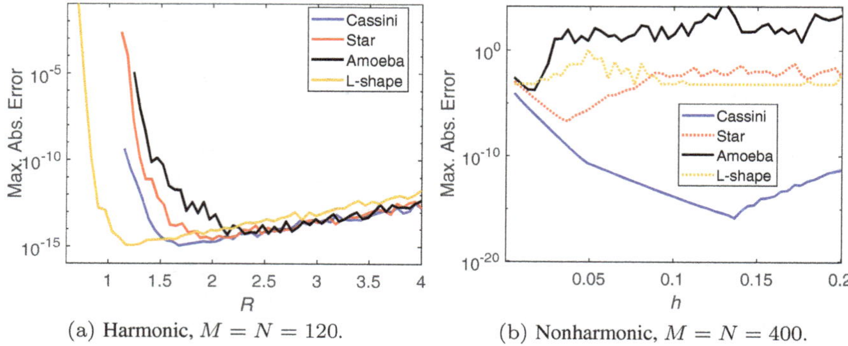

(a) Harmonic, $M = N = 120$.   (b) Nonharmonic, $M = N = 400$.

Figure 7.3. Plot of maximum absolute error $\mathcal{E}_{\max}$ for the four geometries for (a) the harmonic BC case, and (b) the nonharmonic BC case.

These cases are solved for the four geometries shown in Fig. 7.2 and the errors are plotted in Fig. 7.3. As observed in Fig. 7.3(a), for the harmonic BC, there is little difference in terms of optimal accuracy for the different boundary shapes considered. On the other hand, for the nonharmonic BC, the symmetry and smoothness of the domains play an important role in accuracy, and we observe in Fig. 7.3(b) significant differences in performance for different boundaries. For the Cassini-shaped domain, which is smooth and symmetric, machine precision can be achieved; whereas for the amoeba-shaped boundary, which is smooth but asymmetric, and the L-shaped boundary, which is non-smooth, the accuracy is poor. The degrees of smoothness and symmetry of the star-shaped domain lie between the Cassini and amoeba domains; as such, its accuracy lies in between too.

Pattern of Source Distribution: In [538], it was demonstrated that for nonharmonic BC problems, the pattern of distribution of collocation and source points can have a significant impact on accuracy. We shall demonstrate this in the following example.

■ **Example 7.2**

In this example, we consider the amoeba-shaped domain with the nonharmonic BC given in Example 7.1, using uniformly (equal spacing) and nonuniformly (equal polar angle) distributed source and

Table 7.1. Results obtained using nonuniform (equal angle) and uniform (equal spacing) distribution of collocation and source points for the amoeba-shaped domain.

$M = N$	Equal angle			Equal spacing		
	h	$\mathcal{E}_{\max}$	κ	h	$\mathcal{E}_{\max}$	κ
400	0.033	8.8(−4)	2.0(14)	0.027	2.2(−4)	3.3(3)
600	0.027	4.9(−4)	4.7(16)	0.022	7.2(−5)	1.5(4)
800	0.027	1.8(−4)	4.1(18)	0.017	3.2(−5)	1.4(4)
1000	0.025	4.2(−5)	7.1(18)	0.017	1.1(−5)	5.9(4)
1200	0.023	8.8(−6)	7.6(18)	0.015	3.6(−6)	9.1(4)
1400	0.023	8.8(−6)	8.3(19)	0.015	1.2(−6)	3.1(5)

collocation points. The source nodes are placed at a distance h normal to the boundary.

In Table 7.1, we observe little difference in accuracy when using uniform or nonuniform points. We find that when the number of boundary and source points ($M = N$) becomes large, the condition number (κ) increases and the accuracy improves. Meanwhile, the source points need to be closer to the boundary; i.e. h is getting smaller. Note that there is a significant difference in the size of the condition numbers. This means that although the two approaches exhibit similar accuracy, using uniformly distributed boundary points leads to higher stability. However, as already stated, the procedure of obtaining equally spaced boundary collocation points is a non-trivial task. We refer readers to [538] for the construction of an equally spaced boundary collocation point distribution in 2D when the parametric equation of the boundary curve is available. For general domains in 2D and 3D, obtaining such a distribution is an even greater challenge which defeats the purpose of using a meshless method in the first place. In practice, we try to distribute the boundary collocation points as uniformly as possible.

Effect of Linear Solver: Next, we investigate the effect different linear solvers may have on the MFS approximation accuracy. In Table 7.1, we observe the very high condition numbers for the nonuniform distribution of source and collocation points, which can cause an issue for the solver. For a square matrix, the MATLAB® `backslash \` command normally uses Gaussian elimination to solve the general system $Ax = b$. When the matrix

A is highly ill-conditioned, it is better to use a least squares solver and in the MFS, we recommend the use of the MATLAB® least squares solver `lsqminnorm`. Note that the backslash command only aims to minimize the norm of $Ax - b$, whereas `lsqminnorm` also aims to minimize the norm of x.

■ Example 7.3

To investigate the difference between these two solvers, we examine BVP (7.6) in the amoeba-shaped domain with nonharmonic BC, using $M = N = 1200$, and uniformly and nonuniformly distributed source and collocation points.

In Fig. 7.4(a), `lsqminnorm` is shown to perform better than the backslash for the case of nonuniform boundary points. In Fig. 7.4(b), we observe that there is no difference when using either solver for the uniformly spaced case, due to the small condition number. Note that the optimal accuracy achieved with the red curves in Figs. 7.4(a) and 7.4(b) is roughly the same. Hence, one may use a more advanced solver instead of generating uniformly distributed boundary collocation points, which, as mentioned earlier, is not trivial.

Number of Sources: Clearly, better accuracy can be achieved by increasing the number of boundary collocation and source points.

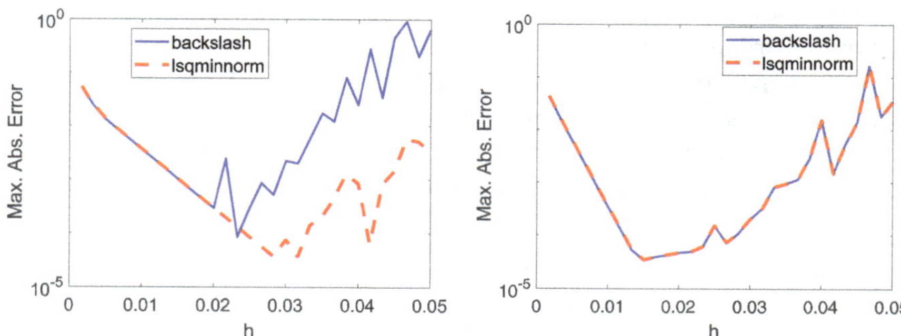

(a) Nonuniform (equal angle) boundary points   (b) Uniform (equal spacing) boundary points

Figure 7.4. Results using different solvers for nonuniform and uniform boundary points for the amoeba-shaped domain with $M = N = 1200$.

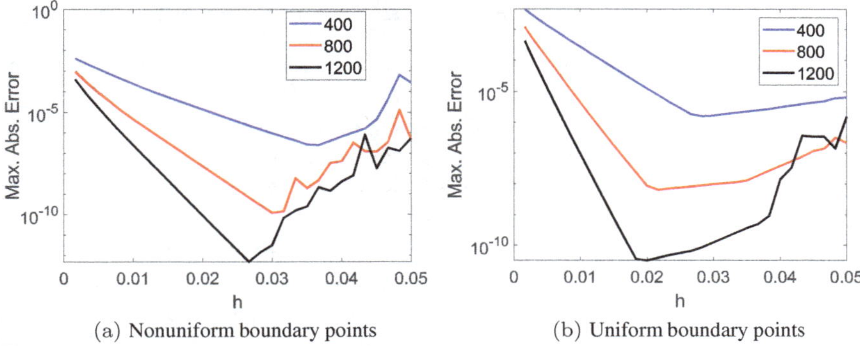

(a) Nonuniform boundary points   (b) Uniform boundary points

Figure 7.5. Results using nonuniform and uniform boundary points with a range of node numbers for the star-shaped domain.

■ Example 7.4

The star-shaped domain with nonharmonic BC is solved using nonuniform and uniform node distributions and a range of node numbers $M = N$.

It can be seen in Fig. 7.5 that contrary to the previous amoeba-shaped domain case, the accuracy obtained in the nonuniform boundary points case, presented in Fig. 7.5(a), is slightly better than that in the uniform boundary points case, shown in Fig. 7.5(b). The condition number sizes are similar to the ones in the amoeba-shape domain case; i.e. the condition number for uniform collocation points is much smaller than for nonuniform collocation points. It has been observed that when the number of collocation and source points becomes large, the accuracy in the star-shaped domain case continues to improve but not so in the case of the L-shaped and amoeba-shaped domains. Clearly, the smoothness and symmetry of the domain boundary greatly affect the accuracy when the BC is nonharmonic.

Optimal Source Location: To find the optimal source location, we apply the LOOCV algorithm described in Section 7.1. By removing the `for` loop in the script `Laplace2D.m` and slightly modifying the code, we obtain `Laplace2D_LOOCV.m` incorporating LOOCV. In Line 11, the subprogram `Costeps` can be found in Section 7.1. The points generator in Line 6 is the same as the one in `Laplace2D.m`. One can slightly modify this code for the nonharmonic BC case.

```matlab
% Solving Laplace equation with harmonic BC's using LOOCV
% Laplace2D_LOOCV.m
1  function Laplace2D_LOOCV
2  clear;  warning off all;
3  u = @(x,y) exp(2*x).*cos(2*y); % Exact solution
   % M = # of boundary points, N= # of source points
4  M=120; N=120; MT=101;  %MT= # of test points
5  L=1; R=5;  %[L, R]: search interval
   %[x,y]:boundary points;[xs,ys]:sources;[xt,yt]: test points
6  [x,y,xs,ys,xt,yt]=coll_pts(M,N,MT);
7  xc=(max(x)-min(x))/2+min(x);
8  yc=(max(x)-min(y))/2+min(y);
9  RHS=u(x,y); %BC
10 ut = u(xt,yt);  % exact solution on the test points
11 R=fminbnd(@(ep) Costeps(ep,[x,y],RHS,[xs,ys],[xc,yc]),L,R);
12 source=R*[xs,ys]+[xc,yc]; %source points
13 DM=pdist2([x,y],source);
14 FS=(1/(2*pi))*log(DM); % MFS interpolation matrix
15 coef=lsqminnorm(FS,RHS);% system solution
16 DMT=pdist2([xt,yt],source);
17 uh = (1/(2*pi))*log(DMT)*coef;  % approximate solution
18 error = norm(uh-ut,inf); % maximum absolute error
19 fprintf('R = %5.3f, Max. Error: %8.3e\n',R,error);
20 end
```

■ Example 7.5

We investigate the amoeba domain BVP with harmonic and non-harmonic BCs using the LOOCV algorithm.

For the harmonic BC problem, the minimum radius of the source circle covering the amoeba is 1.28, hence, we take $L = R_{\min} = 1.28$. In Table 7.2, we present the results for various search intervals. We observe that the estimated R and $\mathcal{E}_{\max}$ are consistent and highly accurate for $M = N = 120$. When fewer source points are taken ($N = 60$), the results are also very sharp. For the other three domains, the results for the harmonic BC using LOOCV are consistent with the plot in Fig. 7.3(a). In general, for harmonic BCs, high accuracy can be easily achieved using different search intervals.

For the nonharmonic BC problem, Fig. 7.3(b) shows that the performance of the MFS is closely related to the smoothness and symmetry of the boundary shape. In Table 7.3, we present the results

Table 7.2. Numerical results for the amoeba domain with harmonic BC using LOOCV.

$[R_{\min}, R_{\max}]$	$M = N = 120$		$M = 120, N = 60$	
	R	$\mathcal{E}_{\max}$	R	$\mathcal{E}_{\max}$
$[1.28, 5]$	2.85	$3.5(-13)$	2.18	$8.0(-14)$
$[1.28, 6]$	3.51	$1.7(-13)$	2.72	$1.8(-13)$
$[1.28, 7]$	3.27	$1.5(-12)$	2.71	$1.7(-13)$
$[1.28, 8]$	3.51	$3.1(-13)$	2.60	$1.4(-13)$
$[1.28, 9]$	4.15	$2.1(-12)$	2.56	$1.9(-13)$

Table 7.3. Numerical results for amoeba domain with non-harmonic BC using LOOCV.

$[h_{\min}, h_{\max}]$	$M = N = 400$		$M = 400, N = 200$	
	h	$\mathcal{E}_{\max}$	h	$\mathcal{E}_{\max}$
$[0, 0.10]$	0.020	$1.0(-3)$	0.067	$2.9(-3)$
$[0, 0.15]$	0.016	$1.0(-4)$	0.093	$4.0(-4)$
$[0, 0.20]$	0.015	$1.2(-4)$	0.124	$2.9(-2)$
$[0, 0.25]$	0.017	$4.2(-4)$	0.129	$1.2(-1)$
$[0, 0.30]$	0.077	$1.5(+1)$	0.185	$9.2(-2)$

for the amoeba domain with nonharmonic BC. Unlike the harmonic BC case, the initial search interval needs to be relatively small. In addition, taking a smaller number of source points has a noticeable impact on the accuracy. In Table 7.4, we present the results for the other three domains using LOOCV with search interval $[0, 0.2]$ and $M = N = 400$.

7.2.2 *Numerical implementation in 3D*

For 3D BVPs, the numerical implementation is similar to that for 2D BVPs, except for the fundamental solution and the generation of the boundary collocation and source points. Moreover, for a general irregular 3D domain, constructing the normal vectors to the surface at given boundary points is challenging. As a result, the generation of source points close to and at the

Table 7.4. Numerical results for various domains with nonharmonic BC using LOOCV with search interval $[0, 0.2]$ and $M = N = 400$.

Domain	h	$\mathcal{E}_{\max}$
Cassini	0.149	3.5($-$15)
Star	0.061	6.3($-$5)
L-shape	0.128	6.7($-$5)

same distance from the boundary, similar to that shown in Fig. 7.1(b) for the 2D case, is difficult. Therefore, for simplicity, we adopt Approaches 1 and 2 proposed in [538]; i.e. we place the sources on a surrounding sphere and on a dilated boundary for the harmonic and nonharmonic BCs, respectively.

In this section, we consider the four 3D boundary shapes shown in Fig. 7.6. The spherical equations $r(\theta, \phi)$ with $0 \leq \theta < \pi, 0 \leq \phi < 2\pi$, for the peanut and bumpy sphere, and the description of other geometries are given as follows:

$$\text{Peanut}: \quad r(\theta) = \sqrt{\cos(2\theta) + \sqrt{1.1 - \sin^2(2\theta)}}, \tag{7.13a}$$

$$\text{Bumpy sphere}: \quad r(\theta, \phi) = 1 + \frac{1}{6}\sin(6\theta)\sin(7\phi), \tag{7.13b}$$

$$\text{Dual-sphere}: \quad \min\left\{\left(x - \frac{3}{4}\right)^2, \left(x + \frac{3}{4}\right)^2\right\} + y^2 + z^2 = 1, \tag{7.13c}$$

$$\text{Bunny domain}: \text{http}://\text{graphics.stanford.edu}/\text{data}/\text{3Dscanrep}/. \tag{7.13d}$$

For a harmonic BC, as was the case in 2D problems, high accuracy can be achieved for all these domains. Hence, we will only focus on the nonharmonic BC case.

■ Example 7.6

We solve the BVP (7.6) in 3D with the Dirichlet BC given as $f(x, y, z) = x^2 y^2 z^2$, which is a nonharmonic function.

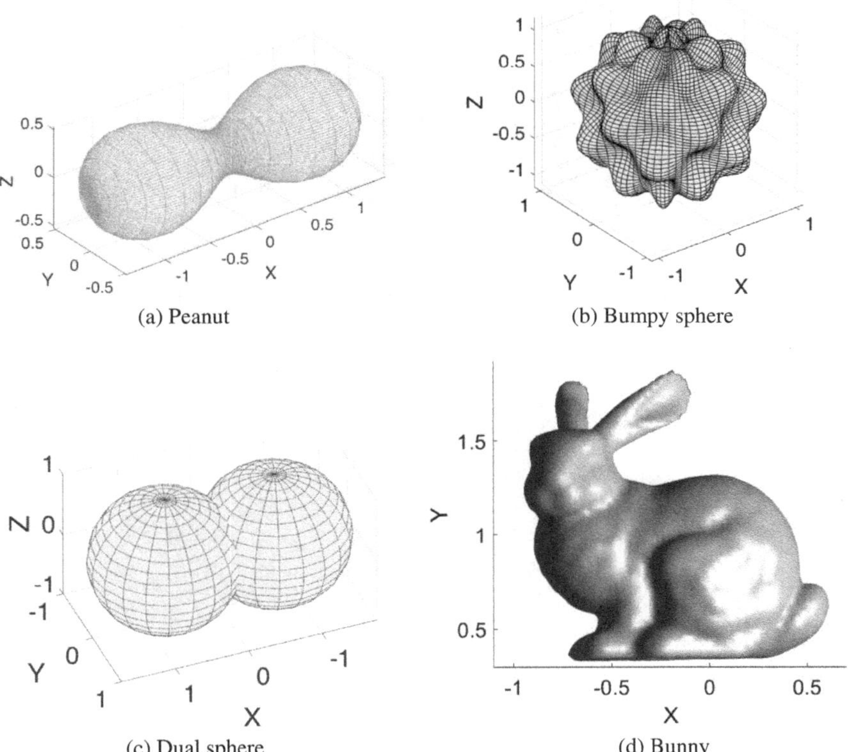

(a) Peanut

(b) Bumpy sphere

(c) Dual sphere

(d) Bunny

Figure 7.6. Profiles of the four 3-D domains.

We take $M = 1000$ boundary collocation points and another $L = 1000$ test points on the boundary different from the collocation points for the peanut, bumpy sphere, and dual sphere problems. For the bunny, we use the cloud data set available at the website of the Stanford Computer Graphics Laboratory which consists of $M = 1877$ boundary collocation points and $L = 453$ test points on the boundary. The coordinates of the boundary points are multiplied by 10 since the original data scale is too small. The solution profile on the boundary surface of the bunny is displayed in Fig. 7.7.

In Figs. 7.8(a)–7.8(c), we present results obtained using Approaches 1 and 2. In the former, R denotes the radius of sphere, and in the latter, η denotes the dilation factor. For the bunny domain,

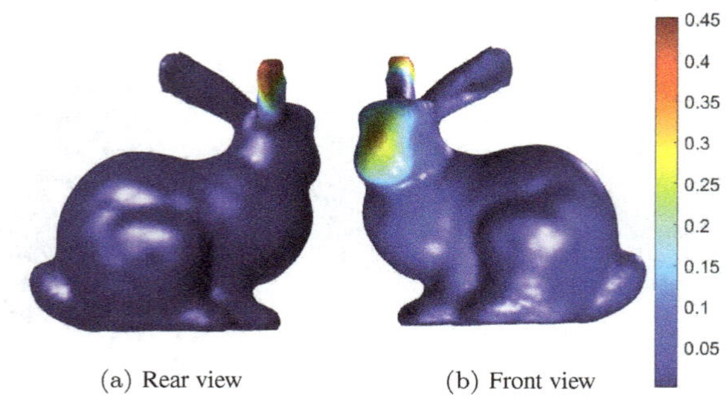

(a) Rear view (b) Front view

Figure 7.7. Solution profile on the bunny boundary surface.

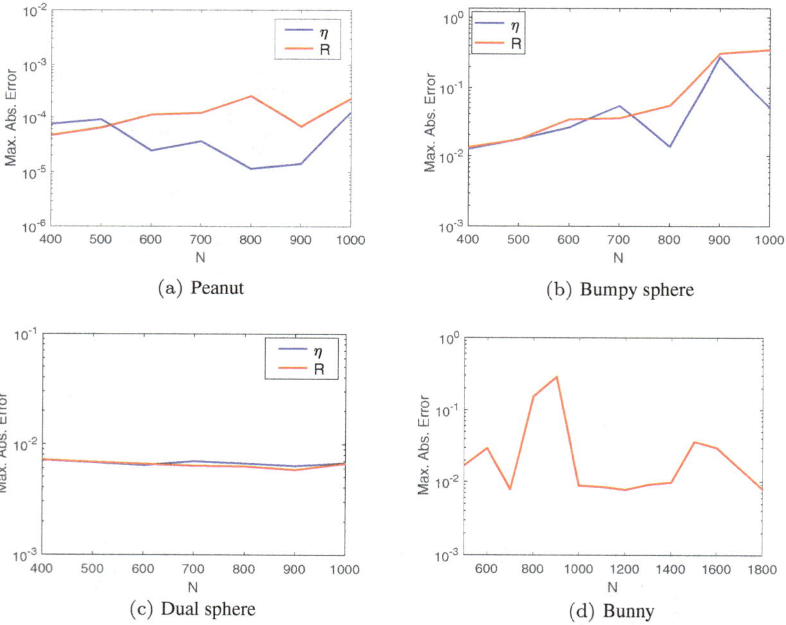

(a) Peanut (b) Bumpy sphere

(c) Dual sphere (d) Bunny

Figure 7.8. Error profiles for four domains with various numbers of source points using LOOCV.

we only use Approach 1, i.e. the source points are placed on a spherical surface. The results in Fig. 7.8 were obtained using LOOCV. We observe that in some cases better accuracy is achieved using fewer source points (N). For the bumpy sphere, the wrinkles on the boundary surface make it difficult to obtain accurate approximations. The bunny domain is asymmetric and not as smooth as the other three domains; thus obtaining satisfactory accuracy is challenging. These results show that the boundary shape has a severe impact on the accuracy of the method.

As seen in Fig. 7.8, a small number of source points can still produce accurate results. When the number of source points is the same as the number of boundary collocation points, special care is required. In this case, the MFS system is square and the backslash command will perform Gaussian elimination. Since the MFS matrix is highly ill-conditioned, this could present a problem as we have observed in the 2D case. As mentioned earlier, we need to force MATLAB®️ to perform the least squares method. Using the least squares solver `lsqminnorm`, as in the 2D case, we obtain consistent results for $N = 500, 1000$ and 1877 as shown in Fig. 7.9 for the case of Bunny domain. Note that when $N < M$, the backslash command will automatically use the least squares solver. In Table 7.5, we compare the optimal results obtained by brute force with those obtained with LOOCV and find that they are consistent. When $M = N$, we can replace `pinv` in Line 5 of `Costeps.m` by `inv` for better efficiency and accuracy.

To visualize the error variation on the boundary, in Fig. 7.10, we plot the absolute errors on the boundary surface for $M = 1877, N = 600$ with $R = 4$.

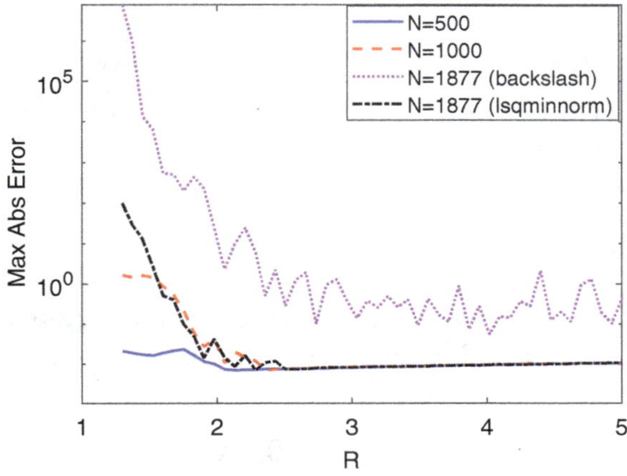

Figure 7.9. Maximum absolute errors for $M = 1877$ boundary collocation points, with various source sphere radii R and numbers N of source points, for the bunny domain.

Table 7.5. Optimal errors for $M = 1877$ and various R and N and the corresponding results using LOOCV. (1877*: `backslash`, 1877**: `lsqminnorm`.)

N	Optimal			LOOCV		
	R	$\mathcal{E}_{max}$	CPU	R	$\mathcal{E}_{max}$	CPU
500	2.13	7.0($-$3)	2.8	1.53	1.7($-$2)	1.5
1000	2.36	7.2($-$3)	12.9	3.47	8.8($-$3)	7.7
1877*	4.02	5.5($-$2)	5.1	1.95	5.5($+$1)	22.9 (`pinv`)
1877**	2.28	7.0($-$3)	46.5	3.54	8.9($-$3)	3.7 (`inv`)

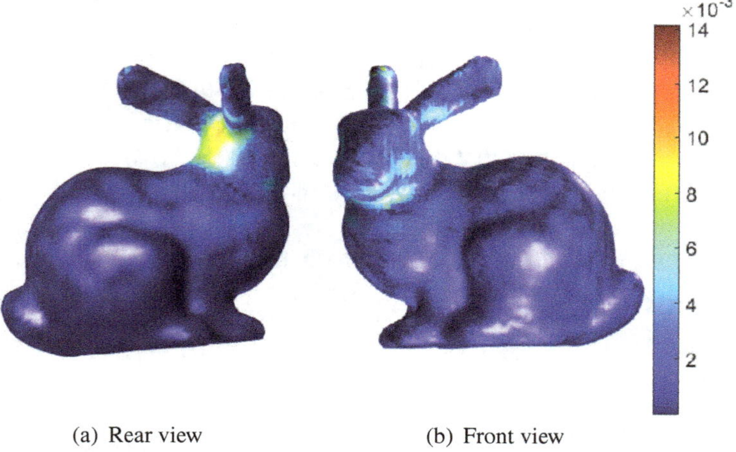

(a) Rear view   (b) Front view

Figure 7.10. Profiles of absolute errors for bunny domain using $M = 1877$, $N = 600$. The error is shown as color indicated in the error bar.

7.3 Helmholtz Equation

We now consider BVPs of the Helmholtz equation in a closed and bounded domain $\Omega \subset \mathbb{R}^d$, $d = 2, 3$, with a Neumann BC:

$$(\Delta + k^2)\phi(x) = 0, \quad x \in \Omega, \tag{7.14a}$$

$$\frac{\partial \phi}{\partial n}(x) = g(x), \qquad x \in \partial\Omega. \tag{7.14b}$$

Note that the above is an interior Helmholtz BVP. The corresponding exterior Helmholtz BVP in which Ω is replaced by $\mathbb{R}^2\backslash\overline{\Omega}$ or $\mathbb{R}^3\backslash\overline{\Omega}$ in 2D or 3D, respectively, will be studied later.

We approximate the solution of BVP (7.14) by

$$\hat{\phi}(\boldsymbol{x}) = \sum_{j=1}^{N} c_j\, G^H(\boldsymbol{x}, \boldsymbol{x}'_j), \quad \boldsymbol{x} \in \overline{\Omega}, \tag{7.15}$$

where $\{\boldsymbol{x}'_j\}_{j=1}^{N}$ are the sources, $\{c_j\}_{j=1}^{N}$ the coefficients, and $G^H(\boldsymbol{x}, \boldsymbol{x}')$ is a fundamental solution of the Helmholtz operator, given in (2.32a) and (2.32b), respectively for 2D and 3D problems. The value of $\partial\phi/\partial n$ is known at the boundary collocation points $\{\boldsymbol{x}_i\}_{i=1}^{M}$, leading to the system of equations

$$\boldsymbol{A}^H\boldsymbol{c} = \boldsymbol{g}, \tag{7.16}$$

where $\boldsymbol{c} = [c_1, c_2, \ldots, c_N]^T$, $\boldsymbol{g} = [g(\boldsymbol{x}_1), g(\boldsymbol{x}_2), \ldots, g(\boldsymbol{x}_M)]^T$, and the matrix $\boldsymbol{A}^H$ is defined from

$$\left(A^H\right)_{i,j} = \frac{\partial G^H}{\partial n(\boldsymbol{x})}(\boldsymbol{x}_i, \boldsymbol{x}'_j), \quad i = 1, \ldots, M; \ j = 1, \ldots, N, \tag{7.17}$$

where

$$\frac{\partial G^H(\boldsymbol{x}, \boldsymbol{x}')}{\partial n(\boldsymbol{x})} = \begin{cases} -\dfrac{k}{4}\dfrac{Y_1(kr)}{r}(\boldsymbol{x} - \boldsymbol{x}') \cdot \boldsymbol{n} & \text{in 2D,} \\[2ex] \dfrac{1}{4\pi}\left(\dfrac{\cos(kr)}{r^3} + k\dfrac{\sin(kr)}{r^2}\right)(\boldsymbol{x} - \boldsymbol{x}') \cdot \boldsymbol{n} & \text{in 3D,} \end{cases} \tag{7.18}$$

and Y_1 denotes the Bessel function of the second kind of order one.

7.3.1 *Numerical implementation in 2D*

In this section, we implement the numerical solution of the Helmholtz BVP by the MFS in 2D.

■ **Example 7.7**

We solve BVP (7.14) for the Helmholtz equation in the amoeba-shaped domain in Fig. 7.2(c) with the Neumann BC (7.14b) corresponding to the exact solution

$$\phi(x, y) = Y_0(k\rho),$$

Table 7.6. Numerical results for the amoeba domain using LOOCV with $M = 200$, $N = 100$, $k = 2$ for various initial search intervals.

$[R_{min}, R_{max}]$	R	$\mathcal{E}_{rms}$
$[1.5, 5]$	3.85	7.1(−16)
$[1.5, 6]$	3.85	6.3(−16)
$[1.5, 7]$	3.63	8.6(−16)
$[1.5, 8]$	3.66	7.1(−16)
$[1.5, 9]$	3.63	1.7(−16)

where $\rho = \sqrt{(x-2)^2 + (y-3)^2}$. The BC is given by

$$\frac{\partial \phi}{\partial n}(x, y) = -\frac{k}{2}Y_1(k\rho)\left[(2x - 4)n_x + (2y + 6)n_y\right],$$

where (n_x, n_y) is the normal vector to the boundary at the point (x, y).

To solve the problem, we take $M = 200$, $N = 100$, $k = 2$, and 653 test points, which are uniformly distributed inside Ω. The results using LOOCV are reported in Table 7.6. The errors are the $\mathcal{E}_{rms}$ defined in (7.11). These results are near machine precision and remain highly consistent when using different search intervals.

The code `Helmholtz_2D_LOOCV.m` used to solve the above problem is listed below. Line 8 generates the points and normal vectors required in the MFS discretization. The code `pts.m` can be found in Appendix C. In Line 14, (0.64, 0.42) is the geometric center of the amoeba. Line 16 defines the difference matrices as shown in (7.8).

```
%   Helmholtz_2D_LOOCV.m
1   function Helmholtz_LOOCV
2   D=@(x,y) sqrt((x-2).^2+(y+3).^2);
3   u=@(x,y,k) bessely(0,k*D(x,y)); % Exact solution
4   ux=@(x,y,k) -(k*(2*x-4).*bessely(1,k*D(x,y)))./(2*D(x,y));
5   uy=@(x,y,k) -(k*(2*y+6).*bessely(1,k*D(x,y)))./(2*D(x,y));
%   M: # of boundary points; N: # of source points;
6   M=200; N=100;  kappa=2;
7   r0=1.5; r1=10; %Search interval of LOOCV
8   [x,y,xs0,ys0,xt,yt,xn,yn]= pts(M,N);
9   MT=length(xt); % MT: # of test points
```

```
10 XN=repmat(xn,1,N); YN=repmat(yn,1,N);%normal derivative
11 rhs=ux(x,y,kappa).*xn+uy(x,y,kappa).*yn; %rhs
12 ut = u(xt,yt,kappa);  % exact solution
%  Search for optimal radius of sources using LOOCV
13 R=fminbnd(@(ep) Costeps(ep,x,y,rhs,xs0,ys0,XN,YN,kappa),r0,r1);
14 xs1=R*xs0+.64;  ys1=R*ys0+.42; %Source points
15 DM=pdist2([x y],[xs1  ys1]);
16 DMX=x-xs1'; DMY=y-ys1'; %Difference Matrix
17 A=-(kappa/4)*(bessely(1,kappa*DM)./DM).*(DMX.*XN+DMY.*YN);
18 coef=A\rhs; % system solution
19 DMT=pdist2([xt yt],[xs1 ys1]);
20 uh = (1/4)*bessely(0,kappa*DMT)*coef;  %Approximate solution
21 rmse=norm(uh-ut,2)/sqrt(MT);   %RMSE error
22 fprintf('M=%4d, N=%4d, R =%5.3f, RMSE=%8.3e\n',M,N,R,rmse)
end
%  LOOCV
23 function ceps = Costeps(ep,x,y,rhs,xs,ys,XN,YN,kappa)
24 xs1=xs*ep+0.64;    ys1=ys*ep+0.42;
25 DM=pdist2([x,y],[xs1,ys1]);
26 DMX=x-xs1';  %Difference Matrix in x
27 DMY=y-ys1';  %Difference Matrix in y
28 A=-(kappa/4)*(bessely(1,kappa*DM)./DM).*(DMX.*XN+DMY.*YN);
29 invA=pinv(A);
30 errorvector=(invA*rhs)./diag(invA);
31 ceps=norm(errorvector);
32 end
```

7.3.2 *Numerical implementation in 3D*

The 3D solution of the Helmholtz BVP is implemented in this section.

■ **Example 7.8**

We now solve BVP (7.14) in 3D for the Bunny domain shown in Fig. 7.6(d). The exact solution is given by

$$\phi(x, y, z) = \frac{\cos(k\rho)}{\rho}, \quad \rho = \sqrt{(x - 2)^2 + (y - 3)^2 + (z - 1)^2},$$

and the Neumann BC (7.14b) is now given by

$$\frac{\partial \phi}{\partial n} = \nabla \phi \cdot \left(n_x, n_y, n_z\right),$$

where (n_x, n_y, n_z) is the normal vector to the boundary.

Table 7.7. Numerical results for the Bunny domain using LOOCV for various initial search intervals.

$[R_{\min}, R_{\max}]$	R	$\mathcal{E}_{\mathrm{rms}}$	$\mathcal{E}_{\max}$
$[1, 5]$	4.056	1.2(−15)	1.4(−14)
$[1, 6]$	4.090	2.2(−15)	3.1(−14)
$[1, 7]$	4.708	2.6(−15)	2.3(−14)
$[1, 8]$	2.239	1.6(−15)	3.3(−14)
$[1, 9]$	2.167	2.2(−15)	3.9(−14)

The results in Table 7.7 were obtained using LOOCV with $M = 1877$, $N = 1000$, $k = 2$. In this table, we observe the consistency in accuracy for different initial search intervals.

In the code `Helmholtz_Bunny_LOOCV.m` below, we implement the LOOCV algorithm to solve the above 3-D Helmholtz problem. The data file `Data3D_Bunny3.mat` in Line 6 contains 1877 boundary points and the normal vectors to the boundary at these points. The data file `bunny_int.mat` in Line 4 contains 1000 interior test points for the evaluation of errors. The source points are distributed on a sphere of radius R and center $(−0.167, 1.0912, 0)$. Line 13 generates 1000 points on the surface of the unit sphere. The subprogram `unit_sphere` in Line 13 can be found in Appendix C. As before, we enlarge the size of the bunny tenfold.

```
%   Helmholtz_Bunny_LOOCV.m
%   Interior Neumann Helmholtz problem
1   function Helmholtz_Bunny_LOOCV
2   D=@(x,y,z) sqrt((x-2).^2+(y+3).^2+(z-1).^2);
3   u=@(x,y,z,k) cos(k*D(x,y,z))./D(x,y,z); % exact solution
4   load('bunny_int.mat','intpts');
5   X=intpts(:,1); Y=intpts(:,2); Z=intpts(:,3);
6   load('Data3D_Bunny3.mat','dsites','normals');
7   dsites=dsites*10;
8   kappa=2;
9   ut = u(X,Y,Z,kappa);   % exact solution
10  xn=normals(:,1); yn=normals(:,2); zn=normals(:,3);
11  x=dsites(:,1); y=dsites(:,2); z=dsites(:,3);
12  M=length(x); N =1000;
13  [xs0,ys0,zs0]=unit_sphere(N);
14  [ux,uy,uz]=uxyz(x,y,z,kappa);
15  rhs=ux.*xn+uy.*yn+uz.*zn;
```

```
16 minep=1; maxep=5; %LOOCV search interval
17 R=fminbnd(@(c) costEps(c,rhs,x,y,z,xs0,ys0,zs0,...
18   xn,yn,zn,kappa),minep,maxep,optimset('TolX',1e-2));
19 xs=R*xs0-0.167; ys=R*ys0+1.0912;  zs=R*zs0;%sources
20 DM=pdist2(dsites, [xs ys zs]);
21 DX=x-xs';  DY=y-ys';   DZ=z-zs';   %Difference Matrix
22 A=(1/(4*pi))*((cos(kappa*DM)./DM.^3)+...
23    (kappa*sin(kappa*DM)./DM.^2)).*(DX.*xn+DY.*yn+DZ.*zn);
24 coef=A\rhs; % Solution of system
25 DMT=pdist2(intpts, [xs ys zs]);
26 uh=-(1/(4*pi))*(cos(kappa*DMT))./DMT*coef; %Appr. solution
27 RMSE=norm(uh-ut,2)/sqrt(length(X));   %RMSE error
28 maxer=norm(uh-ut,inf);  % maximum error
29 fprintf('R =%5.3f,Maxerr=%8.3, RMSE=%8.3e\n',R,maxer,RMSE)
30 end
%  partial derivatives of u in x, y, and z directions
31 function [ux,uy,uz]=uxyz(x,y,z,k)
32 D=sqrt((x-2).^2+(y+3).^2+(z-1).^2);
33 D3=2*D.^3;    D2=2*D.^2;
34 COS=cos(k*D); SIN=sin(k*D);
35 ux= -COS.*(2*x - 4)./D3-(k*SIN.*(2*x - 4))./D2;
36 uy= -COS.*(2*y + 6)./D3-(k*SIN.*(2*y + 6))./D2;
37 uz= -COS.*(2*z - 2)./D3-(k*SIN.*(2*z - 2))./D2;
38 end
%  LOOCV
39 function ceps = costEps(ep,rhs,x,y,z,xs0,ys0,zs0,xn,yn,zn,k)
40 xs=ep*xs0-0.167; ys=ep*ys0+1.0912; zs=ep*zs0; %sources
41 DM=pdist2([x y z], [xs ys zs]);
42 DX=x-xs';  DY=y-ys';   DZ=z-zs';
43 A=(1/(4*pi))*((cos(k*DM)./DM.^3)+(k*sin(k*DM)./DM.^2))...
44    .*(DX.*xn+DY.*yn+DZ.*zn);
45 invA=pinv(A);
46 errorvector = (A\rhs)./diag(invA);
47 ceps = norm(errorvector);
48 end
```

7.3.3 *Implementation for exterior problems*

In this section we consider the exterior Helmholtz BVP corresponding to (7.14), in which Ω is replaced by $\mathbb{R}^2\backslash\overline{\Omega}$ and $\mathbb{R}^3\backslash\overline{\Omega}$ in 2D and 3D problems, respectively.

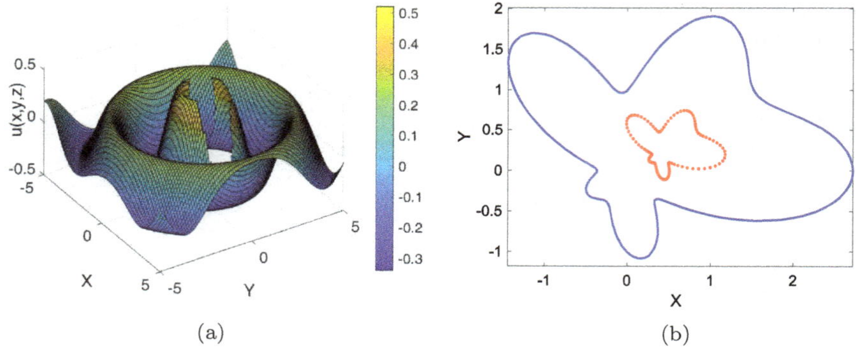

(a)   (b)

Figure 7.11. (a) Profile of exact solution in the extended exterior domain. (b) Typical location of source points (red) inside the amoeba after LOOCV search using $M = 200$, $N = 120$, and $k = 2$.

■ Example 7.9

We solve the exterior 2D Helmholtz problem for the amoeba-shaped geometry in Fig. 7.2(c). The exact solution is given by

$$\phi(x, y) = Y_0(k\rho), \quad \rho = \sqrt{(x - 0.5)^2 + (y - 0.2)^2},$$

and the Neumann BC is evaluated from the above solution. The profile of the exact solution in the square $[-5, 5]^2$ is shown in Fig. 7.11(a).

We apply LOOCV for determining a good location of the source points. For interior problems, the source points can be placed on a circle outside the domain. For exterior problems however, it is difficult to place the source points on a circle when the geometry of the domain is irregular and un-symmetric. We thus place the source points on a contraction of the amoeba boundary by a factor in the range (0,1). To better position the source points towards the center of the domain, after the contraction, we shift the source points as shown in Lines 12 and 25 of the following code `Helmholtz_Ext_2D_LOOCV.m`. A typical final location of the sources (on the contracted and shifted curve) after the application of LOOCV is presented in Fig. 7.11(b). We take $M = 200$, $N = 120$, and $k = 2$ and the test points are placed on an enlarged amoeba boundary with a magnification factor of 3/2 (see Line 41 in the code). In Table 7.8, we present $\mathcal{E}_{\max}$ and $\mathcal{E}_{\mathrm{rms}}$ for various contraction factor search intervals. The results are consistent and highly accurate.

Table 7.8. Numerical results for the 2D exterior Helmholtz problem for the amoeba domain using LOOCV with $M = 200$, $N = 120$, and $k = 2$ for various initial search intervals.

$[\eta_{\min}, \eta_{\max}]$	η	$\mathcal{E}_{\max}$	$\mathcal{E}_{\mathrm{rms}}$
$[0.05, 0.95]$	0.286	2.3(−14)	2.7(−15)
$[0.10, 0.90]$	0.289	3.3(−14)	3.8(−15)
$[0.15, 0.85]$	0.284	1.8(−14)	2.1(−15)
$[0.20, 0.80]$	0.275	6.6(−14)	9.4(−15)

```matlab
%   Helmholtz_Ext_2D_LOOCV.m
%   2D Exterior Helmholtz's equation for amoeba with Neumann BC's
1   function Helmholtz_Ext_2D_LOOCV
2   D=@(x,y) ((x - 1/2).^2 + (y - 1/5).^2).^(1/2);
3   u=@(x,y,k) bessely(0,k*D(x,y)); % Exact solution
4   ux=@(x,y,k) -(k*(2*x-1).*bessely(1,k*D(x,y)))./(2*D(x,y));
5   uy=@(x,y,k) -(k*(2*y-2/5).*bessely(1, k*D(x,y)))./(2*D(x,y));
%   M:# boundary points  N: # source points MT:# test points
6   M=200;   N=120;   MT=101;   kappa=2;
%   Generate boundary points (x,y), source points (xs0,ys0),
%   test points (xt,yt), and normal vectors (xn,yn)
7   [x,y,xs0,ys0,xt,yt,xn,yn]= nodes(M,N,MT);
8   ut = u(xt,yt,kappa);   % exact solution
9   rhs = ux(x,y,kappa).*xn+uy(x,y,kappa).*yn; %RHS
10  minep=0.1;maxep=0.9;  %search interval for LOOCV
11  d=fminbnd(@(c)costEps(c,rhs,x,y,xn,yn,kappa),minep,maxep);
12  xs=d*xs0+.4;   ys=d*ys0+.2;
13  DM=pdist2([x y],[xs  ys]);
14  DMX=x-xs';   DMY=y-ys'; %DifferenceMatrix
15  A=-(kappa/4)*(bessely(1,kappa*DM)./DM).*(DMX.*xn+DMY.*yn);
16  sol=A\rhs; % system solution
17  DMT=pdist2([xt yt],[xs ys]);
18  uh = (1/4)*bessely(0,kappa*DMT)*sol; %approximate solution
19  rmse=norm(uh-ut,2)/sqrt(length(xt));   %RMSE error
20  maxer=norm(uh-ut,inf);  % maximum error
21  fprintf('M=%4d, N=%4d, d =%5.3f\n',M,N,d)
22  fprintf('Max_err=%8.3e, RMSE=%8.3e\n',maxer,rmse)
23  end
%%  LOOCV
24  function ceps = costEps(ep,rhs,x,y,xn,yn,k)
25  xs=ep*x+.4; ys=ep*y+.2; %sources
26  DM=pdist2([x y], [xs ys]);
27  DMX=x-xs';   DMY=y-ys';
```

```
28 A=-(k/4)*(bessely(1,k*DM)./DM).*(DMX.*xn+DMY.*yn);
29 invA=pinv(A);
30 errorvector = (A\rhs)./diag(invA);
31 ceps = norm(errorvector);
32 end
%% points and normal vectors generator
33 function [x,y,xs,ys,xt,yt,xn,yn]= nodes( M,N,MT)
34 r=@(x) (exp(sin(x)).*sin(2*x).^2+exp(cos(x)).*(cos(2*x)).^2);
35 t=(1:M)'/(M)*2*pi;
36 [x, y]=pol2cart(t,r(t));
37 t=(1:N)'/N*2*pi;
38 [xs, ys]=pol2cart(t,r(t));
39 t=(1:MT)'/MT*2*pi;
40 [xt, yt]=pol2cart(t,r(t));
41 xt=3/2*xt;    yt=3/2*yt;
42 dx=gradient(x);    dy=gradient(y);  D=sqrt(dx.^2+dy.^2);
43 xn=dy./D;    yn=-dx./D; %normal derivatives
44 end
```

■ Example 7.10

For the 3D exterior Helmholtz BVP, we consider the double sphere domain (see (7.13c) and Fig. 7.6(c)) with a Neumann BC derived from the exact solution

$$\phi(x, y, z) = \frac{\cos(k\rho)}{\rho}, \quad \rho = \sqrt{(x - 0.2)^2 + (y + 0.3)^2 + (z - 0.1)^2}.$$

Since the geometry of the domain is symmetric, we can simply shrink the domain boundary by multiplying its coordinates by a contraction factor, which can be determined by applying LOOCV. The code for this 3D exterior Helmholtz problem is `Helmholtz_Ext_3D_LOOCV.m`, where 1000 boundary points are generated in Line 3. The code `doublesphere.m` can be found in Appendix C. To estimate the normals for a sparse 3D point cloud, the code `findPointNormals.m` in Line 6 is available in the MATLAB® Central File Exchange [832]. The $L = 1000$ test points are placed on a surface similar to the domain boundary defined by a magnification factor of 2. In this case, the number of source points is taken to be the same as the number of boundary collocation points. A typical source distribution (on the contracted surface) after the application of LOOCV is displayed in Fig. 7.12. In Table 7.9, we present some numerical results for various search intervals and these indicate

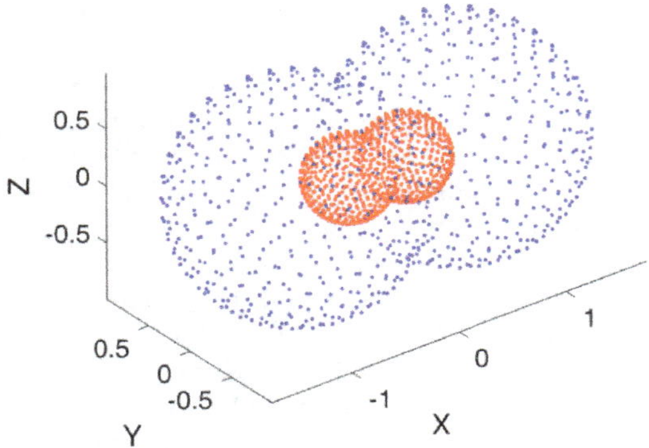

Figure 7.12. Typical location of source points (red) inside the double sphere after LOOCV search using $M = 1000$, $N = 1000$, and $k = 2$.

Table 7.9. Numerical results for the 3-D exterior Helmholtz problem for double sphere domain using LOOCV with $M = 1000$, $N = 1000$, and $k = 2$ for various initial search intervals.

$[\eta_{min}, \eta_{max}]$	η	$\mathcal{E}_{max}$	$\mathcal{E}_{rms}$
[0.05, 0.95]	0.316	1.7(−11)	1.8(−12)
[0.10, 0.90]	0.410	6.8(−10)	4.1(−11)
[0.15, 0.85]	0.366	5.0(−11)	3.7(−12)
[0.20, 0.80]	0.334	1.1(−10)	6.6(−12)

that the optimal position of the sources is at a moderate distance from the boundary.

```
%  Helmholtz_Ext_3D_LOOCV.m
%  3D exterior Helmholtz's equation for double sphere
%  subject to Neumann BC's
1  function Helmholtz_Ext_3D_LOOCV %
2  kappa=2;  n=1000;
3  bdpt=doublesphere(n);   %boundary points generator
4  x=bdpt(:,1); y=bdpt(:,2); z=bdpt(:,3);
5  X=2*x;  Y=2*y;  Z=2*z;  %test points
6  [normals, ~ ] = findPointNormals(bdpt);
7  xn=normals(:,1); yn=normals(:,2); zn=normals(:,3);
8  [ut,ux,uy,uz]=uxyz(X,Y,Z,x,y,z,kappa);
9  rhs=ux.*xn+uy.*yn+uz.*zn;  %RHS: Neumann condition
10 L=0.1;R=0.90;  %search interval
```

```matlab
11 d=fminbnd(@(c) costEps(c,rhs,x,y,z,xn,yn,zn,kappa),L,R);
12 xs=d*x; ys=d*y;  zs=d*z;  %source points
13 DM=pdist2([x y z], [xs ys zs]);   %DistanceMatrix
14 DX=x-xs';   DY=y-ys';   DZ=z-zs'; %DifferenceMatrix
15 A=((cos(kappa*DM)./DM.^3+(kappa*sin(kappa*DM)./DM.^2)).*...
16    (DX.*xn+DY.*yn+DZ.*zn)/(4*pi);  % system matrix
17 coef=A\rhs; % Solution of system
18 DMT=pdist2([X Y Z], [xs ys zs]);
19 uh=-(1/(4*pi))*((1./DMT).*cos(kappa*DMT))*coef;
20 rmse=norm(uh-ut,2)/sqrt(length(X));   %RMSE error
21 maxer=norm(uh-ut,inf);    % maximum error
22 fprintf('M=%4d, d =%5.3f\n',length(x);,d)
23 fprintf('Max_er=%10.3e, RMSE=%10.3e\n',maxer,rmse)
24 end
%%%% Exact solution and partial derivatives
25 function [ut,ux,uy,uz]=uxyz(X,Y,Z,x,y,z,k)
26 DT=sqrt((X-.2).^2+(Y+.3).^2+(Z-.1).^2);
27 ut= cos(k*DT)./DT; %exact solution
28 D=sqrt((x-0.2).^2+(y+0.3).^2+(z-0.1).^2);
29 D2=D.^2; D3=D.^3;   COS=cos(k*D);   SIN=sin(k*D);
30 ux = -COS.*(2*x - 2/5)./(2*D3)-(k*SIN.*(2*x-2/5))./(2*D2);
31 uy = -COS.*(2*y + 3/5)./(2*D3)-(k*SIN.*(2*y+3/5))./(2*D2);
32 uz = -COS.*(2*z - 1/5)./(2*D3)-(k*SIN.*(2*z-1/5))./(2*D2);
33 end
%%% LOOCV
34 function ceps = costEps(ep,rhs,x,y,z,xn,yn,zn,k)
35 xs=ep*x;  ys=ep*y;   zs=ep*z;
36 DM=pdist2([x y z], [xs ys zs]); %DistanceMatrix
37 DX=x-xs';  DY=y-ys';   DZ=z-zs'; %DifferenceMatrix
38 A=((cos(k*DM)./DM.^3)+(k*sin(k*DM)./DM.^2)).*...
39    (DX.*xn+DY.*yn+DZ.*zn)/(4*pi);
40 invA=pinv(A);
41 errorvector = (A\rhs)./diag(invA);
42 ceps = norm(errorvector);
43 end
```

7.4 Biharmonic Equation

We consider the biharmonic equation in $\mathbb{R}^d$, $d = 2, 3$, subject to the BCs
of the first biharmonic problem, as

$$\Delta^2 \phi(\boldsymbol{x}) = 0, \qquad\qquad\qquad \boldsymbol{x} \in \Omega, \qquad\qquad (7.19a)$$

$$\phi(\boldsymbol{x}) = f(\boldsymbol{x}) \ \text{ and } \ \frac{\partial \phi}{\partial n}(\boldsymbol{x}) = g(\boldsymbol{x}), \ \ \boldsymbol{x} \in \partial\Omega. \qquad (7.19b)$$

We approximate the solution of (7.19) by

$$\hat{\phi}(x) = \sum_{j=1}^{N} c_j \, G^L(x, x'_j) + \sum_{j=1}^{N} d_j \, G^B(x, x'_j), \quad x \in \overline{\Omega}, \tag{7.20}$$

where $\{x'_j\}_{j=1}^{N}$ are the sources, and $\{c_j\}_{j=1}^{N}$ and $\{d_j\}_{j=1}^{N}$ are unknown coefficients to be determined. In (7.20), $G^L(x, x')$ is a fundamental solution of the Laplace operator given by (2.23), and $G^B(x, x')$ is a fundamental solution of the biharmonic operator given by (2.156).

The values of ϕ and $\partial\phi/\partial n$ are known at the boundary collocation points $\{x_i\}_{i=1}^{M}$ leading to the $2M \times 2N$ system of equations

$$\left(\begin{array}{c|c} A_{11} & A_{12} \\ \hline A_{21} & A_{22} \end{array} \right) \left(\frac{c}{d} \right) = \left(\frac{f}{g} \right), \tag{7.21}$$

where $c = [c_1, c_2, \ldots, c_N]^T$, $d = [d_1, d_2, \ldots, d_N]^T$, $f = [f(x_1), f(x_2), \ldots, f(x_M)]^T$ and $g = [g(x_1), g(x_2), \ldots, g(x_M)]^T$. Also, for $i = 1, \ldots, M; \ j = 1, \ldots, N$,

$$(A_{11})_{i,j} = G^L(x_i, x'_j), \quad (A_{12})_{i,j} = G^B(x_i, x'_j),$$

$$(A_{21})_{i,j} = \frac{\partial G^L}{\partial n(x)}(x_i, x'_j), \quad (A_{22})_{i,j} = \frac{\partial G^B}{\partial n(x)}(x_i, x'_j), \tag{7.22}$$

where

$$\frac{\partial G^L}{\partial n(x)}(x, x') = \begin{cases} \dfrac{1}{2\pi} \dfrac{1}{r^2} (x - x') \cdot n & \text{in 2D,} \\[2ex] \dfrac{1}{4\pi} \dfrac{1}{r^3} (x - x') \cdot n & \text{in 3D,} \end{cases} \tag{7.23}$$

and

$$\frac{\partial G^B}{\partial n(x)}(x, x') = \begin{cases} \dfrac{1}{8\pi}(1 + 2\log(r))(x - x') \cdot n & \text{in 2D,} \\[2ex] -\dfrac{1}{8\pi} \dfrac{1}{r} (x - x') \cdot n & \text{in 3D.} \end{cases} \tag{7.24}$$

7.4.1 *Numerical implementation in 2D*

In this section, we solve biharmonic BVPs in 2D.

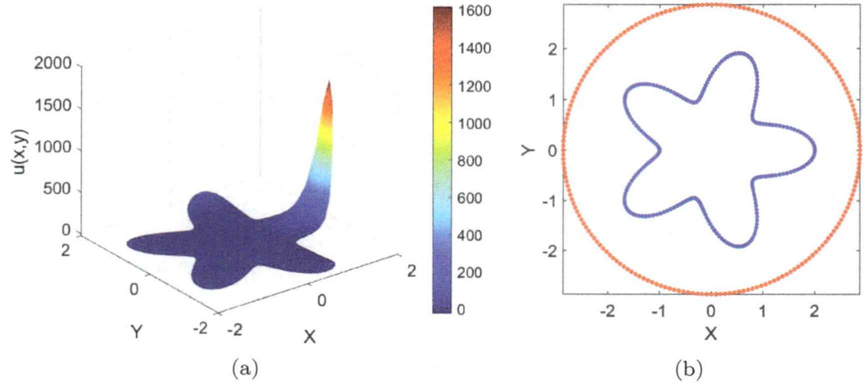

(a) (b)

Figure 7.13. (a) Profile of exact solution. (b) Typical distribution of source points (red) after LOOCV search using $M = 250$ and $N = 200$.

■ Example 7.11

We first consider the 2D biharmonic BVP (7.19) in the interior of the star-shaped domain depicted in Fig. 7.13(b), with the BCs (7.19b) derived from the exact solution

$$\phi(x, y) = (x^2 + y^2) \exp(3x) \cos(3y) + \exp(2y) \sin(2x).$$

We choose $M = 250$, $N = 150$, and $L = 739$ interior test points. The profile of the exact solution is presented in Fig. 7.13(a). The code using LOOCV is `Biharmonic1.m` listed below. In Line 11, the boundary points, source points, test points, and the normal vectors are generated. The code `pts.m` can be found in Appendix C. The highly accurate results obtained using various initial search intervals for the radii of the source circle are listed in Table 7.10 in maximum relative errors $\epsilon_{\max}$ (see (7.12)). The final distribution of the sources (on the source circle) after LOOCV is shown in Fig. 7.13(b).

```
%  Biharmonic1.m
%  Biharmonic equation subject to first biharmonic problem
1  function Biharmonic1
2  clear; warning off all;
3  T=@(x,y) cos(3*y).*exp(3*x);
4  u = @(x,y) T(x,y).*(x.^2+y.^2)+exp(2*y).*sin(2*x);
5  ux = @(x,y) 2*cos(2*x).*exp(2*y) + 2*x.*T(x,y) ...
6               + 3*T(x,y).*(x.^2 + y.^2);
```

Table 7.10. Numerical results for the 2D biharmonic problem for star-shaped domain using LOOCV with $M = 250, N = 200$ for various initial source circle radius search intervals.

$[R_{\min}, R_{\max}]$	R	$\epsilon_{\max}$
$[2, 4]$	2.70	$7.2(-12)$
$[2, 6]$	2.91	$1.3(-11)$
$[2, 8]$	3.09	$1.3(-12)$
$[2, 10]$	2.86	$8.3(-12)$

```
7  uy = @(x,y) 2*sin(2*x).*exp(2*y) + 2*y.*T(x,y) ...
8              - 3*sin(3*y).*exp(3*x).*(x.^2 + y.^2);
9  M=250; N=200;
10 L=2; R=10;  %[L, R]: search interval%%%%
11 [x,y,xs0,ys0,xt,yt,xn,yn]= pts(M,N);
12 rhs=[u(x,y);  ux(x,y).*xn+uy(x,y).*yn]; %BCs
13 d=fminbnd(@(ep) Costeps(ep,x,y,rhs,xs0,ys0,xn,yn),L,R);
14 xs=d*xs0;    ys=d*ys0;  %source circle
15 DM=pdist2([x y],[xs  ys]);
16 DMX=x-xs';   DMY=y-ys'; %DifferenceMatrix
17 G1=log(DM)/(2*pi);
18 G2=DM.^2.*log(DM)/(8*pi);
19 G3=((DMX./DM.^2).*xn+(DMY./DM.^2).*yn)/(2*pi);
20 G4=((DMX.*(1+2*log(DM))).*xn+(DMY.*(1+2*log(DM))).*yn)/(8*pi);
21 A=[G1 G2; G3 G4];  %system matrix
22 sol=lsqminnorm(A,rhs); % system solution
23 DMT=pdist2([xt yt],[xs ys]);
24 uh=[log(DMT)/(2*pi),  (DMT.^2.*log(DMT))/(8*pi)]*sol;
25 ut = u(xt,yt);  % exact solution
26 error = norm(uh-ut,inf)/norm(ut,inf);  %relative error
27 rmse=norm(uh-ut,2)/sqrt(length(xt));   %RMSE error
%  maxer=norm(uh-ut,inf);  % maximum error
28 fprintf('M=%4d,  N=%4d, d =%5.3f\n',M,N,d)
29 fprintf('rel_er=%10.3e, RMSE=%10.3e\n',error,rmse)
30 end
%%%% LOOCV
31 function ceps = Costeps(ep,x,y,rhs,xs0,ys0,xn,yn)
32 xs=xs0*ep;  ys=ys0*ep;
33 DM=pdist2([x,y],[xs,ys]);
34 c1=1/(2*pi);  c2=1/(8*pi);
35 DMX=x-xs'; DMY=y-ys';%DifferenceMatrix(x,xs);
36 G1=c1*log(DM); % system matrix
```

```
37 G2=c2*DM.^2.*log(DM); % system matrix
38 G3=c1*((DMX./DM.^2).*xn+(DMY./DM.^2).*yn);
39 G4=c2*((DMX.*(1+2*log(DM))).*xn+(DMY.*(1+2*log(DM))).*yn);
40 A=[G1 G2; G3 G4];  %system matrix
41 invA=pinv(A);
42 errorvector=(invA*rhs)./diag(invA);
43 ceps=norm(errorvector);
44 end
```

7.4.2 *Numerical implementation in 3D*

Next, we consider a 3D biharmonic BVP.

■ Example 7.12

We solve (7.19a) with the BCs (7.19b) derived from the exact solution

$$\phi(x, y, z) = (x^2 + y^2 + z^2) \cos(x) \sinh(\sqrt{5}y) \sin(2z)$$
$$+ \sin(x) \cosh(\sqrt{5}y) \cos(2z),$$

and Ω is the peanut-shaped domain in Fig. 7.6(a).

We take $M = N = 931$ boundary and source points, and $L = 1001$ interior test points. In this case, the sources are distributed uniformly on an ellipsoid and a typical distribution of these (with x, y, z axes ratio 1:1/2:1/2) after the application of LOOCV can be seen in Fig. 7.14. The results presented in Table 7.11 were obtained with various search

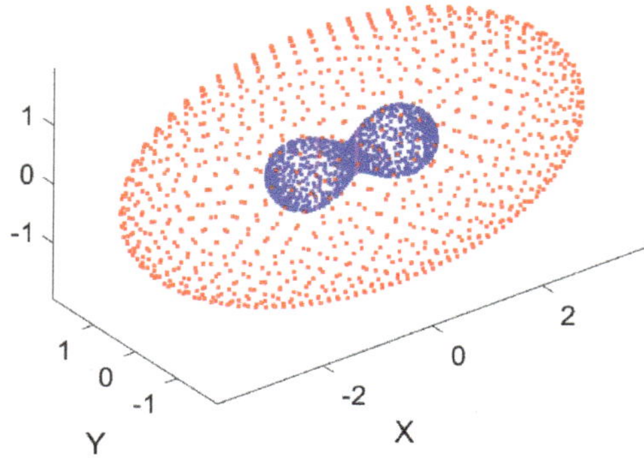

Figure 7.14. Typical final location of source points (red) after LOOCV search with $M = N = 931$.

Table 7.11. Numerical results for the 3D biharmonic problem for the peanut domain using LOOCV with $M = N = 931$ for various search intervals.

$[R_{\min}, R_{\max}]$	R	$\epsilon_{\max}$
$[1.5, 5]$	3.35	$9.6(-10)$
$[1.5, 6]$	3.22	$2.8(-10)$
$[1.5, 7]$	3.29	$4.7(-11)$
$[1.5, 8]$	4.10	$8.7(-11)$

intervals $[R_{\min}, R_{\max}]$ and R denotes the scale of the major axis of the ellipsoid.

The results in Table 7.11 were obtained using the code `Bihar_3D_LOOCV.m`. The subprograms `peanut_nodes.m` in Line 4, `findPointNormals.m` in Line 6, and `unit_sphere.m` are available in Appendix C.

```
%   Bihar_3D_LOOCV.m
%   Solving 3D biharmonic BVP.
%   M: # of boundary points; nt: # of interior test points
1   function Bihar_3D_LOOCV
2   clear; warning off
3   M=931; nt=1001; tic
4   [bdpt,int]=peanut_nodes(M,nt);
5   x=bdpt(:,1); y=bdpt(:,2); z=bdpt(:,3);
6   [normals,~] = findPointNormals(bdpt); %Find normal vectors
7   xn=normals(:,1); yn=normals(:,2); zn=normals(:,3);
8   X=int(:,1); Y=int(:,2); Z=int(:,3); %Interior test points
9   N=M; %N: # of source points
10  [xs0,ys0,zs0]=unit_sphere(N); %source points on unit sphere
11  [u,ux,uy,uz,ut]=uxyz(x,y,z,X,Y,Z);
12  rhs=[u;  ux.*xn+uy.*yn+uz.*zn]; %Forcing term
13  L=1.5; R=8;  %search interval of LOOCV
14  R=fminbnd(@(c) costEps(c,rhs,x,y,z,xn,yn,zn,xs0,ys0,zs0),L,R);
15  xs=R*xs0;  ys=R*ys0/2;   zs=R*zs0/2;
%   Construction of global matrix
16  DM=pdist2([x y z], [xs ys zs]);
17  DX=x-xs';  DY=y-ys';   DZ=z-zs';   %DifferenceMatrix
18  G1=-(1./DM)/(4*pi);
19  G2=-DM/(8*pi);
20  G3=(1./DM.^3).*(DX.*xn+DY.*yn+DZ.*zn)/(4*pi);
```

```matlab
21 G4=-(1./DM).*(DX.*xn+DY.*yn+DZ.*zn)/(8*pi);
22 A=[G1 G2; G3 G4];   % system matrix
23 sol=A\rhs; % system solution
%  Compute the approximate solution and errors
24 DMT=pdist2([X Y Z], [xs ys zs]);
25 uh=[-(1./DMT)/(4*pi), -DMT/(8*pi)]*sol; % approx. solution
26 err = norm(uh-ut,inf)/norm(ut,inf);   %relative error
27 rmse=norm(uh-ut,2)/sqrt(length(X));    %RMSE error
28 fprintf('R=%5.3f,Rerr=%8.3e,RMSE=%8.3e\n',R,err,rmse)
29 end
%%% Exact solution and its partial derivatives
30 function [u,ux,uy,uz,ut]=uxyz(x,y,z,X,Y,Z)
31 S1=sinh(sqrt(5)*y); C1=cosh(sqrt(5)*y);   XYZ2=x.^2+y.^2+z.^2;
32 S2=sinh(5^(1/2)*y); C2=cosh(5^(1/2)*y);
33 S=sin(2*z);  C=cos(2*z);   CX=cos(x);
34 u= CX.*S1.*S.*XYZ2+sin(x).*C1.*C; % exact solution
35 ux= C.*C2.*CX - S.*S2.*sin(x).*XYZ2+ 2*x.*S.*S2.*CX;
36 uy= 5^0.5*C.*S2.*sin(x)+2*y.*S.*S2.*CX+5^0.5*S.*C2.*CX.*XYZ2;
37 uz= 2*C.*S2.*CX.*XYZ2 - 2*S.*C2.*sin(x) + 2*z.*S.*S2.*CX;
38 ut=(cos(X).*sinh(sqrt(5)*Y).*sin(2*Z)).*(X.^2+Y.^2+Z.^2)+...
39    sin(X).*cosh(sqrt(5)*Y).*cos(2*Z); % exact solution
40 end
%% LOOCV
41 function ceps = costEps(ep,rhs,x,y,z,xn,yn,zn,xs,ys,zs)
42 xs=ep*xs;  ys=ep*ys/2;  zs=ep*zs/2;
43 DM=pdist2([x y z], [xs ys zs]);
44 DX=x-xs';  DY=y-ys';   DZ=z-zs';   %DifferenceMatrix
45 G1=-(1./DM)/(4*pi);
46 G2=-DM/(8*pi);
47 G3=(1./DM.^3).*(DX.*xn+DY.*yn+DZ.*zn)/(4*pi);
48 G4=-(1./DM).*(DX.*xn+DY.*yn+DZ.*zn)/(8*pi);
49 A=[G1 G2; G3 G4];  % system matrix
50 invA=inv(A);
51 errorvector = (A\rhs)./diag(invA);
52 ceps = norm(errorvector);
53 end
```

7.4.3 *Alternative biharmonic formulations*

We next present two alternative MFS formulations for the solution of BVP (7.19).

First Alternative Formulation: Based on the formulations of [113] and [908], instead of approximation (7.20), we may approximate the solution of BVP (7.19) by

$$\hat{\phi}(x) = \sum_{j=1}^{2N} c_j \, G^B(x, x'_j), \quad x \in \overline{\Omega}, \tag{7.25}$$

where $\left\{x'_j\right\}_{j=1}^{2N}$ are the sources and $\left\{c_j\right\}_{j=1}^{2N}$ are unknown coefficients to be determined.

We place N sources on each of two distinct pseudo-boundaries $\partial\Omega'_1$ and $\partial\Omega'_2$ surrounding Ω, i.e. the sources $\left\{x'_j\right\}_{j=1}^{N}$ on $\partial\Omega'_1$ and the sources $\left\{x'_j\right\}_{j=N+1}^{2N}$ on $\partial\Omega'_2$. Collocation of the BCs (7.19b) at the boundary collocation points $\{x_i\}_{i=1}^{M}$ yields a system of the form (7.21), where for $i = 1, \ldots, M, \, j = 1, \ldots, N,$

$$(A_{11})_{i,j} = G^B(x_i, x'_j), \quad (A_{12})_{i,j} = G^B(x_i, x'_{N+j}), \tag{7.26}$$

$$(A_{21})_{i,j} = \frac{\partial G^B}{\partial n(x)}(x_i, x'_j), \quad (A_{22})_{i,j} = \frac{\partial G^B}{\partial n(x)}(x_i, x'_{N+j}). \tag{7.27}$$

Figure 7.15(a) shows a typical distribution of source points on two circles $\partial\Omega'_1$ and $\partial\Omega'_2$ in 2D. Alternatively, we can scatter $K \leq 2M$ sources in an

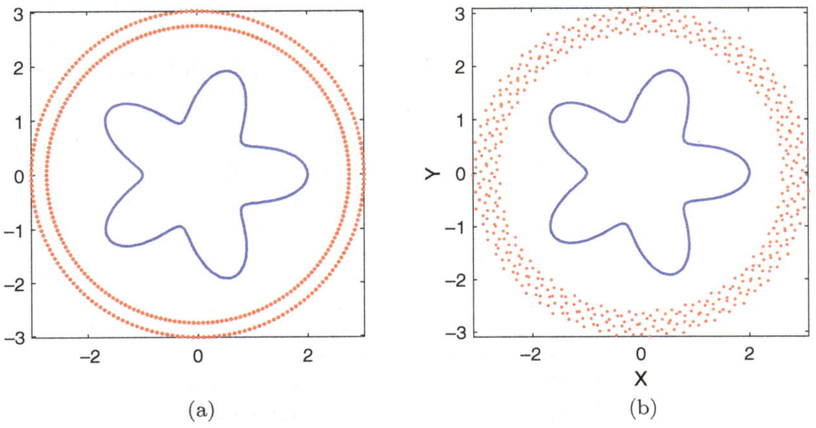

(a)   (b)

Figure 7.15. Typical distributions of source points (red) (a) on two pseudo-boundaries, (b) scattered in an annular region containing $\partial\Omega$.

annular domain containing $\partial\Omega$ and bounded by the boundaries $\partial\Omega_1'$ and $\partial\Omega_2'$ (see Fig. 7.15(b)).

In this case we get a system of the form

$$\left[\frac{\boldsymbol{B}_1}{\boldsymbol{B}_2}\right]\boldsymbol{c} = \boldsymbol{b}, \tag{7.28}$$

where, for $i = 1, \ldots, M$, $j = 1, \ldots, K$,

$$(B_1)_{i,j} = G^B(\boldsymbol{x}_i, \boldsymbol{x}_j'), \quad (B_2)_{i,j} = \frac{\partial G^B}{\partial n(\boldsymbol{x})}(\boldsymbol{x}_i, \boldsymbol{x}_j'). \tag{7.29}$$

■ Example 7.13

We redo the biharmonic problem in Example 7.11 using the alternative approach described above.

The results in Table 7.12 were obtained using the code `Biharmonic2.m` shown below and are compatible with those in Table 7.10 in terms of accuracy. However, with respect to the formulation and ease of implementation, the first alternative formulation presented in this subsection has advantages, especially for higher order polyharmonic equations.

The MATLAB® code that produces the results in the above table is listed below. The points generator in Line 10 is similar to that in `Biharmonic1.m` in Example 7.11 except for the distribution of source points.

Table 7.12. Numerical results for the 2D biharmonic problem using first alternative formulation for various initial search intervals for the first source circle radius.

$[R_{\min}, R_{\max}]$	R	$\epsilon_{\max}$
[2, 4]	2.33	9.5(−13)
[2, 6]	2.41	1.3(−12)
[2, 8]	3.13	1.5(−11)
[2, 10]	2.75	5.0(−12)

```
%  Biharmonic2.m
1  function Biharmonic2
2  T=@(x,y) 3*exp(3*x).*(x.^2 + y.^2);
3  Q=@(x,y) cos(3*y).*exp(3*x);
4  Ex=@(x,y) (x.^2+y.^2).*Q(x,y)+exp(2*y).*sin(2*x);
5  Exx=@(x,y) 2*cos(2*x).*exp(2*y)+2*x.*Q(x,y)+cos(3*y).*T(x,y);
6  Exy=@(x,y) 2*sin(2*x).*exp(2*y)+2*y.*Q(x,y)-sin(3*y).*T(x,y);
7  m=250; % m:# of boundary pts
8  nt=739; % nt:# of test pts
9  ns=200;
%  collocation pts and sources
10 [x,y,xn,yn,xs0,ys0,xt,yt]=star_pts(m,nt,ns);   % All points
11 GR1=Ex(x,y); % RHS Dirichlet BC
12 GR2=Exx(x,y).*xn+Exy(x,y).*yn; % RHS Neumann BC
13 RHS=[GR1; GR2];   % RHS  forcing term
14 Exact=Ex(xt,yt); %exact solution
15 ml=2; mr=4;   %search interval
16 eta1=fminbnd(@(eta) costeps(eta,x,y,xs0,ys0,RHS,xn,yn),...
17       ml,mr,optimset('TolX',1e-2));
18 xs=eta1*xs0; ys=eta1*ys0;
%  Construction of system
19 R=pdist2([x,y],[xs,ys]); % distance matrix
20 G=R.^2.*log(R);
21 Gx=(x-xs').*(1+2*log(R));% xder matrix
22 Gy=(y-ys').*(1+2*log(R));% yder matrix
23 gl=G;% Dirichlet matrix
24 gln=xn.*Gx+yn.*Gy;% Neumann matrix
25 A=[gl;gln]; % Global matrix
26 sol=lsqminnorm(A,RHS); % solution of MFS system
27 R=pdist2([xt,yt],[xs,ys]);  % distance matrix
28 ffa=R.^2.*log(R); % coeff. matrix
29 Appr=ffa*sol; % Approximation at test pts
30 err=norm(Appr-Exact,inf)/norm(Exact,inf);
31 fprintf('maxer=%8.3e, m=%4d\n',err,m)
%  evaluation of solution at test points
%% LOOCV
32 function ceps = costeps(eta,x,y,px,py,b,xn,yn)
33 xs=eta*px; ys=eta*py;
34 dx=x-xs'; dy=y-ys'; % difference matrix;
35 DM=sqrt(dx.^2+dy.^2);% distance matrix
36 G=DM.^2.*log(DM);
37 Gx=dx.*(1+2*log(DM));% x-der matrix
38 Gy=dy.*(1+2*log(DM));% y-der matrix
39 gln=xn.*Gx+yn.*Gy;% Neumann matrix
40 A=[G; gln]; % Global matrix
41 invA=pinv(A);
```

```
42 errorvector=(invA*b)./diag(invA);
43 ceps=norm(errorvector);
```

The following code produces the boundary collocation points and the corresponding normal vectors, source points, and test points.

```
1   function [x,y,xn,yn,xs,ys,xt,yt]=star_pts(m,nt,ns)
2   t=(2*pi/m)*(0:m-1)';
3   rt=(1+cos(5*t/2).^2)/1;
4   [x,y]=pol2cart(t,rt);
5   dx=gradient(x);    dy=gradient(y);
6   xn=dy./sqrt(dx.^2+dy.^2);
7   yn=-dx./sqrt(dx.^2+dy.^2);
8   t=(2*pi/ns)*(0:ns-1)';
9   xr=cos(t);   yr=sin(t);
10  xs=[xr;xr*1.1];  ys=[yr;yr*1.1];
11  p=haltonset(2,"Leap",1500);
12  q=net(p,m*25);
13  q1=q*4-2;
14  in3=inpolygon(q1(:,1),q1(:,2),x,y);
15  xt=q1(in3,1);  yt=q1(in3,2);
16  xt=xt(1:nt);  yt=yt(1:nt);% test pts
17  end
```

Second Alternative Formulation: We may also apply the following MFS approximation which is based on the auxiliary boundary biharmonic single layer representation, see, e.g. [678, 928],

$$u_N(\boldsymbol{x}) = \sum_{j=1}^{N} c_j G^B(\boldsymbol{x}, \boldsymbol{x}'_j) + \sum_{j=1}^{N} d_j \frac{\partial G^B}{\partial n(\boldsymbol{x}')}(\boldsymbol{x}, \boldsymbol{x}'_j), \quad \boldsymbol{x} \in \overline{\Omega}, \qquad (7.30)$$

where $\{\boldsymbol{x}'_j\}_{j=1}^{N}$ are the sources and $\{c_j\}_{j=1}^{N}$ and $\{d_j\}_{j=1}^{N}$ are unknown coefficients to be determined. The sources are placed on a pseudo-boundary $\partial\Omega'$ surrounding Ω and $\boldsymbol{n}(\boldsymbol{x}')$ is the outward unit normal vector to $\partial\Omega'$. Collocation of the BCs (7.19b) yields a system of the form (7.21), where for $i = 1, \ldots, M, j = 1, \ldots, N,$

$$(A_{11})_{i,j} = G^B(\boldsymbol{x}_i, \boldsymbol{x}'_j), \quad (A_{12})_{i,j} = \frac{\partial G^B}{\partial n(\boldsymbol{x}')}(\boldsymbol{x}_i, \boldsymbol{x}'_j), \qquad (7.31)$$

$$(A_{21})_{i,j} = \frac{\partial G^B}{\partial n(\boldsymbol{x})}(\boldsymbol{x}_i, \boldsymbol{x}'_j), \quad (A_{22})_{i,j} = \frac{\partial^2 G^B}{\partial n(\boldsymbol{x})\partial n(\boldsymbol{x}')}(\boldsymbol{x}_i, \boldsymbol{x}'_{N+j}). \quad (7.32)$$

Table 7.13. Numerical results for the 2D biharmonic problem using second alternative formulation for various initial search intervals for the source circle radius.

$[R_{\min}, R_{\max}]$	R	$\epsilon_{\max}$
[2, 4]	2.46	1.9(−12)
[2, 6]	2.92	2.9(−12)
[2, 8]	3.59	5.3(−11)
[2, 10]	3.13	4.5(−12)

The MFS has also been applied to BVPs governed by the polyharmonic equation $\Delta^{\mathcal{N}} u = 0$, $\mathcal{N} \in \mathbb{N}\backslash\{1\}$ in [456, 853], and also [473].

■ Example 7.14

For comparison, we again chose the same example as in Examples 7.11 and 7.13 for the second alternative formulation.

The results presented in Table 7.13 are also compatible with those in Tables 7.10 and 7.12 in terms of accuracy. On the other hand, the formulation and implementation of the current approach are more complicated than in the first alternative formulation. The results in Table 7.13 were obtained using the code `Biharmonic3.m` shown at the end of this example.

The point generator in Line 11 is the same as shown in the code `Biharmonic1.m` in Example 7.11.

```
%  Biharmonic3.m
1  function Biharmonic3
2  warning off all;
3  T=@(x,y) 3*exp(3*x).*(x.^2 + y.^2);
4  Q=@(x,y) cos(3*y).*exp(3*x);
5  Exsol=@(x,y) (x.^2+y.^2).*Q(x,y)+exp(2*y).*sin(2*x);
6  Exsolx=@(x,y) 2*cos(2*x).*exp(2*y)+2*x.*Q(x,y)+cos(3*y).*T(x,y);
7  Exsoly=@(x,y) 2*sin(2*x).*exp(2*y)+2*y.*Q(x,y)-sin(3*y).*T(x,y);
8  nn=250; m=nn; % m:# of bry pts
9  np=739;% np:# of test  pts,
10 ml=2; mr=4;  %search interval
%  collocation pts and sources
11 [x,y,xn,yn,xs0,ys0,xt,yt]=pts(m,np);
```

```
12 n=length(xs0);
13 RHS(1:m,1)=Exsol(x,y); % RHS Dirichlet BC
14 RHS(m+1:2*m,1)=Exsolx(x,y).*xn+Exsoly(x,y).*yn; %Neumann BC
15 Exact=Exsol(xt,yt);
16 eta1=fminbnd(@(eta) costeps(eta,x,y,xs0,ys0,RHS,xn,yn),...
17         ml,mr,optimset('TolX',1e-2));
18 xs=eta1*xs0; ys=eta1*ys0;
19 dx=gradient(xs);      dy=gradient(ys);
20 xns=dy./sqrt(dx.^2+dy.^2);    yns=-dx./sqrt(dx.^2+dy.^2);
%  Construction of system
21 R=pdist2([x,y],[xs,ys]); % distance matrix
22 A11=R.^2.*log(R);
23 Gx=(x-xs').*(1+2*log(R));% xder matrix
24 Gy=(y-ys').*(1+2*log(R));% yder matrix
25 Gxx=2*log(R) + (2*(x-xs').^2)./R.^2 + 1;
26 Gyy=2*log(R) + (2*(y-ys').^2)./R.^2 + 1;
27 Gxy=2*(x-xs').*(y-ys')./R.^2;
28 A12=xns'.*Gx+yns'.*Gy;
29 A21=xn.*Gx+yn.*Gy;
30 A22=xn.*(xns'.*Gxx+yns'.*Gxy)+yn.*(xns'.*Gxy+yns'.*Gyy);
31 A=[A11 -A12; A21 -A22];
32 sol=lsqminnorm(A,RHS);   %solve the MFS system
%  evaluation of solution at test points
33 Rp=pdist2([xt,yt],[xs,ys]); % distance matrix
34 F=Rp.^2.*log(Rp);
35 Fx=(xt-xs').*(1+2*log(Rp));% xder matrix
36 Fy=(yt-ys').*(1+2*log(Rp));% yder matrix
37 ffa=F; % coeff. matrix
38 ffb=-(xns'.*Fx+yns'.*Fy);
39 Appr=ffa*sol(1:n,1)+ffb*sol(n+1:2*n,1);
40 err=norm(Appr-Exact,inf)/norm(Exact,inf);
41 fprintf('maxer=%8.3e, eta1=%5.3f, m=%4d, mr=%3.1f\n',err,eta1,m,mr)
```

The following code is the LOOCV for seeking the optimal radius of the
source circle.

```
%  LOOCV
1  function ceps = costeps(eta,x,y,px,py,b,xn,yn)
2  xs=eta*px; ys=eta*py;
3  R=pdist2([x,y],[xs,ys]); % distance matrix
4  dx=gradient(xs);                dy=gradient(ys);
5  xns=dy./sqrt(dx.^2+dy.^2);    yns=-dx./sqrt(dx.^2+dy.^2);
6  A11=R.^2.*log(R);
7  Gx=(x-xs').*(1+2*log(R));% xder matrix
8  Gy=(y-ys').*(1+2*log(R));% yder matrix
```

```
9   Gxx=2*log(R) + (2*(x-xs').^2)./R.^2 + 1;
10  Gyy=2*log(R) + (2*(y-ys').^2)./R.^2 + 1;
11  Gxy=2*(x-xs').*(y-ys')./R.^2;
12  A12=xns'.*Gx+yns'.*Gy;
13  A21=xn.*Gx+yn.*Gy;
14  A22=xn.*(xns'.*Gxx+yns'.*Gxy)+yn.*(xns'.*Gxy+yns'.*Gyy);
15  A=[A11 -A12; A21 -A22];
16  invA=pinv(A);
17  errorvector=(invA*b)./diag(invA);
18  ceps=norm(errorvector);
```

7.5　Cauchy–Navier Equations

We first consider the Cauchy–Navier equations in $\mathbb{R}^2$ (see also (2.78))

$$\sigma \frac{\partial^2 u_1}{\partial x^2} + \tau \frac{\partial^2 u_2}{\partial x \partial y} + \frac{\partial^2 u_1}{\partial y^2} = 0, \qquad (7.33a)$$

$$\frac{\partial^2 u_2}{\partial x^2} + \tau \frac{\partial^2 u_1}{\partial x \partial y} + \sigma \frac{\partial^2 u_2}{\partial y^2} = 0, \qquad (7.33b)$$

with

$$\sigma = \frac{2 - 2v}{1 - 2v}, \quad \tau = \frac{1}{1 - 2v}, \qquad (7.34)$$

where v is Poisson's ratio. The above equations are subject to the Dirichlet BCs

$$u_1(x) = f_1(x) \quad \text{and} \quad u_2(x) = f_2(x), \quad x \in \partial\Omega, \qquad (7.35)$$

where $\partial\Omega$ is the boundary of the domain Ω.

We approximate the solution (u_1, u_2) of (7.33) and (7.35) by

$$\hat{u}_1(x) = \sum_{j=1}^{N} c_j\, G_{11}(x, x_j') + \sum_{j=1}^{N} d_j\, G_{12}(x, x_j'), \quad x \in \overline{\Omega}, \qquad (7.36a)$$

$$\hat{u}_2(x) = \sum_{j=1}^{N} c_j\, G_{21}(x, x_j') + \sum_{j=1}^{N} d_j\, G_{22}(x, x_j'), \quad x \in \overline{\Omega}, \qquad (7.36b)$$

where $\{x'_j\}_{j=1}^N$ are the sources, and $\{c_j\}_{j=1}^N$ and $\{d_j\}_{j=1}^N$ are unknown coefficients to be determined. In (7.36), the tensor $\{G_{ij}(x, x')\}_{i,j=1,2}$ is a fundamental solution of the 2D Cauchy-Navier operator defined by [510, Chapter 6.2] (see Section 2.8)

$$G_{11}(x, x') = \frac{1}{8\pi\mu(1-v)}\left[-(3-4v)\ln r + \frac{(x-x')^2}{r^2}\right], \qquad (7.37a)$$

$$G_{12}(x, x') = \frac{1}{8\pi\mu(1-v)}\left[\frac{(x-x')(y-y')}{r^2}\right], \qquad (7.37b)$$

$$G_{21}(x, x') = G_{12}(x, x'), \qquad (7.37c)$$

$$G_{22}(x, x') = \frac{1}{8\pi\mu(1-v)}\left[-(3-4v)\ln r + \frac{(y-y')^2}{r^2}\right], \qquad (7.37d)$$

where $r = \|x - x'\|$ and μ is the shear modulus.

The values of u_1 and u_2 are prescribed at the boundary collocation points $\{x_i\}_{i=1}^M$ leading to the $2M \times 2N$ system of equations

$$\left(\begin{array}{c|c}A_{11} & A_{12} \\ \hline A_{21} & A_{22}\end{array}\right)\left(\frac{c}{d}\right) = \left(\frac{f_1}{f_2}\right), \qquad (7.38)$$

where $f_1 = [f_1(x_1), f_1(x_2), \ldots, f_1(x_M)]^T$, $f_2 = [f_2(x_1), f_2(x_2), \ldots, f_2(x_M)]^T$, $c = [c_1, c_2, \ldots, c_N]^T$, and $d = [d_1, d_2, \ldots, d_N]^T$. Also, for $i = 1, \ldots, M$, and $j = 1, \ldots, N$,

$$(A_{k\ell})_{i,j} = G_{k\ell}(x_i, x'_j), \quad k, \ell = 1, 2. \qquad (7.39)$$

We next consider the Cauchy–Navier equations in $\mathbb{R}^3$

$$\sigma\frac{\partial^2 u_1}{\partial x^2} + \frac{\partial^2 u_1}{\partial y^2} + \frac{\partial^2 u_1}{\partial z^2} + \tau\frac{\partial^2 u_2}{\partial x\partial y} + \tau\frac{\partial^2 u_3}{\partial x\partial z} = 0, \qquad (7.40a)$$

$$\frac{\partial^2 u_2}{\partial x^2} + \sigma\frac{\partial^2 u_2}{\partial y^2} + \frac{\partial^2 u_2}{\partial z^2} + \tau\frac{\partial^2 u_1}{\partial x\partial y} + \tau\frac{\partial^2 u_3}{\partial y\partial z} = 0, \qquad (7.40b)$$

$$\frac{\partial^2 u_3}{\partial x^2} + \frac{\partial^2 u_3}{\partial y^2} + \sigma\frac{\partial^2 u_3}{\partial z^2} + \tau\frac{\partial^2 u_1}{\partial x\partial z} + \tau\frac{\partial^2 u_2}{\partial y\partial z} = 0. \qquad (7.40c)$$

The above equations are subject to the Dirichlet BCs

$$u_1(x) = f_1(x), \quad u_2(x) = f_2(x) \text{ and } u_3(x) = f_3(x), \quad x \in \partial\Omega. \quad (7.41)$$

We approximate the solution (u_1, u_2, u_3) of (7.40) and (7.41) by

$$\hat{u}_1(x) = \sum_{j=1}^{N} c_j\, G_{11}(x, x'_j) + \sum_{j=1}^{N} d_j\, G_{12}(x, x'_j) + \sum_{j=1}^{N} e_j\, G_{13}(x, x'_j),$$

$$(7.42a)$$

$$\hat{u}_2(x) = \sum_{j=1}^{N} c_j\, G_{21}(x, x'_j) + \sum_{j=1}^{N} d_j\, G_{22}(x, x'_j) + \sum_{j=1}^{N} e_j\, G_{23}(x, x'_j),$$

$$(7.42b)$$

$$\hat{u}_3(x) = \sum_{j=1}^{N} c_j\, G_{31}(x, x'_j) + \sum_{j=1}^{N} d_j\, G_{32}(x, x'_j) + \sum_{j=1}^{N} e_j\, G_{33}(x, x'_j),$$

$$(7.42c)$$

where $\{x'_j\}_{j=1}^{N}$ are the sources, and $\{c_j\}_{j=1}^{N}$, $\{d_j\}_{j=1}^{N}$, and $\{e_j\}_{j=1}^{N}$ are unknown coefficients to be determined. In (7.42), the tensor $\{G_{ij}(x, x')\}_{i,j=1,2,3}$ is a fundamental solution of the 3D Cauchy–Navier operator defined by [510, Chapter 6.2] (see Section 2.8)

$$G_{11}(x, x') = \frac{1}{16\pi\,\mu(1-v)\,r}\left[(3-4v) + \frac{(x-x')^2}{r^2}\right], \qquad (7.43a)$$

$$G_{12}(x, x') = \frac{1}{16\pi\,\mu(1-v)}\frac{(x-x')(y-y')}{r^3}, \qquad (7.43b)$$

$$G_{13}(x, x') = \frac{1}{16\pi\,\mu(1-v)}\frac{(x-x')(z-z')}{r^3}, \qquad (7.43c)$$

$$G_{22}(x, x') = \frac{1}{16\pi\,\mu(1-v)\,r}\left[(3-4v) + \frac{(y-y')^2}{r^2}\right], \qquad (7.43d)$$

$$G_{23}(x, x') = \frac{1}{16\pi\,\mu(1-v)}\frac{(y-y')(z-z')}{r^3}, \qquad (7.43e)$$

$$G_{33}(x, x') = \frac{1}{16\pi\,\mu(1-v)\,r}\left[(3-4v) + \frac{(z-z')^2}{r^2}\right]. \qquad (7.43f)$$

In the above equations, we note the symmetries $G_{12} = G_{21}$, $G_{13} = G_{31}$ and $G_{23} = G_{32}$.

The values of u_1, u_2 and u_3 are prescribed at the boundary collocation points $\{x_i\}_{i=1}^M$ leading to the $3M \times 3N$ system of equations

$$\left(\begin{array}{c|c|c} A_{11} & A_{12} & A_{13} \\ \hline A_{21} & A_{22} & A_{23} \\ \hline A_{31} & A_{32} & A_{33} \end{array}\right) \left(\begin{array}{c} c \\ \hline d \\ \hline e \end{array}\right) = \left(\begin{array}{c} f_1 \\ \hline f_2 \\ \hline f_3 \end{array}\right), \tag{7.44}$$

where $c = [c_1, c_2, \ldots, c_N]^T$, $d = [d_1, d_2, \ldots, d_N]^T$, $e = [e_1, e_2, \ldots, e_N]^T$, and $f_k = [f_k(x_1), f_k(x_2), \ldots, f_k(x_M)]^T$, $k = 1, 2, 3$. Also, for $i = 1, \ldots, M$, and $j = 1, \ldots, N$,

$$(A_{k\ell})_{i,j} = G_{k\ell}(x_i, x_j'), \quad k, \ell = 1, 2, 3. \tag{7.45}$$

7.5.1 *Numerical implementation in 2D*

In this section, we examine the numerical solution of 2D linear elasticity problems.

■ Example 7.15

We first consider BVP (7.33) and (7.35) in 2D when Ω is the star-shaped domain in Fig. 7.16. The Dirichlet BCs are given based on the exact solution

$$u_1(x, y) = x^2 - cy^2 + xy,$$
$$u_2(x, y) = -cx^2 + y^2 + xy,$$

where $c = (5 - 4\nu)/(2 - 4\nu)$.

We choose $\nu = 0.3$, $\mu = 1$, $M = 100$ boundary collocation points, $N = 50$ source points, and $L = 179$ interior test points. In Table 7.14, we present the maximum relative errors $\epsilon_{\max}$ in u_1 and u_2 obtained using LOOCV with various search intervals, and R denotes the radius of the source circle.

The following code `CauchyNavier_2D_LOOCV.m` solves the above BVP using LOOCV. The subprogram `pts.m` in Line 7 is available in Appendix C.

```
%   CauchyNavier2D_LOOCV.m
%   2D Cauchy-Navier equations in star domain with Dirichlet BC's
1   function CauchyNavier2D_LOOCV
```

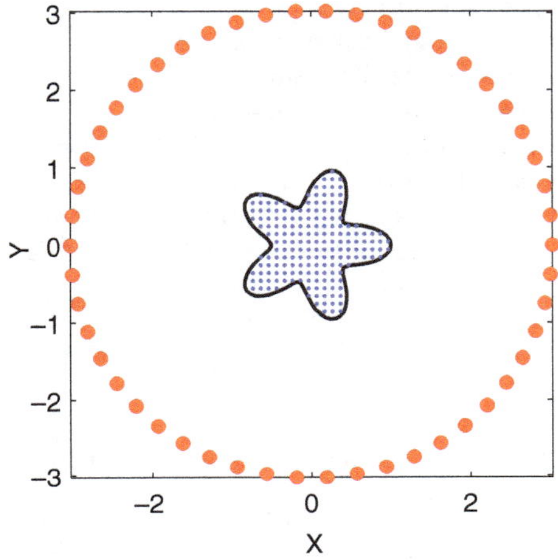

Figure 7.16. Profile of the boundary (black), source (red), and test (blue) points.

Table 7.14. Maximum relative errors using LOOCV with $M = 100$, $N = 50$ for various search intervals.

$[R_{\min}, R_{\max}]$	R	$\epsilon_{\max}(u_1)$	$\epsilon_{\max}(u_2)$
[1.2, 5]	3.39	7.6(−14)	7.6(−14)
[1.2, 6]	3.50	1.0(−13)	7.8(−14)
[1.2, 7]	3.41	6.9(−14)	7.2(−14)
[1.2, 8]	5.47	9.9(−14)	9.0(−14)

```
2  clear; warning off all;
3  u1 = @(x,y,c) x.^2-c*y.^2+x.*y; % Exact solution
4  u2 = @(x,y,c) -c*x.^2+y.^2+x.*y; % Exact solution
5  M=100; N=50; % M:# boundary points; N: # source points
6  nu=0.3;  mu=1;   c=(5-4*nu)/(2-4*nu);
7  [x,y,xs0,ys0,xt,yt,~,~]= pts(M,N);
8  rhs=[u1(x,y,c); u2(x,y,c)];
9  ut1 = u1(xt,yt,c);  % exact solution
10 ut2 = u2(xt,yt,c);  % exact solution
11 L=1.2; R=8;  %search interval for LOOCV
12 d=fminbnd(@(c) costEps(c,rhs,x,y,xs0,ys0,mu,nu),L,R);
13 xs=d*xs0;   ys=d*ys0;
14 [G11,G12,G22]=FS(x,y,xs,ys,mu,nu);
15 A=[G11 G12; G12 G22];   %G21=G12
```

```
16 sol=A\rhs; % system solution
17 [GT11, GT12,GT22]=FS(xt,yt,xs,ys,mu,nu);
18 uh1=[GT11 GT12]*sol;% approximate solution
19 uh2=[GT12 GT22]*sol;% approximate solution
20 error1 = norm(uh1-ut1,inf)/norm(ut1,inf);  %relative error
21 error2 = norm(uh2-ut2,inf)/norm(ut2,inf);  %relative error
22 rmse1=norm(uh1-ut1,2)/sqrt(length(xt));   %RMSE error
23 rmse2=norm(uh2-ut2,2)/sqrt(length(xt));   %RMSE error
24 fprintf('M=%4d,  N=%4d, R =%5.3f\n',M,N,d);
25 fprintf('rel_er_1=%8.3e, RMSE_1=%8.3e\n',error1,rmse1)
26 fprintf('rel_er_2=%8.3e, RMSE_1=%10.3e\n',error2,rmse2)
27 end
%% Matrix of fundamental solutions
28 function [G11, G12, G22]=FS(x,y,xs,ys,mu,nu)
29 DM=pdist2([x y],[xs  ys]);
30 DMX=x-xs';   DMY=y-ys';   %DifferenceMatrix
31 coeff=(1/(8*pi*mu*(1-nu)));
32 G11=coeff*((3-4*nu)*log(1./DM)+(DMX.^2./DM.^2));
33 G12=coeff*((DMX.*DMY)./DM.^2);
34 G22=coeff*((3-4*nu)*log(1./DM)+(DMY.^2./DM.^2));
35 end
%% LOOCV
36 function ceps = costEps(ep,rhs,x,y,xs0,ys0,mu,nu)
37 xs=ep*xs0;   ys=ep*ys0;
38 [G11,G12,G22]=FS(x,y,xs,ys,mu,nu);
39 A=[G11 G12; G12 G22];   %G21=G12
40 invA=pinv(A);
41 errorvector = (A\rhs)./diag(invA);
42 ceps = norm(errorvector);
43 end
```

7.5.2 *Numerical implementation in 3D*

We finally consider the 3D Cauchy–Navier equations.

■ Example 7.16

We solve the 3D Cauchy–Navier equations with Dirichlet BCs in the airplane domain shown in Fig. 7.17. The Dirichlet BCs (7.41) are derived from the exact solution

$$u_1(x, y, z) = \frac{3 - 4v}{\rho} + \frac{(x - 10)^2}{\rho^3},$$

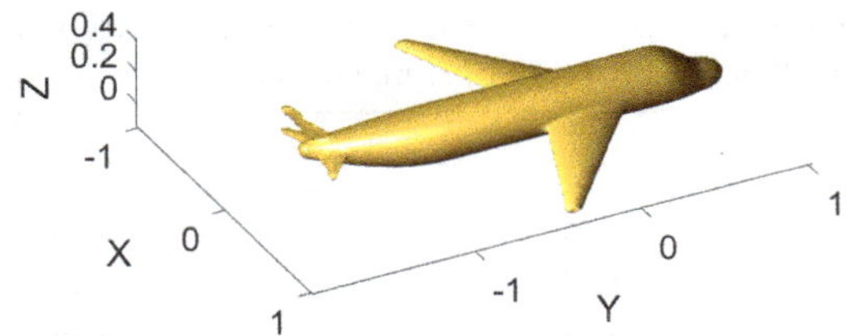

Figure 7.17. Profile of the airplane domain.

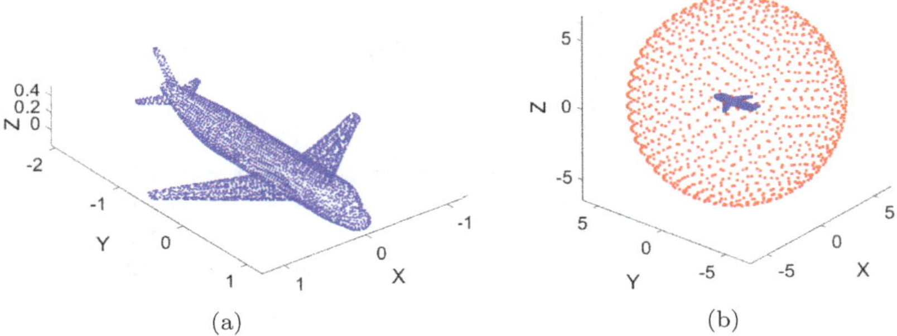

(a)   (b)

Figure 7.18. (a) Profile of 1105 boundary points on the airplane surface. (b) Typical distributions of source points (red) and boundary collocation points (blue).

$$u_2(x, y, z) = \frac{(x - 10)y}{\rho^3},$$

$$u_3(x, y, z) = \frac{(x - 10)z}{\rho^3},$$

where $\rho = \sqrt{(x - 10)^2 + y^2 + z^2}$.

We choose $\nu = 0.3$, $\mu = 1$, $M = N = 1105$ boundary collocation and source points, and $L = 2087$ interior test points; see Fig. 7.18. The geometric center of the airplane $(0, -0.41, 0.1641)$ is set as the center of the source sphere. In Table 7.15, we present the maximum relative errors $\epsilon_{\max}$ obtained using LOOCV with various search intervals.

The code `CauchyNavier_3D_LOOCV.m` solves the above BVP using LOOCV. The subprogram `unit_sphere.m` can be found in Appendix C.

Table 7.15. Maximum relative errors using LOOCV with $M = N = 1105$ for various search intervals.

$[R_{\min}, R_{\max}]$	R	$\epsilon_{\max}(u_1)$	$\epsilon_{\max}(u_2)$	$\epsilon_{\max}(u_3)$
$[1.5, 8]$	6.33	1.2(−13)	3.9(−12)	8.9(−12)
$[1.5, 9]$	6.87	7.6(−14)	5.9(−13)	3.8(−12)
$[1.5, 10]$	6.14	6.4(−14)	1.2(−12)	1.9(−12)
$[1.5, 11]$	5.97	1.4(−13)	1.8(−12)	4.7(−12)

```
%   CauchyNavier_3D_LOOCV.m
%   3D Cauchy-Navier equations in airplane with Dirichlet BC's
1   function CauchyNavier_3D_LOOCV
2   clear;  warning off all;
3   r=@(x,y,z) sqrt((x-10).^2+y.^2+z.^2);
4   u1=@(x,y,z,nu) (3-4*nu)./r(x,y,z)+(x-10).^2./r(x,y,z).^3;
5   u2=@(x,y,z,nu) (x-10).*y./r(x,y,z).^3;    % exact solution
6   u3=@(x,y,z,nu) (x-10).*z./r(x,y,z).^3;    % exact solution
%   Input boundary points (x,y,z) and interior points "int"
7   load airplane.mat x y z int
8   nu=0.3;  mu=1;
9   L=1.5; R=11; %search interval of LOOCV
10  M=length(x);
11  N=M;
12  [xs0,ys0,zs0]=unit_sphere(N); %Unit sphere
13  X=int(:,1); Y=int(:,2); Z=int(:,3);
14  ut1 = u1(X,Y,Z,nu);  % exact solution
15  ut2 = u2(X,Y,Z,nu);  % exact solution
16  ut3 = u3(X,Y,Z,nu);  % exact solution
17  rhs=[u1(x,y,z,nu); u2(x,y,z,nu); u3(x,y,z,nu)];
18  R=fminbnd(@(c) costEps(c,rhs,x,y,z,xs0,ys0,zs0,mu,nu),L,R);
19  xs=R*xs0; ys=R*ys0-0.41; zs=R*zs0+0.1641;
%   Construction of global matrix and rhs
20  [G11, G12, G13, G22,G23,G33]=FS(x,y,z,xs,ys,zs,mu,nu);
21  A=[G11 G12 G13; G12 G22 G23; G13 G23 G33];
22  sol=A\rhs; % system solution
%%%%%%%
23  [GT11, GT12, GT13, GT22,GT23,GT33]=FS(X,Y,Z,xs,ys,zs,mu,nu);
24  uh1=[GT11 GT12 GT13]*sol;
25  uh2=[GT12 GT22 GT23]*sol;
26  uh3=[GT13 GT23 GT33]*sol;
27  err1 = norm(uh1-ut1,inf)/norm(ut1,inf);  %relative error
28  err2 = norm(uh2-ut2,inf)/norm(ut2,inf);  %relative error
29  err3 = norm(uh3-ut3,inf)/norm(ut3,inf);  %relative error
30  rm1=norm(uh1-ut1,2)/sqrt(length(X));    %RMSE error
```

```
31 rm2=norm(uh2-ut2,2)/sqrt(length(X));    %RMSE error
32 rm3=norm(uh3-ut3,2)/sqrt(length(X));    %RMSE error
33 fprintf('M=%4d,   N=%4d,  R =%5.3f\n',M,N,R)
34 fprintf('RMSE1=%8.3e,RMSE2=%8.3e,RMSE3=%8.3e\n',rm1,rm2,rm3)
35 fprintf('Err1=%8.3e,Err2=%8.3e,Err3=%8.3e\n',err1,err2,err3)
36 end
%% Matrix of fundamental solutions
37 function [G11,G12,G13,G22,G23,G33]=FS(x,y,z,xs,ys,zs,mu,nu)
38 coeff=1/(16*pi*mu*(1-nu));
39 DM=pdist2([x y z], [xs ys zs]);
40 DX=x-xs';  DY=y-ys';  DZ=z-zs';  %DifferenceMatrix;
41 G11=coeff*(1./DM).*((3-4*nu)+(DX.^2./DM.^2));
42 G12=coeff*(1./DM).*((DX.*DY)./DM.^2);
43 G13=coeff*(1./DM).*((DX.*DZ)./DM.^2);
44 G22=coeff*(1./DM).*((3-4*nu)+(DY.^2./DM.^2));
45 G23=coeff*(1./DM).*((DY.*DZ)./DM.^2);
46 G33=coeff*(1./DM).*((3-4*nu)+(DZ.^2./DM.^2));
47 end
%% LOOCV
48 function ceps = costEps(ep,rhs,x,y,z,xs0,ys0,zs0,coeff,nu)
49 xs=ep*xs0; ys=ep*ys0-0.41; zs=ep*zs0+0.1641;
50 [G11, G12, G13, G22,G23,G33]=FS(x,y,z,xs,ys,zs,mu,nu);
51 A=[G11 G12 G13; G12 G22 G23; G13 G23 G33];
52 invA=inv(A);
53 errorvector = (A\rhs)./diag(invA);
54 ceps = norm(errorvector);
55 end
```

Chapter 8

The Method of Particular Solutions

For several years after its introduction as a numerical technique, the MFS was confined to solving BVPs governed by homogeneous PDEs. In 1996, Golberg and Chen [335] proposed the extension of the MFS to solving inhomogeneous problems using RBFs to approximate a particular solution of the given governing equation. The procedure for approximating the particular solution is called the MPS, which has been extensively discussed in Chapter 4 and the combined method is known as the MPS-MFS. In this chapter, we present the numerical implementation of the MPS-MFS with MATLAB® codes. Sections 8.2 and 8.3 present the MPS-MFS using the MQ RBF and polynomial basis functions, respectively. In Section 8.6, we employ the fundamental solution of Helmholtz-type operators as the basis function for the particular solution approximation.

8.1 Two-Step MPS-MFS

One of the standard methods for solving BVPs is to split the solution of the governing PDE into a particular solution and a homogeneous one. Since the particular solution is not unique and no BC satisfaction is required, various techniques can be applied to approximate it, as described in Chapter 4, while the homogeneous solution will be approximated by the MFS. The solution procedure of this two-step approach was briefly discussed in Section 4.1. A more detailed description is given below.

369

Let $\Omega \subset \mathbb{R}^d$, $d = 2, 3$, be a closed and bounded domain with boundary $\partial\Omega$. Consider the BVP

$$\mathcal{L}\phi(x) = f(x), \quad x \in \Omega, \tag{8.1a}$$

$$\mathcal{B}\phi(x) = g(x), \quad x \in \partial\Omega, \tag{8.1b}$$

where $\mathcal{L}$ is a linear partial differential operator, $\mathcal{B}$ a boundary differential operator, and f and g are known functions, or unknown functions whose discrete values are provided at a set of points.

The first step in solving (8.1) is to obtain a particular solution ϕ_p satisfying (8.1a) without considering the BC (8.1b); i.e.

$$\mathcal{L}\phi_p(x) = f(x). \tag{8.2}$$

For a general function $f(x)$, we seek to approximate it using certain basis functions, for example, RBFs. For an RBF approximation, we choose a set of RBF centers $\{\xi_j\}_{j=1}^n$ scattered within and outside the domain Ω. We then select a set of collocation points $\{x_i\}_{i=1}^m$, which can be different from the RBF centers. Figure 8.1 gives an illustration of the distribution of the RBF centers (in red) and collocation points (in blue) for a gear-shaped domain. When the RHS $f(x)$ is an explicit function and can be smoothly extended outside the domain, these nodes can cover the inside and

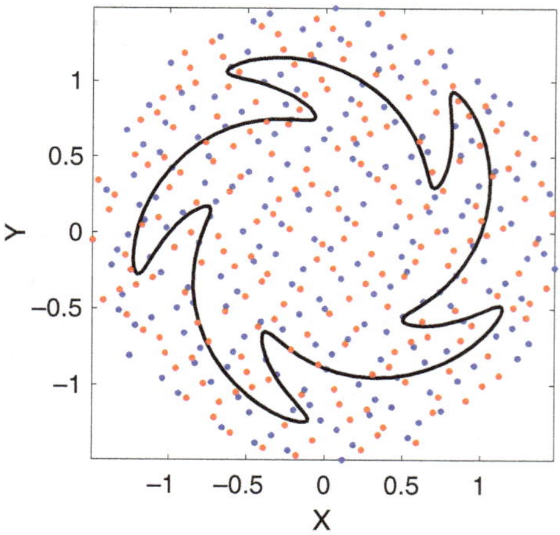

Figure 8.1. The profile of the distribution of collocation (blue dots) and RBF centers (red dots).

outside of Ω, which generally yields a better approximation. Furthermore, for complicated multi-connected domains, especially in 3D, these nodes can be distributed freely covering the domain regardless of the complexity of the given domain. On the other hand, when $f(x)$ is given as data at discrete points in Ω, it is obvious that the collocation points must coincide with these data points. We should also mention that it is a common practice to use the same set of points for the collocation and the RBF centers.

When $f(x)$ is expressed in terms of the RBF $\psi(r)$, the approximation formula is

$$\hat{f}(x) = \sum_{j=1}^{n} d_j \, \psi(r_j), \tag{8.3}$$

where $r_j = \|x - \xi_j\|$. By performing collocation at $\{x_i\}_{i=1}^{n}$, we have

$$\sum_{j=1}^{n} d_j \, \psi(r_{ij}) = f(x_i), \quad i = 1, 2, \ldots, n, \tag{8.4}$$

where $r_{ij} = \|x_i - \xi_j\|$. If the collocation nodes and RBF centers coincide, the above system of equations is symmetric. Once the unknown coefficients d_j have been determined, an approximate particular solution $\hat{\phi}_p$ of (8.2) can be obtained as

$$\hat{\phi}_p(x) = \sum_{j=1}^{n} d_j \, \varphi^p(r_j), \tag{8.5}$$

where $\varphi^p(r)$ is a particular solution satisfying the equation

$$\mathcal{L}\varphi^p(r) = \psi(r). \tag{8.6}$$

The derivation of closed-form particular solutions φ^p for various differential operators $\mathcal{L}$ and basis functions ψ can be found in Chapter 4.

8.2 MPS-MFS by Multiquadric Basis Functions

As the MQ-RBF

$$\psi(r) = \sqrt{r^2 + c^2} \tag{8.7}$$

defined in (3.76) is highly accurate in interpolating a function, it can be chosen to approximate $f(x)$. For $\mathcal{L} \equiv \Delta$ and Δ^2, the closed-form particular solutions φ^L and φ^B have been derived in (4.138) and (4.144), respectively. For other differential operators, however, closed-form particular solutions corresponding to the MQ-RBF are not yet available. To fill in the gap, we shall present a procedure that can construct a particular solution by "reverse engineering". That is, we start from a particular solution of a certain form, and then back derive the basis functions that approximate $f(x)$.

8.2.1 *Polymetaharmonic operators*

We give a demonstration of finding a particular solution for the polymeta-harmonic operator

$$\mathcal{L} \equiv \Delta^2 + \alpha\,\Delta + \beta, \tag{8.8}$$

where α and β are constants. First, we seek particular solutions for the operators Δ and Δ^2 corresponding to a basis function ψ such that

$$\Delta\varphi^L(r) = \psi(r), \tag{8.9}$$

$$\Delta^2\varphi^B(r) \equiv \Delta\Delta(\varphi^B(r)) = \psi(r). \tag{8.10}$$

From (8.9) and (8.10), we have $\Delta(\varphi^B(r)) = \varphi^L(r)$. As stated earlier, ψ is the MQ-RBF defined in (8.7), φ^L and φ^B have been derived and are respectively given in (4.138) and (4.144). We then assume that an approximate particular solution of the operator in (8.8) can be expressed in the form

$$\hat{\phi}_p(x) = \sum_{j=1}^{n} d_j\,\varphi^B(r_j). \tag{8.11}$$

Substituting the above into (8.8), we obtain

$$\mathcal{L}\hat{\phi}_p(x) = \sum_{j=1}^{n} d_j \left[\Delta^2\varphi^B(r_j) + \alpha\,\Delta\varphi^B(r_j) + \beta\,\varphi^B(r_j) \right]$$

$$= \sum_{j=1}^{n} d_j \left[\psi(r_j) + \alpha\,\varphi^L(r_j) + \beta\,\varphi^B(r_j) \right]$$

$$= \hat{f}(x), \tag{8.12}$$

Hence, if we approximate $f(x)$ by a linear combination of the basis function $\Psi(r)$,

$$\hat{f}(x) = \sum_{j=1}^{n} d_j \Psi(r_j), \qquad (8.13)$$

where

$$\Psi(r) = \psi(r) + \alpha \varphi^L(r) + \beta \varphi^B(r), \qquad (8.14)$$

its corresponding particular solution is given by (8.11).

8.2.2 *Helmholtz-type operators*

For the Helmholtz and modified Helmholtz operators

$$\mathcal{L} \equiv \Delta \pm k^2, \qquad (8.15)$$

we seek a particular solution of the form

$$\hat{\phi}_p(x) = \sum_{j=1}^{n} d_j \varphi^L(r_j). \qquad (8.16)$$

Following a procedure similar to (8.12), we find that if we approximate $f(x)$ using the basis function

$$\Psi(r) = \psi(r) \pm k^2 \varphi^L(r), \qquad (8.17)$$

as in (8.13), then (8.16) is its particular solution.

8.2.3 *Ghost point method*

As described in Section 8.1, in the second step of the proposed technique, we solve the following homogeneous equation subject to the updated BC by the MFS:

$$\mathcal{L}\phi_h(x) = 0, \qquad\qquad x \in \Omega, \qquad (8.18a)$$

$$\mathcal{B}\phi_h(x) = g(x) - \mathcal{B}\hat{\phi}_p(x), \quad x \in \partial\Omega. \qquad (8.18b)$$

The approximate homogeneous solution $\hat{\phi}_h$ can thus be computed using the MFS. The approximate solution $\hat{\phi}$ of (8.1) can then be obtained from $\hat{\phi} = \hat{\phi}_h + \hat{\phi}_p$.

The above two-step MPS-MFS has been successfully applied to solving various types of BVPs (see Section 1.7 for a review). In most implementations, the collocation nodes and RBF centers are located in $\overline{\Omega}$. Recently, a new technique in which the RBF centers are placed far outside the domain has been proposed to considerably increase the accuracy of the particular solution [955]. This approach is called the fictitious or ghost point method. The basic idea of this approach resembles that of the MFS where the source points are placed outside the domain due to the singularity of the fundamental solution. As RBFs are nonsingular, we have the option of placing them inside and outside the domain.

The profile of a typical distribution of collocation and fictitious center points is presented in Fig. 8.2. In contrast to the distribution in the traditional MPS-MFS approach shown in Fig. 8.1, the RBF centers or fictitious centers in Fig. 8.2 are distributed far outside the domain. Note that there are various ways of distributing the fictitious points and, in this chapter, their deployment is quite different from that in [955]. Furthermore, we apply the LOOCV algorithm to determine both the optimal shape parameter value for the MQ-RBF approximation and the source location in the MFS.

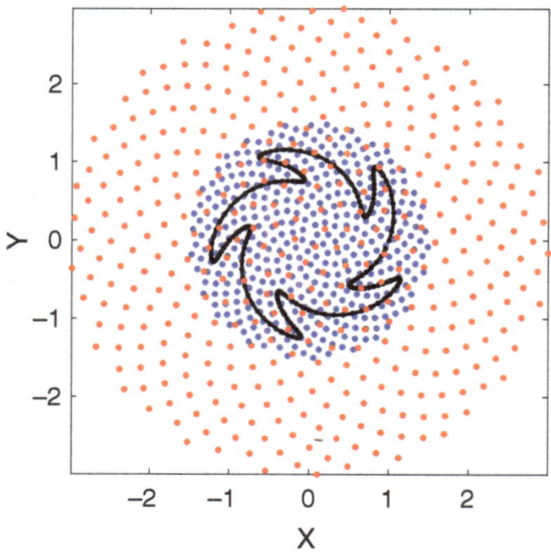

Figure 8.2. The profile of the collocation (blue) and fictitious (red) points covering a gear-shaped domain for the ghost point method.

8.2.4 *Numerical implementation*

We next examine some numerical examples.

Modified Helmholtz Equation: We solve a BVP for the 2D modified Helmholtz equation, where the governing equation and BC are

$$(\Delta - k^2)\phi(x) = f(x), \quad x \in \Omega, \tag{8.19a}$$

$$\phi(x) = g(x), \qquad\qquad x \in \partial\Omega, \tag{8.19b}$$

and Ω is the gear-shaped domain in Fig. 8.2, which is defined by the parametric equation

$$\Omega = \{(x, y) | x = \rho(t)\cos(\theta(t)), \, y = \rho(t)\sin(\theta(t)), 0 \le t \le 2\pi\}, \tag{8.20}$$

where

$$\rho(t) = 2 + \frac{1}{2}\sin(5t), \quad \theta(t) = t + \frac{1}{2}\sin(5t). \tag{8.21}$$

■ **Example 8.1**

We solve (8.19) with $k = 3$ and the functions $f(x, y)$ and $g(x, y)$ are derived from the exact solution

$$\phi(x, y) = e^{x+y}.$$

The profile of the gear-shaped domain Ω and the collocation and fictitious points are shown in Fig. 8.2, where each set is distributed in a circular sunflower-shaped region. The radius of the disk containing the collocation points is 1.5 so that it covers the gear-shaped domain. The fictitious points are extended further beyond the domain and the radius of the circle containing them is denoted by R_g. The MFS source points are distributed on a circle centered at the origin with radius R_s.

To obtain an approximate particular solution in the first step, we take 400 collocation points and the same number of fictitious centers. For the approximation of the homogeneous solution with the MFS, we choose 200 boundary collocation and source points. In Table 8.1, we present results for the optimal shape parameter values c, radii

Table 8.1. Results for the optimal shape parameter and the radius of the MFS source circle using LOOCV with 200 boundary and source points and 400 collocation and fictitious points.

R_g	c	R_s	$\mathcal{E}_{max}$	ϵ_{max}
1.3	0.72	2.83	2.5(−6)	4.5(−7)
2.0	2.45	2.81	3.8(−8)	1.5(−9)
2.5	1.07	2.84	1.5(−11)	2.8(−12)
3.0	1.39	2.82	1.5(−11)	2.7(−12)
3.5	1.26	2.84	1.0(−11)	1.9(−12)
4.0	1.52	2.82	1.3(−11)	2.4(−12)
4.5	1.87	2.81	4.0(−11)	7.3(−12)

of the source circle R_s, maximum absolute errors $\mathcal{E}_{max}$, and maximum relative errors ϵ_{max}, for various sizes of the RBF fictitious circle R_g obtained using LOOCV. We observe that the accuracy increases sharply when R_g becomes larger and remains stable beyond that. In the traditional MPS-MFS where the collocation and RBF centers are all placed inside the domain, we obtain $c = 0.686$, $R_s = 2.81$, $\mathcal{E}_{max} = 2.2(−6)$, and $\epsilon_{max} = 4.1(−7)$. The accuracy of the MPS-MFS improves considerably by simply distributing the RBF centers at appropriate locations outside the domain. The fictitious MQ-RBF centers play a role similar to that of the source points in the MFS, regarding accuracy.

The code `MPS_Loocv_mq.m` used to solve the above problem is listed below. Line 2 shows the closed-form MQ particular solution of the Laplace operator φ^L given in (4.138a). In Line 7, the boundary points, collocation points, fictitious points, and test points are generated. In Lines 13–18, an approximate particular solution is obtained and the BC is updated for calculating the homogeneous solution in the second step. The homogeneous problem is solved in Lines 19–24. The final approximate solution at a set of test points is evaluated in Line 29.

Note that in Lines 16 and 24, the least squares solver `lsqminnorm` was used instead of the linear solver `backslash`. The reason for using `lsqminnorm` was addressed in Section 7.2.1. In Example 8.1, when we used the `backslash` command, we obtained for the first line in Table 8.1

$c = 0.69$, $R_s = 2.84$, $\mathcal{E}_{max} = 7.84(-5)$, and $\epsilon_{max} = 1.39(-5)$, which is one order less accurate than using `lsqminnorm`. The accuracy in Table 8.1 generally drops by one to two orders of magnitude if `backslash` is used.

```
%   MPS_Loocv_mq.m
%   Two-step MPS-MFS for solving modified Helmholtz equation
1   clear; warning off all;
2   mps2 = @ (c,r, mq) mq.*(r.^2+4*c^2)/9-c^3*log(mq+c)/3;
3   rbf=@(r,c) sqrt(r.^2+c^2);
4   forcing=@(x,y,lam) (2-lam^2)*exp(1*x+y);   %forcing term
5   bound_con=@(x,y) exp(x+y);
%   nb: # of boundary points; n:# of RBF coll and centers
6   nb=200;   n=400;   D=3;   lam=3;
%   (x,y): boundary points; (xc,yc):RBF collocation points
%   (xg,yg): fictitious points; (xt,yt): test points
7   [x,y,xc,yc,xg,yg,xt,yt]=gear(nb,n,D);
8   center=[xg, yg];
9   f=forcing(xc,yc,lam);     %Forcing term
10  exact=bound_con(xt,yt);  %exact solution
11  DM1=pdist2([xc,yc],center);
12  DM2=pdist2([x y],center);
%% LOOCV for RBF
13  c = fminbnd(@(c) cost_loocv_mq(c,rbf,f,DM1,lam,mps2),0,4);
14  mq1=sqrt(DM1.^2.+c*c);
15  A1=rbf(DM1,c)-lam^2*mps2(c,DM1,mq1);   %Governing equation
16  coef_rbf=lsqminnorm(A1,f);
17  mq=sqrt(DM2.^2.+c*c);%
18  g=bound_con(x,y)-mps2(c,DM2,mq)*coef_rbf;   %update BC
%% LOOCV for MFS
19  [sx0,sy0]=pol2cart(2*pi*(1:nb)'/nb,1);
20  R=fminbnd(@(d) cost_loocv2(d,x,y,sx0,sy0,g,lam),1.5,5);
21  xs=R*sx0;     ys=R*sy0;   %source points of the MFS
22  DM=pdist2([x,y],[xs,ys]);
23  A2=besselk(0,lam* DM);     % boundary condition
24  coef_mfs= lsqminnorm(A2,g);
%% Test
25  DM5=pdist2([xt yt],center);
26  DM4=pdist2([xt yt],[xs, ys]);
27  mq=sqrt(DM5.^2.+c*c);%
28  A3=mps2(c,DM5,mq);        A4=besselk(0,lam*DM4);
29  approx=A3*coef_rbf+A4*coef_mfs;   %Approximate solution
30  error=norm(exact-approx,inf); % Maximum absolute error
31  rel_err=norm(exact-approx,inf)/norm(exact,inf);
32  fprintf('D =%5.1f, c= %5.3f, R =%5.3f\n',D,c,R);
33  fprintf('Maxerr=%8.3e, Rel_err=%8.3e\n',error,rel_err);
```

The code gear.m below generates the boundary collocation points, RBF collocation points, fictitious points, and test points. Note that the code sunflower in Line 6 can be found in the Appendix C.

```
%   gear.m
1   function [x,y,xc,yc,xg,yg,xt,yt]=gear(nb,n,D)
2   t=2*pi*(1:nb)'/nb;
3   r=(2+0.5*sin(5*t));
4   x=r.*cos(t+0.5*sin(5*t))/2;
5   y=r.*sin(t+0.5*sin(5*t))/2;
6   [xg0,yg0]=sunflower(n, 1);
7   xg=xg0*D;   yg=yg0*D;   % ghost points
8   xc=xg0*1.5; yc=yg0*1.5;
9   [xt,yt]=sunflower(800, 1);
10  xt=xt*1.4; yt=yt*1.4;
11  in=inpolygon(xt,yt,x,y);
12  xt=xt(in); yt=yt(in);
13  end
```

The LOOCV codes for the MQ-RBF and the MFS are listed as follows:

```
%% LOOCV for the shape parameter of MQ-RBF
1   function ceps = cost_loocv_mq(c,rbf,rhs,DM1,lam,mps2)
2   mq1=sqrt(DM1.^2.+c*c);%
3   A=rbf(DM1,c)-lam^2*mps2(c,DM1,mq1);
4   invA=pinv(A);
5   errorvector=(A\rhs)./diag(invA);
6   ceps=norm(errorvector);
7   end
```

```
%% LOOCV for the radius of circle of the MFS
1   function ceps = cost_loocv2(d,x,y,sx,sy,rhs,lam)
2   xs=d*sx; ys=d*sy;
3   DM=pdist2([x,y],[xs,ys]);
4   A=besselk(0,lam* DM);
5   invA=pinv(A);
6   errorvector=(A\rhs)./diag(invA);
7   ceps=norm(errorvector);
8   end
```

Product of Elliptic Operators: We now consider the following 2D fourth-order BVP:

$$\Delta(\Delta - k^2)\phi(x) = f(x), \quad x \in \Omega, \tag{8.22a}$$

$$\phi(x) = g_1(x), \qquad\qquad x \in \partial\Omega, \tag{8.22b}$$

$$\Delta\phi(x) = g_2(x), \qquad\qquad x \in \partial\Omega. \tag{8.22c}$$

The basis function for approximating $f(x)$ is given by (8.14) with $\alpha = -k^2$ and $\beta = 0$, and the approximate particular solution is given by (8.11). The fundamental solution of the above operator is given by (2.241a). In view of the fact that at each of the N boundary collocation points there are two prescribed BCs, we need $2N$ degrees of freedom in the approximation. This can be done by using N points and splitting the terms of the fundamental solution in (2.241a) to obtain the approximation formula

$$\hat{\phi}_h(x, y) = \sum_{j=1}^{N} a_j \ln(r_j) + \sum_{j=1}^{N} b_j K_0(kr_j), \tag{8.23}$$

where $r_j = \|x - x'_j\|$. We notice that each of the above terms is not a standard fundamental solution and is more singular than (2.241a). They nevertheless satisfy the governing equation except at the center; hence they can be used for the MFS approximation.

For the collocation of the second BC, (8.22c), we find that $\Delta \ln r = 0$ for $r \neq 0$, and $\Delta K_0(kr) = k^2 K_0(kr)$. Hence the MFS solution system becomes

$$\begin{bmatrix} \ln r & K_0(kr) \\ 0 & k^2 K_0(kr) \end{bmatrix} \begin{bmatrix} a \\ b \end{bmatrix} = \begin{bmatrix} \widetilde{g}_1 \\ \widetilde{g}_2 \end{bmatrix}, \tag{8.24}$$

in which $\widetilde{g}_1$ and $\widetilde{g}_2$ represent the modified BCs obtained by subtracting the particular solutions.

■ **Example 8.2**

We solve (8.22) with $k = 3$ and the functions $f(x, y)$, $g_1(x, y)$, and $g_2(x, y)$ derived from the exact solution

$$\phi(x, y) = \frac{4}{4 + x + y}.$$

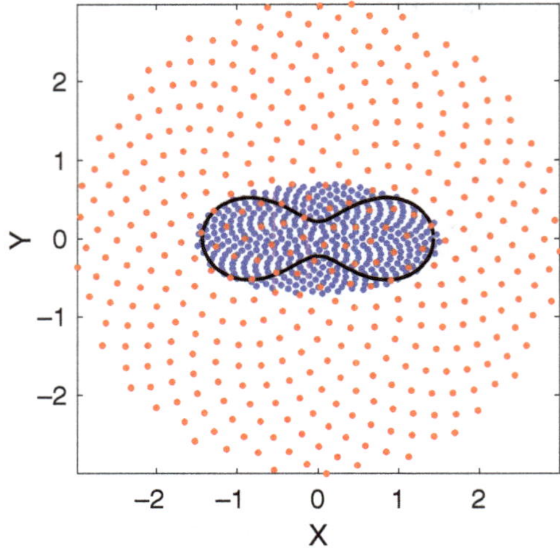

Figure 8.3. The profile of the collocation (blue) and fictitious (red) points covering a peanut-shaped domain.

The profiles of the peanut-shaped domain Ω, collocation points, and fictitious centers are presented in Fig. 8.3. The radius of the sunflower fictitious circle is R_g and the collocation points are distributed in an ellipse covering the peanut. For the approximation of the particular solution, we take 400 collocation points and an equal number of fictitious centers; while in the MFS solution of the homogeneous problem, 200 boundary collocation and source points are selected.

Similar to Example 8.1, we employ the least squares solver `lsqminnorm` to improve the accuracy. In Table 8.2, we present results for the optimal shape parameter c, source circle radius R_s, and accuracy for various R_g, obtained using LOOCV. The LOOCV search interval for the RBF shape parameter is set to $[0, 4]$ and for the radius of the source circle $[1.5, 5]$. Overall, the obtained accuracy is very high.

For a comparison with the traditional MPS-MFS, which places all the collocation and RBF centers inside the domain, we present some results in Table 8.3. It is clear that the accuracy is inferior to the ghost point method.

Table 8.2. Results for the optimal shape parameter and the radius of the MFS source circle obtained using LOOCV with 200 boundary and source points and 400 collocation and fictitious points.

R_g	c	R_s	$\mathcal{E}_{max}$	ϵ_{max}
2.0	1.09	1.85	3.5(−11)	2.1(−11)
2.5	0.94	2.02	8.0(−12)	4.9(−12)
3.0	0.83	2.38	9.6(−12)	5.9(−12)
3.5	0.81	2.02	3.1(−11)	1.9(−11)
4.0	0.65	2.44	4.1(−11)	2.5(−11)
4.5	0.94	1.82	3.0(−10)	1.8(−10)

Table 8.3. The results using the traditional MPS-MFS with different solvers.

c	R_s	$\mathcal{E}_{max}$	ϵ_{max}	Solver
2.13	2.67	2.5(−8)	1.5(−8)	`lsqminnorm`
2.13	2.75	3.5(−4)	2.1(−4)	`backslash`

The results for the above example were produced using the code `MPS_4th_order.m` listed below. Lines 4–6 show the closed-form particular solution of the biharmonic operator.

```
%  MPS_4th_order.m
1  clear; warning off all;  tic
2  rbf=@(r,c) sqrt(r.^2+c.^2);     % MQ
3  mps2=@ (c,r, mq) mq.*(r.^2+4*c^2)/9-c^3*log(mq+c)/3;
4  mps4=@(r,c) 2*c*c/45*(r.^2+c*c).^(3/2)-7*c^4/60*rbf(r,c)+...
5             (2*c^2-5*r.^2)*c^3/60.*log(c+rbf(r,c)) + ...
6             (r.^2+c^2).^(5/2)/225+c^3*r.^2/12;
7  bound_con=@(x,y) 4./(4+x+y);   %Dirichlet BC
8  Del=@(x,y) 16./((4+x+y).^3);   %Laplace BC
9  forcing=@(x,y) 384./(x + y + 4).^5 -9*Del(x,y);
%  nb: # of boundary points;  n:# of collocation points;
10 nb=200;  n=400;  D=3; lam=3;
11 [x,y,xc,yc,xg,yg,xs0,ys0,xt,yt]=cassini(nb,n,D);
12 center=[xg, yg];
13 f=forcing(xc,yc);          %forcing term
14 exact=bound_con(xt,yt);   %exact solution
15 DM1=pdist2([xc, yc],center);
```

```
16 DM2=pdist2([x, y],center);
%  LOOCV for the shape parameter of MQ-RBF
17 c = fminbnd(@(c) cost_4thmps(c,rbf,f,DM1,lam,mps2),0,4);
18 mq=sqrt(DM1.^2+c.*c);
19 A1=rbf(DM1,c)-lam^2*mps2(c,DM1,mq);      %MPS matrix
20 coef=lsqminnorm(A1,f);
21 mq=sqrt(DM2.^2+c.*c);%
22 MPS4=mps4(DM2,c)*coef;
23 g=bound_con(x,y)-MPS4;   %update Dirchlet BC
24 h=Del(x,y)-mps2(c,DM2,mq)*coef;   %update Laplace BC
25 rhs=[g;h];
%  LOOCV for the radius of source circle of the MFS
26 d=fminbnd(@(d) cost_4thmfs(d,x,y,xs0,ys0,rhs,lam),1.5,5);
27 xs=d*xs0; ys=d*ys0;
28 DM3=pdist2([x y],[xs, ys]);
29 A11=log(DM3);  A12=besselk(0,lam*DM3);       %Dirchlet BC
30 A21=0*DM3;  A22=lam^2*besselk(0,lam*DM3);     %Laplace BC
31 A2=[A11 A12; A21 A22];                      %MFS matrix
32 coe=lsqminnorm(A2,rhs);
%%%%%%  Test
33 DM5=pdist2([xt yt],center);
34 DM4=pdist2([xt yt],[xs ys]);
35 A3=mps4(DM5,c);
36 A4=[log(DM4) besselk(0,lam*DM4)];
37 approx=A3*coef+A4*coe;                     %Approximate solution
38 error=norm(exact-approx,inf);        % Maximum absolute error
39 rel_err=norm(exact-approx,inf)/norm(exact,inf);
40 fprintf('D =%5.1f, c= %5.3f, d =%5.3f\n',D,c,d);
41 fprintf('Maxerr=%8.3e, Rel_err=%8.3e\n',error,rel_err);
```

The subprogram for generating points is the following:

```
1  function [x,y,xc,yc,xg,yg,xs0,ys0,xt,yt]=cassini(nb,n,D)
2  r=@(t) sqrt(cos(2*t)+sqrt(1.1-(sin(2*t)).^2));
3  t=2*pi*(1:nb)'/nb;
4  [x,y]=pol2cart(t,r(t)); % boundary points
5  [x0,y0]=sunflower(n, 1);
6  xg=x0*D ;   yg=y0*D;   % Fictitious points
7  xc=x0*1.5;  yc=y0*0.7;  %collocation points of the RBF
8  tt=2*pi/nb*(1:nb)';
9  xs0=cos(tt); ys0=sin(tt); %source points of the MFS
10 [xt,yt]=sunflower(1200, 1);
11 xt=xt*1.4; yt=yt*1.4;
12 in=inpolygon(xt,yt,x,y);
13 xt=xt(in); yt=yt(in);    %Test points
```

The LOOCV subprogram for RBFs is:

```
%   cost_4thmps.m      LOOCV for RBF
1   function ceps = cost_4thmps(c,rbf,rhs,DM1,lam,mps2)
2   mq1=sqrt(DM1.^2.+c*c);%
3   A=rbf(DM1,c)-lam^2*mps2(c,DM1,mq1);
4   invA=pinv(A);
5   errorvector=(A\rhs)./diag(invA);
6   ceps=norm(errorvector);
7   end
```

The LOOCV subprogram for the MFS is:

```
%   cost_4thmfs.m      LOOCV for the MFS
1   function ceps = cost_4thmfs(d,x,y,sx,sy,rhs,lam)
2   xs=d*sx; ys=d*sy;
3   DM=pdist2([x,y],[xs,ys]);
4   A11=log(DM);   A12=besselk(0,lam*DM);        %Dirchlet
5   A21=0*DM;      A22=lam^2*besselk(0,lam*DM); %Laplace BC
6   A=[A11 A12; A21 A22];                        %MFS matrix
7   invA=pinv(A);
8   errorvector=(A\rhs)./diag(invA);
9   ceps=norm(errorvector);
10  end
```

8.3   MPS-MFS by Polynomial Basis Functions

In addition to RBFs, polynomial basis functions also play an important role in the solution of BVPs for PDEs. Particular solutions for various differential operators using polynomial basis functions have been provided in Chapter 4. In this section, we show the effectiveness of the MPS-MFS when using polynomial basis functions for solving second and fourth order BVPs in 2D.

A 2-D polynomial basis of degree (order) s contains the following monomial elements:

$$P_s = \{x^{i-j}y^j : 0 \leq j \leq i, 0 \leq i \leq s\}$$
$$= \{1, x, y, x^2, xy, y^2, \ldots, x^s, x^{s-1}y, x^{s-2}y^2, \ldots, xy^{s-1}, y^s\}. \quad (8.25)$$

We also know that $(s+1)(s+2)/2$ is the number of monomial terms in P_s. To find an approximate particular solution for the operator $\mathfrak{L}$ in (8.1a), we approximate the RHS as

$$\hat{f}(x, y) = \sum_{i=0}^{s} \sum_{j=0}^{i} d_{ij}\, x^{i-j} y^{j}. \tag{8.26}$$

Let $\{(x_k, y_k)\}_{k=1}^{n}$ be a set of points covering the domain, at which the function values are known. By collocating for the function values, we form the system of linear equations

$$\sum_{i=0}^{s} \sum_{j=0}^{i} d_{ij}\, x_k^{i-j} y_k^{j} = f(x_k, y_k), \quad k = 1, 2, \ldots, n. \tag{8.27}$$

If we have more collocation points than the number of polynomial basis functions, i.e., $n \geq (s+1)(s+2)/2$, system (8.27) is overdetermined and can be solved by the least squares method.

Once the coefficients d_{ij} have been determined, the approximate particular solution can be constructed as

$$\hat{\phi}_p(x, y) = \sum_{i=0}^{s} \sum_{j=0}^{i} d_{ij}\, \varphi_{ij}^{p}(x, y), \tag{8.28}$$

where the φ_{ij}^{p} are particular solutions satisfying

$$\mathfrak{L}\varphi_{ij}^{p}(x, y) = x^{i-j} y^{j}, \quad 0 \leq j \leq i, \ 0 \leq i \leq s. \tag{8.29}$$

In Section 4.2, explicit particular solutions of various operators $\mathfrak{L}$ for monomial terms have been derived. In this way, the calculation of an approximate particular solution of (8.1a) is transformed into a data interpolation problem.

8.3.1 *Polymetaharmonic operator*

We consider the polymetaharmonic operator

$$\mathfrak{L} \equiv (\Delta - k^2)(\Delta + k^2). \tag{8.30}$$

Rather than directly deriving particular solutions of the fourth-order differential operator, which is quite tedious [940], we shall try to assemble particular solutions using those of the Helmholtz and modified Helmholtz operators [127], which are available in (4.30) and (4.31).

Our aim is to construct a closed-form particular solution of the equation

$$(\Delta - k^2)(\Delta + k^2)\varphi^P(x, y) = x^m y^n. \tag{8.31}$$

The closed-form particular solutions φ^H and φ^M of the equations

$$(\Delta + k^2)\varphi^H(x, y) = x^m y^n, \tag{8.32}$$

$$(\Delta - k^2)\varphi^M(x, y) = x^m y^n, \tag{8.33}$$

are given in (4.30) and (4.31), respectively. Multiplying both sides of the above equations by $\Delta - k^2$ and $\Delta + k^2$, respectively, we have

$$(\Delta - k^2)(\Delta + k^2)\varphi^H = (\Delta - k^2)x^m y^n, \tag{8.34}$$

$$(\Delta + k^2)(\Delta - k^2)\varphi^M = (\Delta + k^2)x^m y^n. \tag{8.35}$$

Subtracting (8.34) from (8.35), we get

$$(\Delta + k^2)(\Delta - k^2)(\varphi^M - \varphi^H) = 2k^2 x^m y^n. \tag{8.36}$$

Finally, comparing (8.31) and (8.36), we obtain

$$\varphi^P = \frac{1}{2k^2}(\varphi^M - \varphi^H). \tag{8.37}$$

This suggests that φ^P in (8.31) can be constructed as the difference of two lower order particular solutions.

The above procedure for obtaining particular solutions can be extended to other products of differential operators, such as $\Delta(\Delta - k^2)$, $(\Delta - k_1^2)(\Delta + k_2^2)$, etc., and for different basis functions.

8.3.2 *Numerical implementation*

It is well-known that polynomial basis function approximations become notoriously unstable when the order of the polynomial basis is large. Hence, they are not the method of choice for solving BVPs using a global approach as the resultant matrix becomes severely ill-conditioned when the polynomial order increases. This is due to the mixing of the equations corresponding to the BCs with those corresponding to the governing equation in the resultant matrix. As such, various types of pre-conditioning techniques are required. In the MPS-MFS technique, however, the MPS and MFS matrices are separated and the corresponding systems may be solved without ill-conditioning problems.

Modified Helmholtz Operator: For the modified Helmholtz operator in Example 8.1, the particular solution φ^M for the basis function $x^m y^n$ is given by (4.31), which will be used in the following example.

■ Example 8.3

We consider the same BVP as in Example 8.1 involving the modified Helmholtz equation. However, rather than using MQ-RBFs for the approximation of the particular solution, we shall use polynomial basis functions.

Compared to Example 8.1, in this example we use the same number of boundary points $N = 200$ for the MFS solution, but fewer collocation points for the particular solution ($n = 200$ instead of 400). For the approximation of the particular solution, we choose polynomials up to degree $s = 18$, which contain monomials up to $m = 190$ terms. Since the accuracy of the approximation depends mainly on the polynomial order used, using more collocation points has little effect on the accuracy.

In Fig. 8.4, we present the plot of the maximum absolute error $\mathcal{E}_{\max}$ and condition number versus the polynomial order. We observe that the accuracy increases exponentially for cases up to $s = 15$, where the error is of $\mathcal{O}(10^{-14})$, and no further improvement is observed for $s > 15$. The plot of the condition number is inversely proportional to that of the accuracy. This phenomenon is described by the Uncertainty Principle of Schaback [775] (see Section 5.4.3).

In comparison with Example 8.1, in which 400 MQ-RBF terms were used, we observe that the same level of accuracy can be achieved in the polynomial approximation with $s = 12$ using much fewer collocation points. In addition, an LOOCV type search procedure for the optimal shape parameter is not needed. Hence the polynomial approximation seems to be more efficient.

The code `MFS_MPS_POLY.m` for obtaining the results in Fig. 8.4 is listed below. In Line 10, the generation of the boundary, collocation, and test points for the gear-shaped domain is the same as that shown in the code for

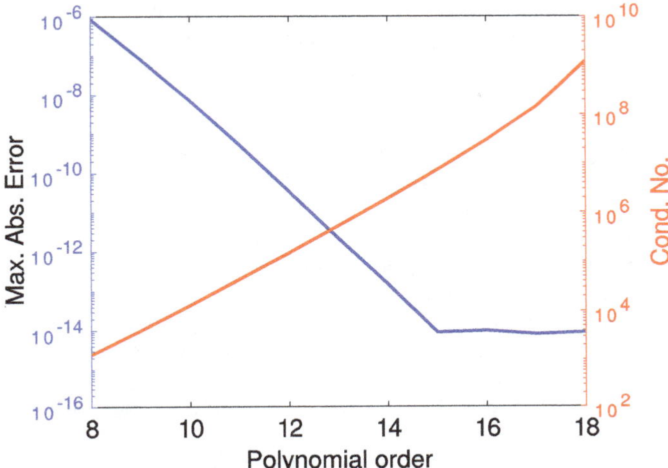

Figure 8.4. The maximum absolute error $\mathcal{E}_{\max}$ and condition number versus the order of polynomial basis functions.

Example 8.1. From Fig. 8.4, we observe that the magnitudes of the condition numbers are moderate; thus special treatment using pre-conditioning techniques is not required. In this example, we also observed that the homogeneous solution is relatively stable with respect to the size of the source circle radius. As a result, we set the radius of the source circle to 3 without using LOOCV to determine it. Lines 29–39 generate the polynomial basis functions, and Lines 40–52 give the closed-form particular solutions. The solution of the Helmholtz equation can be obtained by setting `ep=1` in Line 4 and replacing `besselk` by `bessely` in Lines 19 and 22. Results similar to those presented in Fig. 8.4 can be obtained.

```
%  MFS_MPS_POLY.m
1  function MFS_MPS_POLY
2  warning off all;
3  global lmd
4  ep=-1;   %ep=1 is Helm. eq; ep=-1 is modified Helmh. eq
5  force = @(x,y,lmd,ep) (2+ep*lmd^2)*exp(x+y);
7  boundary_cond = @(x,y)  exp(x+y); % BC
8  order=16;   lmd=3; %order: polynomial order
9  nb=200; n=200;
10 [xb,yb,xc,yc,~,~,xt,yt]=gear(nb,n);% GEAR-SHAPED DOMAIN
```

```matlab
11 f = force(xc,yc,lmd,ep);   % Forcing term of PDE
12 poly_terms=((order+1)*(order+2))/2; %# of polynomial terms
13 [A,B,Bt]=poly(xc,yc,xb,yb,xt,yt,order,poly_terms,ep);
14 coef =  A\f;
15 con=cond(A);
16 g = boundary_cond(xb,yb)-B*coef; %update BC
17 [xs,ys]=pol2cart(2*pi*(1:nb)'/nb,3);
18 DM=pdist2([xb,yb],[xs,ys]);
19 A2=besselk(0,lmd* DM);          % boundary condition
20 coef_mfs= lsqminnorm(A2,g);
21 DM4=pdist2([xt yt],[xs ys]);
22 A4=besselk(0,lmd*DM4);
23 approx=Bt*coef+A4*coef_mfs; %Approximate solution
24 exact = exp(xt+yt);          % Exact solution
25 error = norm((approx-exact), inf); % absolute maximum error
26 rer=norm(exact-approx)/norm(exact);    % relative max error
27 fprintf('n=%4d,nb=%4d,order=%2d,con=%8.2e\n',n,nb,order,con);
28 fprintf('Maxerr: %8.3e, Rel. error: %8.3e\n', error, rer);
%%% Polynomial basis
29 function [A, B, Bt]=poly(x,y,xb,yb,xt,yt,P,size,ep)
30 A=zeros(length(x),size);B=zeros(length(xb),size);
31 Bt=zeros(length(xt),size);   index=0;
32 for i=0:P
33     for j=0:i
34         index=index+1;
35         A(:,index)= x.^(i-j).*y.^j;
36         [T1,T2] = particular(i-j,j,xb,yb,xt,yt,ep);
37         B(:,index)=T1;       Bt(:,index)=T2;
38     end
39 end
%%% Closed-form particular solution of Helmholtz-type equations
40 function [tot1,tot2]=particular(m,n,xb,yb,xt,yt,ep)
41 global lmd
42 tot1=0; tot2=0;
43 for k=0:floor(m/2)
44     temp0=prod(m-2*k+1:m)*xb.^(m-2*k);
45     temp1=prod(m-2*k+1:m)*xt.^(m-2*k);
46     for l=0:floor(n/2)
47         temp=ep*(-ep)^(k+l)*prod(k+1:k+l)/factorial(l)*...
48             prod(n-2*l+1:n)/lmd^(2*k+2*l+2);
49         tot1=tot1+temp*temp0.*yb.^(n-2*l);
50         tot2=tot2+temp*temp1.*yt.^(n-2*l);
51     end
52 end
```

Polymetaharmonic Operator: We consider the 2D BVP for the fourth-order polymetaharmonic operator

$$(\Delta + k^2)(\Delta - k^2)\phi(x) = f(x), \quad x \in \Omega, \tag{8.38a}$$

$$\phi(x) = g_1(x), \qquad\qquad x \in \partial\Omega, \tag{8.38b}$$

$$\Delta\phi(x) = g_2(x), \qquad\qquad x \in \partial\Omega. \tag{8.38c}$$

We notice that the implementation of the second BC (8.38c) requires the application of the Δ operator to the particular solution. Rather than direct differentiation, we perform the following procedure. Subtracting (8.32) from (8.33), we obtain

$$\Delta(\varphi^M - \varphi^H) = k^2(\varphi^M + \varphi^H). \tag{8.39}$$

Hence we find from (8.37) that

$$\Delta\varphi^P(x, y) = \frac{1}{2}(\varphi^M + \varphi^H). \tag{8.40}$$

The BC (8.38c) can thus be applied without differentiation.

The fundamental solution of the operator in (8.38a) is given by (2.226). Similar to (8.23), we split the terms to obtain the approximation

$$\hat{\phi}_h(x, y) = \sum_{j=1}^{N} a_j Y_0(kr_j) + \sum_{j=1}^{N} b_j K_0(kr_j). \tag{8.41}$$

Performing similar operations as in (8.24) for the two BCs, (8.38b) and (8.38c), we obtain the matrix system

$$\begin{bmatrix} \mathbf{Y_0}(kr) & \mathbf{K_0}(kr) \\ -k^2\mathbf{Y_0}(kr) & k^2\mathbf{K_0}(kr) \end{bmatrix} \begin{bmatrix} a \\ b \end{bmatrix} = \begin{bmatrix} \widetilde{g}_1 \\ \widetilde{g}_2 \end{bmatrix}, \tag{8.42}$$

for the homogeneous part of the solution, in which $\widetilde{g}_1$ and $\widetilde{g}_2$ represent the modified BCs after subtracting the particular solutions.

■ Example 8.4

In this example, we take $k = 30$, and f, g_1, and g_2 are given functions derived from the exact solution

$$\phi(x, y) = \sin(\pi x)\cosh(y) - \cos(\pi x)\sinh(y).$$

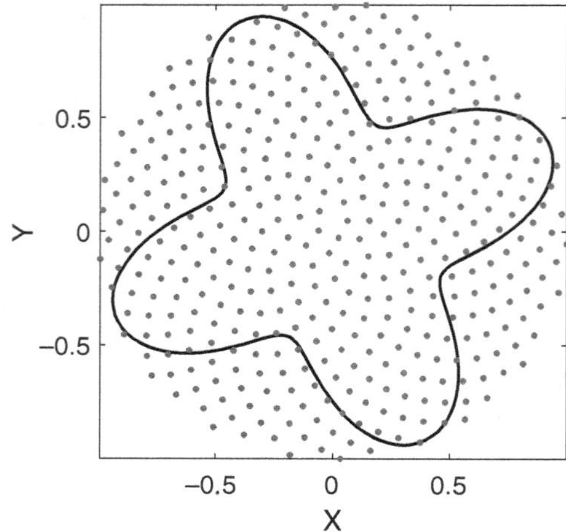

Figure 8.5. The profiles of a four-petal domain and collocation points for the particular solution.

Here Ω is the four-petal domain in Fig. 8.5, where the collocation points covering the domain are also displayed.

Recalling that s is the polynomial basis function order and we take $m = (s+1)(s+2)/2$ to be the corresponding number of polynomial basis functions. To obtain the approximate particular solution, we choose $m + 100$ collocation points for each s. In the MFS, we take $N = 200$ boundary and source points. As was the case in the last example, the accuracy is insensitive to the size of the source circle; hence we fix the radius of the source circle to be $R = 3$ in all tests. Figure 8.6 shows the maximum absolute error $\mathcal{E}_{\max}$ and the condition number versus the polynomial order s. In this case, the error drops up to $s = 21$ and then starts to increase. The error $\mathcal{E}_{\max}$ at $s = 21$ is of $\mathcal{O}(10^{-13})$ which means that the current method is highly accurate, particularly for a fourth order BVP, which is considerably more challenging than a second order one.

The code producing the results in Fig. 8.6 is $\texttt{poly\_4th\_MPSMFS.m}$, which is listed below. Line 14 solves the system of equations (8.27), while Lines 15 and 16 define the particular solutions φ^M and φ^H, respectively. Line 17 defines φ^P and Line 19, $\Delta\varphi^P$. Lines 20 and 21 update the BCs and Lines 26–28 produce the MFS interpolation matrix. The particular

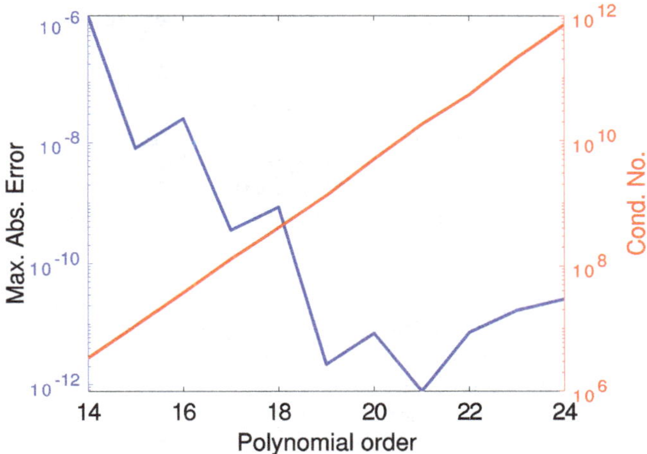

Figure 8.6. The maximum absolute error $\mathcal{E}_{\max}$ and condition number versus the order of polynomial basis functions.

solutions φ^M and φ^H given in (4.30) and (4.31) are generated by the function $\texttt{part1}$ in Lines 56–67.

```
%% poly_4th_MPSMFS.m
1   function poly_4th_MPSMFS
2   global lmd
3   force=@ (x,y,lmd) ((pi^2-1)^2-lmd^2)*(sin(pi*x).*cosh(y)...
4                       -cos(pi*x).*sinh(y));
5   BC1=@ (x,y)-(pi^2-1)*(sin(pi*x).*cosh(y)-cos(pi*x).*sinh(y));
6   BC2=@ (x,y) sin(pi*x).*cosh(y)-cos(pi*x).*sinh(y);
7   analy = @(x,y) sin(pi*x).*cosh(y)-cos(pi*x).*sinh(y);
8   order=20;    lam=30; lmd=lam^2;
9   m=((order+1)*(order+2))/2;
10  n=m+100;   nb=200;
11  [xb,yb,xc,yc,xt,yt]=petal(nb,n);
12  f = force(xc,yc,lmd);   % Forcing term of PDE
13  A=POLY(xc,yc,order,poly_terms);
14  coef1 = A\f;
%% MFS
15  [B21,Bt1]=part1(xb,yb,xt,yt,order,m,-1); %B21=\psi^mh
16  [B22,Bt2]=part1(xb,yb,xt,yt,order,m,1); %B22=\psi^h
17  B2=1/(2*lmd)*(B21-B22); %\Psi(xb,yb)
18  Bt=1/(2*lmd)*(Bt1-Bt2); %\Psi(xt,yt)
19  B1=1/2*(B21+B22); %BC: \Delta Psi=(\psi^mh +\psi^h)/2
20  g1 = BC1(xb,yb)-B1*coef1;    %update Laplace BC
```

```matlab
21 g2 = BC2(xb,yb)-B2*coef1;    %update Dirichlet BC
22 F=[g2;g1];   % RHS
23 t=2*pi*(1:nb)'/nb;
24 [xs,ys]=pol2cart(t,3); %source points
25 DM=pdist2([xb,yb], [xs,ys]);
26 A11=bessely(0,lam*DM);        A12= besselk(0,lam*DM);
27 A21=-lmd*bessely(0,lam*DM);   A22=-lmd*besselk(0,lam*DM);
28 A=[A11,A12;A21,A22]; %  MFS matrix
29 coef = lsqminnorm(A,F);
30 DM1=pdist2([xt,yt],[xs,ys]);
31 approx=bessely(0,lam*DM1)*coef(1:nb)+besselk(0,lam*DM1)...
32         *coef(nb+1:2*nb)+Bt*coef1;  %approximate solution
33 exact = analy(xt,yt);              % exact solution
34 error = norm((approx-exact), inf);
35 rer=norm(exact-approx)/norm(exact);
36 fprintf('n = %3d, nb = %3d, order = %3d\n',n,nb,order);
37 fprintf('Maxerr: %8.3e, Rel_err: %8.3e\n', error,rer);
%%% polynomial basis functions
38 function A=POLY(x,y,P,size)
39 A=zeros(length(x),size); index=0;
40 for i=0:P
41     for j=0:i
42         index=index+1;
43         A(:,index)= x.^(i-j).*y.^j;
44     end
45 end
%%%%%
46 function [A, B]=part1(xb,yb,xt,yt,P,size,ep)
47 A=zeros(length(xb),size); B=zeros(length(xt),size);
48 index=0;
49 for i=0:P
50     for j=0:i
51         index=index+1;
52         A(:,index)= particular1(i-j,j,xb,yb,ep);
53         B(:,index)=particular1(i-j,j,xt,yt,ep);
54     end
55 end
%% closed-form particular solution of Helmholtz-type equations
56 function part=particular1(m,n,x,y,ep)
57 global lmd
58 part=0;
59 FM=factorial(m);  FN=factorial(n);
60 for k=0:floor(m/2)
61     FK=factorial(k);  FK2=factorial(m-2*k);
62     for l=0:floor(n/2)
63         part=part+ep*(-ep)^(k+l)*factorial(k+l)*FM*FN...
```

```
64                   *x.^(m-2*k).*y.^(n-2*l)/(FK*factorial(l)...
65                   *FK2*factorial(n-2*l)*lmd^(k+l+1));
66       end
67 end
%%% points generator: petal.m
68 function [xb,yb,xi,yi,xt,yt]=petal(nb,n)
69 r=@(t) (1.5+sin(4*t)/2)/2;
70 t=2*pi*(1:nb)'/nb;
71 [xb,yb]=pol2cart(t,r(t));
72 [xt,yt]=sunflower(800,1);
73 in=inpolygon(xt,yt,xb,yb);
74 xt=xt(in); yt=yt(in);   %interior test points
75 [xi,yi]=sunflower(n,1); %collocation points
```

8.4 One-Step MPS-MFS

In 2001, Balakrishman and Ramachandran [57] proposed combining the homogeneous and particular solutions for solving nonlinear Poisson problems. A similar technique proposed by Wang and Qin [894] extended the two-step approach to solving Poisson-type problems. In this section, we will combine the RBF and MFS basis functions to form dual basis functions for solving general PDEs. Furthermore, the performance of the one-step MPS-MFS will be enhanced by the use of the ghost point method [972] mentioned in the previous sections.

Let us consider BVP (8.1) with

$$\mathcal{L} \equiv \Delta + \hat{\mathcal{L}}, \tag{8.43}$$

where $\hat{\mathcal{L}}$ is a linear operator of order at most two. For example, we could have

$$\hat{\mathcal{L}} \equiv \alpha(x)\frac{\partial}{\partial x} + \beta(x)\frac{\partial}{\partial y} + \gamma(x), \tag{8.44}$$

where $\alpha(x)$, $\beta(x)$, and $\gamma(x)$ are known functions.

Let $\{x_i\}_{i=1}^n$ be a set of collocation points covering the domain of the problem under consideration and $\{\xi_j\}_{j=1}^n$ be the RBF centers or ghost points shown in Fig. 8.2. These two sets of points are used in the approximation of a particular solution. For the homogeneous solution approximation, we choose the boundary collocation points $\{x_i\}_{i=n+1}^{n+m}$ on $\partial\Omega$ and a set of source points $\{\eta_k\}_{k=1}^m$ outside the domain. The selection of all these points is carried out in basically the same way as in the two-step approach.

In the one-step approach, the solution is approximated by the sum of the particular solution $\hat{\phi}_p$ and the homogeneous solution $\hat{\phi}_h$

$$\phi(x) \simeq \hat{\phi}_p(x) + \hat{\phi}_h(x) = \sum_{j=1}^{n} a_j \psi(r_j) + \sum_{k=1}^{m} c_k G^L(\rho_k), \qquad (8.45)$$

where $r_j = \|x - \xi_j\|$ and $\rho_k = \|x - \eta_k\|$. Note that in (8.45), the basis functions ψ and G^L are RBFs and the fundamental solution of the Laplacian Δ, respectively. The fundamental solution of the differential operator $\mathfrak{L}$ in (8.43) is generally not available. As a result, we consider the Laplacian Δ as the main differential operator and take G^L to be $\ln \rho$ in 2D and $1/\rho$ in 3D. In this section, we take ψ to be the normalized MQ, $\psi = \sqrt{1 + r^2 c^2}$.

Instead of using the closed-form particular solution as in the two-step approximation in Section 8.1, we directly substitute (8.45) into (8.1a) to obtain

$$\sum_{j=1}^{n} a_j \mathfrak{L} \psi(r_j) + \sum_{k=1}^{m} c_k \mathfrak{L} G^L(\rho_k) = f(x), \quad x \in \Omega. \qquad (8.46)$$

BC (8.1b) then becomes

$$\sum_{j=1}^{n} a_j \mathcal{B} \psi(r_j) + \sum_{k=1}^{m} c_k \mathcal{B} G^L(\rho_k) = g(x), \quad x \in \partial\Omega. \qquad (8.47)$$

In 2D, we have

$$\Delta \psi(r) = \frac{c^2(2 + c^2 r^2)}{(1 + c^2 r^2)^{3/2}}, \quad \frac{\partial \psi(r)}{\partial x} = \frac{c^2 x}{\sqrt{1 + c^2 r^2}}, \quad \frac{\partial \psi(r)}{\partial y} = \frac{c^2 y}{\sqrt{1 + c^2 r^2}}, \qquad (8.48)$$

and

$$\frac{\partial G^L(\rho)}{\partial x} = \frac{x}{\rho^2}, \quad \frac{\partial G^L(\rho)}{\partial y} = \frac{y}{\rho^2}. \qquad (8.49)$$

As in the two-step approach, we select n RBF collocation points in an extended region covering the domain Ω and the same number of RBF centers or ghost points (see Fig. 8.2). We also take m boundary points on $\partial\Omega$, and m source points on the auxiliary boundary outside the domain. Collocation yields the following system of linear equations of size $(m + n) \times (m + n)$

$$\begin{bmatrix} \mathfrak{L}\psi(r_{ij}) & \mathfrak{L}G^L(\rho_{ik}) \\ \mathcal{B}\psi(r_{ij}) & \mathcal{B}G^L(\rho_{ik}) \end{bmatrix} \begin{bmatrix} a \\ c \end{bmatrix} = \begin{bmatrix} f \\ g \end{bmatrix}, \qquad (8.50)$$

where

$$a = [a_1, a_2, \ldots, a_n]^T, \quad c = [c_1, c_2, \ldots, c_m]^T,$$
$$f = [f(x_1), f(x_2), \ldots, f(x_n)]^T, \tag{8.51}$$
$$g = [g(x_{n+1}), g(x_{n+2}), \ldots, g(x_{n+m})]^T.$$

Once $\{a_j\}_{j=1}^n$ and $\{c_k\}_{k=1}^m$ are determined by solving (8.50), the approximate solution $\hat{\phi}$ can be obtained from (8.45).

Numerical Implementation: In the numerical implementation, the ghost points for the RBF and source points for the MFS are chosen as in the two-step approach in Section 8.1. One of the challenges of the one-step approach is that we now have three parameters to be determined simultaneously for finding the optimal solution, namely the radii of the ghost and source circles denoted by R_g and R_s, respectively, and the shape parameter of the RBF. To measure the error, we compute the maximum relative error ϵ_{max}.

In 1982, Franke [314] proposed an experimental formula for the estimation of a good shape parameter for the RBF-MQ. This formula has been modified due to the advent of double precision arithmetic in recent decades [502]. The modified Franke formula for the normalized MQ is $c = 0.8\sqrt[4]{N}/D$, where N is the number of collocation points and D is the diameter of the smallest circle/sphere containing the domain. The modified Franke formula provides a rough estimate for the normalized MQ shape parameter. In the next example, we shall apply the modified Franke formula for the initial guess in an adaptive approach for determining the optimal shape parameter value.

■ Example 8.5

We consider the following second-order BVP with variable coefficients:

$$\left(\Delta + \frac{\partial}{\partial y} - x^2 y\right) \phi(x, y) = f(x, y), \quad (x, y) \in \Omega,$$

$$\phi(x, y) = g(x, y), \quad (x, y) \in \partial\Omega_1,$$

$$\frac{\partial\phi}{\partial n}(x, y) = h(x, y), \quad (x, y) \in \partial\Omega_2,$$

where f, g and h are obtained from the exact solution

$$\phi(x, y) = e^{2x} \cos(y^2 + x).$$

The computational domain Ω is the star-shaped domain in Fig. 7.2(b). Note that $\partial\Omega_1$ is the part of the boundary where $y \geq 0$ and $\partial\Omega_2$ the part where $y < 0$.

We choose 600 circular sunflower collocation points, 200 boundary points, and 200 interior test points to measure the error. The radius of the sunflower disk containing the domain is 1.32. Since there are three parameters to be determined, we need to fix two parameters and then plot the error when the third parameter is varied. For observation purposes, we set the radius of the source circle to $R_s = 3$ for the two error plots shown in Fig. 8.7. In Fig. 8.7(a), as we have no way of predicting an appropriate initial guess for R_g, the difference in the results for different radii R_g can be quite significant. In contrast, in the results presented in Fig. 8.7(b), we can use the modified Franke formula ($c = 1.492$) to provide guidance for the selection of a better shape parameter (red curve in Fig. 8.7(b)).

In [972], the main novelty was the introduction of the ghost points to improve the accuracy. The LOOCV algorithm was applied to find a good shape parameter by fixing the radii of the ghost and source circles. The optimal selection of these three parameters is still an open issue. In this example, we propose the following adaptive procedure:

(1) Choose an RBF shape parameter based on the Franke formula and properly guess a source circle radius R_s. Apply LOOCV to select a ghost circle radius R_g.

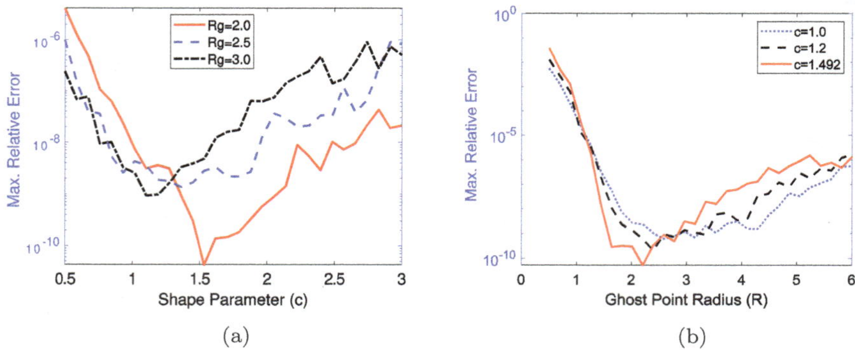

Figure 8.7.   The profiles of maximum relative errors $\epsilon_{\max}$ by fixing the radius of the source circle $R_s = 3$ and (a) scanning shape parameter c for three different radii of ghost circle R_g, and (b) scanning R_g for three different shape parameters.

Table 8.4. Results obtained by the adaptive scheme using an initial guess for the shape parameter from Franke's formula and $R_s = 3$.

Step	Input	Output	ϵ_{max}
1	$c = 1.492,\ R_s = 3$	$R_g = 2.528$	7.8($-$09)
2	$R_g = 2.528,\ R_s = 3$	$c = 0.875$	7.2($-$11)
3	$R_g = 2.528,\ c = 0.875$	$R_s = 3.747$	7.9($-$11)

(2) Apply LOOCV to update the shape parameter in Step 1 by using the R_g produced in Step 1 and the same R_s as in Step 1.

(3) Apply LOOCV to update R_s using the R_g in Step 1 and the shape parameter in Step 2.

■ Example 8.6

We solve the same problem as Example 8.5 by applying the above procedure.

In Table 8.4, we present the results obtained using the above adaptive procedure. In Step 1, $c = 1.492$ was obtained by the Franke formula, and $R_s = 3$ is a reasonable guess. The search intervals in the MATLAB® search function `fminbnd` for c, R_g, and R_s are $[0, 6]$, $[0, 5]$, and $[0, 5]$, respectively. For different search intervals, the results differ slightly. The distribution of the points producing the results in Step 3, Table 8.4, is shown in Fig. 8.8.

On the other hand, if the adaptive procedure starts with a different initial guess for R_g, say $R_g = 2$, we obtain the results presented in Table 8.5. Clearly, there is no improvement in the adaptive process if the initial R_g is not properly selected. However, if the initial guess of R_g is set to 3, then we obtain results compatible with those in Table 8.4. Since the initial selection of R_g is problematic, we suggest adopting the approach where the initial shape parameter value is set using Franke's formula.

We provide the following code for the first step of the adaptive procedure when the initial shape parameter is chosen using Franke's formula. In Line 10, we notice that the number of source points is only half the number of boundary points (`ns=nb/2`). In Line 19, the sunflower collocation points are defined as in Example 8.1. Lines 11–28 generate the various sets of

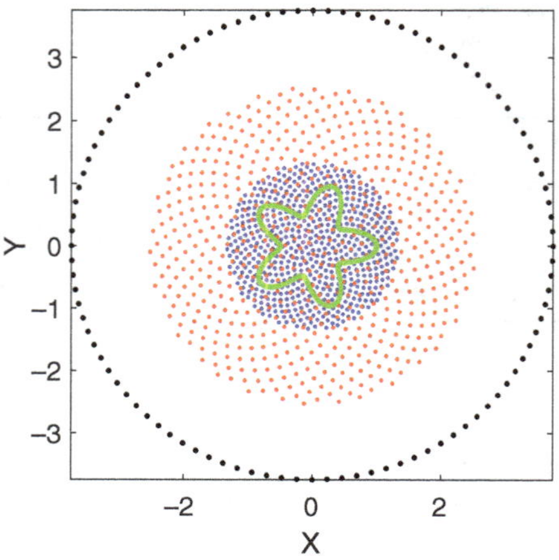

Figure 8.8. Profile of the final distribution of collocation points (blue), ghost points (red), boundary points (green), and source points (black).

Table 8.5. Results using the adaptive scheme with the initial guess $R_g = 2$ and $R_s = 3$.

Step	Input	Output	ϵ_{max}
1	$R_g = 2.0, R_s = 3$	$c = 3.062$	2.2(−07)
2	$c = 3.062, R_s = 3$	$R_g = 3.852$	1.8(−06)
3	$R_g = 3.852, c = 3.062$	$R_s = 4.271$	2.1(−06)

points that are needed for the numerical implementation. In Line 22, we apply the modified Franke formula for the initial estimation of the shape parameter. In Line 35, we construct the block matrix $\mathcal{L}G^L(\rho_{ik})$ in (8.50) while in Line 47, the block matrix $\mathcal{L}\psi(r_{ij})$ is formulated.

```
%   One-step MPS-MFS
1   rbf=@(r,c) sqrt(1+(r*c).^2);
2   Delta= @(r,c) c^2*(2+c^2*r.^2)./(rbf(r,c).^3);
3   phi_r = @ (r,c) c^2./rbf(r,c);
4   f=@(x,y) (3-4*y.^2-x.^2.*y).*exp(2*x).*cos(y.^2+x)-...
5           (2*y+6).*exp(2*x).*sin(y.^2+x);
6   g=@(x,y) exp(2*x).*cos(y.^2+x);          % Exact solution
7   h=@(x,y,nx,ny) exp(2*x).*(2*cos(y.^2+x)-sin(y.^2+x)).*nx-...
8           2*y.*exp(2*x).*sin(y.^2+x).*ny;
9   r=@(t) 1+(cos(5/2*t).^2); %Star-shaped boundary
%   nb: # of boundary points; ni: # of RBF collocation pints
```

```
%   ns: # of source points; sr: radius of source circle
10  nb=200;   ni=600;   ns=nb/2;       sr=3;
11  t=2*pi*(1:nb)'/nb;
12  [xb,yb]=pol2cart(t,r(t)); %boundary points
%   Dirichlet boundary points: (xdb,ydb); Neumann: (xbn,ybn)
13  xb=xb./2;                 yb=yb./2;
14  xbd=xb(1:nb/2);        ybd=yb(1:nb/2);       bdptd=[xbd,ybd];
15  xbn=xb(nb/2+1:nb);    ybn=yb(nb/2+1:nb); bdptn=[xbn,ybn];
16  dx=gradient(xbn);     dy=gradient(ybn); r=sqrt(dx.^2+dy.^2);
17  xn=dy./r;    yn=-dx./r; %normal vectors
18  a1=min(xb);    a2=max(xb);    a3=min(yb);    a4=max(yb);
19  [xi,yi]=sunflower(ni,1);
20  gx=xi; gy=yi;
21  D=sqrt((a2-a1)^2+(a4-a3)^2)/2;
22  c=0.8*ni^(1/4)/(2*D);    % Modified Franke's formula
23  xi=xi*D; yi=yi*D;        % RBF collocation Points
24  [sx,sy]=pol2cart((1:ns)'./ns.*2.*pi,sr);
25  source=[sx,sy];    % Source Points of the MFS
26  xt=xi*.8; yt=yi*0.8;
27  in=inpolygon(xt,yt,xb,yb);
28  xt=xt(in); yt=yt(in);    %test points
29  z1=f(xi,yi); z2=g(xbd,ybd);    z3=h(xbn,ybn,xn,yn);
30  rhs=[z1;z2;z3];   %Forcing term
31  exact=g(xt,yt);   %Exact solution
%% Formulation of the MFS Matrix
32  DM12=pdist2([xi yi],source);
33  DM22=pdist2(bdptd,   source);
34  DM32=pdist2(bdptn,   source);
35  A12=(yi-sy')./(DM12.^2)-xi.^2.*yi.*log(DM12);
36  A22=log(DM22);
37  A32=((xbn-sx').*xn+(ybn-sy').*yn)./(DM32.^2);
38  A2=[A12;A22;A32];   %MFS matrix
%% Particular Solution (MPS)
%   LOOCV for radius of ghost circle Rg
39  RL=0; RR=5;        % Search interval (RL,RR)
40  options=optimset('TolX',1e-2);   % Options for fminbnd
41  Rg=fminbnd(@(r) costeps(r,c,rhs,xi,yi,gx,gy,xbn,ybn,...
42                     xn,yn,A2,bdptd),RL,RR,options);
43  gx=gx*Rg;    gy=gy*Rg;   ghost=[gx,gy];   %ghost points
%   Formulation of RBF Matrices
44  DM11=pdist2([xi yi],ghost);
45  DM21=pdist2(bdptd,   ghost);
46  DM31=pdist2(bdptn,   ghost);
47  A11=Delta(DM11,c)+phi_r(DM11,c).*(yi-gy')-xi.^2.*yi.*rbf(DM11,c);
48  A21=rbf(DM21,c);
49  A31=((xbn-gx').*xn+(ybn-gy').*yn).*phi_r(DM31,c);
```

```
50 A1=[A11;A21;A31];   %RBF matrix
51 A=[A1, A2];
52 coef=lsqminnorm(A,rhs); %least square solver
%  Numerical Test
53 DMT1=pdist2([xt,yt], ghost);    B1=rbf(DMT1,c);
54 DMT2=pdist2([xt,yt], source);   B2=log(DMT2);
55 approx=[B1 B2]*coef;
56 rel_err=norm(approx-exact,inf)/max(exact);
57 fprintf('sr=%4.3f, c=%5.3f, R = %5.3f\n',sr,c,Rg)
58 fprintf('Rel_err = %8.3e \n',rel_err);
```

To perform the LOOCV algorithm in Line 41, we need the following code:

```
1   function eval=costeps(r,c,rhs,xi,yi,gx,gy,xbn,ybn,nx,ny,A2,bdptd)
2   rbf=@(r,c) sqrt(1+(r*c).^2);
3   Delta= @(r,c) c^2*(2+c^2*r.^2)./((1+r.*r*c*c).^(3/2));
4   phi_r = @ (r,c) c^2./((1+r.*r*c*c).^(1/2));
5   gx1=gx*r;  gy1=gy*r; ghost=[gx1,gy1];
6   DM11=pdist2([xi yi],ghost);
7   DM21=pdist2(bdptd,  ghost);
8   DM31=pdist2([xbn,ybn],  ghost);
9   A11=Delta(DM11,c)+phi_r(DM11,c).*(yi-gy1')...
10      -xi.^2.*yi.*rbf(DM11,c);
11  A21=rbf(DM21,c);
12  A31=((xbn-gx1').*nx+(ybn-gy1').*ny).*phi_r(DM31,c);
13  A1=[A11;A21;A31];
14  A=[A1 A2];
15  invA=pinv(A);
16  errorvector=(invA*rhs)./diag(invA);
17  eval=norm(errorvector);
18  end
```

For other types of BVPs such as the ones governed by fourth order PDEs or 3D cases, the one-step MPS-MFS is also highly effective. We refer readers to [972] for further information.

8.5 MPS-MFS by Chebyshev Polynomials

Chebyshev polynomial interpolation has been introduced in Section 3.2.2, and the derivation of particular solutions for various differential operators has been presented in Section 4.4. Due to their spectral convergence properties, Chebyshev polynomials are highly effective for function approximation and solving BVPs. However, their use is confined to rectangular or rectangularly decomposable domains and they lose their attractiveness when solving BVPs in irregular domains. On the other hand, in the evaluation of particular solutions, we have the freedom to choose the interpolation domain without a specified boundary (and thus in a rectangle); hence Chebyshev polynomials can be used effectively. In this section, we employ a technique similar to the one-step MPS-MFS presented in Section 8.4 by replacing RBFs with Chebyshev polynomials, exploiting some of the properties of the latter, mentioned in Sections 3.2.2 and 4.4. In the following, we present the Chebyshev polynomial approximation coupled with the MFS for solving a general BVP [329].

We consider the BVP in 2D

$$\mathcal{L}\phi(x) = f(x), \quad x \in \Omega, \tag{8.52a}$$

$$\mathcal{B}\phi(x) = g(x), \quad x \in \partial\Omega, \tag{8.52b}$$

where $\Omega \subseteq [a, b] \times [c, d]$ is a bounded and closed domain, $\mathcal{B}$ a boundary differential operator, and

$$\mathcal{L} \equiv \Delta + \alpha(x, y)\frac{\partial}{\partial x} + \beta(x, y)\frac{\partial}{\partial y} + \gamma(x, y). \tag{8.53}$$

Before solving (8.52) using Chebyshev polynomials, we rescale the domain $\Omega \subseteq [a, b] \times [c, d]$ to $\hat{\Omega} \subseteq [-1, 1] \times [-1, 1]$. Let $x = (x, y) \in \Omega$ and $(\xi, \eta) \in \hat{\Omega}$. From (3.53) and (3.54), we have

$$x = \alpha_1\xi + \beta_1, \quad \text{where } \alpha_1 = \frac{b - a}{2}, \; \beta_1 = \frac{a + b}{2},$$

$$y = \alpha_2\eta + \beta_2, \quad \text{where } \alpha_2 = \frac{d - c}{2}, \; \beta_2 = \frac{d + c}{2}, \tag{8.54}$$

and

$$\xi = \frac{x - \beta_1}{\alpha_1}, \quad \eta = \frac{y - \beta_2}{\alpha_2}. \tag{8.55}$$

Then, BVP (8.52a) and (8.52b) is transformed from a problem in the (x, y) plane into a problem in the (ξ, η) plane. As such, for every $\phi(x, y) \in \Omega$, we can find a corresponding $\psi(\xi, \eta) \in \hat{\Omega}$ such that $\phi(x, y) = \psi(\xi, \eta)$. Clearly,

$$\frac{\partial \phi}{\partial x} = \frac{1}{\alpha_1}\frac{\partial \psi}{\partial \xi}, \quad \frac{\partial \phi}{\partial y} = \frac{1}{\alpha_2}\frac{\partial \psi}{\partial \eta}, \quad \frac{\partial^2 \phi}{\partial x^2} = \frac{1}{\alpha_1^2}\frac{\partial^2 \psi}{\partial \xi^2}, \quad \frac{\partial^2 \phi}{\partial y^2} = \frac{1}{\alpha_2^2}\frac{\partial^2 \psi}{\partial \eta^2}. \tag{8.56}$$

As in the numerical formulation presented in the previous section for the one-step MPS-MFS, the solution of (8.52) can be approximated by two types of basis functions, namely Chebyshev polynomials and fundamental solutions of the Laplacian; i.e.

$$\hat{\phi}(x, y) = \sum_{j=0}^{m} \sum_{i=0}^{n} a_{ij} T_i(x) T_j(y) + \sum_{k=1}^{N} b_k G^L(\rho_k),\qquad (8.57)$$

where $\rho_k = \|(x, y) - (s_k, t_k)\|$, $(x, y) \in \Omega \cup \partial\Omega$, $(s_k, t_k) \in \partial\hat{\Gamma}$, $\{a_{ij}\}$ and $\{b_k\}$ are unknown coefficients to be determined, $\partial\hat{\Gamma}$ is the auxiliary boundary of Ω, and G^L is the fundamental solution of the Laplacian; i.e., $G^L(\rho) = \ln \rho$.

As we have shown in the last few sections, we can freely embed a given domain into an extended domain of any shape such as a sunflower-shaped domain. In the current context, due to the use of Chebyshev polynomials, it is natural to embed Ω into a rectangular domain with a change of variables.

Since $\Delta G^L(\rho) = 0$ for $(x, y) \in \Omega$, we have

$$\Delta\hat{\phi}(x, y) = \sum_{j=0}^{m} \sum_{i=0}^{n} a_{ij} \Delta(T_i(x) T_j(y)) + \sum_{k=1}^{N} b_k \Delta G^L(\rho_k)$$

$$= \sum_{j=0}^{m} \sum_{i=0}^{n} a_{ij} \Delta(T_i(x) T_j(y)).\qquad (8.58)$$

In addition, we have

$$\frac{\partial\hat{\phi}}{\partial x}(x, y) = \sum_{j=0}^{m} \sum_{i=0}^{n} a_{ij} T_j(y) \frac{d T_i(x)}{dx} + \sum_{k=1}^{N} b_k \frac{\partial G^L}{\partial x}(\rho_k),\qquad (8.59)$$

$$\frac{\partial\hat{\phi}}{\partial y}(x, y) = \sum_{j=0}^{m} \sum_{i=0}^{n} a_{ij} T_i(x) \frac{d T_j(y)}{dy} + \sum_{k=1}^{N} b_k \frac{\partial G^L}{\partial y}(\rho_k).\qquad (8.60)$$

In the above, $\partial G^L(\rho)/\partial x$ and $\partial G^L(\rho)/\partial y$ are given in (8.49). To calculate the derivatives of Chebyshev polynomials, we shall use the following identities. Let T_n denote the Chebyshev polynomial of the first kind and U_n the Chebyshev polynomial of the second kind with degree n. The polynomials T_n and U_n can be obtained by the following recursive relations [312]:

$$T_0(x) = 1, \quad T_1(x) = x, \quad T_{n+1}(x) = 2x\, T_n(x) - T_{n-1}(x), \quad n \in \mathbb{N},$$

$$(8.61a)$$

$$U_0(x) = 1, \quad U_1(x) = 2x, \quad U_{n+1}(x) = 2x\,U_n(x) - U_{n-1}(x), \quad n \in \mathbb{N}.$$
(8.61b)

Furthermore,

$$T_n(x) = U_n(x) - x\,U_{n-1}(x), \quad \text{and} \quad U_n(x) = x\,U_{n-1}(x) + T_n(x).$$
(8.62)

The first and second derivatives of T_n with respect to x are given by

$$T_n' = nU_{n-1}, \quad T_n'' = \frac{n}{x^2 - 1}\big((n+1)T_n - U_n\big),$$

$$T_n''\big|_{x=1} = \frac{1}{3}(n^4 - n^2), \quad T_n''\big|_{x=-1} = \frac{(-1)^n}{3}(n^4 - n^2).$$
(8.63)

By directly substituting (8.58)–(8.60) into (8.52a), we obtain

$$\sum_{j=0}^{m}\sum_{i=0}^{n} a_{ij}\Psi(x, y) + \sum_{k=1}^{N} b_k\Phi(\rho_k) = f(x, y),$$
(8.64)

where

$$\Psi(x, y) = \left[\Delta + \alpha(x, y)\frac{\partial}{\partial x} + \beta(x, y)\frac{\partial}{\partial y} + \gamma(x, y)\right] T_i(x)T_j(y),$$
(8.65)

$$\Phi(\rho_k) = \left[\alpha(x, y)\frac{\partial}{\partial x} + \beta(x, y)\frac{\partial}{\partial y} + \gamma(x, y)\right] G^L(\rho_k).$$
(8.66)

Meanwhile, the BC in (8.52b) becomes

$$\sum_{j=0}^{m}\sum_{i=0}^{n} a_{ij}\mathfrak{B}(T_i(x)T_j(y)) + \sum_{k=1}^{N} b_k\mathfrak{B}G^L(\rho_k) = g(x, y).$$
(8.67)

To achieve spectral convergence using Chebyshev polynomials, we need to collocate at the Gauss–Lobatto quadrature nodes. These are the roots of the degree m Chebyshev polynomial of the first kind, which contribute to the high accuracy of Chebyshev pseudospectral approximations. These points can be obtained using (3.57) in Section 3.2.2. Collocation of the Chebyshev polynomials at the Gauss–Lobatto quadrature nodes and using the standard

MFS procedure on (8.64)–(8.67), yields the $(m \times n + N) \times (m \times n + N)$ system of equations:

$$\begin{bmatrix} \Psi(x, y) & \Phi(\rho_k) \\ \mathcal{B}(T_i(x)T_j(y)) & \mathcal{B}G^L(\rho_k) \end{bmatrix} \begin{bmatrix} \mathbf{a} \\ \mathbf{b} \end{bmatrix} = \begin{bmatrix} \mathbf{f} \\ \mathbf{g} \end{bmatrix}, \qquad (8.68)$$

where $\mathbf{a} = [a_1, a_2, \ldots, a_{m \times n}]^T$, $\mathbf{b} = [b_1, b_2, \ldots, b_N]^T$. Once the coefficients $\{a_{ij}\}$ and $\{b_k\}$ have been calculated by solving the above linear system, the approximate solution $\hat{\phi}$ can be obtained from (8.57).

Numerical Implementation: We first map the given domain Ω onto $\hat{\Omega}$ which is contained in the standard Chebyshev domain $[-1, 1] \times [-1, 1]$. Next, we take the Gauss–Lobatto nodes in both the ξ and η directions in this square domain to be the collocation points for the evaluation of an approximate particular solution. The number of these nodes in the x and y directions are denoted by N_x and N_y, respectively. For the MFS, as usual, we place the sources on an auxiliary circle with a fixed radius with its center at the geometric center of the domain. To measure the error, we compute the maximum relative error $\epsilon_{\max}$.

■ Example 8.7

We examine the Dirichlet BVP [329]

$$(\Delta - |xy|)\,\phi(x, y) = f(x, y), \quad (x, y) \in \Omega,$$

$$\phi(x, y) = g(x, y), \quad (x, y) \in \partial\Omega,$$

where f and g are calculated from the exact solution

$$\phi(x, y) = y\sin(\pi x) + x\cos(\pi y), \quad (x, y) \in \overline{\Omega}.$$

We consider the four irregular domains (Cassini, Star, Amoeba, and L-shape domains) in (7.9). We take 160 boundary points for the Cassini, Star, and L-shaped domains, and 200 boundary points for the Amoeba domain, and the radius of the source circle is set to 3 in all cases. From the results presented in Table 8.6, we observe that the accuracy for all four domains improves rapidly when increasing the number of Chebyshev nodes in each axis direction. Such extraordinary accuracy is due to the spectral convergence of both Chebyshev polynomials and the MFS. In addition, the polynomials can be evaluated easily, and the overall computational efficiency is quite remarkable.

Table 8.6. Maximum relative errors ϵ_{max} for the four domains using various numbers of Chebyshev nodes.

$N_x = N_y$	Cassini	Star	Amoeba	L-shape
14	2.9(−06)	1.1(−04)	4.3(−04)	1.6(−11)
16	1.2(−06)	4.2(−05)	1.3(−04)	1.4(−12)
18	2.1(−08)	2.7(−06)	5.9(−06)	3.6(−15)
20	3.7(−10)	1.3(−07)	4.0(−07)	3.2(−15)
22	7.5(−12)	5.5(−09)	5.9(−09)	2.0(−15)
24	1.3(−13)	4.8(−11)	3.8(−10)	2.4(−15)
26	6.4(−15)	1.1(−12)	3.0(−12)	1.6(−15)
28	8.4(−15)	1.7(−14)	8.4(−14)	3.9(−15)
30	8.4(−15)	3.9(−15)	7.1(−15)	5.0(−15)

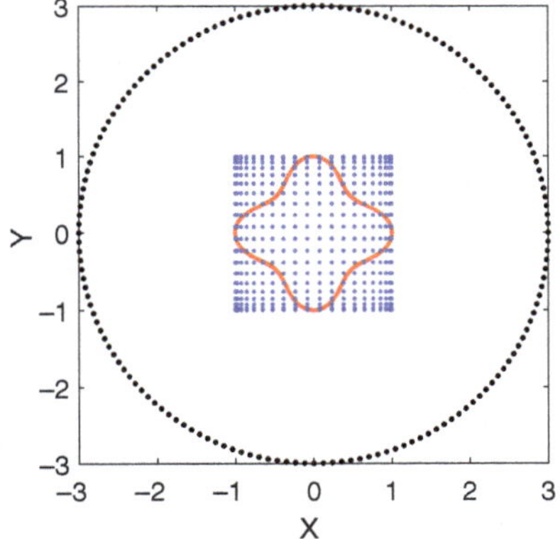

Figure 8.9. Distribution of boundary points (red), Chebyshev collocation points (blue) and source points (black) using $N_x = N_y = 20$ and source circle with radius 3.

If necessary, LOOCV can be applied to determine the optimal radius of the MFS source circle.

Figure 8.9 depicts the Chebyshev collocation points with $N_x = N_y = 20$, 160 Cassini boundary points and the source circle (with radius 3).

The numerical results in Table 8.6 were produced by the following code. In Line 14, bx1 and by1 are the ξ and η shown in (8.55), while the coefficient matrix in (8.68) is constructed in Lines 19–26. Lines 43–54 produce Ψ in (8.68), and Lines 55–63 generate the block matrix $\mathcal{B}(T_i(x)T_j(y))$ in (8.68). Lines 64–72 evaluate the particular solution in (8.57). Note that using the least squares solver lsqminnorm in Line 29 improves the accuracy by two to three orders of magnitude in comparison to using to the standard backslash solver.

```matlab
%   Chebyshev_MFS.m
1   function Chebyshev_MFS
2   rhs = @(x,y) -(pi)^2*(y.*sin(pi*x)+x.*cos(pi*y))-abs(x.*y)...
3               .*(y.*sin(pi*x)+x.*cos(pi*y));
4   exact = @(x,y) y.*sin(pi*x)+x.*cos(pi*y);
5   r=@(x) exp(sin(x)).*sin(2*x).^2+exp(cos(x)).*(cos(2*x)).^2;
6   global  alpha1 alpha2 beta1 beta2
%   nb: # of boundary points; Nx, Ny: # of Gauss-Robatto points
7   nb=160;  Nx=20;   Ny=Nx;  %Input data
8   t=(1:nb)'/nb*2*pi;
9   [bx,by]=pol2cart(t,r(t)); %boundary points
10  a=min(bx); b=max(bx);   c=min(by);     d=max(by);
11  Ax1=min(a,c); Ay1=Ax1;   Ax2=max(b,d);  Ay2=Ax2;
12  alpha1=(Ax2-Ax1)/2;    alpha2=(Ay2-Ay1)/2;
13  beta1=(Ax1+Ax2)/2;     beta2=(Ay1+Ay2)/2;
14  bx1=(bx-beta1)/alpha1;   by1=(by-beta2)/alpha2;
%% (x1,y1): Gauss-Robatto points
15  x=cos((2*(1:Nx)-1)*pi/(2*Nx));
16  [M,N]=meshgrid(x,x);
17  x1=M(:); y1=N(:);      nc=length(x1);
18  [sx, sy]=pol2cart(t,3); %% Generate source points
%% Collocation coefficient matrix Using Chebyshev Polynomial
19  A11=calcCoef(x1,y1,Nx-1,Ny-1);
20  DM=pdist2([x1 y1],[sx sy]);
21  var=repmat(-abs((x1*alpha1+beta1).*(y1*alpha2+beta2)),1,nb);
22  A12=var.*log(DM);
23  A21=PARMAT(bx1,by1,Nx-1,Ny-1);
%% Method of Fundamental solution
24  DM=pdist2([bx1 by1],[sx sy]);
25  A22=log(DM);
26  A=[A11 A12;A21 A22];
27  f=rhs(x1*alpha1+beta1,y1*alpha2+beta2); % Forcing term
28  g=exact(bx1*alpha1+beta1,by1*alpha2+beta2); %Dirichlet BC
29  coef=lsqminnorm(A,[f;g]);
%% Calculate the approximate solution
30  p=haltonset(2);    q=net(p,1000);
31  q(:,1)=q(:,1)*(b-a)+a;   q(:,2)=q(:,2)*(d-c)+c;
32  in = inpolygon(q(:,1),q(:,2),bx,by);
33  testpt = [q(in,1), q(in,2)];
34  xt=testpt(1:300,1); yt=testpt(1:300,2);
35  xt1=(xt-beta1)/alpha1;    yt1=(yt-beta2)/alpha2;
36  DM=pdist2([xt1 yt1], [sx sy]);
37  par=calcSOL(xt1,yt1,coef(1:nc),Nx-1,Ny-1); %particular sol.
```

```matlab
38 hom=log(DM)*coef(nc+1:end); %homogeneous solution
39 zt=hom+par;    %approximate solution
40 z=exact(xt1*alpha1+beta1,yt1*alpha2+beta2); % exact solution
41 dz=max(abs((z-zt)))/max(abs(z)); %Max abs rel error
42 fprintf('Nx=Ny=%3d, Max abs error: %8.3e\n',Nx,dz);
%% Collocation coefficient matrix Using Chebyshev Polynomial
43 function A=calcCoef(x1,y1,Nx,Ny)
44 global alpha1 alpha2 beta1 beta2
45 Nn=(Nx+1)*(Ny+1);  A=zeros(Nn);  Ny1=Ny+1;
46 for k=1:length(x1)
47     [C1,C2]= chebyX2_1(Nx,x1(k));
48     [C3,C4]= chebyX2_1(Ny,y1(k));
49     for i=0:Nx
50         A(k,i*Ny1+1:(i+1)*Ny1)=1/alpha1^2*C2(i+1)*C3(1:Ny1)+...
51         1/alpha2^2*C1(i+1)*C4(1:Ny1)-abs((x1(k)*alpha1+beta1)...
52         .*(y1(k)*alpha2+beta2)).*C1(i+1)*C3(1:Ny+1);
53     end
54 end
%% Matrix of particular solution
55 function A=PARMAT(x1,y1,Nx,Ny)
56 m=length(x1); A=zeros(m,(Nx+1)*(Ny+1));
57 for k=1:m
58     C1= cos((0:Nx)*acos(x1(k)));
59     C2= cos((0:Ny)*acos(y1(k)));
60     A1=repmat(C1,Nx+1,1);   A2=repmat(C2',1,Ny+1);
61     B=A1.*A2;
62     A(k,:)=B(:)';
63 end
%% Evaluation of particular solution
64 function sol=calcSOL(x1,y1,coef,Nx,Ny)
65 m=length(x1); sol=zeros(m,1);
66 for k=1:m
67     C1= cos((0:Nx)*acos(x1(k)));
68     C2= cos((0:Ny)*acos(y1(k)));
69     A1=repmat(C1,Nx+1,1);   A2=repmat(C2',1,Ny+1);
70     B=A1.*A2;
71     sol(k,1)=coef'*B(:); %sol=sum sum coe(i,j)* T(i,x)*T(i,y)
72 end
%% Chebyshev poly and its second derivative
73 function [Tx,Tx2]=chebyX2_1(n,x)
74 Tx2=zeros(n+1,1); Tx=Tx2; Ux=Tx2;
75 Tx(1)=1; Tx(2)=x; Ux(1)=1; Ux(2)=2*x;
76 for i=3:n+1
77     Tx(i)=2*x*Tx(i-1)-Tx(i-2);
78     Ux(i)=2*x*Ux(i-1)-Ux(i-2);
79 end
```

```
80 if x==1
81     Tx2(1:n+1,1)=((0:n).^4-(0:n).^2)/3;
82 else
83     if x==-1
84         Tx2(1:n+1,1)=(-1).^(0:n).*((0:n).^4-(0:n).^2)/3;
85     else
86         for i=0:n
87             Tx2(i+1)=(i/(x*x-1))*((i+1)*Tx(i+1)-Ux(i+1));
88         end
89     end
90 end
```

As already stated, the numerical results presented in this section are remarkably accurate due to the spectral convergence of both the Chebyshev polynomials and the MFS. The limitation in using Chebyshev polynomials for solving BVPs in irregular domains may thus be alleviated by coupling it with the MFS. We refer readers to [329] for extensions of the current approach to solving 3D problems and other types of PDEs.

8.6 Particular Solutions via Fundamental Solutions

In this section, we introduce another technique for constructing particular solutions [18, 455], through the use of fundamental solutions of Helmholtz-type equations, which can be expressed as

$$(\Delta + \lambda)\, G_\lambda(\boldsymbol{x}, \boldsymbol{x}') = \delta(\boldsymbol{x}, \boldsymbol{x}'), \tag{8.69}$$

where $\lambda \in \mathbb{R}$. Clearly, when $\lambda > 0$, (8.69) is the Helmholtz equation with $G_\lambda = G^H$; and when $\lambda < 0$, it is the modified Helmholtz equation with $G_\lambda = G^M$. Both fundamental solutions G^H and G^M are given in Section 2.4.

8.6.1 *Poisson problems*

Let us consider finding a particular solution of the Poisson equation

$$\Delta \phi(\boldsymbol{x}) = f(\boldsymbol{x}). \tag{8.70}$$

We approximate $f(\boldsymbol{x})$ by a summation of fundamental solutions as

$$\hat{f}(\boldsymbol{x}) = \sum_{i=1}^{m} \sum_{j=1}^{N} a_{ij}\, G_{\lambda_i}(r_j), \tag{8.71}$$

where $r_j = \|x - x'_j\|$, $\{x'_1, \ldots, x'_N\}$ are N different source points located on an auxiliary boundary $\partial\Omega'$ outside the solution domain Ω, the set $\{\lambda_1, \ldots, \lambda_m\}$ consists of m distinct nonzero real numbers, and the G_{λ_i} are fundamental solutions corresponding to different values of λ_i. The completeness of the above representation was proved in [18].

A particular solution of (8.70) with the approximated RHS (8.71) is given by

$$\hat{\phi}_p(x) = -\sum_{i=1}^{m}\sum_{j=1}^{N}\frac{a_{ij}}{\lambda_i}G_{\lambda_i}(r_j). \tag{8.72}$$

In the above, each λ_i can be positive or negative, and the G_{λ_i} corresponding to these values should be used. We can verify the above representation by taking the Laplacian of the above equation and obtain

$$\Delta\hat{\phi}_p(x) = -\sum_{i=1}^{m}\sum_{j=1}^{N}\frac{a_{ij}}{\lambda_i}\Delta G_{\lambda_i}(r_j)$$

$$= -\sum_{i=1}^{m}\sum_{j=1}^{N}\frac{a_{ij}}{\lambda_i}[-\lambda_i\, G_{\lambda_i}(r_j)] = \hat{f}(x). \tag{8.73}$$

In the above we have utilized (8.69) and ignored the Dirac delta function, since the singularity stays outside the BVP region. Hence, the particular solution is obtained by dividing each term of (8.71) by $-\lambda_i$. In this way, fundamental solutions are employed to approximate both the homogeneous and particular solutions.

A step by step description of the algorithm is given as follows:

Algorithm I

Step 1: Select a set of collocation points $\{x_k\}_{k=1}^{M}$ covering the domain.

Step 2: Select a set of source points $\{x'_j\}_{j=1}^{N}$ on the auxiliary boundary $\partial\Omega'$.

Step 3: Select a set of nonzero wave numbers $\{k_i\}_{i=1}^{m}$, where $k_i^2 = \lambda_i$.

Step 4: Construct the right hand side vector
$$F = [f(x_1), f(x_2), \ldots, f(x_M)]^T, \quad M > mN.$$

Step 5: Construct the interpolation matrix $A = [A_1\, A_2\, \cdots\, A_m]$, where
$$A_i = [G_{\lambda_i}(r_{kj})]_{1 \le k \le M, 1 \le j \le N}, \quad r_{kj} = \|x_k - x'_j\|, \quad 1 \le i \le m.$$

Step 6: Solve the over-determined system $Aa = F$.

Step 7: Particular solutions can be evaluated using (8.72).

After a particular solution has been determined, the homogeneous solution can be obtained using the MFS as described in the previous sections.

■ Example 8.8

Consider the following Poisson 2D BVP

$$\Delta\phi(x) = f(x), \quad x \in \Omega,$$

$$\phi(x) = g(x), \quad x \in \partial\Omega,$$

where f and g are derived from the exact solution

$$\phi(x, y) = e^{x+y}$$

and Ω is a gear-shaped domain whose parametric equation is given by (8.20).

For the evaluation of the particular solution, we take $M = 700$ collocation points covering the domain, and $N = 60$ source points on a circle of initial radius $R = 3$ centered at (0,0). The radius R will be adjusted in the simulation according to the LOOCV algorithm. Figure 8.10 shows a typical distribution of collocation (blue) and source (red) points.

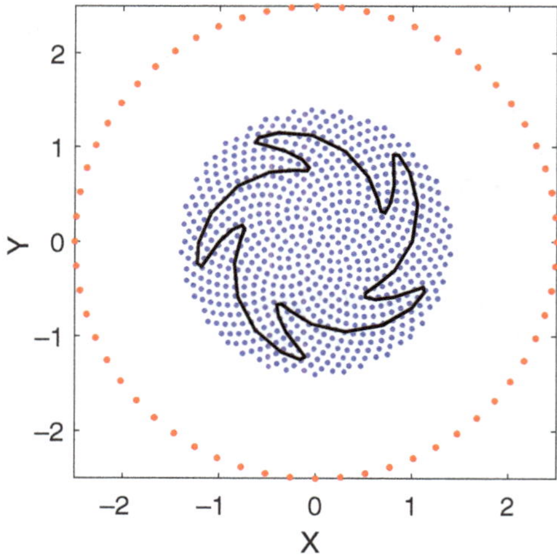

Figure 8.10. Typical distribution of collocation points (blue) and source points (red) for the evaluation of the particular solution.

For the basis functions G_{λ_i} in (8.71), we can use either Y_0 or K_0, or even the nonsingular solution J_0. The wave numbers k_i in Step 3 of Algorithm I were chosen as $\{k_i\}_{i=1}^{10} = \{(9 + 2(i - 1))/9\}_{i=1}^{10}$; i.e. $k_i \in [1, 3]$. The range of the k_i does have a significant impact on the accuracy and extensive experimentation has revealed that smaller k_i values produce better results. Furthermore, we applied LOOCV to determine an appropriate source circle radius R. For the determination of the homogeneous solution, we apply the MFS using the same source points as those employed for the particular solution.

Table 8.7 shows the maximum absolute errors $\mathcal{E}_{\max}$ resulting from simulations using different fundamental solutions, source circle radii, and numbers of basis functions. We found that the fundamental solution K_0 consistently performs better than other fundamental solutions.

The code `MFS_par2.m` that produced the results in Table 8.7 is presented as follows. The points generator for the gear-shaped domain in Line 5 can be found in Example 8.1. Line 8 produces the set of wave numbers in Step 3 of Algorithm I. In Line 11, we search for the optimal radius of the source circle for the MPS using LOOCV. Both the MPS and MFS use the same source points in Line 12. The interpolation matrix for the MPS in Step 5 of Algorithm I is generated in Lines 16–18. Particular solutions on the boundary are evaluated in Lines 22–25, while particular solutions and homogeneous solutions at the test points are evaluated in Lines 34–39.

Table 8.7. Maximum absolute errors $\mathcal{E}_{\max}$ using various fundamental solutions and numbers of basis functions with different wave numbers k_i.

	J_0		Y_0		K_0	
# of k_i	$\mathcal{E}_{\max}$	R	$\mathcal{E}_{\max}$	R	$\mathcal{E}_{\max}$	R
5	4.2(−07)	3.71	2.3(−06)	3.54	9.9(−10)	3.53
6	1.4(−08)	3.66	1.4(−08)	4.57	1.9(−11)	4.67
7	4.8(−10)	3.54	1.8(−08)	2.94	3.4(−11)	5.19
8	5.5(−11)	4.92	5.8(−09)	3.51	1.7(−11)	3.30
9	2.5(−10)	3.53	1.4(−10)	5.19	7.3(−12)	4.48
10	4.9(−11)	3.51	5.9(−11)	4.45	2.4(−11)	4.47

```matlab
%  MFS_par2.m     Poisson's Problem
1  clear; warning off all;  tic
2  forcing=@(x,y) 2*exp(1*x+y);              %forcing term
3  bound_con=@(x,y) exp(1*x+y);  %Dirichlet boundary condition
%  nb: # of boundary points;  n:# of collocation points;
%  ep=1 : Helmholtz FS;  ep=-1: modified Helmholtz FS
4  nb=60;  n=700; ep = -1;
5  [x,y,xc,yc,~,~,xt,yt]=gear(nb,n,1.5);
6  f=forcing(xc,yc);
7  exact=bound_con(xt,yt);     %exact solution
8  wl=linspace(1,3,10);  %wave frequencies
9  N=9;  % # of wave frequency
10 [sx0,sy0]=pol2cart(2*pi*(1:nb)'/nb,1);
11 R=fminbnd(@(d) cost_mfs_mps(d,xc,yc,sx0,sy0,f,wl,N,n,nb),2,6);
12 [xs,ys]=pol2cart(2*pi*(1:nb)'/nb,R);
13 center=[xs,ys]; % source points for the MPS and MFS
14 DM1=pdist2([xc,yc],center);
15 phi=zeros(n,nb*N);  %[M_1 M_2 ... M_p] Interpolation matrix
16 for i=1:N
17     phi(:,(i-1)*nb+1:i*nb)=besselk(0,wl(i)*DM1);
18 end
19 coef_par=phi\f;     %lsqminnorm(phi,f);
20 DM2=pdist2([x, y],center);
21 ps=zeros(nb,1);
22 for i=1:N
23     ps=ps+besselk(0, wl(i)*DM2)*coef_par(1+nb*(i-1):i*nb)...
24     /(-ep*wl(i)^2);
25 end
26 g=bound_con(x,y)-ps;  %updated Dirchlet BC
27 DM=pdist2([x,y],[xs,ys]);
28 A2=log(DM);          % boundary condition
29 coef_mfs= lsqminnorm(A2,g);  %Solve linear system
%%%%%%  Test
30 DM5=pdist2([xt, yt],center);
31 DM4=pdist2([xt, yt],[xs, ys]);
32 nt=length(xt);
33 A3=zeros(nt,1);
34 for i=1:N
35     A3=A3+besselk(0,wl(i)*DM5)*coef_par(1+nb*(i-1):i*nb)...
36     /(-ep*wl(i)^2);
37 end
38 A4=log(DM4);
39 approx=A3+A4*coef_mfs;   %Approximate solution
40 error=norm(exact-approx,inf); % Maximum absolute error
41 fprintf('n =%4d, nb =%4d, N=%2d,d=%5.2f\n',n,nb,N,R);
42 fprintf('Maxerr=%8.3e\n',error); toc
```

The following code searches for the optimal radius of the source circle of the MPS using LOOCV.

```
%   cost_mfs_mps.m
1   function ceps = cost_mfs_mps(d,x,y,sx,sy,rhs,wl,N,n,nb)
2   xs=d*sx; ys=d*sy;
3   =pdist2([x,y],[xs,ys]);
4   A=zeros(n,nb*N);   %[M_1 M_2 ... M_p] Interpolation matrix
5   for i=1:N
6       A(:,(i-1)*nb+1:i*nb)=besselk(0,wl(i)*DM);
7   end
8   invA=pinv(A);
9   errorvector=(A\rhs)./diag(invA);
10  ceps=norm(errorvector);
11  end
```

8.6.2 *Inhomogeneous Helmholtz-type problems*

For the inhomogeneous Helmholtz-type equation

$$(\Delta + \mu)\,\phi(x) = f(x), \tag{8.74}$$

where μ can be positive or negative, a similar technique as that in the preceding section can be applied to obtain a particular solution. First, we approximate the RHS using Helmholtz-type fundamental solutions as basis functions, as in (8.71). An approximate particular solution of (8.74) is then given by

$$\hat{\phi}_p(x) = \sum_{i=1}^{m}\sum_{j=1}^{N} \frac{a_{ij}}{-\lambda_i + \mu}\,G_{\lambda_i}(r_j), \tag{8.75}$$

with $\lambda_i \neq \mu$. We may verify the above statement by applying the Helmholtz operator to it:

$$(\Delta + \mu)\hat{\phi}_p(x) = \sum_{i=1}^{m}\sum_{j=1}^{N} \frac{a_{ij}}{-\lambda_i + \mu}(\Delta + \mu)\,G_{\lambda_i}(r_j)$$

$$= \sum_{i=1}^{m}\sum_{j=1}^{N} \frac{a_{ij}}{-\lambda_i + \mu}\left[(-\lambda_i + \mu)\,G_{\lambda_i}(r_j)\right] = \hat{f}(x). \tag{8.76}$$

Note that in the above we have used (8.69).

■ **Example 8.9**

Consider the 2D BVP

$$(\Delta + k^2)\phi(\pmb{x}) = f(\pmb{x}), \quad \pmb{x} \in \Omega,$$

$$\phi(\pmb{x}) = g(\pmb{x}), \quad \pmb{x} \in \partial\Omega,$$

where f and g are derived from the exact solution

$$\phi(x, y) = \frac{4}{4 + x + y}.$$

The domain Ω is the unit disk with five holes in it, shown in Fig. 8.11(a).

We take $k = 18$ and choose $n_i = 20$ boundary points on each inner circle which has radius equal to 0.2, and take $n_o = 100$ boundary points on the outer circle. We also take m wave numbers λ_i. The number of boundary points is thus $M = 5n_i + n_o$, and the same number of source points N is selected. The number of collocation points which are distributed inside the unit circle is $N \times m$. In this example, we choose J_0 as the basis function for the particular solution approximation. Since J_0 is nonsingular, the source points can be distributed freely inside and outside the domain as shown in Fig. 8.11(b). As such, the source points are selected uniformly inside a circle of radius

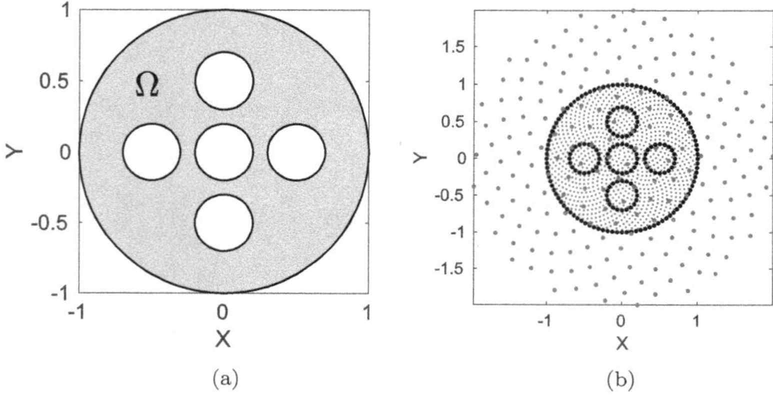

(a)   (b)

Figure 8.11. (a) The unit disk with five holes. (b) Typical distribution of collocation (blue) and source (red) points for the approximation of particular solutions.

Table 8.8. Maximum absolute errors using J_0 as the basis function of the MPS and MFS for various number of wave numbers.

# of wave numbers	$\mathcal{E}_{\max}$	R
5	1.4(−4)	7.65
6	2.2(−6)	4.49
7	1.3(−7)	5.06
8	4.7(−10)	6.82
9	3.0(−11)	5.06
10	1.2(−9)	4.64
11	4.2(−11)	3.17

R containing the domain, as shown in red dots in Fig. 8.11(b), and for the calculation of the maximum absolute error $\mathcal{E}_{\max}$, we took 323 uniformly distributed points inside the domain.

For the evaluation of the homogeneous solution, we apply the MFS using J_0 as the fundamental solution and use the same source points as those in the particular solution approximation. The LOOCV algorithm is used to determine the source circle radius and the computed results are presented in Table 8.8 as maximum absolute errors. We notice that the accuracy improves with the number of k_i terms, but there is no further improvement beyond nine such terms.

The code that produced the results in Table 8.8 can be obtained by slightly revising the code `MFS_par2.m` used in Example 8.8 for solving Poisson problems. All functions `besselk` should be replaced by `besselj` while Lines 24 and 36 should be replaced using (8.75). The points generator for the domain with five holes is not a difficult task and thus no code is provided for this example.

The above solution procedure for solving inhomogeneous second order elliptic equations can be extended to certain polyharmonic differential equations using various kinds of fundamental solutions as the basis functions.

Chapter 9

Matrix Decomposition Algorithms for Axisymmetric BVPs

Matrix decomposition algorithms (MDAs) are fast solution methods that reduce systems resulting from the discretization of multi-dimensional problems to a set of independent 1D problems, in which the coefficient matrices are generally banded, block tridiagonal, or almost block diagonal. Such algorithms have been implemented to linear systems resulting from finite difference, finite element, spectral and orthogonal spline collocation methods. Surveys of MDAs can be found in [85, 86].

MDAs have also been used in MFS solutions of harmonic and biharmonic problems [287, 453, 861, 862] and problems in linear elasticity and thermoelasticity [448, 450]. As MDAs require matrices to have a certain structure to manipulate, such as matrices with circulant patterns, their applications are limited to problems with symmetrical geometries, such as regular polygonal and axisymmetric shapes. In this chapter, we consider MDAs for the solution of BVPs in rotationally symmetric domains in 2D and 3D, governed by elliptic PDEs.

9.1 Preliminaries

In the current chapter, we shall need the definition of a circulant matrix $A \in \mathbb{R}^{N \times N}$, which is a square matrix of the form

$$
A = \begin{pmatrix}
a_1 & a_2 & \ldots & a_N \\
a_N & a_1 & \ldots & a_{N-1} \\
\vdots & \vdots & \ddots & \vdots \\
a_2 & a_3 & \ldots & a_1
\end{pmatrix}, \tag{9.1}
$$

and which is denoted by $A = \mathrm{circ}\,(a_1, a_2, \ldots, a_N)$. The various properties of circulant matrices can be found in [238].

In addition, we shall require the definition of the Kronecker or tensor product of two matrices $\mathbf{A} \in \mathbb{R}^{M_1 \times N_1}$ and $\mathbf{B} \in \mathbb{R}^{M_2 \times N_2}$, which is the matrix $\mathbf{C} \in \mathbb{R}^{M_1 M_2 \times N_1 N_2}$ defined by

$$
\mathbf{C} = \mathbf{A} \otimes \mathbf{B} = \begin{pmatrix}
a_{11}\mathbf{B} & a_{12}\mathbf{B} & \ldots & a_{1N_1}\mathbf{B} \\
a_{21}\mathbf{B} & a_{22}\mathbf{B} & \ldots & a_{2N_1}\mathbf{B} \\
\vdots & \vdots & \ddots & \vdots \\
a_{M_1 1}\mathbf{B} & a_{M_1 2}\mathbf{B} & \ldots & a_{M_1 N_1}\mathbf{B}
\end{pmatrix}. \tag{9.2}
$$

The properties of the Kronecker product may be found in, e.g. [633].

9.2 Laplace Equation

We first consider the Laplace equation in $\mathbb{R}^2$ or $\mathbb{R}^3$ subject to a Dirichlet BC

$$
\Delta \phi(x) = 0, \quad x \in \Omega, \tag{9.3a}
$$

$$
\phi(x) = f(x), \quad x \in \partial\Omega. \tag{9.3b}
$$

where the boundary of the domain Ω is $\partial\Omega$.

9.2.1 *2D implementation*

In 2D we examine the case when Ω is a disk, and for simplicity we shall assume that Ω is the unit disk. Although the BVP (9.3) might appear overly simple in this case, it is useful for problems in more complex domains which can be conformally transformed onto the unit disk; see [452].

We express the solution of BVP (9.3) with the MFS approximation

$$\hat{\phi}(\boldsymbol{x}) = \sum_{j=1}^{N} c_j \, G^L(\boldsymbol{x}, \boldsymbol{x}'_j), \quad \boldsymbol{x} \in \overline{\Omega}, \tag{9.4}$$

where the $\{\boldsymbol{x}'_j\}_{j=1}^{N}$ are the sources located on the auxiliary boundary, and $G^L(\boldsymbol{x}, \boldsymbol{x}'_j)$ is a fundamental solution of the 2D Laplacian operator given by (2.23a). We choose the collocation points $\{\boldsymbol{x}_i\}_{i=1}^{N}$ on the unit circle as

$$(x_i, y_i) = \left(\cos \frac{2(i-1)\pi}{N}, \ \sin \frac{2(i-1)\pi}{N} \right), \quad i = 1, \ldots, N, \tag{9.5}$$

and the sources as

$$(x'_j, y'_j) = R \times \left(\cos \frac{2(j-1+\alpha)\pi}{N}, \ \sin \frac{2(j-1+\alpha)\pi}{N} \right),$$

$$j = 1, \ldots, N, \tag{9.6}$$

where R is the radius of the circle (auxiliary boundary) on which the sources are placed, and $0 \le \alpha < 1$ is a rotational parameter of the sources with respect to the collocation points.

Collocation of the BCs leads to the square system

$$A c = f, \tag{9.7}$$

where the matrix A is circulant [238], and in particular $A = \mathrm{circ}\,(a_1, a_2, \ldots, a_N)$. Following [808], we define a unitary $N \times N$ matrix U_N, whose conjugate is the Fourier matrix

$$U_N^* = \frac{1}{\sqrt{N}} \begin{pmatrix} 1 & 1 & 1 & \cdots & 1 \\ 1 & \omega_N & \omega_N^2 & \cdots & \omega_N^{N-1} \\ 1 & \omega_N^2 & \omega_N^4 & \cdots & \omega_N^{2(N-1)} \\ \vdots & \vdots & \vdots & \cdots & \vdots \\ 1 & \omega_N^{N-1} & \omega_N^{2(N-1)} & \cdots & \omega_N^{(N-1)(N-1)} \end{pmatrix}, \tag{9.8}$$

where $\omega_N = e^{2\pi \mathrm{i}/N}$, $\mathrm{i} = \sqrt{-1}$, and $U_N U_N^* = I$.

From the properties of circulant matrices [238] we have

$$U_N A U_N^* = D, \tag{9.9}$$

and

$$D = \mathrm{diag}(d_1, \ldots, d_N), \quad d_j = \sum_{k=1}^{N} a_k \, \omega_N^{(k-1)(j-1)}. \tag{9.10}$$

We can rewrite system (9.7) as

$$U_N A U_N^* U_N c = U_N f, \tag{9.11}$$

or

$$D \hat{c} = v, \tag{9.12}$$

where

$$\hat{c} = U_N c \quad \text{and} \quad v = U_N f. \tag{9.13}$$

The solution of system (9.12) is

$$\hat{c}_i = \frac{v_i}{d_i}, \quad i = 1, \ldots, N, \tag{9.14}$$

and

$$c = U_N^* \hat{c}. \tag{9.15}$$

Note that all matrix-vector multiplications by the matrices U_N and U_N^*, as well as the calculation of the diagonal elements d_j, $j = 1, \ldots, N$, can be carried out efficiently using FFTs with the MATLAB® commands `fft` and `ifft`.

■ Example 9.1

The code `Laplace.m` listed below solves BVP (9.3) in the unit disk with exact solution $\phi(x, y) = \exp(4x) \cos(4y)$.

The errors were calculated at 101 test points on the boundary. The results in Table 9.1 were obtained for various numbers of collocation points N with the code `Laplace.m`. In the table, R denotes the radius of the source circle (c.f. (9.6)) and this notation will be used throughout this chapter.

Table 9.1. Results using various numbers of collocation points.

N	R	$\epsilon_{\max}$	$\mathcal{E}_{\mathrm{rms}}$
50	3.15	4.2(−12)	7.2(−11)
60	2.10	3.6(−14)	7.3(−13)
70	1.81	6.2(−15)	1.3(−13)
90	1.58	2.0(−15)	4.4(−14)
100	1.49	1.8(−15)	3.4(−14)
300	1.99	1.2(−15)	1.8(−14)

```matlab
   % MATLAB code: Laplace.m
1  function Laplace(N) % N is the number of coll. pts and sources
2  u = @(x,y) exp(4*x).*cos(4*y); % exact solution
3  alfa=0.2; % rotational parameter
4  mp=101; %number of test points
5  ang=2*pi*(1:N)'/N; % angles
6  [xp,yp]=pol2cart(ang,1);
7  [xs0,ys0]=pol2cart(ang+(2*alfa*pi/N),1);
8  ttp=2*pi*(1:mp)'/mp; % angles for test points
9  [xpp,ypp]=pol2cart(ttp,1);
10 b=u(xp,yp); % rhs
11 ut=u(xpp,ypp); %exact solution
12 rs=linspace(1.04,3,30);
13 error1=rs; rmse=rs;
14 for iter=1:30
15     xs=rs(iter)*xs0;    ys=rs(iter)*ys0; % sources
16     DX=xp(1,1)-xs;   DY=yp(1,1)-ys;
17     circ=(1/(2*pi))*log(sqrt(DX.*DX+DY.*DY));
18     d=N*ifft(circ); % diagonal elements
19     ct=(1/sqrt(N))*fft(b)./d; % vector \tilde{c}
20     c=sqrt(N)*ifft(ct); % solution c;
21     DM=pdist2([xpp, ypp],[xs,ys]);
22     A=(1/(2*pi))*log(DM); % Fundamental solution
23     uh=real(A*c); %approximation
24     error1(iter)= norm(uh-ut,inf)/norm(ut,inf);
25     rmse(iter)=norm(uh-ut,2)/sqrt(length(xpp));
26 end
27 [err,n1]=min(error1);
28 [RMSE,~]=min(rmse);
29 fprintf('N =%4d, ext rad = %6.2f\n',N,rs(n1));
30 fprintf('rel error = %8.4e | rmse = %8.4e,\n',err,RMSE);
```

9.2.2 *3D implementation*

We first define N boundary collocation points P_n, $n = 1, \ldots, N$, on the meridian 2D curve $\partial \Upsilon$. None of these points lies on the z-axis, as they would then remain invariant under rotation in the azimuthal direction leading to a singular matrix. These boundary collocation points are described by their planar Cartesian coordinates

$$P_n = (r_n, z_n), \quad n = 1, \ldots, N. \tag{9.16}$$

The collocation points on the boundary $\partial \Omega$ of the axisymmetric domain Ω are generated by rotating these points about the z-axis. We first define M angles

$$\theta_m = \frac{2\pi \, (m - 1)}{M}, \quad m = 1, \ldots, M. \tag{9.17}$$

The collocation points $\{(x_{mn}, y_{mn}, z_{mn})\}_{m=1,n=1}^{M,N}$ are then

$$x_{mn} = r_n \cos\left(\theta_m + \frac{2\pi \, \alpha_n}{M}\right), \quad y_{mn} = r_n \sin\left(\theta_m + \frac{2\pi \, \alpha_n}{M}\right), \quad z_{mn} = z_n, \tag{9.18}$$

where $m = 1, \ldots, M$, $n = 1, \ldots, N$, and the parameters $\{\alpha_n\}_{n=1}^{N} \in [-1/2, 1/2]$ correspond to rotations of the collocation points and may be used to produce more uniform distributions. The sources $\{(x'_{mn}, y'_{mn}, z'_{mn})\}_{m=1,n=1}^{M,N}$ are defined by a magnification R of the collocation points about the origin, as

$$\left(x'_{mn}, y'_{mn}, z'_{mn}\right) = R \times (x_{mn}, y_{mn}, z_{mn}), \quad m = 1, \ldots, M, \; n = 1, \ldots, N. \tag{9.19}$$

The MFS approximation is now

$$\hat{\phi}(x) = \sum_{m,n=1}^{M,N} c_{mn} \, G^L(x, x'_{mn}), \quad x \in \overline{\Omega}, \tag{9.20}$$

where $G^L(x, x')$ is a fundamental solution of the 3D Laplace operator, given in (2.23b).

    Another important quantity of interest in the discretization of the solids under consideration is the outward normal unit vector $n(x, y, z)$ to the boundary surface $\partial\Omega$. In order to calculate it, we first find the outward normal vector to the boundary $\partial\Upsilon$, denoted by $n(r, z) = (n_r, n_z)$. So, if the outward normal vector to $\partial\Upsilon$ at the boundary point $P_n = (r_n, z_n)$ is $n(r_n, z_n) = (n_{r_n}, n_{z_n})$, then the outward normal vector to $\partial\Omega$ at the boundary point (x_{mn}, y_{mn}, z_{mn}) is given by

$$n(x_{mn}) = (n_x, n_y, n_z)$$

$$= \left(n_{r_n} \cos\left(\theta_m + \frac{2\pi \alpha_n}{M} \right), n_{r_n} \sin\left(\theta_m + \frac{2\pi \alpha_n}{M} \right), n_{z_n} \right). \quad (9.21)$$

Collocation of BC (9.3b) leads to the system

$$Ac = \begin{pmatrix} A_{1,1} & A_{1,2} & \cdots & A_{1,N} \\ A_{2,1} & A_{2,2} & \cdots & A_{2,N} \\ \vdots & \vdots & \ddots & \vdots \\ A_{N,1} & A_{N,2} & \cdots & A_{N,N} \end{pmatrix} \begin{pmatrix} c_1 \\ c_2 \\ \vdots \\ c_N \end{pmatrix} = \begin{pmatrix} f_1 \\ f_2 \\ \vdots \\ f_N \end{pmatrix} = f. \quad (9.22)$$

Each of the $M \times M$ submatrices $A_{k,\ell}$, $k, \ell = 1, \ldots, N$, is circulant and defined by $A_{k,\ell} = \operatorname{circ}\left(a_{k\ell_1}, a_{k\ell_2}, \ldots, a_{k\ell_N} \right)$, and each of the c_ℓ and f_k are $M \times 1$ subvectors.

Following [810], and with U_M defined as in (9.8) with N replaced by M, we can rewrite (9.22) as

$$(I_N \otimes U_M) \begin{pmatrix} A_{1,1} & A_{1,2} & \cdots & A_{1,N} \\ A_{2,1} & A_{2,2} & \cdots & A_{2,N} \\ \vdots & \vdots & \ddots & \vdots \\ A_{N,1} & A_{N,2} & \cdots & A_{N,N} \end{pmatrix} (I_N \otimes U_M^*) (I_N \otimes U_M) \begin{pmatrix} c_1 \\ c_2 \\ \vdots \\ c_N \end{pmatrix}$$

$$= (I_N \otimes U_M) \begin{pmatrix} f_1 \\ f_2 \\ \vdots \\ f_N \end{pmatrix}, \quad (9.23)$$

where I_N is the $N \times N$ identity matrix, or

$$\begin{pmatrix} D_{1,1} & D_{1,2} & \cdots & D_{1,N} \\ D_{2,1} & D_{2,2} & \cdots & D_{2,N} \\ \vdots & \vdots & \ddots & \vdots \\ D_{N,1} & D_{N,2} & \cdots & D_{N,N} \end{pmatrix} \begin{pmatrix} \hat{c}_1 \\ \hat{c}_2 \\ \vdots \\ \hat{c}_N \end{pmatrix} = \begin{pmatrix} \hat{f}_1 \\ \hat{f}_2 \\ \vdots \\ \hat{f}_N \end{pmatrix}, \quad (9.24)$$

where

$$\hat{c}_n = U_M c_n, \quad \hat{f}_n = U_M f_n, \quad n = 1, \ldots, N. \tag{9.25}$$

The matrices $D_{k\ell}, k, \ell = 1, \ldots, N$, are diagonal with

$$D_{k\ell} = \mathrm{diag}\left(d_{k\ell_1}, d_{k\ell_2}, \ldots, d_{k\ell_M}\right), \quad d_{k\ell_j} = \sum_{m=1}^{M} a_{k\ell_m} \omega_M^{(m-1)(j-1)}. \tag{9.26}$$

The solution of system (9.24) can thus be decomposed into the solution of the M, $N \times N$ systems

$$E_m x_m = y_m, \quad m = 1, \ldots, M, \tag{9.27}$$

where

$$(E_m)_{k,\ell} = D_{k,\ell_m}, \quad k, \ell = 1, \ldots, N, \tag{9.28}$$

and

$$(x_m)_n = (\hat{c}_n)_m, \quad (y_m)_n = (\hat{f}_n)_m, \quad n = 1, \ldots, N. \tag{9.29}$$

Having obtained the vectors $x_m, m = 1, \ldots, M$, we can recover the vectors $\hat{a}_n, n = 1, \ldots, N$, and, subsequently, the vector c from

$$c = \begin{pmatrix} c_1 \\ c_2 \\ \vdots \\ c_N \end{pmatrix} = \left(I_N \otimes U_M^*\right) \hat{c} = \begin{pmatrix} U_M^* \hat{c}_1 \\ U_M^* \hat{c}_2 \\ \vdots \\ U_M^* \hat{c}_N \end{pmatrix}. \tag{9.30}$$

Again, all matrix-vector multiplications by the matrices U_M and U_M^* as well as the calculation of diagonal elements $d_{k\ell_j}, k, \ell = 1, \ldots, N$, $j = 1, \ldots, M$, can be carried out efficiently using FFTs.

■ Example 9.2

We consider the Laplace equation with the Dirichlet BC (9.3b) in a 3D peanut-shaped domain using the MDA described in Section 9.2.2 where the exact solution is given by

$$\phi(x, y, z) = \cos(x) \sinh\left(\sqrt{5}\, y\right) \sin(2z).$$

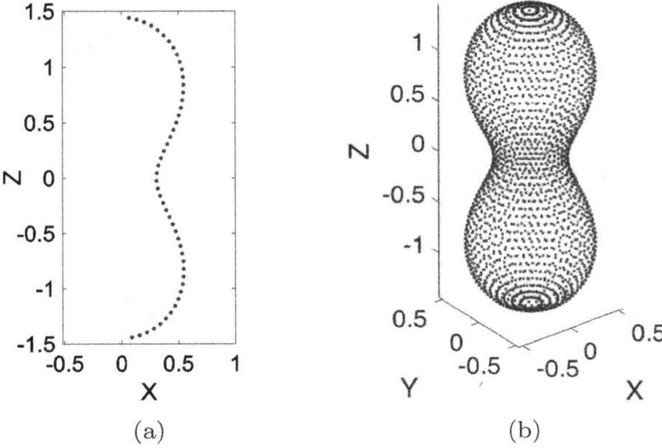

Figure 9.1. (a) Profile of 50 boundary collocation points on the meridian 2D curve $\partial\Upsilon$. (b) Profile of 2,500 circular boundary points on $\partial\Omega$.

To produce the distributed points P_n in (9.16), the code `equibdd_points.m` is used and available in Appendix C. In Fig. 9.1(a), we show 50 evenly distributed boundary points on the meridian 2D curve $\partial\Upsilon$. By appropriately rotating these points about the z-axis, see (9.18), 2500 boundary points are generated on the boundary $\partial\Omega$ of the axisymmetric domain Ω as shown in Fig. 9.1(b).

The code `Laplace1_3D_AxisymA_peanut.m` solves the BVP in question for $M = N = 120$ using the proposed MDA. Lines 4–11 are used to generate the boundary points on the surface of the peanut domain. In Line 16 the source points are determined by multiplying the boundary point coordinates by a magnification factor. Standard MDA procedures are shown in Lines 17–31. The results of the code are presented in Fig. 9.2 with the accuracy tested on a set of 1,000 randomly selected boundary points (`xt`,`yt`,`zt`) as shown in Line 2. Alternatively, these test points can be generated using the code `peanut_nodes` listed in Appendix C. In Table 9.2, we present the results obtained using various numbers of collocation points. Note that the CPU time for each case is obtained for a loop of 30 iterations. From this table it is clear that accuracy is not an issue and that we can handle problems with a large number of collocation points. We also observe that the optimal magnification factor R is inversely proportional to the number of collocation points.

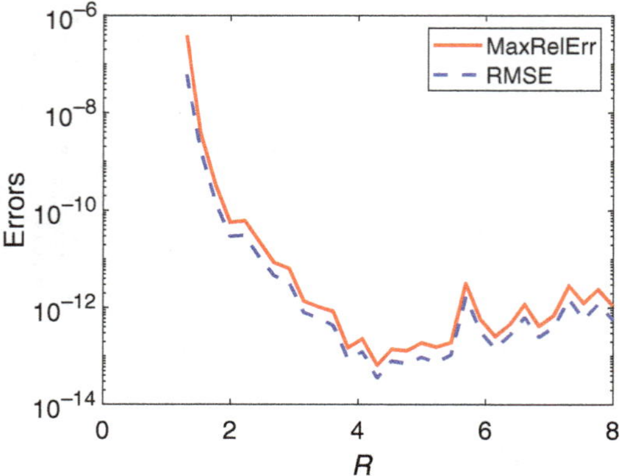

Figure 9.2. The maximum relative error ϵ_{max} and RMSE $\mathcal{E}_{rms}$ for the 3D Laplace BVP using the proposed MDA with $M = N = 120$.

Table 9.2. Results using various numbers of collocation points.

$M = N$	R	ϵ_{max}	$\mathcal{E}_{rms}$	CPU (secs)
80	7.77	3.6(−11)	2.0(−11)	2.0
100	7.54	7.2(−13)	3.8(−13)	3.4
150	3.15	7.0(−14)	4.2(−14)	8.3
200	2.14	3.8(−14)	1.8(−14)	16.6
250	2.42	4.5(−14)	2.7(−14)	28.9
300	1.99	4.8(−14)	2.7(−14)	52.5

```
%   Laplace1_3D_AxisymA_peanut.m
%   n: no of collocation points on generating curve
%   m: no collocation points in azimuthal direction
1   u=@(x,y,z) cos(x).*sinh(sqrt(5)*y).*sin(2*z);%exact solution
2   load peanut_pts.mat    xt yt zt    % boundary test points
3   n=120; m=120;
4   [xp, yp]=equibdd_points(m);
5   alfa=(1-(-1).^(1:n)')/5; %rotational parameter for coll pts
6   ang1=2*pi/m*(0:m-1)';    %Angles in azimuthal direction
7   ang1=repmat(ang1,1,n);   alfa=repmat(alfa',m,1);
8   t= ang1+(2*alfa*pi/m);
9   x=xp'.*cos(t);            x=x(:); %Vectorization of coll pts
10  y=xp'.*sin(t);            y=y(:);
11  z=repmat(yp',m,1);  z=z(:);
12  rhs=u(x,y,z);        %right hand side
13  ut = u(xt,yt,zt);  % exact solution on test points
```

```matlab
14 d=linspace(1.3,8,30);   error1=d; rmse=d;
15 for iter=1:30
16     source=d(iter)*[x,y,z];    % sources
17     b=reshape(rhs,m,n);
18     b=transpose(fft(b)/sqrt(m));   % fft of rhs
19     circul=zeros(n,n,m);
20     k=(0:n-1)*m+1;
21     DM=pdist2([x(k), y(k), z(k)], source);
22     Mat=-(1/(4*pi))./DM; % relevant parts of global matrix
23     for ii=1:n              %first row of circulant matrices
24         circul(ii,:,:)=(reshape(Mat(ii,:),m,n))';
25     end
26     dd=ifft(circul,[],3)*m;   % diagonal matrices
27     sol=zeros(m,n);
28     for ii=1:m
29         sol(ii,:)=dd(:,:,ii)\b(:,ii);%solution of subsystems
30     end
31     sol=(ifft(sol)*sqrt(m));  % fft to recover solution
%  Calculation of solution at test points
32     DMT=pdist2([xt yt zt], source);
33     uh=-(1/(4*pi))*(1./DMT)*sol(:);   %approximate solution
34     error1(iter) = norm(uh-ut,inf)/norm(ut,inf);
35     rmse(iter)=norm(uh-ut,2)/sqrt(length(xt));
36 end
37 [err, n1]=min(error1);
38 [RMSE,~]=min(rmse);
39 fprintf('N=%4d, d =%5.3f\n',n*m,d(n1))
40 fprintf('rel_er=%10.3e, RMSE=%10.3e\n',err,RMSE)
41 figure (1) % plot of errors vs magnification factor
42 semilogy(d,error1,'-r',d,rmse,'--b')
43 legend('MaxRelErr','RMSE')
44 xlabel('\it d'); ylabel('Errors')
45 set(gca,'FontSize',16)
```

The selection of a suitable magnification factor for the auxiliary boundary is crucial in obtaining good accuracy. As demonstrated in Chapter 7, the LOOCV algorithm provides an effective way for the identification of a good magnification factor. However, when the number of collocation points becomes large, LOOCV is inefficient. One of the attractive features of MDAs is that it enables us to solve large-scale problems. Typically, when using MDAs, M systems of equations of order N need to be solved. Clearly, it is not cost-effective to apply LOOCV to each of the M systems and determine a good magnification factor for each. Instead, we apply LOOCV to a randomly chosen system among the M systems. As such, the same (thus obtained) magnification factor will then be used for all the other systems. The above MDA script can be slightly modified by removing the `for` loop and then adding the following LOOCV script:

```
1   function ceps = Costeps(d,x,y,z,rhs,m,n,k)
    % [x,y,z]: boundary collocation points
    % k: a randomly selected circular matrix.
2   source=d*[x,y,z];    %sources
    % MDA matrices
3   b=reshape(rhs,m,n);
4   b=transpose(fft(b)/sqrt(m)); % fft of rhs
5   K=(0:n-1)*m+1;
6   DM=pdist2([x(K), y(K), z(K)], source);
7   Mat=-(1/(4*pi))./DM; % relevant parts of global matrix
8   circul=zeros(n,n,m);
9   for ii=1:n
10       circul(ii,:,:)=(reshape(Mat(ii,:),m,n))';
11  end
12  dd=ifft(circul,[],3)*m;   % diagonal matrices
13  A=dd(:,:,k); RHS=b(:,k);
14  invA=pinv(A);
15  errorvector=(invA*RHS)./diag(invA);
16  ceps=norm(errorvector);
17  end
```

The following MATLAB® function `fminbnd` needs to be added after Line 13 in the code `Laplace1_3D_AxisymA_peanut.m`:

```
k= 8;  % A randomly selected circular matrix.  k < m.
d=fminbnd(@(ep) Costeps(ep,x,y,z,rhs,m,n,k),MIN,MAX, ...
    optimset('TolX',1e-3));
```

Note that `[MIN,MAX]` denotes the search interval using `fminbnd`.

From Table 9.3, we notice that the results are very close to the optimal accuracy as shown in Table 9.2. When the number of collocation points becomes large, the optimal magnification factor is expected to get smaller, and thus the search interval should be gradually reduced for achieving better results. Meanwhile, the CPU times in this table are much smaller than those reported in Table 9.2, without compromising the accuracy.

Table 9.3. Results using LOOCV.

$M = N$	[MIN, MAX]	R	$\epsilon_{\max}$	$\mathcal{E}_{\mathrm{rms}}$	CPU (secs)
80	[1.3, 9]	7.18	7.0(−11)	4.4(−11)	0.24
100	[1.3, 9]	7.18	1.0(−12)	5.3(−13)	0.44
150	[1.3, 6]	3.10	3.8(−14)	2.2(−14)	1.25
200	[1.3, 5]	2.70	9.8(−14)	4.5(−14)	2.89
250	[1.3, 3]	2.04	7.7(−14)	4.0(−14)	5.42
300	[1.3, 3]	1.95	9.3(−14)	5.4(−14)	8.35

9.3 Helmholtz Equation

We consider the Helmholtz equation in $\mathbb{R}^2$ or $\mathbb{R}^3$ in the closed bounded domain Ω, subject to the Neumann BC

$$(\Delta + k^2)\phi(x) = 0, \quad x \in \Omega, \tag{9.31a}$$

$$\frac{\partial \phi}{\partial n}(x) = g(x), \qquad x \in \partial\Omega. \tag{9.31b}$$

9.3.1 *2D implementation*

We examine the case when Ω is the unit disk. The solution of BVP (9.31) is approximated by

$$\hat{\phi}(x) = \sum_{j=1}^{N} c_j \, G^H(x, x_j'), \quad x \in \overline{\Omega}, \tag{9.32}$$

where $G^H(x, x')$ is a fundamental solution of the 2D Helmholtz operator, given by (2.32a). The discretization details are the same as those described in Section 9.2.1, leading to the system

$$A c = g, \tag{9.33}$$

where the matrix A is again block circulant. The solution of system (9.33) is identical to that of (9.7), see also [446].

Table 9.4. Results using various numbers of collocation points.

N	R	$\epsilon_{\max}$	$\mathcal{E}_{\mathrm{rms}}$
30	3.80	$1.1(-9)$	$1.5(-10)$
40	4.20	$3.0(-12)$	$4.3(-13)$
50	3.60	$1.5(-15)$	$1.9(-16)$
80	3.12	$1.2(-15)$	$1.8(-16)$
100	1.96	$1.3(-15)$	$2.0(-16)$
300	1.84	$1.6(-15)$	$2.5(-16)$

■ Example 9.3

The code `Helmholtz.m` listed below solves BVP (9.31) in the unit disk with exact solution

$$\phi(x, y) = Y_0 \left(k\sqrt{(x - 2)^2 + (y + 3)^2}\right)$$

using the MDA of Section 9.2.1 for increasing values of R (Line 22).

This code is similar to the code for Example 9.1 except for the use of a Neumann BC (9.31b) instead of a Dirichlet BC, see Line 19, and also the fundamental solution (see Line 26 for its normal derivative on the boundary).

The results in Table 9.4 were obtained using the code `Helmholtz_MDA.m`.

```
%   MATLAB code: Helmholtz_MDA.m
%   N is the number of coll. pts and sources
1   function Helmholtz_MDA(N)
2   u = @(x,y,kappa) bessely(0,kappa*sqrt((x-2).^2+(y+3).^2));
3   syms x y c
4   uxd=diff(u(x,y,c),x);
5   ux=matlabFunction(uxd,'vars',{x,y,c});
6   uyd=diff(u(x,y,c),y);
7   uy=matlabFunction(uyd,'vars',{x,y,c});
8   clear x y c
9   alfa=0.2; % rotational parameter
10  kappa=2;
11  mp=101; %number of test points
```

```
12 tt=2*pi*(0:N-1)'/N; % angles
13 [xp,yp]=pol2cart(tt,1); %boundary collocation points
14 [xs0,ys0]=pol2cart(tt+(2*alfa*pi/N),1); %Source points
15 ttp=2*pi*(0:mp-1)'/mp; % angles for test points
16 [xpp,ypp]=pol2cart(ttp,1); %boundary collocation points
17 xn=xp;  yn=yp; %normal derivatives
18 XN=xn(1,1); YN=yn(1,1); %normal derivative matrices
19 b(1:N,1)=ux(xp,yp,kappa).*xn+uy(xp,yp,kappa).*yn; %rhs
20 error=zeros(50,1); rmse=error; rs=error;
21 for iter=1:50
22      rs(iter)=1+iter*0.04; % radius of pseudo-boundary
23      xs=rs(iter)*xs0;    ys=rs(iter)*ys0; % sources
24      DX=xp(1,1)-xs;  DY=yp(1,1)-ys;
25      DM=sqrt(DX.*DX+DY.*DY);   DMX=DX.*XN+DY.*YN
%  first row of circulant matrix
26      circ=-(kappa/4)*(bessely(1,kappa*DM)./DM).*DMX;
27      d=N*ifft(circ); % diagonal elements
28      ct=(1/sqrt(N))*fft(b)./d;   % vector \tilde{c}
29      sol=sqrt(N)*ifft(ct); % solution c;
30      dxp=xpp*ones(1,N)-ones(mp,1)*xs';
31      dyp=ypp*ones(1,N)-ones(mp,1)*ys';
32      A=(1/4)*bessely(0,kappa*sqrt(dxp.*dxp+dyp.*dyp));
33      uh=A*sol; %approximation
34      ut=u(xpp,ypp,kappa); %exact solution
35      error(iter) = norm(uh-ut,inf)/norm(ut,inf);
36      rmse(iter)=norm(uh-ut,2)/sqrt(length(xpp));
37 end
38 [err,n1]=min(error);
39 [RMSE,~]=min(rmse);
40 fprintf('N =%4d, ext rad = %6.2f\n',N,rs(n1));
41 fprintf('rel error = %8.4e | rmse = %8.4e,\n',err,RMSE);
```

9.3.2 *3D implementation*

The domain Ω is now an axisymmetric domain constructed as described in Section 9.2.2, see (9.16)–(9.21). The discretization details are the same as those described in Section 9.2.2, leading to a system of the form (9.33), where the matrix A is block circulant. The MFS approximation of the solution is of the form (9.20), where G^L is replaced by the fundamental solution of the 3D Helmholtz operator, given by (2.32b). The MDA solution of system (9.33) is identical to that of (9.22).

Table 9.5. Results of the Helmholtz equation with $k = 2$ for various $M = N$ using LOOCV.

$M = N$	[MIN, MAX]	R	ϵ_{max}	$\mathcal{E}_{rms}$	CPU (secs)
80	[1, 8]	4.20	1.9(-14)	2.1(-15)	0.33
100	[1, 8]	4.63	9.7(-14)	1.0(-14)	0.63
150	[1, 8]	4.60	5.8(-14)	6.6(-15)	2.3
200	[1, 8]	2.65	5.7(-14)	6.5(-15)	2.9
250	[1, 3]	2.27	5.7(-14)	7.2(-15)	9.3
300	[1, 3]	1.47	8.2(-14)	1.1(-14)	14.7

■ Example 9.4

We consider the Helmholtz BVP (9.31) in an ellipsoidal domain where the exact solution is given by

$$\phi(x, y, z) = \frac{\cos\left(k\sqrt{(x - 2)^2 + (y + 3)^2 + (z - 1)^2}\right)}{\sqrt{(x - 2)^2 + (y + 3)^2 + (z - 1)^2}}.$$

The scale of the ellipsoidal axes in the coordinates x, y and z is 1:1:0.5.

The following code solves the above BVP using LOOCV for the case $M = N = 100$ and $k = 2$. In Table 9.5, we present the results for various $M = N$. The source points are placed on a similar ellipsoid with magnification factor R. The results show the proposed MDA is very efficient and highly accurate for solving axisymmetric problems with a large number of collocation points.

```
%   MATLAB code: Helmholtz1_3D_AxisymA_LOOCV.m
1   function Helmholtz1_3D_AxisymA_LOOCV
2   warning off
3   global kappa  xn yn zn
4   u=@(x,y,z,c) (1./sqrt((x-2).^2+(y+3).^2+(z-1).^2)).*...
5       cos(c*sqrt((x-2).^2+(y+3).^2+(z-1).^2));%exact solution
6   syms x y z c
7   uxd=diff(u(x,y,z,c),x);
8   ux=matlabFunction(uxd,'vars',{x,y,z,c}); % x-derivative
9   uyd=diff(u(x,y,z,c),y);
```

```matlab
10 uy=matlabFunction(uyd,'vars',{x,y,z,c}); %y-derivative
11 uzd=diff(u(x,y,z,c),z);
12 uz=matlabFunction(uzd,'vars',{x,y,z,c}); % z-derivative
13 clear x y z c
14 n=100; % number of collocation points on generating curve
15 m=100; %number of collocation points in azimuthal direction
16 kap=2;  % kappa
% Generation of col pts and normals on ellipsoid generator
17 [xp, yp, xN, yN]=ellipsoid(1,.5,n);
18 alfa(1:n,1)=(1-(-1).^(1:n))/5;
19 ang1(1:m,1)=2*pi/m*(0:m-1); %angles in azimuthal direction
20 ang1=repmat(ang1,1,n);  alfa=repmat(alfa',m,1);
21 t= ang1+(2*alfa*pi/m);
22 x=xp'.*cos(t);        x=x(:);      % Vectorization of coll pts
23 y=xp'.*sin(t);        y=y(:);
24 z=repmat(yp',m,1);    z=z(:);
25 xn=xN'.*cos(t);       xn=xn(:);%Vectorization of normal coords
26 yn=xN'.*sin(t);       yn=yn(:);
27 zn=repmat(yN',m,1);    zn=zn(:);
28 xs0=x;   ys0=y;      zs0=z;     %initial sources
29 [xt,yt,zt]=sphere(30);
30 xt=xt(:); yt=yt(:); zt=zt(:)/2;
31 rhs=ux(x,y,z,kap).*xn+uy(x,y,z,kap).*yn+...
32     uz(x,y,z,kap).*zn; %Neumann boundary condition
33 ut = u(xt,yt,zt,kap);  % exact solution
34 k=8; % k is randomly chosen in [1, n]
35 d=fminbnd(@(ep) Costeps(ep,x,y,z,rhs,m,n,k),1,6,...
36     optimset('TolX',1e-2));
37 xs=d*xs0; ys=d*ys0; zs=d*zs0;%sources
% MDA
38 b=reshape(rhs,m,n);
39 b=transpose(fft(b)/sqrt(m)); % fft of rhs
40 circul=zeros(n,n,m); t=1+(0:n-1)'*m;
41 XN=xn(t);   YN=yn(t);    ZN=zn(t);
42 DM=pdist2([x(t),y(t),z(t)],[xs,ys,zs]); %Distance Matrix
43 DMX=x(t)-xs'; DMY=y(t)-ys'; DMZ=z(t)-zs';%Difference Matrix
44 Mat=(1/(4*pi))*((cos(kap*DM)./DM.^3)+(kap*sin(kap*DM)./...
45     DM.^2)).*(DMX.*XN+DMY.*YN+DMZ.*ZN);% global matrix
46 for ii=1:n
47     circul(ii,:,:)=(reshape(Mat(ii,:),m,n)).';
48 end
49 dd1=ifft(circul,[],3)*m;  % diagonal matrices
50 sol=zeros(m,n);
51 for ii=1:m
52     sol(ii,:)=dd1(:,:,ii)\b(:,ii); % solution of subsystems
53 end
```

```
54 sol=(ifft(sol)*sqrt(m)); % fft to recover solution
55 sol=sol(:);
%  Calculation of solution at test points on sphere surface
56 DMT=pdist2([xt,yt,zt],[xs,ys,zs]);
57 uh=-(1/(4*pi))*((1./DMT).*cos(kap*DMT))*sol;%approx solution
58 err1= norm(uh-ut,inf)/norm(ut,inf);   %relative error
59 err2= norm(uh-ut,2)/sqrt(length(xt)); %RMSE error
60 fprint('m=%4d, n=%4d, d=%5.3f\n',m,n,d)
61 fprintf('rel_er=%10.3e, RMSE=%10.3e\n',err1,err2)
%%%%  LOOCV
62 function ceps = Costeps(d,x,y,z,rhs,m,n,k)
63 global kap xn yn zn
64 xs=d*x; ys=d*y; zs=d*z;
65 b=reshape(rhs,m,n);
66 b=transpose(fft(b)/sqrt(m)); % fft of rhs
67 circul=zeros(n,n,m);
68 t=1+(0:n-1)'*m;
69 XN=xn(t);   YN=yn(t);    ZN=zn(t);
70 DM=pdist2([x(t),y(t),z(t)],[xs,ys,zs]);
71 DMX=x(t)-xs';    DMY=y(t)-ys';    DMZ=z(t)-zs';
72 Mat=(1/(4*pi))*((cos(kap*DM)./DM.^3)+(kap*sin(kap*DM)./...
73     DM.^2)).*(DMX.*XN+DMY.*YN+DMZ.*ZN);% global matrix
74 for ii=1:n
%  first row of circulant matrices
75     circul(ii,:,:)=(reshape(Mat(ii,:),m,n)).';
76 end
77 dd=ifft(circul,[],3)*m;  % diagonal matrices
78 A=dd(:,:,k); RHS=b(:,k);
79 invA=pinv(A);
80 errorvector=(invA*RHS)./diag(invA);
81 ceps=norm(errorvector);
```

9.4 Biharmonic Equation

We consider the biharmonic equation in $\mathbb{R}^2$ or $\mathbb{R}^3$ subject to the first biharmonic problem BCs

$$\Delta^2\phi(x) = 0, \qquad\qquad x \in \Omega, \qquad\qquad (9.34a)$$

$$\phi(x) = f(x), \quad \text{and} \quad \frac{\partial\phi}{\partial n}(x) = g(x), \quad x \in \partial\Omega. \qquad (9.34b)$$

9.4.1 *2D implementation*

We examine the case when Ω is the unit disk. The discretization details are the same as those in Section 9.2.1. The solution of (9.34) is approximated by

$$\hat{\phi}(x) = \sum_{j=1}^{N} c_j \, G^L(x, x'_j) + \sum_{j=1}^{N} d_j \, G^B(x, x'_j), \quad x \in \overline{\Omega}, \qquad (9.35)$$

where $G^L(x, x')$ is a fundamental solution of the Laplacian operator defined in (2.23a) and $G^B(x, x')$ is a fundamental solution of the biharmonic operator given by (2.156a). Collocation of the BCs now leads to the $2N \times 2N$ system of equations

$$\left(\begin{array}{c|c} A^{11} & A^{12} \\ \hline A^{21} & A^{22} \end{array}\right) \left(\begin{array}{c} c \\ d \end{array}\right) = \left(\begin{array}{c} f \\ g \end{array}\right), \qquad (9.36)$$

where each $A^{\mu\nu}$, $\mu, \nu = 1, 2$, is circulant with $A^{\mu\nu} = \mathrm{circ}\left(a_1^{\mu\nu}, a_2^{\mu\nu}, \ldots, a_N^{\mu\nu}\right)$. Following [809], we can rewrite (9.36) as

$$(I_2 \otimes U_N) \left(\begin{array}{c|c} A^{11} & A^{12} \\ \hline A^{21} & A^{22} \end{array}\right) (I_2 \otimes U_N^*) \, (I_2 \otimes U_N) \left(\begin{array}{c} c \\ d \end{array}\right)$$

$$= (I_2 \otimes U_N) \left(\begin{array}{c} f \\ g \end{array}\right), \qquad (9.37)$$

or

$$\left(\begin{array}{c|c} D^{11} & D^{12} \\ \hline D^{21} & D^{22} \end{array}\right) \left(\begin{array}{c} \hat{c} \\ \hat{d} \end{array}\right) = \left(\begin{array}{c} \hat{f} \\ \hat{g} \end{array}\right), \qquad (9.38)$$

where

$$\hat{c} = U_N c, \quad \hat{d} = U_N d, \quad \hat{f} = U_N f, \quad \hat{g} = U_N g, \qquad (9.39)$$

and the matrices $D^{\mu\nu}$, $\mu, \nu = 1, 2$, are diagonal with

$$D^{\mu\nu} = \mathrm{diag}\left(d_1^{\mu\nu}, d_2^{\mu\nu}, \ldots, d_N^{\mu\nu}\right), \quad d_j^{\mu\nu} = \sum_{m=1}^{N} a_m^{\mu\nu} \omega_N^{(m-1)(j-1)}. \qquad (9.40)$$

The solution of system (9.38) is thus

$$\hat{c}_j = \frac{\hat{f}_j d_j^{22} - \hat{g}_j d_j^{12}}{d_j^{11} d_j^{22} - d_j^{21} d_j^{12}}, \quad \hat{d}_j = \frac{\hat{g}_j d_j^{11} - \hat{f}_j d_j^{21}}{d_j^{11} d_j^{22} - d_j^{21} d_j^{12}}, \quad j = 1, \ldots, N, \quad (9.41)$$

from which we can recover the solution of (9.36) from

$$c = U_N^* \hat{c} \quad \text{and} \quad d = U_N^* \hat{d}. \quad (9.42)$$

Again, all matrix–vector multiplications by the matrices U_N and U_N^* as well as the calculation of the diagonal elements $d_j^{\mu\nu}$, $\mu, \nu = 1, 2$, $j = 1, \ldots, N$, can be carried out efficiently using FFTs.

■ Example 9.5

The code `Biharmonic.m` listed below solves BVP (9.34) in the unit disk with exact solution

$$\phi(x, y) = \exp(3x) \cos(3y)(x^2 + y^2) + \exp(2y) \sin(2x).$$

Using the MDA described above, the results in Table 9.6 were obtained using the code `Biharmonic.m` shown below. The errors were calculated at 101 boundary points.

```
%   MATLAB code: Biharmonic.m
%   N is the # of boundary and source points
1   function Biharmonic(N)
2   u=@(x,y) exp(3*x).*cos(3*y).*(x.^2+y.^2)+exp(2*y).*sin(2*x);
3   syms x y
4   uxd=diff(u(x,y),x);
5   ux=matlabFunction(uxd,'vars',{x,y});
```

Table 9.6. Results using various numbers of collocation points.

N	R	$\epsilon_{\max}$	$\varepsilon_{\mathrm{rms}}$
50	2.24	1.4(−12)	1.1(−11)
60	1.83	2.8(−13)	2.0(−12)
70	1.71	6.9(−14)	5.8(−13)
100	1.38	2.1(−14)	1.4(−13)
200	1.19	8.6(−15)	5.0(−14)

```
6  uyd=diff(u(x,y),y);
7  uy=matlabFunction(uyd,'vars',{x,y});
8  clear x y
9  mp=101;  % # of test points
10 alfa=0.2;   coe1=1/(2*pi); coe2=1/(2*pi);
11 ang=2*pi*(1:N)'/N; % angles
12 [xp,yp]=pol2cart(ang,1);
13 [xs0,ys0]=pol2cart(ang+(2*alfa*pi/N),1);
14 ttp=2*pi*(1:mp)'/mp; % angles for test points
15 [xpp,ypp]=pol2cart(ttp,1);
16 xn=xp;  yn=yp; %normal derivatives
17 XN=xn(1,1);YN=yn(1,1); %normal derivative matrices
18 rhs1=u(xp,yp); %rhs Dirichlet
19 rhs2=ux(xp,yp).*xn+uy(xp,yp).*yn;% rhs Neumann
20 rhs=[rhs1; rhs2];
21 RHS1=(1/sqrt(N))*fft(rhs(1:N,1)); % modification of rhs
22 RHS2=(1/sqrt(N))*fft(rhs(N+1:2*N,1));% modification of rhs
23 d=linspace(1.04,4,80);   error=d; rmse=d;
24 for iter=1:80
25     xs=d(iter)*xs0;  ys=d(iter)*ys0;
26     DX=xp(1,1)-xs;   DY=yp(1,1)-ys;
27     DM=sqrt(DX.*DX+DY.*DY);
28     circ11=coe1*log(DM);
29     circ12=coe2*DM.^2.*log(DM);
30     circ21=coe1*((DX./DM.^2).*XN+(DY./DM.^2).*YN);
31     circ22=coe2*((DX.*(1+2*log(DM))).*XN+(DY.*(1+2*log(DM))).*YN);
32     d11=N*ifft(circ11); % diagonal elements
33     d12=N*ifft(circ12); % diagonal elements
34     d21=N*ifft(circ21); % diagonal elements
35     d22=N*ifft(circ22); % diagonal elements
36     ct1=(RHS1.*d22-RHS2.*d12)./(d11.*d22(1:N,1)-d12.*d21);
37     ct2=(RHS2.*d11-RHS1.*d21)./(d11.*d22-d12.*d21);
38     c1=sqrt(N)*ifft(ct1); % recovery of solution c;
39     c2=sqrt(N)*ifft(ct2); % recovery of solution c;
40     DMT=pdist2([xpp ypp],[xs ys]);
41     uh = coe1*log(DMT)*c1+coe2*(DMT.^2.*log(DMT))*c2;
42     ut = u(xpp,ypp); % exact solution
43     error(iter) = norm(uh-ut,inf)/norm(ut,inf);  %relative error
44     rmse(iter)=norm(uh-ut,2)/sqrt(length(xpp));   %RMSE error
45 end
46 [err,n1]=min(error);
47 [RMSE,~]=min(rmse);
48 fprintf('N =%4d, ext rad = %6.2f\n',N,d(n1));
49 fprintf('rel error = %8.4e, rmse = %8.4e,\n',err,RMSE);
```

9.4.2   3D implementation

The domain Ω is now an axisymmetric domain constructed as described in Section 9.2.2, see (9.16)–(9.21). The discretization details are the same as those described in Section 9.2.2. The MFS approximation is

$$\hat{\phi}(\mathbf{x}) = \sum_{m,n=1}^{M,N} c_{mn} G^L(\mathbf{x}, \mathbf{x}'_{mn}) + \sum_{m,n=1}^{M,N} d_{mn} G^B(\mathbf{x}, \mathbf{x}'_{mn}), \quad \mathbf{x} \in \overline{\Omega}, \quad (9.43)$$

where $G^L(\mathbf{x}, \mathbf{x}')$ is a fundamental solution of the 3D Laplacian operator (2.23b), and $G^B(\mathbf{x}, \mathbf{x}')$ is a fundamental solution of the 3D biharmonic operator (2.156b). Collocation of the BCs leads to a system of the form

$$\left(\begin{array}{c|c} \mathbf{A}^{11} & \mathbf{A}^{12} \\ \hline \mathbf{A}^{21} & \mathbf{A}^{22} \end{array}\right) \left(\frac{\mathbf{c}}{\mathbf{d}}\right) = \left(\frac{\mathbf{f}}{\mathbf{g}}\right), \tag{9.44}$$

where

$$\begin{aligned} \mathbf{c} &= [\mathbf{c}_1,\ \mathbf{c}_2,\ \ldots,\ \mathbf{c}_N]^T, \quad \mathbf{d} = [\mathbf{d}_1,\ \mathbf{d}_2,\ \ldots,\ \mathbf{d}_N]^T, \\ \mathbf{f} &= [\mathbf{f}_1,\ \mathbf{f}_2,\ \ldots,\ \mathbf{f}_N]^T, \quad \mathbf{g} = [\mathbf{g}_1,\ \mathbf{g}_2,\ \ldots,\ \mathbf{g}_N]^T, \end{aligned} \tag{9.45}$$

and

$$\mathbf{A}^{\mu\nu} = \begin{pmatrix} \mathbf{A}_{1,1}^{\mu\nu} & \mathbf{A}_{1,2}^{\mu\nu} & \cdots & \mathbf{A}_{1,N}^{\mu\nu} \\ \mathbf{A}_{2,1}^{\mu\nu} & \mathbf{A}_{2,2}^{\mu\nu} & \cdots & \mathbf{A}_{2,N}^{\mu\nu} \\ \vdots & \vdots & \ddots & \vdots \\ \mathbf{A}_{N,1}^{\mu\nu} & \mathbf{A}_{N,2}^{\mu\nu} & \cdots & \mathbf{A}_{N,N}^{\mu\nu} \end{pmatrix}, \quad \mu, \nu = 1, 2. \tag{9.46}$$

Each of the $M \times M$ submatrices $\mathbf{A}_{k,\ell}^{\mu\nu}$, $k, \ell = 1, \ldots, N$, $\mu, \nu = 1, 2$, is circulant and defined by $\mathbf{A}_{k,\ell}^{\mu,\nu} = \mathrm{circ}\left(a_{k\ell_1}^{\mu,\nu}, a_{k\ell_2}^{\mu,\nu}, \ldots, a_{k\ell_N}^{\mu,\nu}\right)$.

Following [287] we rewrite (9.44) as

$$(\mathbf{I}_2 \otimes \mathbf{I}_N \otimes \mathbf{U}_M) \left(\begin{array}{c|c} \mathbf{A}^{11} & \mathbf{A}^{12} \\ \hline \mathbf{A}^{21} & \mathbf{A}^{22} \end{array}\right) (\mathbf{I}_2 \otimes \mathbf{I}_N \otimes \mathbf{U}_M^*) (\mathbf{I}_2 \otimes \mathbf{I}_N \otimes \mathbf{U}_M) \left(\frac{\mathbf{c}}{\mathbf{d}}\right)$$

$$= (\mathbf{I}_2 \otimes \mathbf{I}_N \otimes \mathbf{U}_M) \left(\frac{\mathbf{f}}{\mathbf{g}}\right), \tag{9.47}$$

or

$$\left(\begin{array}{c|c} \mathbf{D}^{11} & \mathbf{D}^{12} \\ \hline \mathbf{D}^{21} & \mathbf{D}^{22} \end{array}\right) \left(\frac{\hat{\mathbf{c}}}{\hat{\mathbf{d}}}\right) = \left(\frac{\hat{\mathbf{f}}}{\hat{\mathbf{g}}}\right), \tag{9.48}$$

where

$$\hat{c} = [\hat{c}_1, \ \hat{c}_2, \ \ldots ; \hat{c}_N]^T, \quad \hat{d} = [\hat{d}_1, \ \hat{d}_2, \ \ldots, \ \hat{d}_N]^T,$$
$$\hat{f} = [\hat{f}_1, \ \hat{f}_2, \ \ldots, \ \hat{f}_N]^T, \quad \hat{g} = [\hat{g}_1, \ \hat{g}_2, \ \ldots, \ \hat{g}_N]^T,$$

$$\hat{c}_n = U_M c_n, \ \ \hat{d}_n = U_M d_n, \ \ \hat{f}_n = U_M f_n, \ \ \hat{g}_n = U_M g_n, \tag{9.50}$$

for $n = 1, \ldots, N$, and

$$D^{\mu\nu} = \begin{pmatrix} D_{1,1}^{\mu\nu} & D_{1,2}^{\mu\nu} & \cdots & D_{1,N}^{\mu\nu} \\ D_{2,1}^{\mu\nu} & D_{2,2}^{\mu\nu} & \cdots & D_{2,N}^{\mu\nu} \\ \vdots & \vdots & \ddots & \vdots \\ D_{N,1}^{\mu\nu} & D_{N,2}^{\mu\nu} & \cdots & D_{N,N}^{\mu\nu} \end{pmatrix}, \quad \mu, \nu = 1, 2. \tag{9.51}$$

The matrices $D_{k,\ell}^{\mu\nu}, k, \ell = 1, \ldots, N, \ \mu, \nu = 1, 2$, are diagonal with

$$D_{k,\ell}^{\mu\nu} = \text{diag}\left(d_{k\ell_1}^{\mu\nu}, d_{k\ell_2}^{\mu\nu}, \ldots, d_{k\ell_M}^{\mu\nu}\right), \quad d_{k\ell_j}^{\mu\nu} = \sum_{m=1}^{M} a_{k\ell_m}^{\mu\nu} \omega_M^{(m-1)(j-1)}. \tag{9.52}$$

The solution of system (9.48) can thus be decomposed into the solution of $M, 2N \times 2N$ systems

$$\left(\begin{array}{c|c} E_m^{11} & E_m^{12} \\ \hline E_m^{21} & E_m^{22} \end{array}\right) \begin{pmatrix} x_m^1 \\ x_m^2 \end{pmatrix} = \begin{pmatrix} y_m^1 \\ y_m^2 \end{pmatrix}, \quad m = 1, \ldots, M, \tag{9.53}$$

where

$$\left(E_m^{\mu\nu}\right)_{k,\ell} = D_{k,\ell_m}^{\mu\nu}, \quad k, \ell = 1, \ldots, N, \ \mu, \nu = 1, 2, \tag{9.54}$$

and

$$(x_m^1)_n = (\hat{c}_n)_m, \ \ (x_m^2)_n = (\hat{d}_n)_m, \ \ (y_m^1)_n = (\hat{f}_n)_m, \ \ (y_m^2)_n = (\hat{g}_n)_m, \tag{9.55}$$

for $n = 1, \ldots, N$. Having obtained the vectors $x_m^\mu, \ m = 1, \ldots, M,$ $\mu = 1, 2$, we can recover the vectors $\hat{c}_n, \ \hat{d}_n, \ n = 1, \ldots, N$, and, subsequently, the vectors c and d from

$$c = \begin{pmatrix} c_1 \\ c_2 \\ \vdots \\ c_N \end{pmatrix} = \left(I_N \otimes U_M^* \right) \hat{c} = \begin{pmatrix} U_M^* \hat{c}_1 \\ U_M^* \hat{c}_2 \\ \vdots \\ U_M^* \hat{c}_N \end{pmatrix}, \tag{9.56}$$

$$d = \begin{pmatrix} d_1 \\ d_2 \\ \vdots \\ d_N \end{pmatrix} = \left(I_N \otimes U_M^* \right) \hat{d} = \begin{pmatrix} U_M^* \hat{d}_1 \\ U_M^* \hat{d}_2 \\ \vdots \\ U_M^* \hat{d}_N \end{pmatrix}. \tag{9.57}$$

Again, all matrix–vector multiplications by the matrices U_M and U_M^* as well as the calculation of the diagonal elements $d_{k\ell_j}^{\mu\nu}$, $k, \ell = 1, \ldots, N$, $j = 1, \ldots, M$, can be carried out efficiently using FFTs.

■ Example 9.6

The code for solving the biharmonic BVP (9.34) in an ellipsoid with scale 0.5:0.5:1 in the x, y and z axes can be found in the following code for the case $M = N = 100$ with the exact solution

$$\phi(x, y, z) = \cos(x) \sinh(\sqrt{5}\,y) \sin(2z)(x^2 + y^2 + z^2)$$
$$+ \sin(x) \cosh(\sqrt{5}\,y) \cos(2z).$$

A typical profile of the ellipsoid with the collocation points is displayed in Fig. 9.3.

In Fig. 9.4, we present the error plot for $M = N = 100$ obtained by brute force. It is known that for the biharmonic case, the error plot is flat when R becomes large. In this case, the optimal R is 1.90. As a result, we choose a much smaller search interval when using LOOCV. In Table 9.7, we present results for various $M = N$. In contrast to the Laplace case in Table 9.2, the search interval [MIN, MAX] is much smaller in the biharmonic case and achieves better results. As we have shown earlier, the magnification factor R is expected to be smaller when the collocation points become denser and the condition number increases.

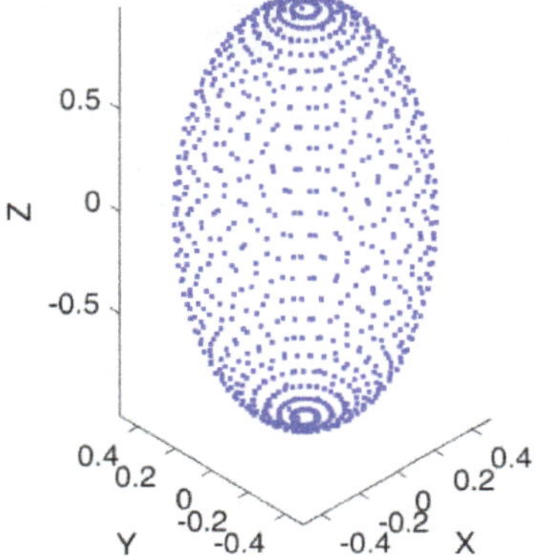

Figure 9.3. Typical profile of circular points on the ellipsoid.

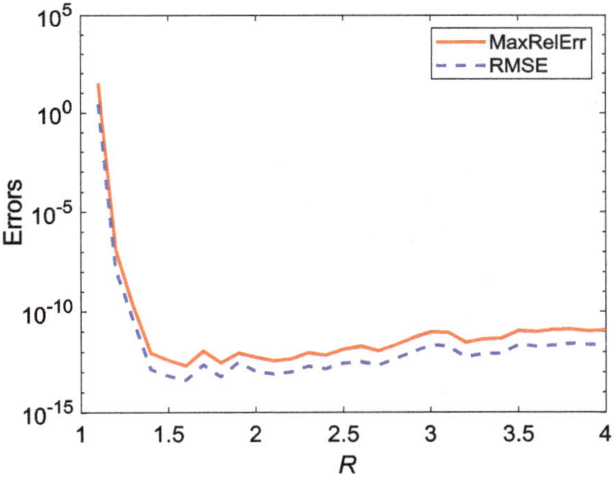

Figure 9.4. Errors for 3D biharmonic BVP using the proposed MDA with $M = N = 100$.

Table 9.7. Results of the biharmonic equation for various $M = N$ using LOOCV.

$M = N$	[MIN, MAX]	R	$\epsilon_{\max}$	$\mathcal{E}_{\mathrm{rms}}$	CPU (secs)
100	[1, 3]	2.06	2.6(-12)	4.6(-13)	1.1
150	[1, 3]	1.77	4.1(-13)	7.2(-14)	3.5
200	[1, 2]	1.68	3.1(-13)	5.8(-14)	6.5
250	[1, 2]	1.69	9.3(-13)	1.5(-13)	17.5
300	[1, 2]	1.34	7.4(-13)	1.3(-13)	23.8

```matlab
%  MATLAB code: Biharmonic1_3D_Axisym_LOOCV.m
1  function Biharmonic1_3D_Axisym_LOOCV
2  warning off
3  global xn yn zn coe1 coe2
4  u=@(x,y,z) (cos(x).*sinh(sqrt(5)*y).*sin(2*z)).*...
5     (x.^2+y.^2+z.^2)+sin(x).*cosh(sqrt(5)*y).*cos(2*z);
6  syms x y z
7  uxd=diff(u(x,y,z),x);
8  ux=matlabFunction(uxd,'vars',{x,y,z}); % x-derivative
9  uyd=diff(u(x,y,z),y);
10 uy=matlabFunction(uyd,'vars',{x,y,z}); %y-derivative
11 uzd=diff(u(x,y,z),z);
12 uz=matlabFunction(uzd,'vars',{x,y,z}); % z-derivative
13 clear x y z
14 m=100;  n=100;
15 L=1; R=3;    %Search interval of LOOCV
16 coe1=-1/(4*pi);   coe2=-1/(8*pi);
17 [xp, yp, xN, yN]=ellipsoid(.5,1,n);
18 alfa(1:n,1)=(1-(-1).^(1:n))/5; %rotational parameter
19 the1=2*pi/m;
20 ang1(1:m,1)=the1*(0:m-1); %angles in azimuthal direction
21 ang1=repmat(ang1,1,n);   alfa=repmat(alfa',m,1);
22 t= ang1+(2*alfa*pi/m);
23 x=xp'.*cos(t);        x=x(:);  % Vectorization of coll pts
24 y=xp'.*sin(t);        y=y(:);
25 z=repmat(yp',m,1);    z=z(:);
26 xn=xN'.*cos(t);       xn=xn(:);  % normal coords
27 yn=xN'.*sin(t);       yn=yn(:);
28 zn=repmat(yN',m,1);   zn=zn(:);
29 f=u(x,y,z); %rhs Dirichlet
30 g=ux(x,y,z).*xn+uy(x,y,z).*yn+uz(x,y,z).*zn;% rhs Neumann
31 [xt,yt,zt]=sphere(30);
32 xt=xt(:); yt=yt(:); zt=zt(:)/2;
33 ut = u(xt,yt,zt);  % exact solution
34 d=fminbnd(@(ep) Costeps(ep,x,y,z,f,g,m,n),L,R,...
35    optimset('TolX',1e-2));
36 sint1=zeros(m*n,1); sint2=sint1;
37 xs=d*x ; ys=d*y ;   zs=d*z ;%sources
38 f=reshape(f,m,n);
39 rhs1=(fft(f)/sqrt(m));    rhs1=rhs1(:);
40 g=reshape(g,m,n);
41 rhs2=(fft(g)/sqrt(m));    rhs2=rhs2(:);
42 [circ11,circ12,circ21,circ22]=circular(x,y,z,d,m,n);
43 dd11=ifft(circ11,[],3)*m; dd12=ifft(circ12,[],3)*m;
44 dd21=ifft(circ21,[],3)*m; dd22=ifft(circ22,[],3)*m;
45 for i=1:m
```

```
46      rrh(1:n,1)=rhs1(i+(0:n-1)*m);
47      rrh(n+1:2*n,1)=rhs2(i+(0:n-1)*m);
48      E=[dd11(:,:,i) dd12(:,:,i); dd21(:,:,i) dd22(:,:,i)];
49      ssh=E\rrh;                  % solution of subsystems
50      sint1(i+(0:n-1)*m)=ssh(1:n);
51      sint2(i+(0:n-1)*m)=ssh(n+1:2*n);
52 end
53 R1=reshape(sint1,m,n);
54 sol1=(ifft(R1)*sqrt(m)); sol1=sol1(:);
55 R1=reshape(sint2,m,n);
56 sol2=(ifft(R1)*sqrt(m)); sol2=sol2(:);
%  Calculation of solution at test points on surface
57 DMT=pdist2([xt,yt,zt],[xs,ys,zs]);
58 uh=coe1*(1./DMT)*sol1+coe2*DMT*sol2;%approximate solution
59 err1= norm(uh-ut,inf)/norm(ut,inf);   %relative error
60 err2=norm(uh-ut,2)/sqrt(length(xt));   %RMSE error
61 fprintf('m=%4d, n=%4d, d=%5.3f, L=%3d, R=%3d\n',m,n,d,L,R)
62 fprintf('rel_er=%10.3e, RMSE=%10.3e\n',err1,err2)
%%% Circular matrices
63 function [c11,c12,c21,c22]=circular(x,y,z,d,m,n)
64 global xn yn zn
65 xs=d*x; ys=d*y; zs=d*z;
66 coe1=-1/(4*pi); coe2=-1/(8*pi);
67 c11=zeros(n,n,m);c12=c11;c21=c11;c22=c11;
68 for i=1:n
69     K=1+(i-1)*m;
70     XN=xn(K); YN=yn(K); ZN=zn(K);
71     dx=x(K)- xs;  dy=y(K)- ys;  dz=z(K)- zs;
72     DM=sqrt(dx.^2+dy.^2+dz.^2);
73     c11(i,:,:)= (reshape(coe1./DM,m,n))';
74     c12(i,:,:)=(reshape(coe2*DM,m,n))';
75     MAT=-coe1./DM.^3.*(dx.*XN+dy.*YN+dz.*ZN);
76     c21(i,:,:)= (reshape(MAT,m,n))';
77     MAT=coe2./DM.*(dx.*XN+dy.*YN+dz.*ZN);
78     c22(i,:,:)= (reshape(MAT,m,n))';
79 end
%%%% LOOCV
80 function ceps = Costeps(d,x,y,z,f,g,m,n)
81 f=reshape(f,m,n);
82 rhs1=(fft(f)/sqrt(m)); % fft of rhs
83 g=reshape(g,m,n);
84 rhs2=(fft(g)/sqrt(m)); % fft of rhs
85 rhs1=rhs1(:);   rhs2=rhs2(:);
86 [circ11,circ12,circ21,circ22]=circular(x,y,z,d,m,n);
87 dd11=ifft(circ11,[],3)*m; dd12=ifft(circ12,[],3)*m;
88 dd21=ifft(circ21,[],3)*m; dd22=ifft(circ22,[],3)*m;
```

```
89 K=8;
90 A=[dd11(:,:,K) dd12(:,:,K); dd21(:,:,K) dd22(:,:,K)];
91 RHS=[rhs1(K+(0:n-1)*m); rhs2(K+(0:n-1)*m)];
92 invA=pinv(A);
93 errorvector=(invA*RHS)./diag(invA);
94 ceps=norm(errorvector);
```

Chapter 10

The Localized Method of Fundamental Solutions

As discussed in earlier chapters, simplicity of implementation and high accuracy are the main attractions of the MFS. Its shortcomings, however, include the uncertainty in the optimal selection of source points, particularly for complicated multi-connected domains, and the difficulty in solving large-scale physical problems that contain a large number of collocation points, resulting in ill-conditioned matrix systems. To alleviate these issues, the localized MFS (LMFS) was recently introduced in [291], and further developed in [356, 358, 588, 740, 741, 822, 960, 961]. Note that MDAs for the LMFS have been developed in [148].

The basic concepts of the LMFS have been briefly introduced in Section 6.8. In this chapter, we further explore the ideas behind it and present its application to homogeneous Laplacian and biharmonic BVPs, with MATLAB® codes.

10.1 Laplacian BVPs

In this section, we consider the following BVP in a closed and bounded domain:

$$\Delta\phi(x) = 0, \qquad x \in \Omega, \tag{10.1a}$$

$$\phi(x) = f(x), \quad x \in \partial\Omega_D, \tag{10.1b}$$

$$\frac{\partial\phi}{\partial n}(x) = g(x), \quad x \in \partial\Omega_N, \tag{10.1c}$$

where f and g are known functions, $x \in \mathbb{R}^d$, $d = 2, 3$, $\partial\Omega = \partial\Omega_D \cup \partial\Omega_N$, and $\partial\Omega_D \cap \partial\Omega_N = \emptyset$.

For convenience, we describe the LMFS for the Laplacian BVP (10.1) in the 2D case. We select N_i and N_b interior and boundary points, respectively, yielding a total number of collocation points/nodes $N = N_i + N_b$, which we shall denote by $\{x_i\}_{i=1}^N$. For each node x_i we select its m_i nearest neighboring nodes $\{x_k^{(i)}\}_{k=1}^{m_i}$. These nodes, along with their center point x_i, will lie inside a sub-domain $\Omega^{(i)} \subset \Omega$; see Fig. 10.1. Moreover, we select $n_i \leq m_i$ sources $\{\xi_j^{(i)}\}_{j=1}^{n_i}$ outside $\Omega^{(i)}$. If x_i is an interior point then the solution ϕ satisfies the Laplace equation in $\Omega^{(i)}$ and can therefore be approximated by the LMFS expansion

$$\phi^{(i)}(x) = \sum_{j=1}^{n_i} c_j^{(i)} G(\rho_j), \quad x \in \Omega^{(i)}, \tag{10.2}$$

where $\{c_j^{(i)}\}_{j=1}^{n_i}$ are the unknown coefficients, and $G(\rho_j) = \ln \rho_j$ is the fundamental solution of the 2D Laplace operator with $\rho_j = \|x - \xi_j^{(i)}\|$ and $x = (x, y)$, $\xi_j^{(i)} = \left(\xi_j^{x(i)}, \xi_j^{y(i)}\right)$. In Fig. 10.1(b), the red dots denote the collocation points in $\Omega^{(i)}$ and the black dots the source points outside the sub-domain. The sources are uniformly distributed on a circle $\partial\Omega_s^{(i)}$, which is centered at x_i with radius $R_s^{(i)}$. Note that $R_s^{(i)} = R \times \max_{1 \leq i \leq N} \left(\lambda^{(i)}\right)$, where $R > 1$ is a parameter, which should be manually determined by the

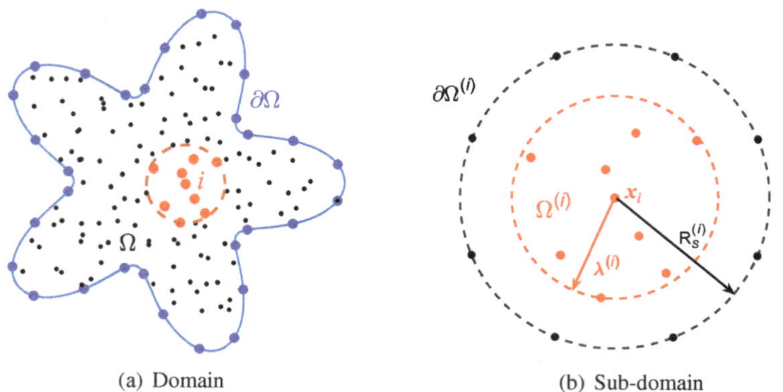

(a) Domain

(b) Sub-domain

Figure 10.1. The profiles of (a) computational domain, boundary and interior points, (b) collocation and source points in a local domain.

user, and $\lambda^{(i)}$ is the radius of the smallest circle containing all the $m_i + 1$ nodes in the sub-domain $\Omega^{(i)}$ (see Fig. 10.1(b)).

We next collocate (10.2) at the m_i nearest nodes $\{x_k^{(i)}\}_{k=1}^{m_i}$ of x_i, and obtain the local system of equations

$$\boldsymbol{\phi}^{(i)} = \boldsymbol{H}\boldsymbol{c}^{(i)}, \tag{10.3}$$

where $\boldsymbol{\phi}^{(i)} = \left[\phi_1^{(i)}, \phi_2^{(i)}, \ldots, \phi_{m_i}^{(i)}\right]^T$ is the vector of the (unknown) approximations to ϕ at the m_i nodes and $\boldsymbol{c}^{(i)} = \left[c_1^{(i)}, c_2^{(i)}, \ldots, c_{n_i}^{(i)}\right]^T$ is the vector of unknown coefficients. The $m_i \times n_i$ coefficient matrix $\boldsymbol{H}$ is defined from

$$\left(H_{kj}\right)_{k,j=1}^{m_i,n_i} = \ln \rho_{kj}, \tag{10.4}$$

where $\rho_{kj} = \|x_k^{(i)} - \xi_j^{(i)}\|$. From (10.3), we have

$$\boldsymbol{c}^{(i)} = \boldsymbol{H}^{-1}\boldsymbol{\phi}^{(i)}. \tag{10.5}$$

Note that the matrix $\boldsymbol{H}$ is highly ill-conditioned and can be non-square if the number of source points is chosen to be less than the number of collocation nodes ($n_i \leq m_i$) in the sub-domain. As such, in the numerical experiments of this chapter, the Moore-Penrose pseudoinverse $\boldsymbol{H}^{-1}$ is obtained by the MATLAB® function `pinv` which is a truncated singular value decomposition (TSVD) with tolerance set to 10^{-8}. Clearly, $\boldsymbol{H}^{-1}$ is a matrix of order $n_i \times m_i$.

From (10.2), the numerical solution at the node x_i can be written in the form

$$\phi_i := \phi(x_i) = \sum_{j=1}^{n_i} c_j^{(i)} \ln \rho_{ij} = \boldsymbol{h}^{(i)T}\boldsymbol{c}^{(i)}$$

$$= \boldsymbol{h}^{(i)T} \boldsymbol{H}^{-1}\boldsymbol{\phi}^{(i)} = \sum_{k=1}^{m_i} w_k^{(i)}\phi_k^{(i)}, \tag{10.6}$$

where $\boldsymbol{h}^{(i)} = \left[\ln \rho_{i1}, \ln \rho_{i2}, \ldots, \ln \rho_{in_i}\right]^T$ is the vector of fundamental solutions at x_i, and $\{w_k^{(i)}\}_{k=1}^{m_i}$ are the weighting coefficients, which are dependent on the spatial coordinates of the $m_i + 1$ nodes in $\Omega^{(i)}$ and the corresponding sources. Equation (10.6) provides a relationship between the unknown approximations of ϕ at x_i, denoted by ϕ_i, and those at the surrounding nodes $\{x_k^{(i)}\}_{k=1}^{m_i}$, by $\boldsymbol{\phi}^{(i)}$, in the sub-domain $\Omega^{(i)}$.

We can apply the above procedure to each node $\boldsymbol{x}_i$ in the computational domain Ω, and obtain a system of equations

$$\phi_i - \sum_{k=1}^{m_i} w_k^{(i)} \phi_k^{(i)} = 0, \quad i = 1, 2, \ldots, N_i. \tag{10.7}$$

We note in the above equation that each local unknown $\phi_k^{(i)}$ corresponds to a global unknown, as all local nodes were selected from a set of global nodes. We may view (10.7) as similar to the FDM, in which the unknown at a center node is equated to the weighted average of unknowns at the neighboring nodes on a grid. However, we should stress the following differences. First, the global nodes of the LMFS are not confined to a rectilinear grid like the FDM, and are generally scattered. Second, the approximation (10.7) exactly satisfies the Laplace equation by virtue of using the fundamental solution as a basis function; while for the FDM, the fulfillment of the governing equation is enforced only to within the truncated higher order terms in the Taylor series.

For the Dirichlet BC (10.1b), we have

$$\phi_i = f(\boldsymbol{x}_i), \quad i = N_i + 1, \ldots, N_i + N_{b_1}, \tag{10.8}$$

where N_{b_1} is the number of nodes on $\partial\Omega_D$. For the Neumann BC, we can differentiate (10.6) with respect to x at a boundary node $\boldsymbol{x}_i$, as

$$\frac{\partial \phi}{\partial x}(\boldsymbol{x}_i) = \sum_{j=1}^{n_i} c_j^{(i)} \frac{\partial \ln \rho_j}{\partial x}(\boldsymbol{x}_i) = \boldsymbol{h}_x^{(i)T} \boldsymbol{c}^{(i)}$$

$$= \boldsymbol{h}_x^{(i)T} \boldsymbol{H}^{-1} \boldsymbol{\phi}^{(i)} = \sum_{k=1}^{m_i} w_k^{x(i)} \phi_k^{(i)}, \tag{10.9}$$

where

$$\boldsymbol{h}_x^{(i)} = \left[\frac{\partial \ln \rho_1}{\partial x}(\boldsymbol{x}_i), \ \frac{\partial \ln \rho_2}{\partial x}(\boldsymbol{x}_i), \ \ldots, \ \frac{\partial \ln \rho_{n_i}}{\partial x}(\boldsymbol{x}_i) \right]^T. \tag{10.10}$$

Similarly, we find

$$\frac{\partial u}{\partial y}(\boldsymbol{x}_i) = \sum_{k=1}^{m_i} w_k^{y(i)} \phi_k^{(i)}. \tag{10.11}$$

To enforce the Neumann BC (10.1c) on $\partial\Omega_N$, we have

$$\frac{\partial\phi}{\partial n}(\boldsymbol{x}_i) = n_{x_i}\frac{\partial\phi}{\partial x}(\boldsymbol{x}_i) + n_{y_i}\frac{\partial\phi}{\partial y}(\boldsymbol{x}_i)$$

$$= \sum_{k=1}^{m_i}\left(w_k^{x(i)}n_{x_i} + w_k^{y(i)}n_{y_i}\right)\phi_k^{(i)}$$

$$= g(\boldsymbol{x}_i), \quad i = N_i + N_{b_1} + 1, \ldots, N, \qquad (10.12)$$

where (n_{x_i}, n_{y_i}) is the unit outward normal vector at the boundary point $\boldsymbol{x}_i$.

The assembly of equations (10.7)–(10.12) yields the global system

$$A\boldsymbol{\phi} = b = \begin{pmatrix} \mathbf{0} \\ f \\ g \end{pmatrix}, \qquad (10.13)$$

where A is an $N \times N$ sparse matrix, $\boldsymbol{\phi} = [\phi_1, \phi_2, \ldots, \phi_N]^T$ is the vector of approximate solutions at all the nodes, and b contains the right-hand sides of (10.7)–(10.12), i.e.

$$\mathbf{0} = \underbrace{[0, 0, \ldots, 0]^T}_{N_i}, \qquad (10.14a)$$

$$f = \left[f(\boldsymbol{x}_{N_i+1}), f(\boldsymbol{x}_{N_i+2}), \ldots, f(\boldsymbol{x}_{N_i+N_{b_1}})\right]^T, \qquad (10.14b)$$

$$g = \left[g(\boldsymbol{x}_{N_i+N_{b_1}+1}), g(\boldsymbol{x}_{N_i+N_{b_1}+2}), \ldots, g(\boldsymbol{x}_N)\right]^T. \qquad (10.14c)$$

To solve the sparse matrix system (10.13) efficiently, the matrix A needs to be stored appropriately. We will provide a MATLAB® code and show how to solve (10.13) efficiently in the next section.

10.1.1 *Numerical implementation in 2D*

For the 2D case, the source points are located on a circle with radius R centered at the center node $\boldsymbol{x}_i$ of the local domain. As before, we denote by N_b and N_i the numbers of boundary and interior points, respectively.

In most BVPs describing physical problems, the exact solution is not known. One way of assessing the accuracy of a numerical method is to calculate the approximation on a set of boundary points different from the collocation points. The error can then be assessed by comparison with the given BC. To facilitate the error calculation, we treat this set of boundary points as a part of the interior points following (10.7), so no additional test points are needed.

■ Example 10.1

To demonstrate the LMFS procedure, we first consider a simple BVP for the 2D Laplace equation with a harmonic Dirichlet BC

$$f(x, y) = \sinh(x)\cos(y),$$

for the four-petal domain in Fig. 10.2. The parametric equation of the domain is given by

$$r(\theta) = 1 + \cos^2(2\theta), \quad 0 \le \theta < 2\pi.$$

For the interior points, we choose Halton quasi-random points, which are uniformly distributed over the domain. As already stated, the radius of the source circle is chosen as $R = \max(D)$ where $\max(D)$ is the maximum distance from the center of the sub-domain to its neighboring points and R is a magnifying factor.

In Table 10.1, we present the maximum absolute error for various N_i, N_b, and R, with $m_i = 30$ and $n_i = 15$. In our tests, we obtained

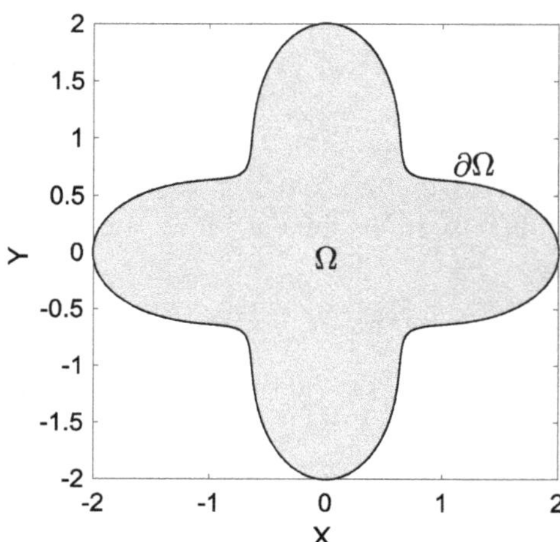

Figure 10.2. The profile of a four-petals domain.

Table 10.1. Harmonic BC: Maximum absolute error for various N_i and N_b with $m_i = 30$ and $n_i = 15$.

(N_i, N_b)	$\mathcal{E}_{\max}\ (R = 10)$	$\mathcal{E}_{\max}\ (R = 2)$
(1000, 200)	3.5(−8)	5.3(−7)
(2000, 400)	6.8(−8)	4.0(−8)
(3000, 600)	2.7(−8)	1.8(−7)
(4000, 800)	1.9(−8)	1.2(−7)
(5000, 1000)	2.2(−8)	1.6(−7)

Table 10.2. Nonharmonic BC problem. Maximum absolute error for various N_i and N_b with $m_i = 30$ and $n_i = 15$.

(N_i, N_b)	$\mathcal{E}_{\max}\ (R = 10)$	$\mathcal{E}_{\max}\ (R = 2)$
(1000, 200)	2.4(−2)	9.9(−3)
(2000, 400)	1.2(−3)	2.5(−4)
(3000, 600)	3.6(−4)	7.2(−5)
(4000, 800)	2.4(−4)	2.6(−6)
(5000, 1000)	1.8(−5)	3.1(−7)

better accuracy using $R = 10$ rather than $R = 2$. This is consistent with the results presented in Chapter 7, where it was recommended that the source points should be placed far away from the collocation points. For a localized method using a relatively small number of collocation points in each sub-domain, the accuracy achieved is excellent.

Next, we consider the same problem with the nonharmonic Dirichlet BC

$$f(x, y) = \cos(x + y) + x^3 + y^3 + 1.$$

The maximum absolute errors in Table 10.2 were obtained using various N_i and N_b, with $m_i = 30$ and $n_i = 15$. In contrast to the results obtained for the harmonic BC in Table 10.1, the accuracy is much better when using $R = 2$ than when using $R = 10$. This observation is also consistent with the results in Chapter 7, according to which, for nonharmonic BCs, the source points should be placed close to the collocation points. In Fig. 10.3, we present the profile of the maximum

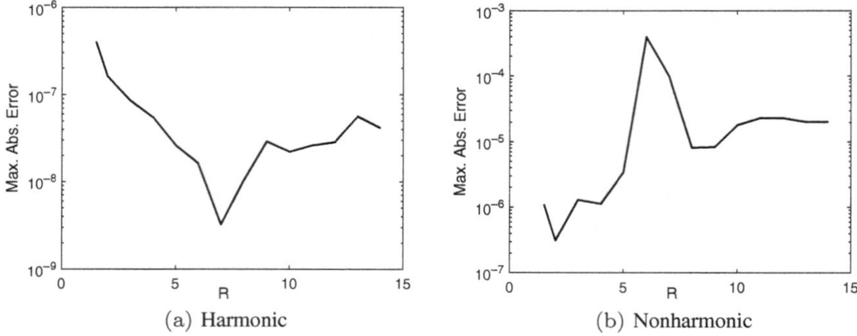

(a) Harmonic   (b) Nonharmonic

Figure 10.3. The maximum absolute errors with respect to R using $N_i = 5000$ and $N_b = 1000$ for harmonic and nonharmonic BCs.

absolute errors versus R for both the harmonic and nonharmonic BC cases using $N_i = 5,000$ and $N_b = 1,000$. From these figures, a larger R is preferable for the harmonic BC case while a smaller R yields better accuracy for the nonharmonic BC case.

The results in the above figure and tables were produced by the code `localMFS.m` listed below. Lines 8–12 generate 1,600 interior points while Line 13 includes the additional boundary test points as part of the interior points. The error is measured on these points as shown in Line 39. In Line 15, we employ an efficient MATLAB® build-in KD-tree search algorithm to determine the m_i nearest neighboring points for each node. Lines 16–18, and 22 produce the source points in each sub-domain while Lines 27–30 store the nonzero entries of the sparse matrix for the interior points. Line 34 stores the nonzero BC and Line 35 assembles the sparse matrix from the three arrays B1, B2, and B3.

```
%   localMFS.m
1   clear; warning off all;
2   v =@(x,y) sinh(x).*cos(y); %harmonic boundary condition
3   r =@(t) (1+cos(4*t/2).^2);
4   ns = 30; ns1=15; R = 10;  tic
5   nb=400; nii=1600; ni=nii+nb; n=nb+ni;
6   t=2*pi*(1:nb)'/nb;
7   [bx,by]=pol2cart(t,r(t));
8   p=haltonset(2,'Skip',1500); q=net(p,50000);
9   pts=q*(max(bx)-min(bx))-(max(bx)-min(bx))/2;
10  in=inpolygon(pts(:,1),pts(:,2),bx,by);
11  pts=pts(in,:); pts=pts(1:nii,:);
```

```
12 xii=pts(1:nii,1);    yii=pts(1:nii,2);
%  Boundary points (bx,by) is included as interior points
13 xi=[bx;xii];   yi=[by;yii];  %interior points
14 x = [xi;bx];   y =[yi; by];  %All collocation points
15 [idx] = knnsearch([x,y],[x,y],'k',ns+1); %KD-tree search
16 dm = ((x-x(idx(:,ns+1))).^2+(y-y(idx(:,ns+1))).^2).^(1/2);
17 [sx0,sy0]=pol2cart(2*pi*(1:ns1)'/ns1, 1); %unit circle
18 rx=R*max(dm)*sx0; ry=R*max(dm)*sy0; %Source circle
19 B1=zeros(1,(ns+1)*ni+nb); B2=B1; B3=B1;    i1=0;
20 for i = 1:ni
21     tpx = x(idx(i,2:ns+1),1); tpy = y(idx(i,2:ns+1),1);
22     sx = x(i)+rx;  sy = y(i)+ry; %source circle
23     DM = pdist2([tpx tpy],[sx sy]);
24     MFS = log(DM); %local matrix
25     DM1 = pdist2([x(i), y(i)],[sx sy]);
26     mfs = log(DM1);
%  Sparse matrix storage for interior points
27     B1(i1+1:i1+ns+1)=i;
28     B2(i1+ns+1)=i; B3(i1+ns+1)=1; %center of the sub-domain
29     B2(i1+1:i1+ns)=idx(i,2:ns+1);
30     B3(i1+1:i1+ns)=-mfs*pinv(MFS,10^(-8));
31     i1=i1+ns+1;
32 end
33 K=i1+1:i1+nb;
34 B1(K)=ni+1:n; B2(K)=ni+1:n; B3(K)=1; %Boundary storage
35 A=sparse(B1,B2,B3,n,n);
36 b = [zeros(ni,1); v(bx,by)];  % RHS
37 u = A\b;  %Approximate solutions
38 exact = v(xi(1:nb),yi(1:nb));  %Exact solutions
39 err = max(abs(u(1:nb)-exact));
40 fprintf('ni=%5d, nb=%3d, ns=%2d, ns1=%2d\n',ni,nb,ns,ns1);
41 fprintf('Maxerr = %8.3e\n',err); toc
```

◼ Example 10.2

We next consider the 2D Laplace equation with mixed BCs in a multiply-connected domain which is a unit square containing 25 circular holes shown in Fig. 10.4 [291]. The radius of each circular hole is 0.05. A Dirichlet BC is imposed on the four sides of the square, while Neumann BCs are imposed on the boundaries of the 25 circular holes. The BCs are derived from the exact solution

$$\phi(x, y) = \cos(x)\sinh(y) + x^2 - y^2 + xy + 1.$$

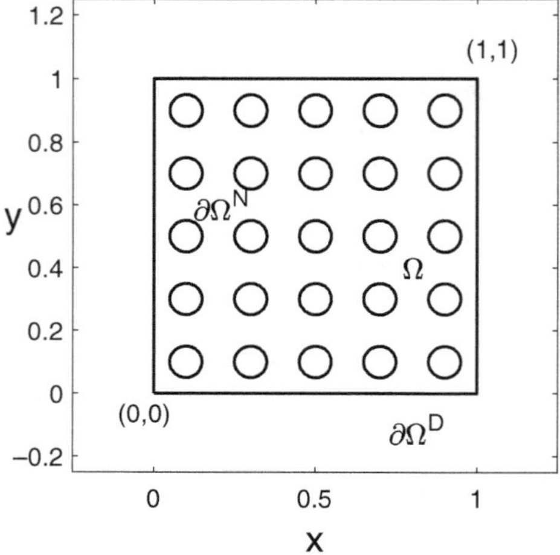

Figure 10.4. The profile of 2D multi-connected domain with 25 holes inside a unit square.

Table 10.3. Harmonic BC problem. Maximum absolute error for various N_i and N_b with $m_i = 30$ and $n_i = 15$.

(N_i, N_b)	$\mathcal{E}_{\max}$ $(R = 8)$	$\mathcal{E}_{\max}$ $(R = 16)$
(1400, 1450)	7.2(−6)	9.3(−6)
(2100, 2030)	2.6(−6)	5.2(−7)
(2600, 2900)	3.0(−6)	2.5(−7)
(3600, 3400)	1.3(−5)	2.4(−7)
(4600, 4230)	3.6(−5)	7.8(−7)
(7600, 5600)	1.1(−5)	7.6(−7)

In Table 10.3, we present the results for $R = 8$ and $R = 16$ for various N_i and N_b with $m_i = 30$ and $n_i = 15$. Since harmonic BCs are imposed, we obtain better results when R is large. In this example, we have demonstrated that the LMFS is capable of handling complicated multi-connected domains with a large number of collocation points, yielding excellent accuracy.

10.1.2 *Numerical implementation in 3D*

For the 3D cases, the source points are located on the surface of a sphere.

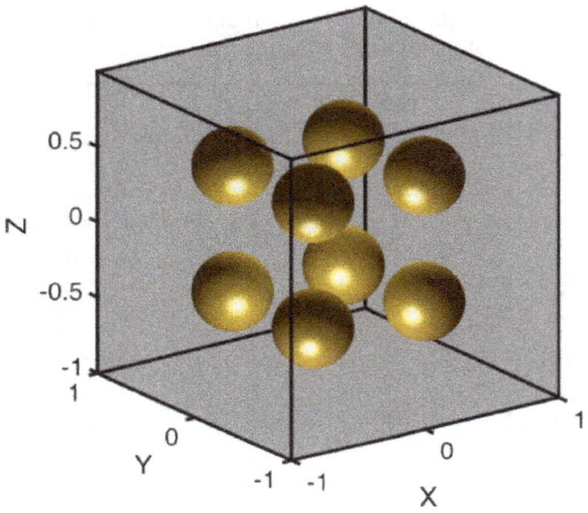

Figure 10.5. The profile of 3D multi-connected domain with eight spheres inside a cube.

■ Example 10.3

We consider a 3D BVP with a Dirichlet BC. The domain is a cube containing eight hollow spheres shown in Fig. 10.5. The radius of each hollow sphere is equal to 1/4 and 536 boundary test points are selected and treated as part of the interior points. The error is computed at these selected boundary points.

For the harmonic BC,

$$f(x, y, z) = x^2 + y^2 - 2z^2,$$

the maximum absolute errors $\mathcal{E}_{\max}$ obtained for various N_i and N_b with $R = 16$ are shown in Table 10.4. We observe that the obtained accuracy using $m_i = 40$ is more stable than for $m_i = 30$. In this example, we choose $n_i = m_i$. Note that the resultant matrix is very sparse and, as a result, the computational cost is low. The CPU time required for each case presented in Table 10.4 is just a few seconds. For example, for $(N_i, N_b) = (1536, 4568)$, the nonzero entries in the resultant sparse matrix are just 0.18% of the total entries.

For the nonharmonic BC

$$f(x, y, z) = x^2 + y^2 + z^2,$$

the maximum absolute errors for various N_i and N_b with $R = 16$ are presented in Table 10.5. In contrast to the harmonic BC case,

Table 10.4. Harmonic BC problem. Maximum absolute error for various N_i and N_b with $R = 16$.

	$\mathcal{E}_{max}$	
(N_i, N_b)	$(m_i = n_i = 30)$	$(m_i = n_i = 40)$
(1336, 1416)	6.1(−7)	9.3(−8)
(1536, 2152)	7.6(−7)	6.1(−8)
(1536, 2378)	9.1(−7)	8.3(−8)
(2036, 2788)	2.0(−5)	8.6(−8)
(2036, 3178)	2.9(−5)	6.7(−8)
(2536, 3978)	3.4(−5)	7.4(−8)
(1536, 4568)	3.8(−7)	3.9(−8)
(1336, 5848)	2.9(−6)	5.9(−8)

Table 10.5. Nonharmonic BC problem. Maximum absolute error for various N_i and N_b with $R = 16$.

	$\mathcal{E}_{max}$	
(N_i, N_b)	$(m_i = n_i = 30)$	$(m_i = n_i = 40)$
(1536, 1978)	2.8(−2)	1.5(−2)
(1536, 3336)	7.3(−3)	4.1(−3)
(2036, 4568)	1.5(−3)	1.6(−3)
(2536, 5848)	1.9(−3)	1.4(−3)
(3536, 7176)	8.4(−3)	5.4(−3)

we notice that there is little difference in terms of accuracy when using $m_i = n_i = 30$ or 40.

In Fig. 10.6(a), we observe better accuracy for large R for the harmonic BC case using $N_i = 2536$, $N_b = 5048$. For the nonharmonic BC case, the selection of R is more unpredictable and it should be neither too small nor too large.

The codes required to obtain the results of the above two tables can be easily reproduced by slightly modifying the 2D code `localMFS.m`. Only the script for generating the collocation points of the 3D domain is provided. In the following code, `NI` denotes the number of interior points, `NBC` the number of boundary points on each face of the cube, `NBH` the number of boundary points on each sphere, `rh` the radius of the hollow sphere,

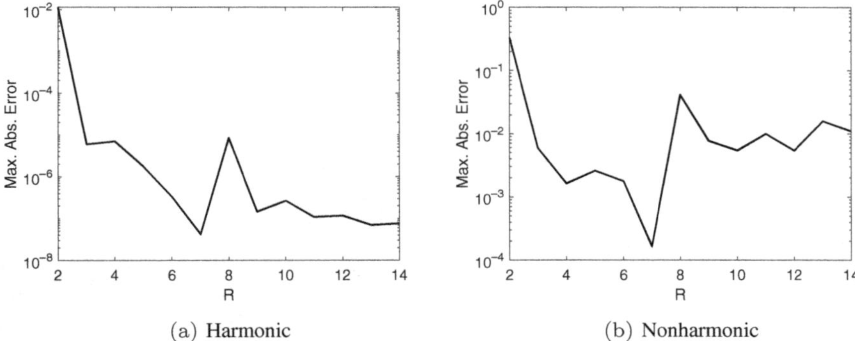

Figure 10.6.   Maximum absolute errors with respect to R using $N_i = 2536$ and $N_b = 5048$ for 3D harmonic and nonharmonic BCs.

$\texttt{ns}$ the number of collocation points in each sub-domain. The number of source points is $\texttt{ns}$. For the output arguments, $(\texttt{xi},\texttt{yi},\texttt{zi})$ denotes the interior points, $(\texttt{xb},\texttt{yb},\texttt{zb})$ the boundary points, and $(\texttt{x0},\texttt{y0},\texttt{z0})$ the m_i points on the unit sphere. Note that the source points of each sub-domain can be generated in a way similar to that described in Lines 17, 18, and 22, of the code $\texttt{localMFS.m}$.

```
1   function [xi,yi,zi,xb,yb,zb,x0,y0,z0]=cube(NI,NBC,NBH,rh,ns)
%   NBH points on a unit sphere.
2   dlong = pi*(3-sqrt(5));
3   long = 0;
4   dz = sqrt(1)*2.0/NBH;
5   z = sqrt(1) - dz/2;
6   x0=zeros(NBH,1); y0=x0; z0=x0;
7   for k = 0 : NBH-1
8       r = sqrt(1-z.^2);
9       y0(k+1)=cos(long)*r;
10      z0(k+1)=sin(long)*r;
11      x0(k+1)=z;
12      z = z - dz;
13      long = long + dlong;
14  end
%   ns points on a unit sphere
15  long = 0;
16  dz = sqrt(1)*2.0/ns;
17  zz = sqrt(1) - dz/2;
18  x0=zeros(ns,1); y0=x0; z0=x0;
19  for k = 0 : ns-1
```

```matlab
20    r = sqrt(1-zz.^2);
21    y0(k+1)=cos(long)*r;
22    z0(k+1)=sin(long)*r;
23    x0(k+1)=zz;
24    zz = zz - dz;
25    long = long + dlong;
26 end
%%%%% setting down the sphere centres in cube
27 gap=0.1; % gap from cube edges
28 n=(2-2*gap)/(2*rh); % max # of sphere on every direction
29 a = linspace(-1-rh,1+rh,floor(n)+1);
30 [X,Y,Z] = meshgrid(a(2:end-1)); %sphere of hollow
31 xc=X(:); yc=Y(:);zc=Z(:); % sphere centers
%%%%% moving reference sphere to centres
32 NH=length(xc);
33 xbh=zeros(NH*NBH,1);ybh=xbh;zbh=xbh;
34 for i=1:length(xc)
35     xbh((i-1)*NBH+1:i*NBH)=rh*x0+xc(i);
36     ybh((i-1)*NBH+1:i*NBH)=rh*y0+yc(i);
37     zbh((i-1)*NBH+1:i*NBH)=rh*z0+zc(i);
38 end
%% interior points picking
39 pp=haltonset(3,'skip',100); p=net(pp,5*NI);
40 xp=2*p(:,1)-1;  yp=2*p(:,2)-1;  zp=2*p(:,3)-1;
41 j=1; xi1=zeros(5*NI,1); yi1=xi1; zi1=xi1;
42 for ii=1:length(xp)
43     d=sqrt((xp(ii)-xc).^2+(yp(ii)-yc).^2+(zp(ii)-zc).^2);
44     if d>rh
45         xi1(j)=xp(ii);  yi1(j)=yp(ii);    zi1(j)=zp(ii);
46         j=j+1;
47     end
48 end
49 in=xi1==0; xi1(in)=[]; yi1(in)=[]; zi1(in)=[];
50 xi=xi1(1:NI); yi=yi1(1:NI);   zi=zi1(1:NI);
%% cube boundary
51 x1=-1;x2=1;y1=-1;y2=1;z1=-1;z2=1;nb1=NBC;nb2=NBC;nb3=NBC;
52 a1=linspace(x1,x2,nb1);
53 a2=linspace(y1,y2,nb2);
54 a3=linspace(z1,z2,nb3);
55 [X,Y,Z]=ndgrid(a1,a2,a3);
56 Px=X(:); Py=Y(:); Pz=Z(:);
57 id = Px==x1 |Px==x2 |Py==y1 |Py==y2 |Pz==z1 |Pz==z2;
58 xbc=Px(id);  ybc=Py(id);    zbc=Pz(id);
59 xb=[xbc;xbh]; yb=[ybc;ybh]; zb=[zbc;zbh];%collecting boundary
```

10.2 Biharmonic BVPs

In this section, we consider the biharmonic BVP

$$\Delta^2 \phi(x) = 0, \qquad x \in \Omega, \tag{10.15a}$$

$$\phi(x) = f(x), \qquad x \in \partial\Omega, \tag{10.15b}$$

$$\frac{\partial \phi}{\partial n}(x) = g(x), \quad x \in \partial\Omega. \tag{10.15c}$$

The LMFS procedure for solving biharmonic problems is similar to that for Laplacian problems and we shall use the same notation. As before, for each node x_i we select its m_i nearest neighboring nodes $\{x_k^{(i)}\}_{k=1}^{m_i}$ in the sub-domain $\Omega^{(i)} \subset \Omega$. We also select $n_i \leq m_i/2$ sources $\{\xi_j^{(i)}\}_{j=1}^{n_i}$ outside $\Omega^{(i)}$. These are uniformly distributed on a circle in 2D or a sphere in 3D with center x_i, containing the sub-domain $\Omega^{(i)}$. If x_i is an interior point, then the solution ϕ satisfies the biharmonic equation in $\Omega^{(i)}$ and can therefore be approximated by

$$\phi^{(i)}(x) = \sum_{j=1}^{n_i} c_j^{(i)} \rho_j^2 \ln \rho_j + \sum_{j=1}^{n_i} d_j^{(i)} \ln \rho_j, \tag{10.16}$$

where $\{c_j^{(i)}\}_{j=1}^{n_i}$ and $\{d_j^{(i)}\}_{j=1}^{n_i}$ are coefficients to be determined. We apply (10.16) at each of the m_i nodes $\{x_k^{(i)}\}_{k=1}^{m_i}$, which yields

$$\phi^{(i)} = K\chi^{(i)}, \tag{10.17}$$

where $\chi^{(i)} = \left[c_1^{(i)}, c_2^{(i)}, \ldots, c_{n_i}^{(i)}, d_1^{(i)}, d_2^{(i)}, \ldots, d_{n_i}^{(i)}\right]^T$. The elements of the $m_i \times n_i$ matrix K can be determined from (10.16), and it follows that

$$\chi^{(i)} = K^{-1}\phi^{(i)}. \tag{10.18}$$

This shows that the strengths of the sources, $\chi^{(i)}$, can be expressed in terms of the unknown physical variables $\phi^{(i)}$. Consequently, the approximation to ϕ and its derivatives at the node x_i can be obtained from (10.16) and (10.18) as follows

$$\phi_i := \phi(x_i) = \sum_{j=1}^{n_i} c_j^{(i)} \rho_{ij}^2 \ln \rho_{ij} + \sum_{j=1}^{n_i} d_j^{(i)} \ln \rho_{ij}$$

$$= \sum_{k=1}^{m_i} v_k^{(i)} \phi_k^{(i)}, \tag{10.19}$$

$$\frac{\partial \phi}{\partial x}(x_i) = \sum_{j=1}^{n_i} c_j^{(i)} \frac{\partial \rho_j^2 \ln \rho_j}{\partial x}(x_i) + \sum_{j=1}^{n_i} d_j^{(i)} \frac{\partial \ln \rho_j}{\partial x}(x_i)$$

$$= \sum_{k=1}^{m_i} v_k^{x(i)} \phi_k^{(i)}, \tag{10.20}$$

where $\{v_k^{(i)}\}_{k=1}^{m_i}$ and $\{v_k^{x(i)}\}_{k=1}^{m_i}$ are the weighting coefficients. Similarly,

$$\frac{\partial \phi}{\partial y}(x_i) = \sum_{k=1}^{m_i} v_k^{y(i)} \phi_k^{(i)}, \tag{10.21}$$

where $\{v_k^{y(i)}\}_{k=1}^{m_i}$ are the weighting coefficients.

Enforcing the governing equation (10.15a) and BCs (10.15b) and (10.15c) and applying (10.19)–(10.21) yields the following sparse global system of linear equations with order $(N + N_b) \times N$:

$$\phi_i - \sum_{k=1}^{m_i} v_k^{(i)} \phi_k^{(i)} = 0, \quad i = 1, 2, \ldots, N_i, \tag{10.22}$$

$$\phi_i = f(x_i), \quad i = N_i + 1, \ldots, N_i + N_b, \tag{10.23}$$

$$\frac{\partial \phi}{\partial n}(x_i) = n_{x_i} \frac{\partial \phi}{\partial x}(x_i) + n_{y_i} \frac{\partial \phi}{\partial y}(x_i)$$

$$= \sum_{k=1}^{m_i} \left(v_k^{x(i)} n_{x_i} + v_k^{y(i)} n_{y_i} \right) \phi_k^{(i)}$$

$$= g(x_i), \quad i = N_i + N_b + 1, \ldots, N_i + 2N_b. \tag{10.24}$$

The above sparse system can be solved efficiently and the approximate solutions at the given interior and boundary collocation nodes can be obtained.

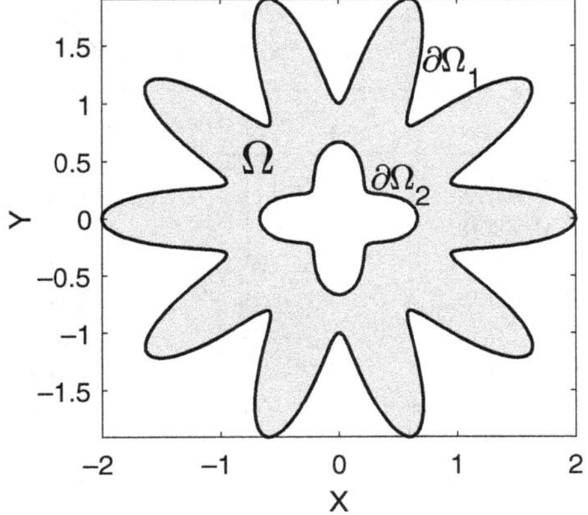

Figure 10.7. The profile of a 2D multi-connected domain.

For numerical implementation, we shall use the same notation as in the previous section.

■ Example 10.4

We consider the biharmonic BVP (10.15) in 2D where the computational domain is the star-shaped domain in Fig. 10.7. The parametric equations for the outer ($\partial\Omega_1$) and inner ($\partial\Omega_2$) boundaries are

$$r_1(\theta) = 1 + \cos^2(5\theta), \quad r_2(\theta) = \frac{1}{2}[1 + \cos^2(2\theta)],$$

respectively, and $\partial\Omega = \partial\Omega_1 \cup \partial\Omega_2$. The BCs are derived from the biharmonic exact solution

$$\phi(x, y) = \frac{x}{2}[\cos(x)\sinh(y) + \sin(x)\cosh(y)] + 1.$$

The numbers of boundary collocation points on $\partial\Omega_1$ and $\partial\Omega_2$ are denoted by N_{b_1} and N_{b_2}, respectively. In Table 10.6, we present the maximum absolute error obtained using various N_i and N_{b_1}, with $N_{b_2} = 200$, $m_i = 30$, $n_i = 12$, and $R = 5, 10$. From this table, we observe that when N_{b_1} gets larger with $N_i \leq 3100$, the accuracy drops. After adding more interior points, the accuracy remains stable for various values of N_{b_1}. Moreover, we observe little difference when using $R = 5$ or 10.

Table 10.6. Biharmonic BC problem. Maximum absolute error for various N_i, N_{b_1} and N_{b_2} with $m_i = 30$ and $n_i = 12$.

(N_i, N_{b_1}, N_{b_2})	$\mathcal{E}_{\max}$ $(R = 5)$	$\mathcal{E}_{\max}$ $(R = 10)$
(2700, 500, 200)	8.2(−7)	4.2(−7)
(2900, 700, 200)	8.4(−7)	1.2(−6)
(3000, 800, 200)	9.5(−6)	7.3(−6)
(3100, 900, 200)	1.1(−5)	7.1(−6)
(5900, 700, 200)	1.9(−7)	3.8(−7)
(6000, 800, 200)	7.6(−8)	1.7(−7)
(6100, 900, 200)	1.0(−7)	1.7(−7)
(6200, 1000, 200)	4.2(−7)	1.7(−7)

Table 10.7. Non-biharmonic BC problem. Maximum absolute error for various N_i, N_{b_1} and N_{b_2} with $m_i = 30$ and $n_i = 12$.

(N_i, N_{b_1}, N_{b_2})	$\mathcal{E}_{\max}$ $(R = 5)$	$\mathcal{E}_{\max}$ $(R = 10)$
(2700, 500, 200)	4.6(−5)	5.0(−5)
(2900, 700, 200)	5.4(−5)	6.3(−5)
(3000, 800, 200)	3.7(−5)	2.7(−5)
(3100, 900, 200)	6.2(−6)	4.0(−5)
(3200, 1000, 200)	3.6(−5)	5.6(−5)
(9000, 800, 200)	1.7(−6)	1.7(−6)
(9100, 900, 200)	1.7(−6)	1.7(−6)
(9200, 1000, 200)	4.4(−6)	5.7(−6)
(11300, 1100, 200)	6.9(−7)	7.3(−7)

Next, we consider the same geometry but with a nonbiharmonic Dirichlet BC given by

$$\phi(x, y) = \exp\left(\frac{x + y}{2}\right) + \sin(x) + \cos(y).$$

The Neumann BC is derived from the above Dirichlet BC. As in the corresponding Laplacian case, we measure the error on the boundary by choosing additional boundary points, which are treated as interior nodes in the implementation of the method. As shown in Table 10.7, the accuracy can be improved by increasing the number of interior points. Little difference is observed for different source locations.

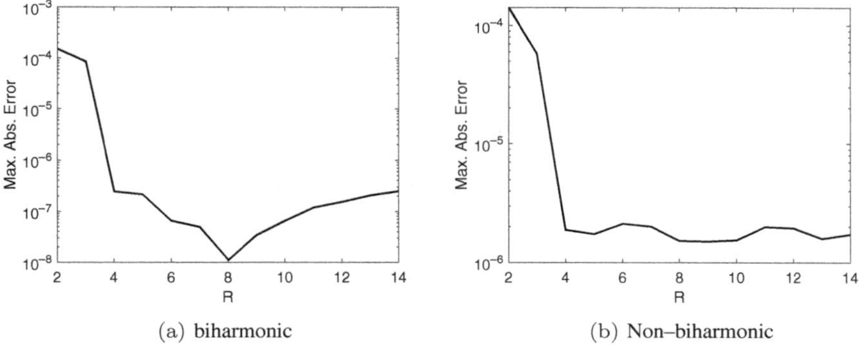

(a) biharmonic   (b) Non–biharmonic

Figure 10.8. Maximum absolute errors with respect to R using $N_i = 7000$ and $N_b = 1000$ for biharmonic and nonbiharmonic BCs in 2D.

In Fig. 10.8, we present the accuracy for varying R using $N_i = 7000$ and $N_b = 1000$ for the cases of biharmonic and nonbiharmonic BCs. In both instances, good accuracy is achieved by choosing the source points sufficiently far from the collocation points.

The code LMFS_bih.m which is listed below produces the results in Table 10.6. In Line 9, we add the extra boundary points as a part of the interior points. Lines 35–51 implement the construction of the sparse matrix, while in Line 56 the errors are evaluated at the extra boundary points. We also provide the code for generating the interior and boundary points, and normal vectors on the boundary. The code is quite similar to the corresponding Laplacian code. The results in Table 10.7 can be obtained by simply changing Lines 2–5 to nonbiharmonic BCs.

```
%   LMFS_bih.m
1   function LMFS_bih
%   biharmonic boundary condition
2   u=@(x,y) x/2.*(cos(x).*sinh(y)+sin(x).*cosh(y))+1;
3   ux =@(x,y) 1/2.*(cos(x).*sinh(y)+sin(x).*cosh(y)) ...
4           +x/2.*(-sin(x).*sinh(y)+cos(x).*cosh(y));
5   uy =@(x,y) x/2.*(cos(x).*cosh(y)+sin(x).*sinh(y));
6   ns = 30; ns1=12; rt = 5; tic
7   nb1=1000; nb2=200; nb=nb1+nb2; nii=2000; ni=nii+nb; n=ni+nb;
8   [xii,yii,xb,yb,nx,ny]=Neu_cond_pts(nb1,nb2,nii);
9   xi=[xii;xb]; yi=[yii; yb];
10  x=[xb;xi]; y=[yb;yi];
11  [idx] = knnsearch([x,y],[x,y],'k',ns+1);
```

```matlab
12 dm = sqrt((x-x(idx(:,ns+1))).^2+(y-y(idx(:,ns+1))).^2);
13 v=u(x,y); vx=ux(x,y); vy=uy(x,y);
14 t=2*pi*(1:ns1)'/ns1;
15 rx = rt*max(dm)*cos(t); ry = rt*max(dm)*sin(t);
16 w = zeros(n,ns); wx=w; wy=w;
17 for i = 1:n
18     tpx = x(idx(i,2:ns+1)); tpy = y(idx(i,2:ns+1));
19     sx = x(i)+rx; sy = y(i)+ry;
20     DM = pdist2([tpx, tpy],[sx sy]);
21     G = [DM.*DM.*log(DM), log(DM)]; %local matrix
22     DM1 = pdist2([x(i), y(i)],[sx sy]);
23     DM2 = DM1.^2;
24     invMFS = pinv(G,10^(-8));
25     if i>nb    %interior points
26         mfs =[DM2.*log(DM1),log(DM1)];
27         w(i,:) = mfs*invMFS;
28     else
29         Dx = (x(i)-sx)';  tax = [Dx.*(2*log(DM1)+1),Dx./DM2];
30         Dy = (y(i)-sy)';  tay = [Dy.*(2*log(DM1)+1),Dy./DM2];
31         wx(i,:) = tax*invMFS; % Neumann BC
32         wy(i,:) = tay*invMFS;
33     end
34 end
35 C1 = zeros( nb*ns+ni*(ns+1)+nb,1); C2=C1; C3=C1;
36 C1(1:nb) = (1:nb); C2(1:nb) = (1:nb); C3(1:nb) = 1; %Dirichlet
37 ic = nb;
38 for i = 1:nb %Neumann boundary
39     C1(ic+1:ic+ns) = i+nb;
40     C2(ic+1:ic+ns) = idx(i,2:ns+1);
41     C3(ic+1:ic+ns) = nx(i)*wx(i,1:ns)+ny(i)*wy(i,1:ns);
42     ic = ic+ns;
43 end
44 for i = nb+1:n    %interior
45     C1(ic+1:ic+ns+1) = i+nb;
46     C2(ic+ns+1)=i;   C3(ic+ns+1)=1;
47     C2(ic+1:ic+ns) = idx(i,2:ns+1);
48     C3(ic+1:ic+ns)= -w(i,1:ns);
49     ic = ic+ns+1;
50 end
51 A = sparse(C1,C2,C3,n+nb,n);
52 b=zeros(n+nb,1);
53 b(1:nb) = v(1:nb);
54 b(nb+1:2*nb) = nx(1:nb).*vx(1:nb) + ny(1:nb).*vy(1:nb);
55 u = A\b;
56 err = max(abs(u(n-nb:n)-v(n-nb:n)));
57 fprintf('ni=%5d, nb=%3d, ns=%2d,R=%2d\n',ni,nb,ns,rt);
```

```matlab
58 fprintf('err = %8.3e\n',err); toc
   %%%%%%%%%%%%%%%%%%%%%%%%%%%%%%%%%%%
59 function [xi,yi,xb,yb,xn,yn]= Neu_cond_pts(nb1,nb2,ni)
60 r=@(t)  (1+cos(10*t/2).^2); %10 leafs
61 r1=@(t) (1+cos(4*t/2).^2);  %4 leafs
62 t=2*pi*(1:nb1)'/nb1;
63 xb1=r(t).*cos(t);  yb1=r(t).*sin(t);
64 dx=gradient(xb1);  dy=gradient(yb1);
65 xn1=dy./sqrt(dx.^2+dy.^2);  yn1=-dx./sqrt(dx.^2+dy.^2);
66 t=2*pi*(1:nb2)'/nb2;
67 xb2=r1(t).*cos(t)/3;  yb2=r1(t).*sin(t)/3;
68 dx=gradient(xb2); dy=gradient(yb2);
69 xn2=dy./sqrt(dx.^2+dy.^2);  yn2=-dx./sqrt(dx.^2+dy.^2);
70 xn=[xn1; xn2]; yn=[yn1; yn2]; %Normal vectors on boundary
71 a=min(xb1); b=max(xb1); c=min(yb1); d=max(yb1);
72 q=haltonset(2); pts=net(q,4*ni);
73 pts(:,1)=pts(:,1)*(b-a)+a;
74 pts(:,2)=pts(:,2)*(d-c)+c;
75 idx = inpolygon(pts(:,1),pts(:,2),xb1,yb1);
76 intnode=pts(idx,:);
77 xi=intnode(:,1); yi=intnode(:,2);  %interior points
78 idx1 = inpolygon(xi,yi,xb2,yb2);
79 xi(idx1)=[]; yi(idx1)=[];
80 xi=xi(1:ni); yi=yi(1:ni); %interior points
81 xb=[xb1;xb2]; yb=[yb1;yb2]; %boundary points
```

Chapter 11

Inverse Problems

The BVPs we have so far studied are known as direct problems. The advantages the MFS has, when applicable, over other techniques make it an ideal method for dealing with another class of problems known as inverse problems, see [528]. In inverse problems, some of the data describing a BVP is missing, rendering their solution more challenging as they are ill-posed. From a numerical point of view, this means that small uncertainties in the data may lead to large errors in the approximation.

In recent years, the MFS has been used extensively for the solution of various types of inverse problems as described in the review papers [458, 459]. Inverse problems may be classified into various categories, see [528, Section 1.1], namely Cauchy problems, initial value problems, source problems, boundary coefficient problems and inverse geometric problems. We shall concentrate on the first and last categories, namely Cauchy problems and inverse geometric problems, for which the MFS has been used the most.

11.1 Cauchy Problems

In these problems, also known as inverse BVPs, the BCs on part of the boundary $\partial\Omega$ (which is known), say Γ_1, are under-specified while, to compensate, on the remaining part of the boundary $\partial\Omega\backslash\Gamma_1$ they are over-specified.

11.1.1 *Laplace equation*

A typical Cauchy problem for the Laplace equation in $\mathbb{R}^2$ or $\mathbb{R}^3$ consists of

$$\Delta\phi(x) = 0, \quad x \in \Omega, \tag{11.1a}$$

subject to the BCs (known as Cauchy data)

$$\phi(x) = f(x) \quad \text{and} \quad \frac{\partial\phi}{\partial n}(x) = g(x), \quad x \in \Gamma_1, \tag{11.1b}$$

while no BCs are provided on the remainder of the domain boundary $\partial\Omega\backslash\Gamma_1$.

The MFS was first used for the solution of BVP (11.1) in 2D in [393], and in 3D in [900].

■ **Example 11.1**

We consider the Cauchy problem (11.1) in 2D examined in [393], in which $\Omega = (0, 1) \times (0, 1)$ and Γ_1 is the lower side of the unit square, i.e. $\Gamma_1 = [0, 1] \times \{0\}$. The BCs (11.1b) are derived from the exact solution

$$\phi(x, y) = x^3 - 3xy^2 + e^{2y}\sin(2x) - e^x\cos(y).$$

The M collocation points $\{x_i\}_{i=1}^{M}$ are uniformly distributed on the side Γ_1 while the N sources $\{x'_j\}_{j=1}^{N}$ are uniformly distributed on a circle with radius R and center $(0.5, 0.5)$. The solution of BVP (11.1) is approximated by

$$\hat{\phi}(x) = \sum_{j=1}^{N} c_j\, G^L(x, x'_j), \quad x \in \overline{\Omega}.$$

Collocation of BCs (11.1b) at the collocation points $\{x_i\}_{i=1}^{M}$ leads to a $2M \times N$ linear system of the form

$$A c = f,$$

which may be solved by least squares. The maximum relative error $\epsilon_{\max}$ and RMSE $\mathcal{E}_{\mathrm{rms}}$ (see (7.11) and (7.12) for definitions) are calculated on 25 quasi-random points in Ω (see code `Cauchy.m` below), and the results obtained for varying values of R and for different $M = N$ are presented in Fig. 11.1.

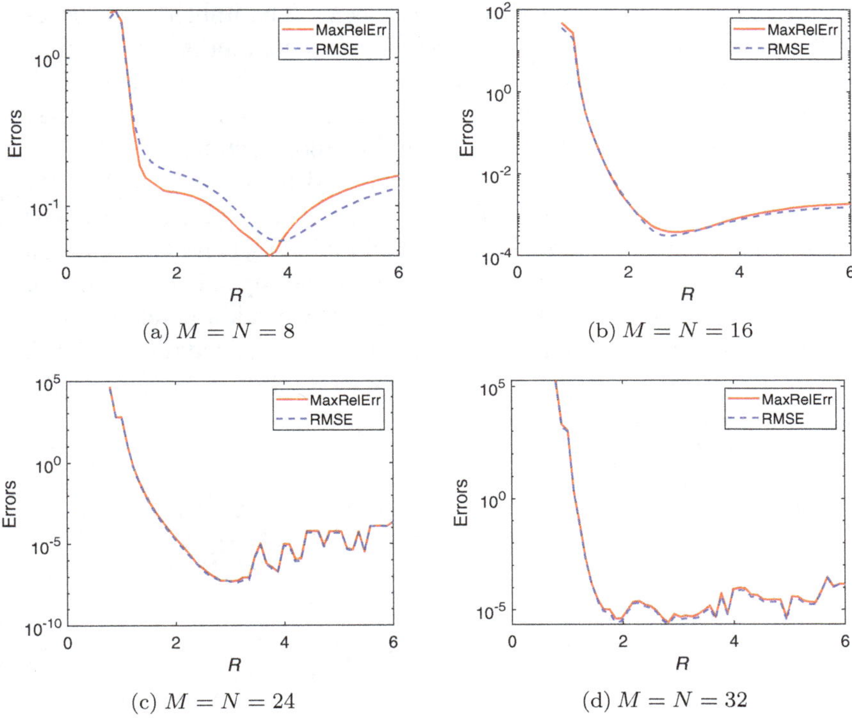

Figure 11.1. Results for varying R and different $M = N$.

To study the stability of the method we consider, instead of the BC $\partial\phi/\partial n = g$ on Γ_1 in (11.1b), the noisy data

$$\frac{\partial\phi}{\partial n}(x) = (1 + p\varrho_m)g(x_m), \quad m = 1, \ldots, M; \ x_m \in \Gamma_1,$$

where p is the percentage noise added and $[\varrho_1, \varrho_2, \ldots, \varrho_M]^T$ is a random noisy variable vector in $[-1, 1]$ constructed using the MATLAB® command `-1+2*rand(1,M)`.

System (11.1) is ill-conditioned and its solution requires the use of a regularization method. We shall use Tikhonov's regularization method [836] in which instead of (11.1), the following least squares functional is minimized in a least squares sense

$$\|Ac - f\|^2 + \lambda\|c\|^2,$$

where $\lambda > 0$ is a regularization parameter. The implementation of Tikhonov's regularization is achieved using the routine `tikhonov` from Hansen's regularization tools package [365], see also [364]. Moreover, the optimal value of the regularization parameter may be selected using the L-curve criterion [366, Sections 4.7, 5.3]. In Fig. 11.2, we present the L-curves obtained for noise $p = 1\%, 3\%$ and 5% with $R = 10$ and $M = N = 24$ (using `tikhonov`). The corresponding approximation $\hat{\phi}$ (blue line) in (11.1) for $p = 1\%$ and the exact solution (red line) are plotted on the top side of the square $[0, 1] \times \{1\}$ for different values of the regularization parameter λ in Fig. 11.3. We observe that improved results are obtained for $\lambda = 10^{-6}$.

```
%   MATLAB code: Cauchy.m
1   warning off
%   M: # collocation points on Gamma_1, N: # sources
2   u=@(x,y) x.^3-3*x.*y.^2+exp(2*y).*sin(2*x)-exp(x).*cos(y);
3   ux=@(x,y) 3*x.^2-exp(x).*cos(y)+2*cos(2*x).*exp(2*y)-3*y.^2;
4   uy=@(x,y) exp(x).*sin(y) - 6*x.*y + 2*sin(2*x).*exp(2*y);
5   M=16; N=16;
6   coe=1/(2*pi);
7   x=linspace(0,1,M); x=x'; y(1:M,1)=0;
8   xn(1:M,1)=0; yn(1:M,1)=-1;   % normal on Gamma_1
9   p=haltonset(2);
10  q=net(p,25);
```

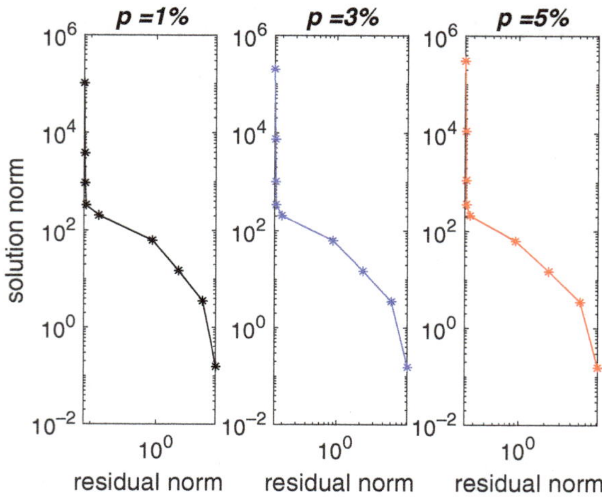

Figure 11.2. L-curves for noise $p = 1\%$, 3% and 5%.

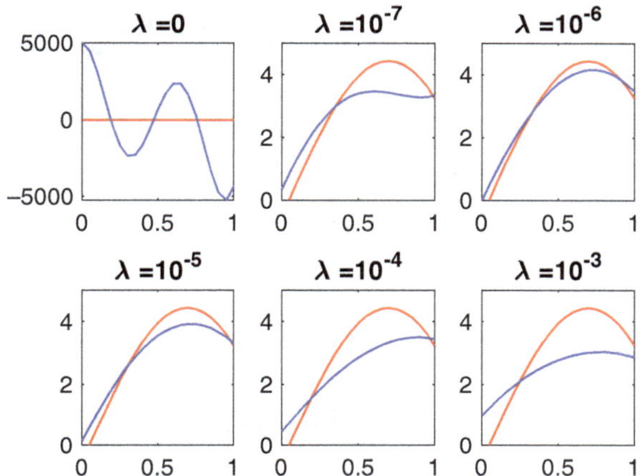

Figure 11.3. Results for noise $p = 1\%$ on side $y = 1$ for different λ. The red line is the exact solution and blue line the approximate solution.

```
11 xt=q(:,1); yt=q(:,2);  % Halton test points in unit square
12 [x0,y0]=pol2cart((1:M)'/M*2*pi,1.0);
13 f=u(x,y);    g=ux(x,y).*xn+uy(x,y).*yn;
14 rhs=[f; g];    %Boundary conditions
15 ut=u(xt,yt);  %Exact solution
16 R=linspace(0.8,6,50); error=zeros(50,1); rmse=zeros(50,1);
17 for k=1:50
18     xs=x0*R(k)+1/2;  ys=y0*R(k)+1/2;  %Source points
19     DM=pdist2([x,y],[xs,ys]);
20     ff=coe*log(DM);
21     DMX=x-xs'; DMY=y-ys';  % Difference Matrix
22     ffn=((DMX./DM.^2).*xn+(DMY./DM.^2).*yn)*coe;
23     G=[ff; ffn];    % approximation matrix
24     sol=G\rhs;       % solution of MFS system
25     DMT=pdist2([xt,yt],[xs,ys]);
26     ffa=coe*log(DMT);
27     uh=ffa*sol; % Evaluation of solution
28     error(k) = norm(uh-ut,inf)/norm(ut,inf); %relative error
29     rmse(k)=norm(uh-ut,2)/sqrt(length(xt));   %RMSE error
30 end
31 [err,n1]=min(error);
32 [RMSE,~]=min(rmse);
33 fprintf('M =%3d, N =%3d, R = %6.2f\n',M,N,R(n1));
34 fprintf('rel error = %8.4e, rmse = %8.4e\n',err,RMSE);
35 figure (1)
36 semilogy(R,error,'-r',R,rmse,'--b','LineWidth',2)
```

```
37 legend('MaxRelErr','RMSE');
38 xlabel('\it R');   ylabel('Errors');
39 set(gca,'FontSize',16)
```

11.1.2 *Helmholtz equation*

A typical Cauchy problem for the Helmholtz equation in $\mathbb{R}^2$ or $\mathbb{R}^3$ is

$$(\Delta + k^2)\phi(x) = 0, \ x \in \Omega, \tag{11.2a}$$

subject to the BCs

$$\phi(x) = f(x) \quad \text{and} \quad \frac{\partial \phi}{\partial n}(x) = g(x), \quad x \in \Gamma_1, \tag{11.2b}$$

and no BCs on $\partial\Omega \backslash \Gamma_1$. The MFS was first used for the solution of BVP (11.2) in 2D in [617], and in 3-D in [616].

11.1.3 *Biharmonic equation*

In this case there are several Cauchy problems for the biharmonic equation in $\mathbb{R}^2$ or $\mathbb{R}^3$. For example we could have

$$\Delta^2 \phi(x) = 0, \quad x \in \Omega, \tag{11.3a}$$

subject to BCs

$$\phi(x) = f(x), \ \frac{\partial \phi}{\partial n}(x) = g(x),$$

$$\Delta\phi(x) = v(x), \ \frac{\partial \Delta \phi}{\partial n}(x) = w(x), \quad x \in \Gamma_1, \tag{11.3b}$$

and no BCs on $\partial\Omega \backslash \Gamma_1$.

The MFS was first used for the solution of BVP (11.3) and other combinations of BCs in 2D in [618].

11.1.4 *Cauchy–Navier equations*

We consider the BVP for the Cauchy–Navier equations in $\mathbb{R}^2$

$$\left(\frac{2-2v}{1-2v}\right)\frac{\partial^2 u_1}{\partial x^2} + \left(\frac{1}{1-2v}\right)\frac{\partial^2 u_2}{\partial x \partial y} + \frac{\partial^2 u_1}{\partial y^2} = 0, \tag{11.4a}$$

$$\frac{\partial^2 u_2}{\partial x^2} + \left(\frac{1}{1-2v}\right)\frac{\partial^2 u_1}{\partial x \partial y} + \left(\frac{2-2v}{1-2v}\right)\frac{\partial^2 u_2}{\partial y^2} = 0, \qquad (11.4\text{b})$$

for $x \in \Omega$, subject to the BCs

$$\begin{aligned}
u_1(x) = \overline{u}_1(x), \ \ u_2(x) = \overline{u}_2(x), \\
t_1(x) = \overline{t}_1(x), \ \ t_2(x) = \overline{t}_2(x), \quad x \in \Gamma_1,
\end{aligned} \qquad (11.4\text{c})$$

where $\overline{u}_1, \overline{u}_2, \overline{t}_1$, and $\overline{t}_2$ are known functions, and (t_1, t_2) are the tractions and no BCs are given on $\partial\Omega \backslash \Gamma_1$.

Such problems were solved using the MFS in [613, 620] and their counterpart in $\mathbb{R}^3$ was considered in [615]. The MFS was also used for related Cauchy problems in thermoelasticity in $\mathbb{R}^2$ and $\mathbb{R}^3$ in [464, 622] and [624], respectively.

11.2 Inverse Geometric Problems

In these problems, unlike Cauchy problems, the location and shape of part of the boundary $\partial\Omega$, say Γ_1, is unknown while, to compensate, on the remaining (fixed and known) part of the boundary $\partial\Omega \backslash \Gamma_1$ the BCs are over-specified. On Γ_1 a specific BC is imposed. There are two types of inverse geometric problems [459], namely BVPs in which Ω is simply-connected and Γ_1 is part of $\partial\Omega$, and BVPs in which Ω is doubly-connected (or multiply-connected) and Γ_1 is an internal boundary. It is noteworthy that, unlike Cauchy problems, the discretization of inverse geometric problems leads to nonlinear systems of equations.

11.2.1 *Laplace equation*

A typical inverse geometric problem for Laplace's equation in $\mathbb{R}^2$ or $\mathbb{R}^3$ consists of

$$\Delta\phi(x) = 0, \quad x \in \Omega, \qquad (11.5\text{a})$$

subject to the BCs

$$\phi(x) = 0, \quad x \in \Gamma_1, \qquad (11.5\text{b})$$

and

$$\phi(x) = f(x) \quad \text{and} \quad \frac{\partial \phi}{\partial n}(x) = g(x), \quad x \in \partial\Omega \setminus \Gamma_1. \tag{11.5c}$$

Here Ω is doubly-connected and Γ_1 is an internal boundary, the shape and size of which are to be determined.

We apply the MFS to (11.5) in $\mathbb{R}^2$ following [462] and assume that the exterior boundary $\partial\Omega \setminus \Gamma_1$ is the unit circle and that the unknown boundary Γ_1 is a smooth, star-like curve with respect to the origin with coordinates

$$(x, y) = r(\vartheta) \, (\cos \vartheta, \sin \vartheta), \quad \vartheta \in [0, 2\pi), \tag{11.6}$$

where r is a smooth 2π-periodic function. The N collocation points on Γ_1 will thus be

$$x_i = (x_i, y_i) = r_i \, (\cos \vartheta_i, \sin \vartheta_i), \quad \vartheta_i = \frac{2\pi (i - 1)}{N}, \quad i = 1, \ldots, N, \tag{11.7}$$

where the radii $r_i = r(\vartheta_i)$, $i = 1, \ldots, N$, are unknown. The corresponding source points are placed on an auxiliary boundary Γ_1' which is a contraction of Γ_1 as follows

$$x_j' = \left(x_j', y_j' \right) = \eta_{\text{int}} r(\vartheta_j) \left(\cos \vartheta_j, \sin \vartheta_j \right),$$

$$\vartheta_j = \frac{2\pi (j - 1)}{N}, \quad j = 1, \ldots, N, \tag{11.8}$$

where the contraction parameter η_{int} is taken in the interval $(0, 1)$. We also place M collocation points on the unit circle $(\partial\Omega \setminus \Gamma_1)$, i.e.

$$x_i = (x_i, y_i) = (\cos \vartheta_i, \sin \vartheta_i),$$

$$\vartheta_i = \frac{2\pi (i - N - 1)}{M}, \quad i = N + 1, \ldots, N + M, \tag{11.9}$$

and N sources on an auxiliary boundary surrounding the unit circle

$$x_j' = \left(x_j', y_j' \right) = \eta_{\text{ext}} \left(\cos \vartheta_j, \sin \vartheta_j \right),$$

$$\vartheta_j = \frac{2\pi (j - N - 1)}{N}, \quad j = N + 1, \ldots, 2N, \tag{11.10}$$

where the dilation parameter η_{ext} is taken to be larger than 1.

We approximate the solution of BVP (11.5) with the MFS approximation

$$\hat{\phi}(x) = \sum_{j=1}^{2N} c_j \, G^L(x, x_j'), \quad x \in \overline{\Omega}. \tag{11.11}$$

The unknowns to be determined are the coefficients $c = (c_1, c_2, \ldots, c_{2N})$ in (11.11) and the radii $r = (r_1, r_2, \ldots, r_N)$ in (11.7). These are determined by imposing BCs (11.5b)–(11.5c) in a least-squares sense by minimizing the functional

$$S(c, r) := \sum_{i=1}^{N} \left[\hat{\phi}(c, x'; x_i) \right]^2 + \sum_{i=N+1}^{N+M} \left[\hat{\phi}(c, x'; x_i) - f(x_i) \right]^2$$

$$+ \sum_{i=N+1}^{N+M} \left[\frac{\partial \hat{\phi}}{\partial n}(c, x'; x_i) - g(x_i) \right]^2, \tag{11.12}$$

where $x' = (x_1', x_2', \ldots, x_{2N}')$. The least-squares minimization of functional (11.12) is carried out using the MATLAB® optimization toolbox routine `lsqnonlin`. This routine offers the option of either providing the Jacobian of the nonlinear system or not.

■ Example 11.2

Here we consider reconstructing a circular rigid inclusion with boundary Γ_1 of radius $1/2$ where the Dirichlet data in (11.5c) on the unit circle $\Omega \backslash \Gamma_1$ is taken as

$$\phi(\vartheta) = e^{-\cos^2 \vartheta}, \quad \vartheta \in [0, 2\pi).$$

The Neumann data in (11.5c) is simulated by solving the direct BVP (11.5a) with the Dirichlet BCs in (11.5b) and (11.5c), using the MFS with $M = N = 500$ and $\eta_{\text{int}} = 3/4$, $\eta_{\text{ext}} = 4/3$.

In order to avoid committing an inverse crime, the inverse solver is applied using $N = 56$, $M = 96$, $\eta_{\text{int}} = 4/5$, $\eta_{\text{ext}} = 3/2$. We choose the initial vector of unknowns $x_0 = (c_0, r_0)^T = (0, 0.1)^T$.

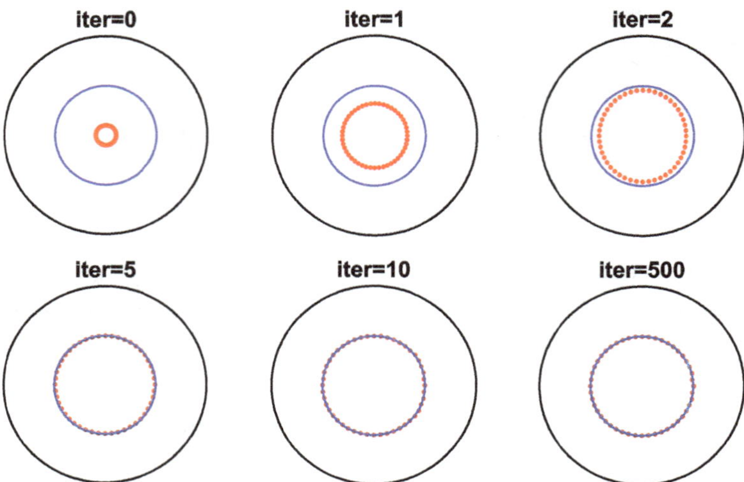

Figure 11.4. Results for various numbers of iterations.

In Fig. 11.4, we present the results obtained for different numbers of iterations, and it can be seen that the solution becomes accurate after as few as five iterations.

The code `Inverse(mm).m` used to solve the above problem is shown below.

```
%   MATLAB code: Inverse(mm).m
1   function Inverse(mm)        % mm is the number of iterations
2   global    n m coe etai etae can san cam sam g gn
3   warning off
4   load results96; % load results of simulation
5   n=48;   m=96;
6   g = results96(:,1)';   gn = results96(:,2)';
7   r0=1/10; % initial circle
8   etai=4/5; % contraction coefficient
9   etae=3/2; % dilation coefficient
10  coe=1/(2*pi);
11  [can, san]=pol2cart(2*pi/n*(0:n-1)',1);
12  [cam, sam]=pol2cart(2*pi/m*(0:m-1)',1);
%   Initialization prior to call to lsqnonlin
13  x0(1,2*n+1:3*n)=r0; %initial radii
14  lb(1,1:2*n)=-100000;     ub(1,1:2*n)=100000;
15  lb(1,2*n+1:3*n)=0.01;     ub(1,2*n+1:3*n)=1;
16  opts2=optimoptions(@lsqnonlin, 'MaxFunEvals',200000, ...
```

```
17 'MaxIter',mm,'TolFun',1.e-14,'TolX',1.e-14);
18 x= lsqnonlin(@f1,x0,lb,ub,opts2);
%  calculation of largest deviation from correct bry
19 ra(1:n,1)=x(2*n+1:3*n);  % final radii
%  plot the recovered bry as well as the correct one
20 [xx2,yy2]=pol2cart(0:0.02:2*pi,1);
21 xx=ra.*can;                yy=ra.*san;   % approx bry
22 xx1=0.5*can;               yy1=0.5*san;  % exact bry
23 plot(xx,yy,'+r',xx1,yy1,'.b',xx2,yy2,'-k','LineWidth',2);
24 axis off;    axis equal;
%%%%%%%%%%%%%%%%%%%%%%%%%%%
25 function f=f1(x)    %  function f
26 global    n m coe etai etae can san cam sam g gn
27 f=zeros(n+2*m,1);  % function f
28 c(1:2*n,1)=x(1,1:2*n);         % unknown coefficients
29 ra(1:n,1)=x(1,2*n+1:3*n);   %unknown radii
30 ri=etai*ra;  % interior sources
31 xp(1:n,1)=ra.*can;   yp(1:n,1)=ra.*san;  % interior coll pts
32 xp(n+1:m+n,1)=cam;   yp(n+1:m+n,1)=sam;   % exterior coll pts
33 xs1=ri.*can;        ys1=ri.*san;       %interior sources
34 xs2=etae*can;    ys2=etae*san;% exterior sources
35 xs=[xs1; xs2];   ys=[ys1; ys2];
36 xn(n+1:m+n,1)=cam;   yn(n+1:m+n,1)=sam;% normal der
37 dx=xp-xs';   dy=yp-ys';    %Difference Matrix
38 DM=pdist2([xp yp],[xs ys]);   %Distance Matrix
39 G1=coe*log(DM);
40 G2=coe*(xn.*dx+yn.*dy)./(dx.^2+dy.^2);
41 app=G1*c(1:2*n,1);     % MFS approximation
42 appn=G2*c(1:2*n,1);  % normal der of MFS approximation
43 f1=app(1:n,1);       % BC on Gamma_1
44 f2=app(n+1:m+n,1)-g(1:m)';  % Dirichlet BC
45 f3=appn(n+1:m+n,1)-gn(1:m)';  % Neumann BC
46 f=[f1; f2; f3];
```

To test the stability of the method, the Neumann data in (11.5c) is contaminated with noise in the following way

$$g^{\varepsilon}(x_i) = (1 + \rho_i\, p)\, g(x_i), \quad i = N+1, \ldots, N+M. \qquad (11.13)$$

In (11.13) p represents the percentage of noise added to the Neumann boundary data on $\partial\Omega\backslash\Gamma_1$ and ρ_j is a pseudo-random noisy variable drawn from a uniform distribution in $[-1, 1]$. In such a case Tikhonov regularization is required and instead of minimizing (11.12), one needs to minimize the functional

$$
S(\boldsymbol{c}, \boldsymbol{r}) := \sum_{i=1}^{N} \left[\hat{\phi}(\boldsymbol{c}, \boldsymbol{x}'; \boldsymbol{x}_i) \right]^2 + \sum_{i=N+1}^{N+M} \left[\hat{\phi}(\boldsymbol{c}, \boldsymbol{x}'; \boldsymbol{x}_i) - f(\boldsymbol{x}_i) \right]^2
$$

$$
+ \sum_{i=N+1}^{N+M} \left[\frac{\partial \hat{\phi}}{\partial n}(\boldsymbol{c}, \boldsymbol{x}'; \boldsymbol{x}_i) - g^{\varepsilon}(\boldsymbol{x}_i) \right]^2 + \lambda_1 |\boldsymbol{c}|^2
$$

$$
+ \lambda_2 \sum_{i=2}^{N} (r_i - r_{i-1})^2 , \tag{11.14}
$$

where λ_1 and $\lambda_2 \geq 0$ are regularization parameters that need to be prescribed.

We now solve the same problem as before, except that a noise level of $p = 5\%$ is added to the Neumann BC in (11.5c) as in (11.13). In Figs. 11.5 and 11.6, we present the reconstructed curves after 500 iterations with various regularization parameters, λ_1 when $\lambda_2 = 0$, and λ_2 when $\lambda_1 = 0$, respectively. From these figures we observe that regularization leads to stable reconstructions.

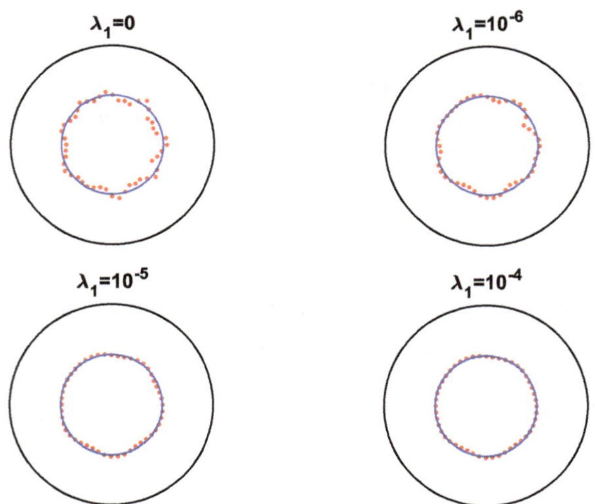

Figure 11.5. Results for noise $p = 5\%$ and regularization with λ_1.

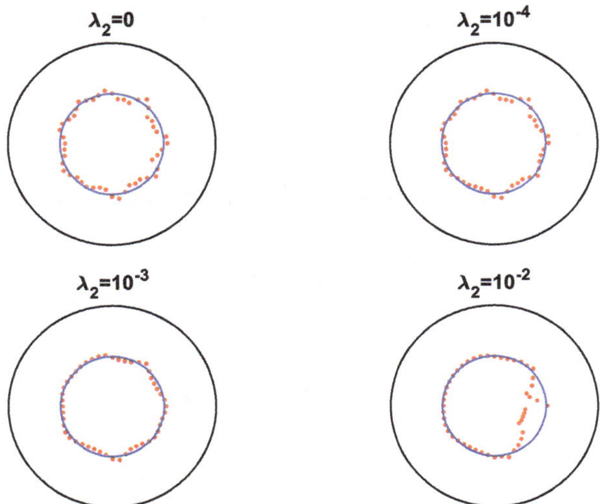

Figure 11.6. Results for noise $p = 5\%$ and regularization with λ_2.

The above solution requires a slight modification of the code `Inverse(mm).m`, in particular the last few lines become:

```
cc(1:2*n,1)=c(1:2*n,1); % coefficients
rr(1:n,1)=ra(1:n,1);% radii
rr1(1:n-1,1)=rr(2:n,1)-rr(1:n-1,1);% radii difference
term1=norm(cc); % first regularization term
term2=norm(rr1); % second regularization term
f1=app(1:n,1); % BC on Gamma_1
f2=app(n+1:m+n,1)-g(1:m)';% Dirichlet BC on circle
% Noisy Neumann BC on circle
f3=appn(n+1:m+n,1)-(ones(1,m)+p*psr(1,1:m))'.*gn(1:m)';
f4=sqrt(m1)*term1;%first regularization term, m1 is lambda_1
f5=sqrt(m2)*term2;%second regularization term, m2 is lambda_2
f=[f1;f2;f3;f4;f5];
```

Problems in which the coordinates of the center of Γ_1 and the contraction and dilation parameters η_{int}, η_{ext} are unknown as well as cases with more complex boundaries Γ_1 maybe found in [462]. Also, in [462] BVP (11.5) is examined where BC (11.5b) is replaced by $\partial\phi/\partial n = 0$. The problem corresponding to BVP (11.5) in 3D is considered in [463]. Numerical experiments when the exact Jacobian is provided in `lsqnonlin` may be found in [471].

11.2.2 *Helmholtz equation*

A typical inverse geometric problem for the Helmholtz equation in $\mathbb{R}^2$ or $\mathbb{R}^3$ consists of

$$(\Delta\phi + k^2)\phi(x) = 0, \quad x \in \Omega, \tag{11.15a}$$

subject to the BCs

$$\phi(x) = f_1(x), \quad x \in \Gamma_1, \tag{11.15b}$$

and

$$\phi(x) = f_2(x) \quad \text{and} \quad \frac{\partial\phi}{\partial n}(x) = g(x), \quad x \in \partial\Omega\backslash\Gamma_1, \tag{11.15c}$$

As in (11.5a) Ω is doubly-connected and Γ_1 is an internal boundary, the shape and size of which are to be determined. Using the approximation (11.11) for $\hat{\phi}$ with G^L replaced with G^H, and a similar discretization as the one described in Section 11.2.1 for the 2D case (see also [457]) the counterpart of (11.12) for the determination of the unknown coefficients and radii (11.7) is given by

$$S(c, r) := \sum_{i=1}^{N}\left[\hat{\phi}(c, x'; x_i) - f_1(x_i)\right]^2 + \sum_{i=N+1}^{N+M}\left[\hat{\phi}(c, x'; x_i) - f_2(x_i)\right]^2$$

$$+ \sum_{i=N+1}^{N+M}\left[\frac{\partial\hat{\phi}}{\partial n}(c, x'; x_i) - g(x_i)\right]^2. \tag{11.16}$$

The case when the data is contaminated with noise, as well as cases with complex shapes of the unknown boundaries to be recovered, were considered in [457].

11.2.3 *Biharmonic equation*

There are several inverse geometric problems for the biharmonic equation in $\mathbb{R}^2$ or $\mathbb{R}^3$. For example we could have , see [958]

$$\Delta^2\phi(x) = 0, \quad x \in \Omega, \tag{11.17a}$$

subject to the BCs

$$\phi(x) = f_1(x), \quad x \in \Gamma_1, \tag{11.17b}$$

and

$$\phi(x) = f_2(x), \quad \frac{\partial \phi}{\partial n}(x) = g(x),$$

$$\Delta\phi(x) = v(x), \quad \frac{\partial \Delta\phi}{\partial n}(x) = w(x), \quad x \in \partial\Omega\backslash\Gamma_1. \tag{11.17c}$$

With an appropriate discretization and the MFS approximation in 2D

$$\hat{\phi}(x) = \sum_{j=1}^{N} c_j\, G^L(x, x_j') + \sum_{j=1}^{N} d_j\, G^B(x, x_j'), \quad x \in \overline{\Omega}, \tag{11.18}$$

the counterpart of functional (11.12) becomes

$$
\begin{aligned}
S(c, r) := & \sum_{i=1}^{N} \left[\hat{\phi}(c, d, x'; x_i) - f_1(x_i) \right]^2 \\
& + \sum_{i=N+1}^{N+M} \left[\hat{\phi}(c, d, x'; x_i) - f_2(x_i) \right]^2 \\
& + \sum_{i=N+1}^{N+M} \left[\frac{\partial \hat{\phi}}{\partial n}(c, d, x'; x_i) - g(x_i) \right]^2 \\
& + \sum_{i=N+1}^{N+M} \left[\Delta\hat{\phi}(c, d, x'; x_i) - v(x_i) \right]^2 \\
& + \sum_{i=N+1}^{N+M} \left[\frac{\partial \Delta\hat{\phi}}{\partial n}(c, d, x'; x_i) - w(x_i) \right]^2.
\end{aligned}
\tag{11.19}
$$

11.2.4 *Cauchy–Navier equations*

We consider the BVP for the Cauchy–Navier equations in $\mathbb{R}^2$ consisting of (11.4) and the BCs

$$u_1(x) = 0, \ u_2(x) = 0, \quad x \in \Gamma_1, \tag{11.20a}$$

$$u_1(x) = \overline{u}_1(x), \quad u_2(x) = \overline{u}_2(x),$$

$$t_1(x) = \overline{t}_1(x), \quad t_2(x) = \overline{t}_2(x), \quad x \in \partial\Omega\backslash\Gamma_1, \tag{11.20b}$$

where $\overline{u}_1, \overline{u}_2, \overline{t}_1$, and $\overline{t}_2$ are known functions. Employing a discretization as the one described in Section 11.2.1 and the approximations for the displacements (u_1, u_2)

$$\hat{u}_1(x) = \sum_{j=1}^{N} c_j\, G_{11}(x, x'_j) + \sum_{j=1}^{N} d_j\, G_{12}(x, x'_j), \tag{11.21a}$$

$$\hat{u}_2(x) = \sum_{j=1}^{N} c_j\, G_{21}(x, x'_j) + \sum_{j=1}^{N} d_j\, G_{22}(x, x'_j), \tag{11.21b}$$

for $x \in \overline{\Omega}$, and analogous approximations for the tractions (see [461]), the coefficients $\{c_j\}_{j=1}^{N}$, $\{d_j\}_{j=1}^{N}$ and the radii $r_i = r(\vartheta_i)$, $i = 1, \ldots, N$, in (11.7) are determined by minimizing the functional

$$S(c, r) := \sum_{k=1}^{2} \sum_{i=1}^{N} \left[\hat{u}_k(c, d, x'; x_i)\right]^2$$

$$+ \sum_{k=1}^{2} \sum_{i=N+1}^{N+M} \left[\hat{u}_k(c, d, x'; x_i) - \overline{u}_k(x_i)\right]^2$$

$$+ \sum_{k=1}^{2} \sum_{i=N+1}^{N+M} \left[\hat{t}_k(c, d, x'; x_i) - \overline{t}_k(x_i)\right]^2. \tag{11.22}$$

Numerical examples for various shapes to be reconstructed as well as the counterpart of (11.22) for noisy data and appropriate regularization may be found in [461]. The inverse geometric problem corresponding to (11.4) with (11.20) in 3D was studied in [467] while a 2D inverse geometric problem in thermoelasticity was considered in [465].

Different MFS approaches for the solution of inverse geometric problems, which do not involve the minimization of functionals such as (11.14), may be found in [21, 24, 472].

11.3 Other Inverse Problems

11.3.1 *Inverse source problems*

In this case the governing equation is inhomogeneous with an unknown source term and, to compensate, the BCs are overspecified on part of the boundary. The simplest such example is the following BVP for the Poisson equation

$$\Delta\phi(x) = F(x), \quad x \in \Omega, \tag{11.23a}$$

subject to the BCs

$$\phi(x) = f(x) \quad \text{and} \quad \frac{\partial\phi}{\partial n}(x) = g(x), \ x \in \partial\Omega, \tag{11.23b}$$

where the unknown source term F belongs to some special classes of functions that ensure its unique identifiability. BVP (11.23) was solved using the MFS under certain assumptions on F in [20].

11.3.2 *Inverse boundary coefficient problems*

Now, a coefficient describing a BC on part of the boundary is unknown and, to compensate, the BCs are overspecified on the remaining part of the boundary. A simple such example is the following BVP for the Laplace equation:

$$\Delta\phi(x) = 0, \quad x \in \Omega, \tag{11.24a}$$

subject to the BCs

$$\phi(x) = f(x) \quad \text{and} \quad \frac{\partial\phi}{\partial n}(x) = g(x), \ x \in \Gamma_1, \tag{11.24b}$$

and

$$\alpha\phi(x) + \frac{\partial\phi}{\partial n}(x) = h(x), \ x \in \partial\Omega\backslash\Gamma_1, \tag{11.24c}$$

where the coefficient α in (11.24c) is unknown. Such a problem was solved using the MFS in [870].

11.3.3 *Inverse initial/boundary value problems*

In a typical inverse initial value problem for the 1D heat equation we have

$$\frac{\partial \phi}{\partial t}(x, t) = \frac{\partial^2 \phi}{\partial x^2}(x, t), \quad (x, t) \in (0, 1) \times (0, T), \tag{11.25a}$$

subject to the initial condition

$$\phi(x, 0) = \bar{\phi}(x) \quad \text{on } [0, 1], \tag{11.25b}$$

and the BCs

$$\phi(1, t) = f(t) \quad \text{and} \quad \frac{\partial \phi}{\partial x}(1, t) = g(t) \quad \text{on } [0, T]. \tag{11.25c}$$

The inverse heat conduction problem (11.25) was first solved with the MFS in [391] and its counterparts in 2D and 3D in [394, 395], respectively. The MFS has also been used in [425, 426] for solving the backward heat conduction problem given by (11.24a) subject to Dirichlet (or Neumann) BCs and final time measurement of $\phi|_{t=T}$.

11.4 Related Methods and Problems

11.4.1 *Free boundary problems*

Free boundary problems are closely related to inverse geometric problems. In such problems, as is the case with inverse geometric problems, the geometry of part of the boundary is unknown and needs to be determined as part of the solution. To compensate for this, BCs are overspecified on the unknown part of the boundary. As an example we consider the free boundary problem in 2D

$$\Delta \phi(x) = 0, \quad x \in \Omega, \tag{11.26a}$$

subject to the BCs

$$\mathcal{B}_1 \phi(x) = g_1(x), \quad \mathcal{B}_2 \phi(x) = g_2(x), \quad x \in \Gamma_1, \tag{11.26b}$$

and

$$\phi(x) = f(x), \quad x \in \partial\Omega \backslash \Gamma_1, \tag{11.26c}$$

where $\mathcal{B}_1$ and $\mathcal{B}_2$ are operators describing the BCs on the unknown (free) boundary Γ_1. As in inverse geometric problems, the MFS discretization of

BVP (11.26), because part of the boundary is unknown, leads to a nonlinear system of equations. Free boundary problems similar to (11.26) were solved using the MFS in, e.g. [444, 932], while potential flow problems past deformable bodies using the MFS was investigated in [772, 774]. In [728] a free boundary Stokes flow problem was solved using the MFS. The MFS was also applied to Signorini problems which are a specific class of free boundary problems in [94, 466, 727]. It is noteworthy that in [466] both direct and inverse Signorini problems were studied. Finally, the MFS has been applied to Stefan problems in [130].

11.4.2 *Related methods*

The singular boundary method which is a variant of the MFS was employed for the solution of inverse Cauchy problems in [959] and of free boundary problems in [150]. Moreover, Trefftz collocation methods [492] have been used extensively for the solution of inverse Cauchy linear problems, see, e.g. [501, 575], and the solution of inverse geometric problems, see, e.g. [290]. The boundary knot method was applied to inverse Cauchy problems in [418]. More recently, the LMFS has been used for inverse Cauchy problems in [890] and the localized boundary knot method has been employed for the solution of inverse Cauchy problems in [929].

Chapter 12

Geometric Modeling Using the MFS

Computer geometric modeling is a computer/software-based mathematical model that can generate an object's geometry. Its importance to industrial applications in computer-aided design (CAD), computer-aided manufacturing (CAM), and computer graphics cannot be overstated. Geometric modeling is generally classified into solid, surface, and wireframe modeling. In this chapter, we address surface modeling and particularly surface construction and reconstruction using a set of data points.

There are two mathematical forms that are commonly used to generate curves and surfaces, the parametric form [293] and the implicit form [875], with the more popular one being the former. In the parametric form, the spatial coordinates x of a surface in 3D can be defined by a parametric equation

$$x(s, t) = (x(s, t), y(x, t), z(s, t)), \quad (s, t) \in [a_1, a_2] \times [b_1, b_2]. \quad (12.1)$$

In the implicit form, the mathematical model is given by an implicit function

$$F(x, y, z) = c. \tag{12.2}$$

In the above, when $c = 0$, the equation defines the target surface, and when c is varied, a family of surfaces is generated. Various methods for constructing curves, surfaces, and other geometric objects can be found in the many computer graphics books [294, 709].

487

With the parametric form, it is easy to generate points on curves and surfaces. However, in many industrial applications, such as robotics and CAM, it is important to determine the position of a point relative to a curve or surface. This can be a non-trivial task in the parametric form, since it requires the solution of nonlinear parametric equations. The implicit form represented by (12.2), on the other hand, enables one to easily identify the position of a point relative to a given curve/surface by merely checking the sign of $F(x)$.

Over the years, there has been a fair amount of interest in representing 3D objects by implicit surface modeling [91, 120, 865, 875], in which the reconstructed surface can be represented by a single function. These methods, however, still face challenges when complex objects are encountered. To overcome these difficulties, Tankelevich *et al.* [829, 830] proposed a potential field (PF) method that represents the implicit function by harmonic functions, or solutions of certain PDEs, such as the biharmonic and Helmholtz equations. The method requires the solution of BVPs and the MFS is undoubtedly the most efficient tool for solving such problems. In this chapter, we give a demonstration of using the PF/MFS for the construction of complex curves and surfaces in 2D and 3D.

12.1 PF/MFS for Surface Construction

In the PF method [829], we define a BVP in a doubly connected domain Ω as

$$\mathcal{L}\phi(x) = 0, \quad x \in \Omega, \tag{12.3}$$

with BCs

$$\phi(x) = f(x), \quad x \in \partial\Omega_I, \tag{12.4a}$$

$$\phi(x) = g(x), \quad x \in \partial\Omega_E, \tag{12.4b}$$

in which $\mathcal{L}$ is a linear second-order partial differential operator, $\partial\Omega_E$ and $\partial\Omega_I$ are, respectively, the external and internal boundaries, and g and f are known functions. The internal boundary is intended to be the generating boundary, see Fig. 12.1 for an illustration.

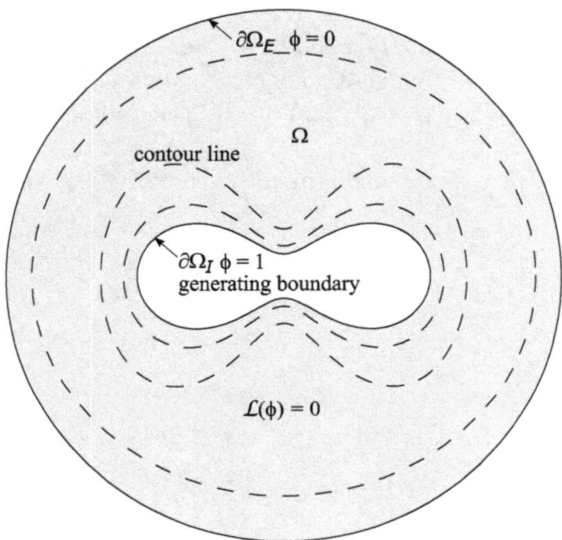

Figure 12.1. A doubly connected region Ω with internal boundary $\partial\Omega_I$ and external boundary $\partial\Omega_E$.

An implicit function is defined as

$$F(\boldsymbol{x}) = \phi(\boldsymbol{x}) - f(\boldsymbol{x}). \tag{12.5}$$

We observe that on the internal boundary, $F(\boldsymbol{x}) = 0$. Hence, $\partial\Omega_I$ is the generating boundary. Typically, we assign the constant boundary values $f = 1$ and $g = 0$. If the operator $\mathcal{L}$ is the one that obeys the maximum principle (see Section 6.7.1), then the equipotential lines/surfaces (level sets) $\phi(\boldsymbol{x}) = c$, for $0 < c < 1$, fall between these two boundaries and morph from the inner surface to the outer one (see Fig. 12.1).

As has been demonstrated in this book, the MFS is among the most efficient methods for solving the BVP defined in (12.3) and (12.4). Once the problem has been solved, an approximation of $\phi(\boldsymbol{x})$ is given by an explicit formula as a summation of fundamental solutions, see (5.96), with known coefficients c_j. It is then easy to construct the level set surfaces of the explicit function $F(\boldsymbol{x})$.

Tankelevich *et al.* [829, 830] utilized the Laplacian and biharmonic equations in the PF/MFS. As in the standard MFS, an auxiliary boundary is needed for the deployment of fundamental solutions. To circumvent the need of creating an auxiliary boundary, Chen *et al.* [146] improved the method by using nonsingular fundamental solutions that can be directly placed on the boundary (see Section 6.1.4). The procedure is described as follows.

For the BVP, a polymetaharmonic operator (see Section 2.12) is considered:

$$\Delta(\Delta - k^2)\phi(x) = 0, \quad x \in \Omega, \tag{12.6}$$

with BCs

$$\phi(x) = 1, \qquad x \in \partial\Omega_I, \tag{12.7a}$$

$$\phi(x) = 0, \qquad x \in \partial\Omega_E, \tag{12.7b}$$

$$\frac{\partial\phi}{\partial n}(x) = g(x), \quad x \in \partial\Omega, \tag{12.7c}$$

where $\partial\Omega = \partial\Omega_I \cup \partial\Omega_E$, and we are letting $g(x)$ be undetermined. We note in the above that because (12.6) is a fourth-order PDE, two BCs are needed. For simplicity, the outer boundary $\partial\Omega_E$ is taken to be a rectangle.

To solve BVP (12.6) and (12.7) with the MFS, we take the collocation points $\{x_i\}_{i=1}^{N_1} \in \partial\Omega_I$, and $\{x_i\}_{i=N_1+1}^{N} \in \partial\Omega_E$. Since there are two BCs at each collocation point, we need in principle $2N$ sources with the same number of undetermined coefficients. As in the biharmonic equation, the operator in (12.6) has two types of fundamental solutions: that of the product of the harmonic and metaharmonic operators (see (2.241))

$$G_1(r) = -\frac{1}{2\pi k^2}[\mathrm{K}_0(kr) + \ln r] \quad (2\mathrm{D}), \tag{12.8a}$$

$$= -\frac{1}{4\pi k^2 r}\left(e^{-kr} - 1\right) \quad (3\mathrm{D}), \tag{12.8b}$$

and that of the modified Helmholtz operator, (2.35). We note that K_0 can be expanded as follows:

$$\mathrm{K}_0(kr) = -\gamma - \ln\left(\frac{k}{2}\right) - \ln r + \mathcal{O}(r), \tag{12.9}$$

where $\gamma \approx 0.5772156649$ is Euler's constant. By taking the limit of G_1 in (12.8), we have

$$\lim_{r \to 0} G_1(r) = \frac{1}{2\pi k^2} \left[\ln\left(\frac{k}{2}\right) + \gamma \right] \quad \text{(2D)}, \tag{12.10a}$$

$$= \frac{1}{4\pi k} \quad \text{(3D)}, \tag{12.10b}$$

thus G_1 is nonsingular. In a normal procedure, we should write an approximation formula that distributes both types of fundamental solutions similar to (7.20) for the biharmonic case. Here, however, we distribute only one type and approximate the solution as

$$\hat{\phi}(x) = \sum_{j=1}^{N} c_j G_1(x, x_j), \quad x \in \overline{\Omega}, \ x_j \in \partial\Omega. \tag{12.11}$$

We refer the reader to [830] for further details on how to avoid using the singular fundamental solution and BC (12.7c). In (12.11), c_j are undetermined coefficients and x_j are the collocation points. Here, we allow the source and collocation points to coincide because G_1 is nonsingular. We then use (12.11) to collocate for the BCs (12.7a) and (12.7b) at the N nodes. The system can then be solved for c_j. In the numerical implementation, we simply ignore the BC (12.7c) due to the theoretical justification shown in [830]. In this improved PF/MFS, the need for an auxiliary boundary is eliminated, and the MFS matrix is symmetric and positive definite.

12.2 Design and Construction of Times Roman Font

In this section, we demonstrate the construction of Times Roman font.

■ Example 12.1

We first consider the design and construction of the letter **S** in the Times Roman font.

To generate the inner boundary points on $\partial\Omega_I$, we use the data points from Table VII in [829], which consist of 13 Bézier curves and six straight lines. Each Bézier curve can be obtained using four control points. Such curves are widely applied in the design of computer fonts

Table 12.1. Coordinates of control points of 13 Bézier curves and two straight lines.

Type	(x_1, y_1)	(x_2, y_2)	(x_3, y_3)	(x_4, y_4)
Straight line 1	$(72, -13)$	$(42, 199)$		
Bézier curve 1	$(65, 199)$	$(84, 152)$	$(135, 22)$	$(270, 22)$
Bézier curve 2	$(270, 22)$	$(350, 22)$	$(390, 79)$	$(390, 133)$
Bézier curve 3	$(390, 133)$	$(390, 190)$	$(374, 224)$	$(227, 310)$
Bézier curve 4	$(227, 310)$	$(135, 363)$	$(71, 407)$	$(71, 506)$
Bézier curve 5	$(71, 620)$	$(162, 676)$	$(252, 676)$	$(312, 676)$
Bézier curve 6	$(252, 676)$	$(312, 676)$	$(366, 642)$	$(391, 642)$
Bézier curve 7	$(391, 642)$	$(419, 642)$	$(424, 665)$	$(426, 676)$
Straight line 2	$(447, 676)$	$(469, 463)$		
Bézier curve 8	$(444, 463)$	$(427, 545)$	$(363, 635)$	$(258, 635)$
Bézier curve 9	$(258, 635)$	$(208, 635)$	$(157, 604)$	$(157, 542)$
Bézier curve 10	$(157, 542)$	$(157, 392)$	$(491, 372)$	$(491, 167)$
Bézier curve 11	$(491, 167)$	$(491, 82)$	$(423, -14)$	$(280, -14)$
Bézier curve 12	$(280, -14)$	$(204, -14)$	$(155, 20)$	$(125, 20)$
Bézier curve 13	$(125, 20)$	$(104, 20)$	$(94, 3)$	$(94, -13)$

and computer graphics. In [146], as shown in Table 12.1, 13 sets of control points and only two sets of straight line points are required for the generation of the inner boundary points.

Next, we describe a step-by-step procedure on how to design and produce the inner boundary collocation points. In Fig. 12.2(a), we present two samples of Bézier curves. The blue and green Bézier curves are obtained using four red and four black control points, respectively, which can be found in curves 1 and 10 in Table 12.1. Figure 12.2(b) displays the 4×13 control points in red dots that produce all 13 Bézier curves for the design of the letter **S**. The collocation points on the two straight lines can be generated easily. More specifically, the points in Line 1 in Table 12.1 can be generated as follows:

```
xL1=linspace(72,42,n);   xL1=xL1';
yL1=linspace(-13,199,n); yL1=yL1';
```

where n is the number of points.

For illustration, let us consider the four black control points in Fig. 12.2(a), whose coordinates are available in curve 10 in Table 12.1. The 2D Bézier curve can be obtained by the following code:

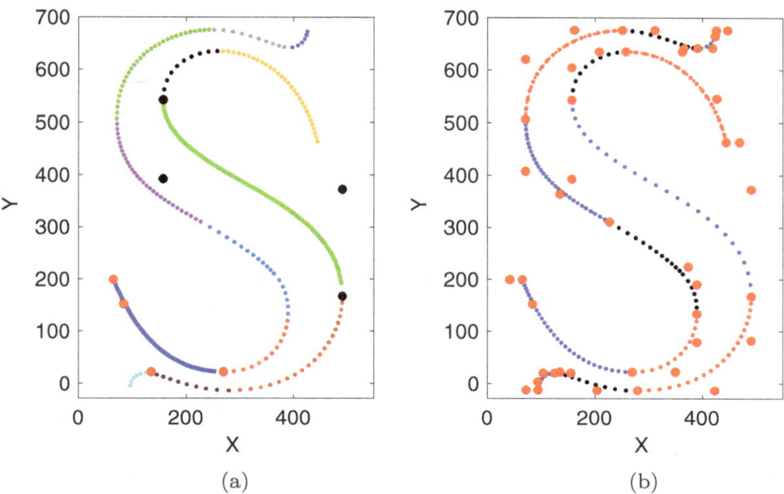

Figure 12.2. (a) Four red control points for the blue Bézier curve and four black control points for the green Bézier curve. (b) All control points for the letter **S**.

```
1 syms t
2 B=bernsteinMatrix(3,t);
3 P=[157, 542; 157, 392 ; 491, 372 ; 491, 167];
4 beziercurve=simplify(B*P);
5 fplot(beziercurve(1), beziercurve(2), [0, 1]);
```

Note that `bernsteinMatrix` is a MATLAB® built-in function. The Bernstein matrix is also called the Bézier matrix and can be used to construct Bézier curves. Line 3 lists the four black control points in Fig. 12.2(a). In Line 4, we obtain the following parametric equations of the third-order Bézier curve in Fig. 12.2(a):

```
x(t) = - 668*t^3 + 1002*t^2 + 157
y(t) = - 315*t^3 + 390*t^2 - 450*t + 542
```

Line 5 plots the Bézier curve (green line) in Fig. 12.2(a).

The following code can be used to generate the collocation points on the exterior boundary $\partial\Omega_E$. In the code, `x1` and `x2` are the left and right ends of the rectangle in the x direction, `y1` and `y2` are upper and lower ends of the rectangle's coordinates in the y direction, and `n1` and `n2` are, respectively, the number of subdivisions in the x and

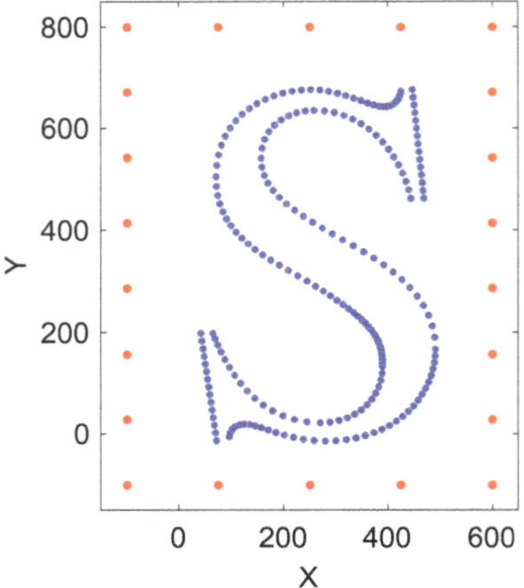

Figure 12.3. The profile of 22 exterior (red) and 218 interior (blue) = boundary collocation points.

y directions. The number and the location of the exterior boundary points can also be chosen freely without affecting the final result.

```
1 function [x,y]=rectangular_pts(x1,x2,y1,y2,n1,n2)
2 a1=linspace(x1,x2,n1);
3 a2=linspace(y1,y2,n2);
4 [X, Y]=ndgrid(a1,a2);
5 id = X==x1 | X==x2 | Y==y1 | Y==y2;
6 x=X(id);   y=Y(id);
```

Using the procedure shown above, we can produce Fig. 12.3, which contains 218 interior and 22 exterior boundary collocation points. With these boundary points, we are in a position to construct the letter **S** in Times Roman font.

The parameter k in the fundamental solution plays an important role in the visual quality of the reproduction of the letter, as the condition number of the MFS matrix depends on the value of kr. Hence, for a large r, a small k is taken and vice versa. To measure the quality of the production of the letter, we choose a different set of 314 inner boundary points as test points. In Table 12.2, we present the maximum absolute errors $\mathcal{E}_{\max}$ and the condition numbers for various values of k. A good quality letter **S** can be constructed with $0.001 \leq k \leq 0.1$. In terms of visual effect, little difference can be detected when using any

Table 12.2. Maximum absolute errors $\mathcal{E}_{\max}$ and condition numbers using various values of k for the construction of the letter **S**.

k	$\mathcal{E}_{\max}$	Cond #
0.0001	6.4(−03)	1.6(11)
0.001	6.4(−03)	9.1(08)
0.005	6.6(−03)	3.1(07)
0.02	7.6(−03)	1.8(06)
0.1	8.1(−03)	7.4(04)
0.2	9.6(−03)	1.9(04)

Figure 12.4. The profile of the constructed letter **S** using $k = 0.02$.

k in this range. For $k \geq 0.2$, however, the visual quality of the letter **S** is poor. Figure 12.4 shows the constructed letter **S** using $k = 0.02$. Figure 12.5(a) depicts the 3D profile of the solution by setting $\hat{\phi}$ to different values. Figure 12.5(b) gives the contour plot for $\hat{\phi} = 0.99, 1$, and 1.01.

The code `MFS_S.m` constructs the letter **S**. The generation of the outer and inner boundary collocation points is carried out in Lines 2–8. The standard MFS solution procedure is shown in Lines 9–15, while in Lines 16–19, we create a bounding box containing the inner boundary collocation points. The solution at 101×101 meshgrid points in the bounding box is evaluated in Lines 20–21. The construction of the

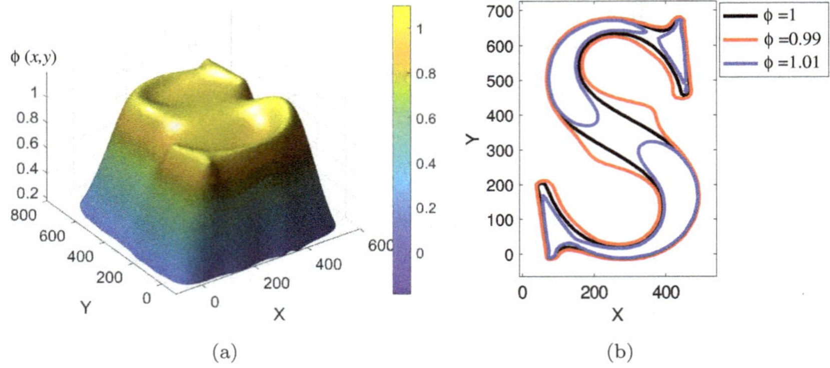

Figure 12.5. For $k = 0.02$, (a) solution profile in the extended domain and (b) contour plot for various $\hat{\phi} = 0.99, 1, 1.01$.

final plot of the letter **S** with yellow face color and red background can be found in Lines 23–29.

```
%   MFS_S.m
%   Design and construction of letter S
1   clear;
2   n1=15;% number of points on some curves
3   n2=15;  % number of points on two straight lines
4   lam=0.02;
5   [x1,y1]=S_curve(n1,n2); % Inner boundary points
6   [xo,yo]=rectangular_pts(-100,600,-100,800,5,8);
7   ni=length(x1); no=length(xo);
8   x=[x1;xo]; y=[y1;yo]; % Boundary collocation points
9   g1=ones(ni,1);   % Inner boundary condition
10  g2 = zeros(no,1); % Outer boundary condition
11  f = [g1; g2]; %Right hand side
12  DM1 = pdist2([x y], [x y]);
13  A = besselk(0,lam*DM1)+log(DM1); %MFS Interpolation
    matrix
14  A(DM1==0)=-log(lam/2)-0.57721566;
15  coe = A\f;
16  L=100;    %meshgrid in each axis direction
17  b = max(x)+50; a = min(x)-50; c=min(y)-50;
    d=max(y )+50;
18  h = a:(b-a)/L:b; k = c:(d-c)/L:d;
```

```
19 [X, Y] = meshgrid(h,k);   % Bounding box
20 DM = pdist2([X(:) Y(:)], [x y]);
21 u=(besselk(0,lam*DM)+log(DM))*coe;%solutions in
   bounding box
22 u = reshape(u,L+1,L+1);
23 figure (1)
24 pts=[-8 -64; 541 -64; 541 727; -8 727;  -8 -64];
25 plot(pts(:,1),pts(:,2));
26 fill(pts(:,1),pts(:,2),'r'); hold on;
27 T=contour(X,Y,u,[1 1]); T=T'; T(1,:)=[];
28 fill(T(:,1),T(:,2),'y','LineWidth',3);
29 axis equal; axis off;  hold off
```

■ Example 12.2

The 3D version of the Times Roman font of the letter **S** can be constructed in a way similar to the 2D technique presented above.

For the design of the boundary collocation points, we can add an extra layer of the inner boundary points of the letter **S** above the other, as shown in Fig. 12.6(a). Let $\{(x, y)\}$ represent the set of boundary points of the letter **S**. Then $\{(x, y, 0), (x, y, 40)\}$ are the 3D inner boundary points that are separated 40 units apart in the vertical direction, as shown in the red and blue dots in Fig. 12.6(a). For the outer boundary points (black dots), we use the following script to produce the boundary box points. In Line 1, x1, x2, y1, y2, z1, and z2 are the left and right end coordinates of the box in the x, y, and z directions, respectively, and n1, n2, and n3 are the numbers of points on each axis. The black dots in Fig. 12.6(a) were obtained using n1 and n2 = 4, and n3 = 2.

```
% box_pts.m  Production of points on a rectangular box
1 function [x,y,z]=box_pts(x1,x2,y1,y2,z1,z2,n1,n2,n3)
2 a1=linspace(x1,x2,n1);
3 a2=linspace(y1,y2,n2);
4 a3=linspace(z1,z2,n3);
5 [X,Y,Z]=ndgrid(a1,a2,a3);
6 id = X==x1 | X==x2 | Y==y1 | Y==y2 | Z==z1 | Z==z2;
7 x=X(id);  y=Y(id);   z=Z(id);
```

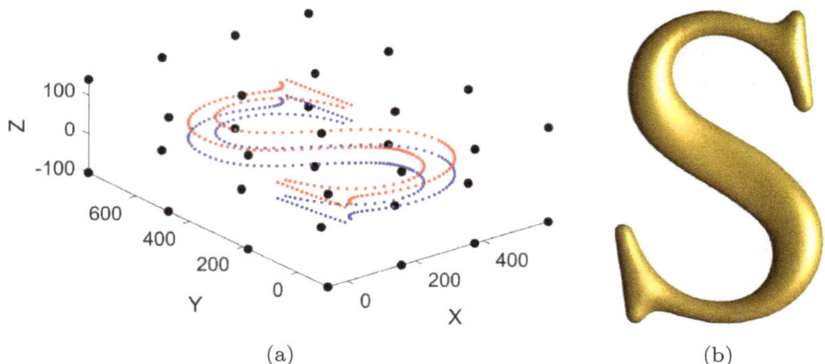

(a)   (b)

Figure 12.6.   (a) Distribution of 436 inner and 32 outer boundary collocation points; (b) profile of the constructed letter **S** in Times Roman font in 3D.

The code for the construction of the 3D case is quite similar to the 2D one. Instead of using the contour plot in the 2D case, we use the MATLAB® command `isosurface` to recover the 3D surface of the letter **S**. Figure 12.6(b) shows the 3D version of the letter **S** using $k = 0.001$.

■ **Example 12.3**

Next, we consider the construction of the joint letters **MFS**.

The control points defining the letters **M** and **F** can be obtained from Tables V and VI in [829]. The profiles of these control points are shown in Figs. 12.7(a) and 12.7(b). Following the steps for the construction of the letter **S** shown above, we can similarly obtain the outer (42 red dots) and inner (750 blue dots) boundary points of the letters **MFS**, as shown in Fig. 12.8(a). Figure 12.8(b) depicts the contour plot at various solution levels. The final plot of the 2D color version of the letters **MFS** using $k = 0.1$ is presented in Fig. 12.9. Note that it is much easier to construct a single letter than the three-letter word **MFS**. In this case, the selection of the parameter k is more challenging due to the geometric complexity of the curves involved.

We can extend the letters of **MFS** from 2D to 3D by adding an additional level of inner collocation points, as shown in Fig. 12.8(a), with 40 units in the vertical direction. Figure 12.10 shows 1600 interior boundary collocation points (red and blue dots) and 102 outer

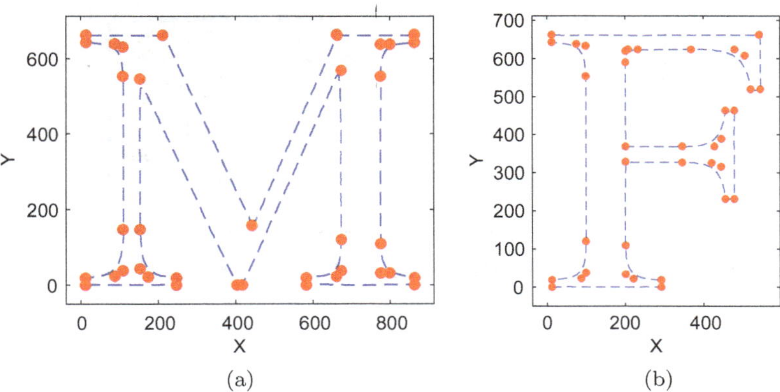

Figure 12.7. Profiles of (a) 25 control points of the letter **M** and (b) 23 control points of the letter **F**.

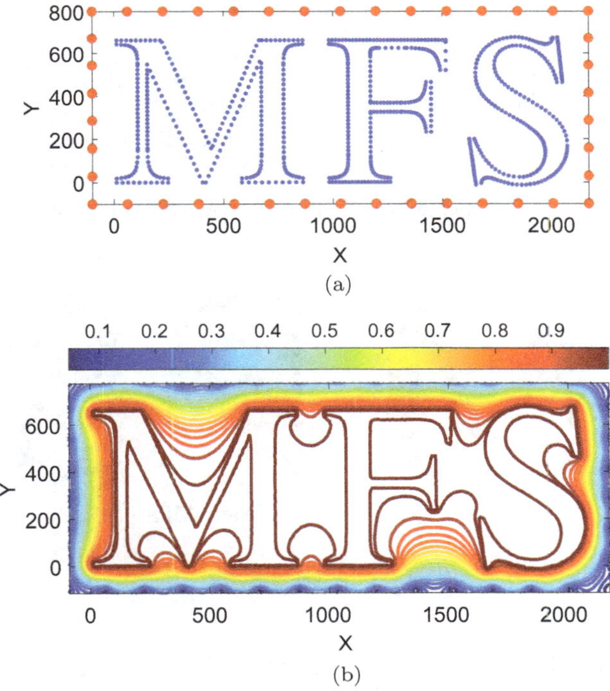

Figure 12.8. (a) Distribution of boundary collocation points; (b) profile of the contour plot.

Figure 12.9. Profile of the constructed 2D letters **MFS** using $k = 0.1$.

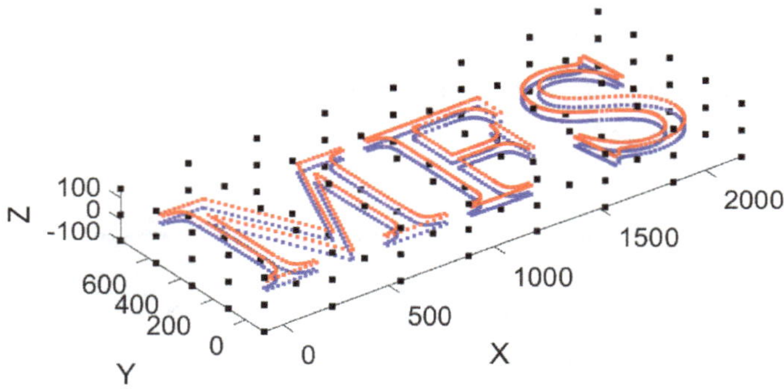

Figure 12.10. Distribution of boundary collocation points of the letter **MFS** in 3D.

Figure 12.11. Profile of the 3D version of the letters **MFS** in Times Roman font using $k = 0.02$.

boundary collocation points (black dots). Following the construction procedure for the letter of **S** in the 3D case, the final reconstruction of the letters **MFS** in Times Roman font in 3D using $k = 0.02$ is presented in Fig. 12.11. For further details, we refer interested readers to [146].

12.3 3D Surface Reconstruction

In this section, we consider the construction of a surface from a set of 3D point-cloud data. This has been the subject of considerable research in computer graphics and CAD. Over the years, there has been a lot of interest in representing 3D objects by implicit surface modeling [91, 120, 830, 875], in which the reconstructed surface can be represented by a single function.

Tankelevich *et al.* [830] proposed the reconstruction of a 3D implicit surface by solving both harmonic and biharmonic BVPs using the MFS. As a result, no additional off-surface points are required. Both of these two models work satisfactorily for simple geometric surfaces. However, for more complicated geometries containing sharp corner edges, spurious surfaces often appear. As such, additional supporting points near the surface outside the domain are required to remove these spurious surfaces. Such a procedure is quite tedious and not desirable. In [968], an improved PDE model was proposed to alleviate the problem of spurious surfaces without adding extra off-surface points. Let $\Omega \subseteq \mathbb{R}^3$ be a bounded and closed domain with boundary $\partial\Omega$. The following modified bi-Helmholtz equation (see Section 2.12.1) as a PDE model for the approximation of an implicit surface is considered in [968]

$$(\Delta - k^2)^2 \phi(x) = 0, \quad x \in \Omega, \tag{12.12a}$$

$$\phi(x) = 1, \qquad\qquad x \in \partial\Omega, \tag{12.12b}$$

$$\frac{\partial \phi}{\partial n}(x) = g(x), \qquad x \in \partial\Omega, \tag{12.12c}$$

where $g(x)$ is unspecified. Compared to the biharmonic model [830], we shall see that the parameter k plays an important role in the removal of unwanted spurious surfaces.

Similar to the formulation in (12.6)–(12.11), the solution of BVP (12.12) can be approximated by a linear combination of the fundamental solution G as follows:

$$\hat{\phi}(x) = \sum_{j=1}^{N} c_j\, G(x, x_j), \quad x \in \overline{\Omega}, \ x_j \in \partial\Omega, \tag{12.13}$$

where G is given by (2.222) as

$$G = \frac{1}{8\pi k} e^{-kr}, \quad \text{where } r = \|x - x_j\|. \tag{12.14}$$

Note that since G is nonsingular, the source points can coincide with the boundary collocation points. As mentioned in Section 12.1, the BC (12.12c) can be avoided. For details, we refer the reader to [830].

■ Example 12.4

We first consider the construction of the surface of the Stanford Bunny in 3D. The boundary cloud data points are available from the Stanford 3D scanning repository [812].

In the numerical implementation, as shown in Fig. 12.12(a), 8171 boundary cloud data points were used. For $k = 12$, a spurious surface appeared behind the ears (see Fig. 12.12(b)). However, for $k = 30$, the spurious surface disappeared (see Fig. 12.12(c)). When the value of k becomes too large, part of the surface starts to crack and fall apart (see Fig. 12.12(d) for the case $k = 65$).

Based on a series of experiments, a visually acceptable Bunny without spurious surfaces can be reconstructed by taking k in the range [20, 40]. To determine an appropriate value of k for a given set of cloud data points $\{(x_i, y_i, z_i)\}_{i=1}^n$, we denote by $d_{\min}$ the average minimum distance expressed as

$$d_{\min} = \frac{1}{n} \sum_{i=1}^n \min_{1 \le j \le n,\ i \ne j} \left(\| (x_i, y_i, z_i) - (x_j, y_j, z_j) \| \right).$$

In the following code, Lines 4 and 5 are used to find $d_{\min}$. For the current data set, we find $d_{\min} = 0.0021$. This implies that we can expect an accurate reconstruction of the Bunny for $kr \in [0.042, 0.084]$. To ensure an accurate reconstruction, it is recommended to choose the middle value of kr which is 0.063. In Line 6, we use this value to give an estimate of k for other data sets. The MFS solution procedure can be coded in just four lines, as shown in Lines 9–12. The bounding box containing the Bunny is found in Lines 13–17.

The most time-consuming part of the process is the repeated evaluation of all the meshgrid points in Lines 20–23. The MATLAB® command `isosurface` extracts isosurface data from the volume data `pf` in Line 25 at the isosurface value `pf = 1`. This means the isosurface connects points that have a certain specific value in much the same way contour lines connect points of equal elevation, as shown in the 2D cases in the previous section. For the case `neval = 100` in

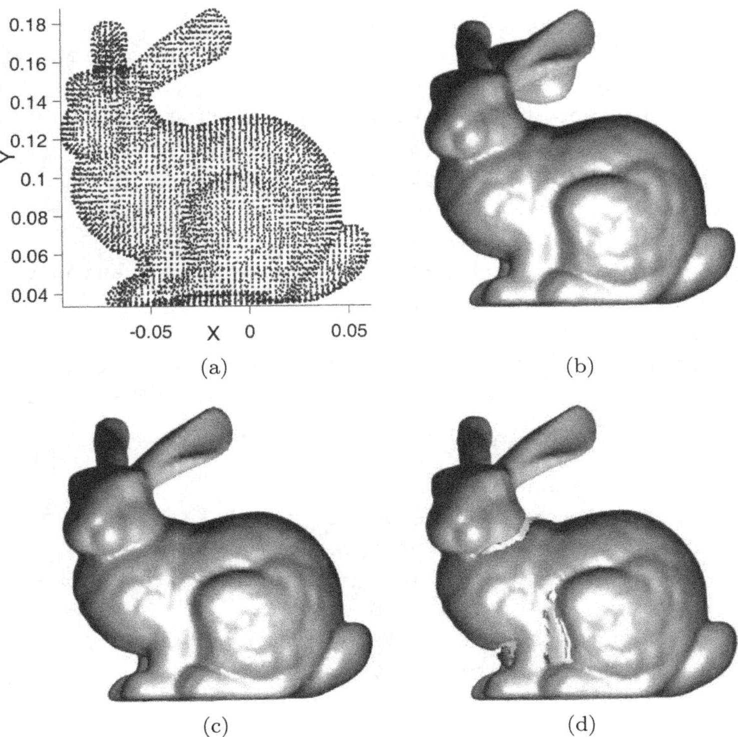

Figure 12.12. Reconstruction of the Stanford Bunny. Profiles of (a) 8171 boundary cloud data points; (b) reconstructed surface using $k = 12$ (observe spurious surface around the ears); (c) $k = 30$ (spurious surface removed); and (d) $k = 65$ (other parts of surface start to break up).

Line 3, one million evaluation points are required. The visual quality of the Bunny can be adjusted by choosing a higher value of `neval`:

```
%   MFS_3D_graphics.m
1   load('Data3D_Bunny2'); tic
2   m=size(dsites,1);       % Number of cloud data points
3   neval=100;   % # evaluation points in each axis direction
4   [~,D]=knnsearch(dsites, dsites,'k',2);
5   d_min=sum(D(:,2))/m;
6   lambda=0.063/d_min;                     % Parameter
7   PS=neval*100;
8   number = floor((neval^3)/PS);
9   RHS=ones(m,1);                          % boundary condition
```

```
10 DM=pdist2(dsites,dsites);              % Distance Matrix
11 MFS=exp(-lambda*DM);                   % MFS matrix
12 coef=MFS\RHS;
13 MIN=min(dsites,[],1); MAX=max(dsites,[],1); % Bounding Box
14 xd=linspace(MIN(1),MAX(1),neval);
15 yd=linspace(MIN(2),MAX(2),neval);
16 zd=linspace(MIN(3),MAX(3),neval);
17 [xe,ye,ze]=meshgrid(xd,yd,zd);
18 epoints=[xe(:),ye(:),ze(:)];          % Evaluation points
19 pf=zeros(length(xe(:)),1);
20 for k=1:number
21     DM = pdist2(epoints(((k-1)*PS+1):k*PS,:),dsites);
22     pf(((k-1)*PS+1):k*PS,1) = exp(-lambda*DM) * coef;
23 end
24 pf=reshape(pf,neval,neval,neval);
   % Plot the isosurface
25 pfit=patch(isosurface(xe,ye,ze,pf,1.0));
26 isonormals(xe,ye,ze,pf,pfit)
27 pfit.FaceLighting='gouraud';  pfit.EdgeColor='none';
28 pfit.FaceColor=[0.94 0.94 0.94];
29 light('Position',[1 0 1],'Style','infinite');
30 axis equal;   axis off;   view([0,90]);
```

■ Example 12.5

In this example, we consider a more complicated geometric model, which is known as the skeleton. The cloud data set is available from the Large Geometric Models Archive at Georgia Institute of Technology [327].

In this case, 21822 data cloud points are available (see Fig. 12.13(a)). Following the procedure described above, we choose $k = 18.5$. Figure 12.13(b) shows the final 3D surface recovery from the given cloud data points in Fig. 12.13(a). The above script `MFS_3D_graphic.m` can be adopted by simply changing the input data in Line 1.

■ Example 12.6

In this example, we consider the reconstruction of the Stanford Dragon [812], which is an even more challenging problem due to the geometric complexity of the surface.

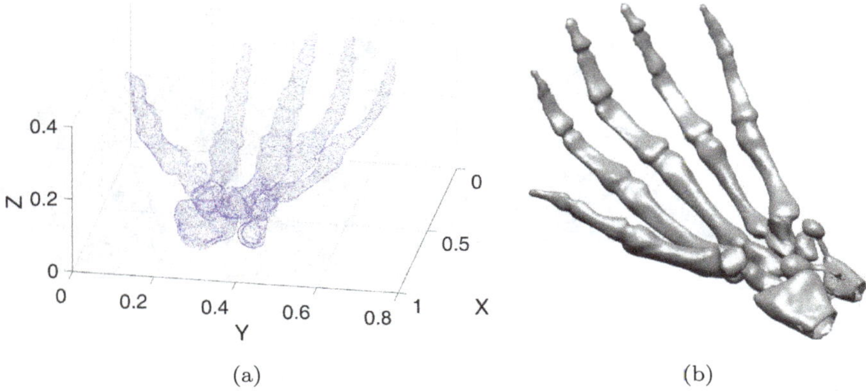

Figure 12.13. Reconstruction of the skeleton hand. Profiles of (a) 21822 cloud data points and (b) reconstructed surface with $k = 18.5$.

We chose 22803 cloud data points (see Fig. 12.14(a)). To speed up the computation, we used a cluster computer with 20 parallel cores. To perform parallel computing, we slightly modified the above code by replacing Lines 19–23 by

```
xe1=reshape(xe,PS,number);
ye1=reshape(ye,PS,number);
ze1=reshape(ze,PS,number);
pf=zeros(length(xe(:)),1);
parfor k=1:number
        DM=pdist2([xe1(:, k),ye1(:,k),ze1(:,k)], dsites);
        pf(:,k)=exp(-lambda*DM)*coef;
end
```

Using the code `MFS_3D_graphics.m` directly, we obtained from Line 6 an estimate of $k = 46$. The performance, however, was not satisfactory. As shown in Fig. 12.14(b), spurious surfaces appeared inside the mouth. We increased the k value to 55 to suppress the spurious surfaces (see Fig. 12.14(c)), which produced some effect, but the surface persisted. If we continued to increase k, the spurious surfaces could be further reduced, but at the same time other parts of the surface would split. Hence, for this complicated geometric surface, adjusting the value of k alone was not sufficient to produce a satisfactory surface.

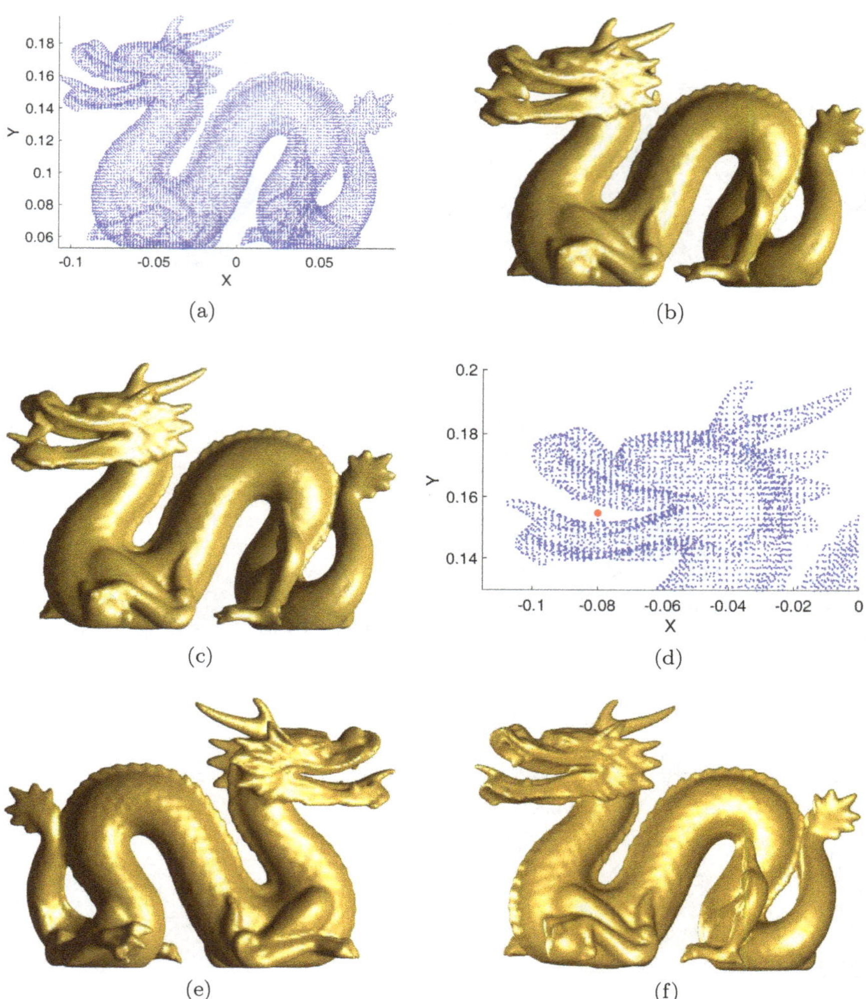

Figure 12.14. Reconstruction of the Stanford Dragon. Profiles of (a) 22803 cloud data points, (b) reconstructed surface with $k = 46$ (observe spurious surface inside the mouth), (c) with $k = 55$ (spurious surface reduced, but not gone), (d) adding a data point (red dot) inside the mouth to improve the algorithm, (e) final profile rear view, and (f) front view.

To overcome this issue, additional points can be added outside the domain in the neighborhood where the spurious surface occurs [830]. For solving (12.12), we note that $\phi(x) \geq 1$ for $x \in \Omega$. By continuity,

$\phi(x) < 1$ for points outside Ω. We hence add a point outside the domain at $(-0.08, 0.155, 0)$, shown as the red dot in Fig. 12.14(d), and assign a ϕ value less than 1 to it. For simplicity, we assign this value as 0. This can be done by replacing Line 2 by

```
dsites=[dsites; [-0.08, 0.155, 0]];  m=size(dsites,1);
```

Meanwhile, we can replace Line 9 by

```
RHS=[ones(m-1,1);  0];
```

To further increase the quality of the reconstructed Dragon, we used `neval=150` in Line 3. The number of evaluation points was $150^3 = 3,375,000$ and it took 147 seconds for a cluster computer with 20 CPU processors to complete the task. Figures 12.14(e) and 12.14(f) show the rear and front sides of the Dragon.

Appendix A

List of Acronyms

Abbreviations

BC	Boundary condition
BVP	Boundary value problem
CAD	Computer-aided design
CAM	Computer-aided manufacturing
CSRBF	Compactly supported RBF
FFT	Fast Fourier transform
GA	Gaussian RBF
IC	Initial condition
IMLS	Interpolating MLS
IMQ	Inverse multiquadric RBF
LHS	Left-hand side
LS	Least squares
MLS	Moving least squares
MQ	Multiquadric RBF
NDIF	Non-dimensional dynamic influence function
ODE	Ordinary differential equation
PDE	Partial differential equation
PU	Partition of unity
RBF	Radial basis function
RHS	Right hand side
RMS	Root mean square
RMSE	Root mean square error
SVD	Singular value decomposition
WLS	Weighted least squares

Acronyms of Numerical Method

ACA	Adaptive cross approximation
AEM	Analog equation method
ANM	Asymptotic numerical method
BCM	Boundary cloud method
BEFM	Boundary element-free method
BEM	Boundary element method
BFM	Boundary face method
BIEM	Boundary integral equation method
BKM	Boundary knot method
BNM	Boundary node method
BPIM	Boundary point interpolation method
CHIEF	Combined Helmholtz integral equation formulation
DDM	Displacement discontinuity method
DEM	Diffuse element method
DQ	Differential quadrature method
DRBEM	Dual reciprocity BEM
EFG	Element free Galerkin method
FCM	Finite cloud method
FDM	Finite difference method
FEM	Finite element method
FMM	Fast multipole method
FPM	Finite point method
FVM	Finite volume method
GBNM	Galerkin boundary node method
GDQ	Generalized differential quadrature
GFD	Generalized finite difference
GFEM	Generalized FEM
HAM	Homotopy analysis method
LBIE	Local boundary integral equation
LMFS	Localized MFS
LOOCV	Leave-one-out cross validation
LRBFCM	Local RBF collocation method
LRPIM	Local radial point interpolation method
LSCM	Least squares collocation method
MDA	Matrix decomposition algorithm
MFS	Method of fundamental solutions
MFVM	Meshless FVM

MLPG	Meshless local Petrov–Galerkin method
MMFS	Modified method of fundamental solutions
MPM	Material point method
MPS	Method of particular solutions
MPS-MFS	Combined MFS and MPS method
MWR	Method of weighted residuals
PIM	Picard iteration method
PF	Potential field method
PIC	Particle-in-cell
PUFEM	Partition of unity FEM
RBFCM	RBF collocation method
RBF-DQ	RBF differential quadrature method
RBF-FD	RBF finite difference method
RKPM	Reproducing kernel particle method
RMM	Regularized meshless method
SBM	Singular boundary method
SFEM	Smoothed FEM
SPH	Smoothed particle hydrodynamics
TM	Trefftz method
TCM	Trefftz collocation method
XFEM	Extended FEM

Appendix B

List of Fundamental Solutions

In Tables B.1 and B.2, we list a few of the most frequently used fundamental solutions. The full presentation is found in Chapter 2.

Table B.1. Fundamental solutions for second order differential operators $\mathcal{L}$.

$\mathcal{L}$	$\mathbb{R}^2$	$\mathbb{R}^3$
Δ	$\dfrac{1}{2\pi}\ln r$	$-\dfrac{1}{4\pi r}$
Δ^2	$-\dfrac{1}{8\pi}r^2\ln r$	$\dfrac{r}{8\pi}$
$\Delta - k^2$	$-\dfrac{1}{2\pi}\mathrm{K}_0(kr)$	$-\dfrac{e^{-kr}}{4\pi r}$
$\Delta + k^2$	$\dfrac{1}{4}\mathrm{Y}_0(kr)$	$-\dfrac{\cos(kr)}{4\pi r}$

Table B.2. Fundamental solutions for fourth order differential operators $\mathfrak{L}$.

$\mathfrak{L}$	$\mathbb{R}^2$	$\mathbb{R}^3$
$\left(\Delta - k^2\right)^2$	$\dfrac{r}{4\pi k}\mathrm{K}_1(kr)$	$\dfrac{e^{-kr}}{8\pi k}$
$\left(\Delta + k^2\right)^2$	$\dfrac{r}{8k}\mathrm{Y}_1(kr)$	$-\dfrac{\sin(kr)}{8\pi k}$
$\Delta\left(\Delta - k^2\right)$	$-\dfrac{1}{2\pi k^2}\left[\mathrm{K}_0(kr) + \ln r\right]$	$-\dfrac{e^{-kr} - 1}{4\pi k^2 r}$
$\Delta\left(\Delta + k^2\right)$	$-\dfrac{1}{4k^2}\left[\mathrm{Y}_0(kr) - \dfrac{2}{\pi}\ln r\right]$	$\dfrac{\cos(kr) - 1}{4\pi k^2 r}$
$\Delta^2 - k^4$	$-\dfrac{1}{8k^2}\left[\mathrm{Y}_0(kr) + \dfrac{2}{\pi}\mathrm{K}_0(kr)\right]$	$\dfrac{\cos(kr) - e^{-kr}}{8\pi k^2 r}$

Appendix C

Additional MATLAB® Codes

In this appendix, we give a list of MATLAB® codes that are commonly used in various chapters in Part II.

The code `sunflower.m` generates uniformly distributed points inside a unit circle. The code can be found in the following website:

https://stackoverflow.com/questions/28567166/uniformly-distribute-x-points-inside-a-circle

```matlab
function [x,y]=sunflower(n, alpha) %  example: n=500, alpha=2
b = round(alpha*sqrt(n));         % number of boundary points
phi = (sqrt(5)+1)/2;              % golden ratio
x=zeros(n,1); y=x;
for k=1:n
    r = radius(k,n,b);
    theta = 2*pi*k/phi^2;
    x(k)=r*sin(theta); y(k)=r*cos(theta);
end
end
function r = radius(k,n,b)
    if k>n-b
        r = 1;            % put on the boundary
    else
        r = sqrt(k-1/2)/sqrt(n-(b+1)/2);
    end
end
```

The following code generates approximately uniformly distributed boundary points on an amoeba boundary and is available from the MathWorks website

http://blogs.mathworks.com/steve/2012/07/06/walking-along-a-path/

```
function [x,y]= equibdd_points(n) % n: number of
 boundary points
r =@(x)  exp(sin(x)).*sin(2*x).^2+exp(cos(x)).*
 (cos(2*x)).^2;%
t=linspace(0,2*pi,n); t=t';
[xv, yv]=pol2cart(t,r(t));
xy = [xv yv];
d = diff(xy,1);
dist_vertex_to_vertex = hypot(d(:,1), d(:,2));
cumulative_dist = [0;cumsum(dist_vertex_to_vertex,1)];
num_points =n +1;
dist_steps = linspace(0, cumulative_dist(end),
 num_points);
bdd_point = interp1(cumulative_dist, xy, dist_steps);
bdd_point(end,:)=[];
x=bdd_point(:,1); y=bdd_point(:,2);
```

The code below generates the boundary points on the double sphere in Example 7.10.

```
function  bdpt=doublesphere(nb)
%nb= # of boundary points on double sphere
bdpt=zeros(nb,3);
dlong = pi*(3-sqrt(5));
long = 0;
dz = sqrt(1)*1.75*2.0/nb;
z = 1.75*sqrt(1) - dz/2;
for k = 0 : 0.5*nb-1
    r = sqrt(1-(z-0.75).^2);
    bdpt(k+1,2)=cos(long)*r;
    bdpt(k+1,3)=sin(long)*r;
    bdpt(k+1,1)=z;
    z = z - dz;
    long = long + dlong;
end
```

```
for k = 0.5*nb : nb-1
    r = sqrt(1-(z+0.75).^2);
    bdpt(k+1,2)=cos(long)*r;
    bdpt(k+1,3)=sin(long)*r;
    bdpt(k+1,1)=z;
    z = z - dz;
    long = long + dlong;
end
```

The following code generates the boundary points, source points, test points, and normal vectors for the star-shaped domain in Example 7.11. For other domains, r in Line 2 can be changed accordingly.

```
function [x,y,xs0,ys0,xt,yt,xn,yn]= pts(M,N)
r =@(x) (1+(cos(5/2*x).^2));  %star-shaped domain
t=(1:M)'/(M)*2*pi;
[x, y]=pol2cart(t,r(t));
t=(1:N)'/N*2*pi;
[xs0, ys0]=pol2cart(t,1); %points on unit circle
[X,Y]=meshgrid(min(x):0.1:max(x),min(y):0.1:max(y));
in=inpolygon(X(:),Y(:),x,y);
xt=X(in); yt=Y(in);
dx=gradient(x);   dy=gradient(y);   D=sqrt(dx.^2+dy.^2);
xn=dy./D;              yn=-dx./D; %normal vectors
end
```

The following code generates the boundary points and interior test points for the 3D peanut domain in Example 7.12.

```
function [bdpt, int]=peanut_nodes(nB,ni)
rho1= @(t) sqrt(cos(2*t)+sqrt(1.1-(sin(2*t).^2)));
 % peanut
q=haltonset(2);
ang=net(q,nB+500);
theta=ang(:,1)*pi;     %0<theta<pi
phi=ang(:,2)*2*pi;     %0<phi<2*pi
xb=rho1(theta).*cos(theta);
yb=rho1(theta).*sin(theta).*cos(phi);
zb=rho1(theta).*sin(theta).*sin(phi);
id=find(xb>-0.25 & xb<0.25);
```

```matlab
ix=length(id);
N=floor(ix*2/3);
xb(id(1:N))=[];   yb(id(1:N))=[];   zb(id(1:N))=[];
bdpt=[xb,yb,zb];
bdpt=bdpt(1:nB,:);
%%%%% generate interior points
m=size(xb,1);
[~,D]=knnsearch([xb,yb,zb], [xb,yb,zb],'k',2);
d_min=sum(D(:,2))/m;
lambda=0.063/d_min;
% The MFS
rhs=ones(m,1);
DM=pdist2([xb,yb,zb], [xb,yb,zb]);
HELM=exp(-lambda*DM);
coe=HELM\rhs;
a1=min(xb);   a2=max(xb);   a3=min(yb);
a4=max(yb);   a5=min(zb);   a6=max(zb);
p=haltonset(3); q=net(p,ni*3);
x=q(:,1)*(a2-a1)+a1;
y=q(:,2)*(a4-a3)+a3;
z=q(:,3)*(a6-a5)+a5;
DM=pdist2([x,y,z],[xb,yb,zb]);
pf=exp(-lambda*DM)*coe;
in=pf>1;
x=x(in); y=y(in); z=z(in);
int=[x(1:ni),y(1:ni),z(1:ni)];
end
```

The following code generates the uniformly distributed points on the surface of the unit sphere in Examples 7.8, 7.12, and 7.14.

```matlab
function [xx,yy,zz]=unit_sphere(nB)
dt = pi*(3-sqrt(5));
t = 0;
dz = 2/nB;
z = 1 - dz/2;
xx=zeros(nB,1); yy=xx; zz=xx;
for k = 1 : nB
    r = sqrt(1-z.^2);
```

```
      yy(k,1)=cos(t)*r;
      zz(k,1)=sin(t)*r;
      xx(k,1)=z;
      z = z - dz;
      t = t + dt;
end
end
```

The following code generates the boundary points, interior points, ghost points of the RBF, and source points of the MFS, as shown in Fig. 8.11(b).

```
function [x,y, xt,yt, xc,yc, xs,ys]=fiveholes(ni,m,n)
%ni: # of boundary points for each interior circle
%m:  # of boundary points for the exterior circle
%n:  # of source points
%(x,y): boundary points
%(xt,yt): interior test points
%(xc,yc): interior collocation points
%(xs,ys): source points
ri=0.2; ro=1;
t1=(1:ni)'/ni*2*pi;          t2=(1:m)'/m*2*pi;
[x2,y2]=pol2cart(t1,ri);
f1(:,1)=x2;       f1(:,2)=y2;
f2(:,1)=x2+0.5; f2(:,2)=y2;
f3(:,1)=x2-0.5; f3(:,2)=y2;
f4(:,1)=x2;       f4(:,2)=y2+0.5;
f5(:,1)=x2;       f5(:,2)=y2-0.5;
[x2,y2]=pol2cart(t2,ro); %outside collocation points
f0(:,1)=x2; f0(:,2)=y2;
[xt,yt]=sunflower(400,1);
in=inpolygon(xt,yt,f1(:,1),f1(:,2));
xt(in)=[]; yt(in)=[];
in=inpolygon(xt,yt,f2(:,1),f2(:,2));
xt(in)=[]; yt(in)=[];
in=inpolygon(xt,yt,f3(:,1),f3(:,2));
xt(in)=[]; yt(in)=[];
in=inpolygon(xt,yt,f4(:,1),f4(:,2));
xt(in)=[]; yt(in)=[];
in=inpolygon(xt,yt,f5(:,1),f5(:,2));
```

```matlab
xt(in)=[]; yt(in)=[];
coll=[f0;f1;f2;f3;f4;f5];
x=coll(:,1); y=coll(:,2);
nb=length(coll(:,1));
[xc,yc]=sunflower(n, 1);
[xs,ys]=sunflower(nb, 1);
```

References

[1] Abou-Seada, M. S. and E. Nasser (1968). "Digital computer calculation of the electric potential and field of a rod gap." *Proceedings of IEEE* **56**(5): 813–820.

[2] Abousleiman, Y. and A. H. D. Cheng (1994). "Boundary element solution for steady and unsteady Stokes flow." *Computer Methods in Applied Mechanics and Engineering* **117**(1–2): 1–13.

[3] Abramowitz, M. and I. A. Stegun (1965). *Handbook of Mathematical Functions: With Formulas, Graphs, and Mathematical Tables*. Dover.

[4] Achenbach, J. D. (1973). *Wave Propagation in Elastic Solids*. North-Holland.

[5] Advanpix LLC (2022). *Advanpix*. https://www.advanpix.com.

[6] Ahmadabadi, M. N., M. Arab and F. M. M. Ghaini (2009). "The method of fundamental solutions for the inverse space-dependent heat source problem." *Engineering Analysis with Boundary Elements* **33**(10): 1231–1235.

[7] Aifantis, E. C. (1980). "On the problem of diffusion in solids." *Acta Mechanica* **37**(3): 265–296.

[8] Ainsworth, M. and J. T. Oden (1997). "A posteriori error estimation in finite element analysis." *Computer Methods in Applied Mechanics and Engineering* **142**(1): 1–88.

[9] Ala, G., G. Fasshauer, E. Francomano, S. Ganci and M. McCourt (2015). "The method of fundamental solutions in solving coupled boundary value problems for M/EEG." *SIAM Journal on Scientific Computing* **37**(4): B570–B590.

[10] Ala, G., G. E. Fasshauer, E. Francomano, S. Ganci and M. J. McCourt (2017). "An augmented MFS approach for brain activity reconstruction." *Mathematics and Computers in Simulation* **141**: 3–15.

[11] Al-Hosani, K., S. Fadhil and A. El-Zafrany (1999). "Fundamental solution and boundary element analysis of thick plates on Winkler foundation." *Computers and Structures* **70**(3): 325–336.

[12] Aliabadi, F. M. H. (2002). *The Boundary Element Method, Vol. 2: Applications in Solids and Structures*. Wiley.

[13] Almansi, E. (1899). "Sull'integrazione dell'Equazione differenziale $\Delta^2 = 0$." *Annali di Matematica Pura ed Applicata, Series III* **2**(1): 1–51.

[14] Altiero, N. J. and S. D. Gavazza (1980). "Unified boundary-integral equation method." *Journal of Elasticity* **10**(1): 1–9.

[15] Aluru, N. R. (2000). "A point collocation method based on reproducing kernel approximations." *International Journal for Numerical Methods in Engineering* **47**(6): 1083–1121.

[16] Aluru, N. R. and G. Li (2001). "Finite cloud method: a true meshless technique based on a fixed reproducing kernel approximation." *International Journal for Numerical Methods in Engineering* **50**(10): 2373–2410.

[17] Alves, C. J. S. and A. L. Silvestre (2004). "Density results using Stokeslets and a method of fundamental solutions for the Stokes equations." *Engineering Analysis with Boundary Elements* **28**(10): 1245–1252.

[18] Alves, C. J. S. and C. S. Chen (2005). "A new method of fundamental solutions applied to nonhomogeneous elliptic problems." *Advances in Computational Mathematics* **23**(1–2): 125–142.

[19] Alves, C. J. S. and S. S. Valtchev (2005). "Numerical comparison of two meshfree methods for acoustic wave scattering." *Engineering Analysis with Boundary Elements* **29**(4): 371–382.

[20] Alves, C. J. S., M. J. Colaço, V. M. A. Leitão, N. F. M. Martins, H. R. B. Orlande and N. C. Roberty (2008). "Recovering the source term in a linear diffusion problem by the method of fundamental solutions." *Inverse Problems in Science and Engineering* **16**(8): 1005–1021.

[21] Alves, C. J. S. and N. F. M. Martins (2008). "Reconstruction of inclusions or cavities in potential problems using the MFS, in C. S. Chen, A. Karageorghis, eds., *The Method of Fundamental Solutions—A Meshless Method*, Dynamic Publishers Inc., Atlanta, pp. 55–73.

[22] Alves, C. J. S. (2009). "On the choice of source points in the method of fundamental solutions." *Engineering Analysis with Boundary Elements* **33**(12): 1348–1361.

[23] Alves, C. J. S. and P. R. S. Antunes (2009). "The method of fundamental solutions applied to the calculation of eigensolutions for 2D plates." *International Journal for Numerical Methods in Engineering* **77**(2): 177–194.

[24] Alves, C. J. S. and N. F. M. Martins (2009). "The direct method of fundamental solutions and the inverse Kirsch-Kress method for the reconstruction of elastic inclusion or cavities." *Journal of Integral Equations and Applications* **21**: 153–178.

[25] Alves, C. J. S. and P. R. S. Antunes (2019). "Determination of elastic resonance frequencies and eigenmodes using the method of fundamental solutions." *Engineering Analysis with Boundary Elements* **101**: 330–342.

[26] Amano, K. (1994). "A charge simulation method for the numerical conformal mapping of interior, exterior and doubly-connected domains." *Journal of Computational and Applied Mathematics* **53**(3): 353–370.

[27] Amano, K. (1998). "A charge simulation method for numerical conformal mapping onto circular and radial slit domains." *SIAM Journal of Scientific Computing* **19**(4): 1169–1187.

[28] Amini, S. and D. T. Wilton (1986). "An investigation of boundary element methods for the exterior acoustic problem." *Computer Methods in Applied Mechanics and Engineering* **54**(1): 49–65.

[29] Anderson, T. G., M. Bonnet, L. M. Faria and C. Pérez-Arancibia (2024). "Construction of polynomial particular solutions of linear constant-coefficient partial differential equations." *Computers & Mathematics with Applications* **162**: 94–103.

[30] Ang, W. T., D. L. Clements and N. Vahdati (2003). "A dual-reciprocity boundary element method for a class of elliptic boundary value problems for non-homogeneous anisotropic media." *Engineering Analysis with Boundary Elements* **27**(1): 49–55.

[31] Antunes, P. R. S. (2011). "Numerical calculation of eigensolutions of 3D shapes using the method of fundamental solutions." *Numerical Methods for Partial Differential Equations* **27**(6): 1525–1550.

[32] Antunes, P. R. S. and É. Oudet (2017). "Numerical minimization of Dirichlet Laplacian eigenvalues of four-dimensional geometries." *SIAM Journal on Scientific Computing* **39**(3): B508–B521.

[33] Antunes, P. R. S. (2018). "Reducing the ill conditioning in the method of fundamental solutions." *Advances in Computational Mathematics* **44**(1): 351–365.

[34] Argyris, J. H., I. Fried and D. W. Scharpf (1968). "The TUBA family of plate elements for the matrix displacement method." *Aeronautical Journal* **72**: 701–709.

[35] Askour, O., A. Tri, B. Braikat, H. Zahrouni and M. Potier-Ferry (2018). "Method of fundamental solutions and high order algorithm to solve nonlinear elastic problems." *Engineering Analysis with Boundary Elements* **89**: 25–35.

[36] Atkinson, K. E. (1985). "The numerical evaluation of particular solutions for poissons-equation." *IMA Journal of Numerical Analysis* **5**(3): 319–338.

[37] Atkinson, K. E. (1989). *An Introduction to Numerical Analysis*, 2nd edn. Wiley.

[38] Atluri, S. N. and T. Zhu (1998). "A new meshless local Petrov-Galerkin (MLPG) approach in computational mechanics." *Computational Mechanics* **22**(2): 117–127.

[39] Atluri, S. N., H. G. Kim and J. Y. Cho (1999). "A critical assessment of the truly meshless local Petrov-Galerkin (MLPG), and local boundary integral equation (LBIE) methods." *Computational Mechanics* **24**(5): 348–372.

[40] Atluri, S. N. and T. L. Zhu (2000). "The meshless local Petrov-Galerkin (MLPG) approach for solving problems in elasto-statics." *Computational Mechanics* **25**(2–3): 169–179.

[41] Atluri, S. N. and S. Shen (2001). "The meshless local Petrov-Galerkin (MLPG) method: A simple and less-costly alternative to the finite element and boundary element methods." *Computer Modeling in Engineering and Sciences* **3**(1): 11–51.

[42] Atluri, S. N. and S. Shen (2002). "The meshless local Petrov-Galerkin (MLPG) method: A simple and less-costly alternative to the finite element and boundary element methods." *Computer Modeling in Engineering and Sciences* **3**(1): 11–51.

[43] Atluri, S. N., Z. D. Han and A. M. Rajendran (2004). "A new implementation of the Meshless Finite Volume Method, through the MLPG-"mixed" approach." *Computer Modeling in Engineering and Sciences* **6**(6): 491–514.

[44] Augustin, M. A. (2015). *A Method of Fundamental Solutions in Poroelasticity to Model the Stress Field in Geothermal Reservoirs*. Birkhäuser.

[45] Azis, M. I. and D. L. Clements (2001). "A boundary element method for anisotropic inhomogeneous elasticity." *International Journal of Solids and Structures* **38**(32): 5747–5763.

[46] Azis, M. I., D. L. Clements and W. S. Budhi (2003). "A boundary element method for the numerical solution of a class of elliptic boundary value problems for anisotropic inhomogeneous media." *ANZIAM Journal* **44**(E): C79–C95.

[47] Azis, M. I. and D. L. Clements (2014). "On some problems concerning deformations of functionally graded anisotropic elastic materials." *Far East Journal of Mathematical Sciences* **87**(2): 173–184.

[48] Azis, M. I. (2017). "Fundamental solutions to two types of 2D boundary value problems of anisotropic materials." *Far East Journal of Mathematical Sciences* **101**(11): 2405–2420.

[49] Azis, M. I., I. Solekhudin, M. H. Aswad and A. R. Jalil (2020). "Numerical simulation of two-dimensional modified Helmholtz problems for anisotropic functionally graded materials." *Journal of King Saud University—Science* **32**(3): 2096–2102.

[50] Azis, M. I., M. Abbaszadeh, M. Dehghan and I. Solekhudin (2021). "A boundary-only integral equation method for parabolic problems of

another class of anisotropic functionally graded materials." *Materials Today Communications* **26**: 101956.

[51] Babuška, I. and W. C. Rheinboldt (1978). "A-posteriori error estimates for the finite element method." *International Journal for Numerical Methods in Engineering* **12**(10): 1597–1615.

[52] Babuška, I. and M. R. Dorr (1981). "Error-estimates for the combined h-version and p-version of the finite-element method." *Numerische Mathematik* **37**(2): 257–277.

[53] Babuška, I., B. A. Szabo and I. N. Katz (1981). "The p-version of the finite-element method." *SIAM Journal on Numerical Analysis* **18**(3): 515–545.

[54] Babuška, I. and J. M. Melenk (1997). "The partition of unity method." *International Journal for Numerical Methods in Engineering* **40**(4): 727–758.

[55] Balakrishnan, K. and P. A. Ramachandran (1999). "A particular solution Trefftz method for non-linear Poisson problems in heat and mass transfer." *Journal of Computational Physics* **150**(1): 239–267.

[56] Balakrishnan, K. and P. A. Ramachandran (2000). "The method of fundamental solutions for linear diffusion-reaction equations." *Mathematical and Computer Modelling* **31**(2–3): 221–237.

[57] Balakrishnan, K. and P. A. Ramachandran (2001). "Osculatory interpolation in the method of fundamental solution for nonlinear Poisson problems." *Journal of Computational Physics* **172**(1): 1–18.

[58] Ball, K., N. Sivakumar and J. D. Ward (1992). "On the sensitivity of radial basis interpolation to minimal data separation distance." *Constructive Approximation* **8**(4): 401–426.

[59] Banoczi, J. M., N. C. Chiu, G. E. Cho and I. C. F. Ipsen (1998). "The lack of influence of the right-hand side on the accuracy of linear system solution." *SIAM Journal on Scientific Computing* **20**(1): 203–227.

[60] Bardenhagen, S. G., J. U. Brackbill and D. Sulsky (2000). "The material-point method for granular materials." *Computer Methods in Applied Mechanics and Engineering* **187**(3): 529–541.

[61] Barnett, A. H. and T. Betcke (2008). "Stability and convergence of the method of fundamental solutions for Helmholtz problems on analytic domains." *Journal of Computational Physics* **227**(14): 7003–7026.

[62] Barnett, D. M. and J. Lothe (1973). "Synthesis of the sextic and the integral formalism for dislocations, Greens function and surface waves in anisotropic elastic solids." *Physica norvegica* **7**: 13–19.

[63] Bassi, F. and S. Rebay (1997). "High-order accurate discontinuous finite element solution of the 2D Euler equations." *Journal of Computational Physics* **138**(2): 251–285.

[64] Bathe, K. J. (2014). *Finite Element Procedures*, 2nd edn. K.J. Bathe, Watertown, MA.

[65] Baxter, B. J. C. (1992). "The asymptotic cardinal function of the multiquadratic $\varphi(r) = (r^2 + c^2)^{1/2}$ as $c \to \infty$." *Computers and Mathematics with Applications* **24**(12): 1–6.

[66] Bear, L. R., I. J. LeGrice, G. B. Sands, N. A. Lever, D. S. Loiselle, D. J. Paterson, L. K. Cheng and B. H. Smaill (2018). "How accurate is inverse electrocardiographic mapping? A systematic in vivo evaluation." *Circulation: Arrhythmia and Electrophysiology* **11**(5): e006108.

[67] Bebendorf, M. (2000). "Approximation of boundary element matrices." *Numerische Mathematik* **86**(4): 565–589.

[68] Bebendorf, M. and S. Rjasanow (2003). "Adaptive low-rank approximation of collocation matrices." *Computing* **70**(1): 1–24.

[69] Becache, E., J. C. Nedelec and N. Nishimura (1993). "Regularization in 3-D for anisotropic elastodynamic crack and obstacle problems." *Journal of Elasticity* **31**: 25–46.

[70] Becker, R. and R. Rannacher (2001). "An optimal control approach to a posteriori error estimation in finite element methods." *Acta Numerica* **10**: 1–102.

[71] Belkin, M., J. Sun and Y. Wang (2008). "Discrete Laplace operator on meshed surfaces." *Proceedings of the Annual Symposium on Computational Geometry*: 278–287.

[72] Bellman, R. and J. Casti (1971). "Differential quadrature and long-term integration." *Journal of Mathematical Analysis and Applications* **34**(2): 235–238.

[73] Bellman, R. (1973). *Methods of Nonlinear Analysis, Vol. II.* Academic Press.

[74] Belytschko, T., Y. Y. Lu and L. Gu (1994). "Element-free Galerkin methods." *International Journal for Numerical Methods in Engineering* **37**(2): 229–256.

[75] Belytschko, T., Y. Krongauz, D. Organ, M. Fleming and P. Krysl (1996). "Meshless methods: An overview and recent developments." *Computer Methods in Applied Mechanics and Engineering* **139**(1–4): 3–47.

[76] Benito, J. J., F. Ureña and L. Gavete (2001). "Influence of several factors in the generalized finite difference method." *Applied Mathematical Modelling* **25**(12): 1039–1053.

[77] Berger, J. R. and A. Karageorghis (1999). "The method of fundamental solutions for heat conduction in layered materials." *International Journal for Numerical Methods in Engineering* **45**(11): 1681–1694.

[78] Berger, J. R. and A. Karageorghis (2001). "The method of fundamental solutions for layered elastic materials." *Engineering Analysis with Boundary Elements* **25**(10): 877–886.

[79] Berger, J. R., A. Karageorghis and P. A. Martin (2007). "Stress intensity factor computation using the method of fundamental solutions: Mixed-mode

problems." *International Journal for Numerical Methods in Engineering* **69**(3): 469–483.

[80] Berger, J. R. and A. Karageorghis (2013). "Galerkin formulations of the method of fundamental solutions." *Advances in Applied Mathematics and Mechanics* **5**(4): 423–441.

[81] Bergman, S. and M. Schiffer (1953). *Kernel Functions and Elliptic Differential Equations in Mathematical Physics*. Academic Press.

[82] Bert, C. W. and M. Malik (1996). "Differential quadrature method in computational mechanics: A review." *Applied Mechanics Reviews* **49**(1): 1–28.

[83] Beskos, D. E. (1997). "Boundary element methods in dynamic analysis: Part II (1986–1996)." *Applied Mechanics Reviews* **50**(3): 149–197.

[84] Bettess, P. (1981). "Operation counts for boundary integral and finite element methods." *International Journal for Numerical Methods in Engineering* **17**(2): 306–308.

[85] Bialecki, B. and G. Fairweather (1993). "Matrix decomposition algorithms for separable elliptic boundary value problems in two space dimensions." *Journal of Computational and Applied Mathematics* **46**(3): 369–386.

[86] Bialecki, B., G. Fairweather and A. Karageorghis (2011). "Matrix decomposition algorithms for elliptic boundary value problems: A survey." *Numerical Algorithms* **56**(2): 253–295.

[87] Bialecki, R. and A. J. Nowak (1981). "Boundary value problems in heat conduction with nonlinear material and nonlinear boundary conditions." *Applied Mathematical Modelling* **5**(6): 417–421.

[88] Bin-Mohsin, B. and D. Lesnic (2011). "The method of fundamental solutions for Helmholtz-type equations in composite materials." *Computers and Mathematics with Applications* **62**(12): 4377–4390.

[89] Biot, M. A. (1941). "General theory of three-dimensional consolidation." *Journal of Applied Physics* **12**(2): 155–164.

[90] Björck, Å. and V. Pereyra (1970). "Solution of Vandermonde systems of equations." *Mathematics of Computation* **24**: 893–903.

[91] Bloomenthal, J. (1977). *Introduction of Implicit Surfaces*. Morgan Kaufmann.

[92] Boag, A., Y. Leviatan and A. Boag (1988). "Analysis of acoustic scattering from fluid cylinders using a multifilament source model." *Journal of the Acoustical Society of America* **83**(1): 1–8.

[93] Bobet, A. and H. H. Einstein (1998). "Numerical modeling of fracture coalescence in a model rock material." *International Journal of Fracture* **92**(3): 221.

[94] Bogomolny, A. (1985). "Fundamental-solutions method for elliptic boundary value problems." *SIAM Journal on Numerical Analysis* **22**(4): 644–669.

[95] Boley, B. A. (1956). "A method for the construction of Green's functions." *Quarterly of Applied Mathematics* **14**(3): 149–157.

[96] Boley, B. A. and J. H. Weiner (1960). *Theory of Thermal Stresses.* Wiley.

[97] Bonnet, G., A. Corfdir and M. T. Nguyen (2014). "On the solution of exterior plane problems by the boundary element method: A physical point of view." *Engineering Analysis with Boundary Elements* **38**: 40–48.

[98] Bonnet, M., G. Maier and C. Polizzotto (1998). "Symmetric galerkin boundary element methods." *Applied Mechanics Reviews* **51**(11): 669–704.

[99] Boresi, A. P., K. P. Chong and J. D. Lee (2010). *Elasticity in Engineering Mechanics*, 3rd edn. Wiley.

[100] Bornemann, F., D. Laurie, S. Wagon and J. Waldvogel (2004). *The SIAM 100-Digit Challenge: A Study in High-Accuracy Numerical Computing.* SIAM.

[101] Boselli, F., D. Obrist and L. Kleiser (2013). "A meshless boundary method for Stokes flows with particles: Application to canalithiasis." *International Journal for Numerical Methods in Biomedical Engineering* **29**(11): 1176–1191.

[102] Bouhamama, O., M. Potse, L. Bear and L. Weynans (2023). "A patchwork method to improve the performance of current methods for solving the inverse problem of electrocardiography." *IEEE Transactions on Biomedical Engineering* **70**(1): 55–66.

[103] Boyd, J. P. (2000). *Chebyshev and Fourier Spectral Methods*, 2nd edn. Dover.

[104] Brackbill, J. U. and H. M. Ruppel (1986). "FLIP: A method for adaptively zoned, particle-in-cell calculations of fluid flows in two dimensions." *Journal of Computational Physics* **65**(2): 314–343.

[105] Brebbia, C. A. and R. Butterfield (1978). "Formal equivalence of direct and indirect boundary element methods." *Applied Mathematical Modelling* **2**(2): 132–134.

[106] Brebbia, C. A. and D. Nardini (1983)."Dynamic analysis in solid mechanics by an alternative boundary element procedure." *International Journal of Soil Dynamics and Earthquake Engineering* **2**(4): 228–233.

[107] Brebbia, C. A. and D. Nardini (1986). "Solution of parabolic and hyperbolic time dependent problems using boundary elements." *Computers and Mathematics with Applications* **12**(5, Part 2): 1061–1072.

[108] Brebbia, C. A. and J. Dominguez (1992). *Boundary Elements. An Introductory Course.* McGraw-Hill.

[109] Breitkopf, P., A. Rassineux and P. Villon (2002). "An introduction to moving least squares meshfree methods." *Revue Européenne des Éléments Finis* **11**(7–8): 825–867.

[110] Buhmann, M. D. (2003). *Radial Basis Functions: Theory and Implementations.* Cambridge University Press.

[111] Burges, C. J. C. (1998). "A tutorial on support vector machines for pattern recognition." *Data Mining and Knowledge Discovery* **2**(2): 121–167.

[112] Burgess, G. and E. Mahajerin (1984). "A comparison of the boundary element and superposition methods." *Computers and Structures* **19**(5): 697–705.

[113] Burgess, G. and E. Mahajerin (1985). "A numerical method for laterally loaded thin plates." *Computer Methods in Applied Mechanics and Engineering* **49**: 1–15.

[114] Burt, P. and E. Adelson (1983). "The Laplacian pyramid as a compact image code." *IEEE Transactions on Communications* **31**(4): 532–540.

[115] Burton, A. J. and G. F. Miller (1971). "The application of integral equation methods to the numerical solution of some exterior boundary-value problems." *Proceedings of the Royal Society of London. A. Mathematical and Physical Sciences* **323**(1553): 201–210.

[116] Butler, J. L. (1970). "Solution of acoustical-radiation problems by boundary collocation." *Journal of the Acoustical Society of America* **48**(1B): 325–336.

[117] Cai, W. and F. Wang (2020). "Numerical investigation of three-dimensional Hausdorff derivative anomalous diffusion model." *Fractals* **28**(02): 2050020.

[118] Canelas, A. and B. Sensale (2010). "A boundary knot method for harmonic elastic and viscoelastic problems using single-domain approach." *Engineering Analysis with Boundary Elements* **34**(10): 845–855.

[119] Canuto, C., A. Quarteroni, M. Y. Hussaini and T. A. Zang (1988). *Spectral Methods in Fluid Dynamics*. Springer.

[120] Carr, J. C., R. K. Beatson, J. B. Cherrie, T. J. Mitchell, W. R. Fright, B. C. McCallum and T. R. Evans (2001). "Reconstruction and representation of 3D objects with radial basis functions." *Proceedings of the ACM SIGGRAPH Conference on Computer Graphics*. 67–76.

[121] Carslaw, H. S. and J. C. Jaeger (1959). *Conduction of Heat in Solids*, 2nd edn. Oxford University Press.

[122] Cecil, T., J. Qian and S. Osher (2004). "Numerical methods for high dimensional Hamilton-Jacobi equations using radial basis functions." *Journal of Computational Physics* **196**(1): 327–347.

[123] Chan, K. S., P. Karasudhi and S. L. Lee (1974). "Force at a point in the interior of a layered elastic half space." *International Journal of Solids and Structures* **10**(11): 1179–1199.

[124] Chan, T. F. and D. E. Foulser (1988). "Effectively well-conditioned linear systems." *SIAM Journal on Scientific and Statistical Computing* **9**(6): 963–969.

[125]   Chang, J. R., W. Yeih and J. T. Chen (1999). "Determination of the natural frequencies and natural modes of a rod using the dual BEM in conjunction with the domain partition technique." *Computational Mechanics* **24**(1): 29–40.

[126]   Chang, J. Y., C. C. Tsai and D. L. Young (2019). "Homotopy method of fundamental solutions for solving nonlinear heat conduction problems." *Engineering Analysis with Boundary Elements* **108**: 179–191.

[127]   Chang, W., C. S. Chen and W. Li (2018). "Solving fourth order differential equations using particular solutions of Helmholtz-type equations." *Applied Mathematics Letters* **86**: 179–185.

[128]   Chang, Y. P., C. S. Kang and D. J. Chen (1973). "Use of fundamental Green's functions for solution of problems of heat-conduction in anisotropic media." *International Journal of Heat and Mass Transfer* **16**(10): 1905–1918.

[129]   Chang, Y. P. (1977). "Analytical solution for heat conduction in anisotropic media in infinite, semi-infinite, and two-plane-bounded regions." *International Journal of Heat and Mass Transfer* **20**(10): 1019–1028.

[130]   Chantasiriwan, S. (2006). "Methods of fundamental solutions for time-dependent heat conduction problems." *International Journal for Numerical Methods in Engineering* **66**(1): 147–165.

[131]   Chantasiriwan, S., B. T. Johansson and D. Lesnic (2009). "The method of fundamental solutions for free surface Stefan problems." *Engineering Analysis with Boundary Elements* **33**(4): 529–538.

[132]   Chen, B., W. Chen and X. Wei (2018). "Characterization of elastic parameters for functionally graded material by a meshfree method combined with the NMS approach." *Inverse Problems in Science and Engineering* **26**(4): 601–617.

[133]   Chen, C. S. (1995). "The method of fundamental solutions for non-linear thermal explosions." *Communications in Numerical Methods in Engineering* **11**(8): 675–681.

[134]   Chen, C. S., M. A. Golberg and Y. C. Hon (1998). "The method of fundamental solutions and quasi-Monte-Carlo method for diffusion equations." *International Journal for Numerical Methods in Engineering* **43**(8): 1421–1435.

[135]   Chen, C. S. and Y. F. Rashed (1998). "Evaluation of thin plate spline based particular solutions for Helmholtz-type operators for the DRM." *Mechanics Research Communications* **25**(2): 195–201.

[136]   Chen, C. S., Y. F. Rashed and M. A. Golberg (1998). "A mesh-free method for linear diffusion equations." *Numerical Heat Transfer Part B–Fundamentals* **33**(4): 469–486.

[137] Chen, C. S., C. A. Brebbia and H. Power (1999). "Dual reciprocity method using compactly supported radial basis functions." *Communications in Numerical Methods in Engineering* **15**(2): 137–150.

[138] Chen, C. S., M. A. Golberg and A. S. Muleshkov (1999). "The numerical evaluation of particular solutions for Poisson's equation—A revisit." In: *Boundary Element Methods XXI.* (eds.) C. A. Brebbia and H. Power, WIT Press: 313–322.

[139] Chen, C. S., M. Ganesh, M. A. Golberg and A. H. D. Cheng (2002). "Multilevel compact radial functions based computational schemes for some elliptic problems." *Computers and Mathematics with Applications* **43**(3–5): 359–378.

[140] Chen, C. S., H. A. Cho and M. A. Golberg (2006). "Some comments on the ill-conditioning of the method of fundamental solutions." *Engineering Analysis with Boundary Elements* **30**(5): 405–410.

[141] Chen, C. S., S. Lee and C. S. Huang (2007). "Derivation of particular solutions using chebyshev polynomial based functions." *International Journal of Computational Methods* **4**(1): 15–32.

[142] Chen, C. S., S. Y. Reutskiy and V. Y. Rozov (2009). "The method of the fundamental solutions and its modifications for electromagnetic field problems." *Computer Assisted Mechanics and Engineering Sciences* **16**(1): 21–33.

[143] Chen, C. S., C. M. Fan and P. H. Wen (2011). "The method of approximate particular solutions for solving elliptic problems with variable coefficients." *International Journal of Computational Methods* **8**(3): 545–559.

[144] Chen, C. S., C. M. Fan and P. H. Wen (2012). "The method of approximate particular solutions for solving certain partial differential equations." *Numerical Methods for Partial Differential Equations* **28**(2): 506–522.

[145] Chen, C. S., A. Karageorghis and Y. Li (2016). "On choosing the location of the sources in the MFS." *Numerical Algorithms* **72**(1): 107–130.

[146] Chen, C. S., L. Amuzu, K. Acheampong and H. Zhu (2021). "Improved geometric modeling using the method of fundamental solutions." *Engineering Analysis with Boundary Elements* **130**: 49–57.

[147] Chen, C.-S., A. Noorizadegan, D. L. Young and C. S. Chen (2023). "On the determination of locating the source points of the MFS using effective condition number." *Journal of Computational and Applied Mathematics* **423**: 114955.

[148] Chen, C. S., A. Karageorghis and M. Lei (2024). "Local MFS matrix decomposition algorithms for elliptic BVPs in annuli." *Numerical Mathmatics: Theory, Methods and Applications* **17**(1): 93–120.

[149] Chen, C. W., D. L. Young, C. C. Tsai and K. Murugesan (2005). "The method of fundamental solutions for inverse 2D Stokes problems." *Computational Mechanics* **37**(1): 2–14.

[150] Chen, F., B. Zheng, J. Lin and W. Chen (2020). "Numerical solution of steady-state free boundary problems using the singular boundary method." *Advances in Applied Mathematics and Mechanics* **13**(1): 163–175.

[151] Chen, I. L. (2006). "Using the method of fundamental solutions in conjunction with the degenerate kernel in cylindrical acoustic problems" *Journal of the Chinese Institute of Engineers* **29**(3): 445–457.

[152] Chen, J. T. and H. K. Hong (1999). "Review of dual boundary element methods with emphasis on hypersingular integrals and divergent series." *Applied Mechanics Reviews* **52**(1): 17–33.

[153] Chen, J. T., J. H. Lin, S. R. Kuo and Y. P. Chiu (2001). "Analytical study and numerical experiments for degenerate scale problems in boundary element method using degenerate kernels and circulants." *Engineering Analysis with Boundary Elements* **25**(9): 819–828.

[154] Chen, J. T., J. H. Lin, S. R. Kuo and S. W. Chyuan (2001). "Boundary element analysis for the Helmholtz eigenvalue problems with a multiply connected domain." *Proceedings of the Royal Society A: Mathematical, Physical and Engineering Sciences* **457**(2014): 2521–2546.

[155] Chen, J. T., S. R. Kuo and J. H. Lin (2002). "Analytical study and numerical experiments for degenerate scale problems in the boundary element method for two-dimensional elasticity." *International Journal for Numerical Methods in Engineering* **54**(12): 1669–1681.

[156] Chen, J. T., C. F. Lee, I. L. Chen and J. H. Lin (2002). "An alternative method for degenerate scale problems in boundary element methods for the two-dimensional Laplace equation." *Engineering Analysis with Boundary Elements* **26**(7): 559–569.

[157] Chen, J. T., W. C. Chen, S. R. Lin and I. L. Chen (2003). "Rigid body mode and spurious mode in the dual boundary element formulation for the Laplace problems." *Computers and Structures* **81**(13): 1395–1404.

[158] Chen, J. T., S. R. Lin and K. H. Chen (2005). "Degenerate scale problem when solving Laplace's equation by BEM and its treatment." *International Journal for Numerical Methods in Engineering* **62**(2): 233–261.

[159] Chen, J., C. Wu, K. Chen and Y. Lee (2006). "Degenerate scale for the analysis of circular thin plate using the boundary integral equation method and boundary element methods." *Computational Mechanics* **38**(1): 33–49.

[160] Chen, J. T. and W. C. Shen (2007). "Degenerate scale for multiply connected Laplace problems." *Mechanics Research Communications* **34**(1): 69–77.

[161] Chen, J. T. and A. C. Wu (2007). "Null-field approach for the multi-inclusion problem under antiplane shears." *Journal of Applied Mechanics* **74**(3): 469–487.

[162] Chen, J. T., C. S. Wu, Y. T. Lee and K. H. Chen (2007). "On the equivalence of the Trefftz method and method of fundamental solutions for Laplace and biharmonic equations." *Computers and Mathematics with Applications* **53**(6): 851–879.

[163] Chen, J. T., Y. T. Lee, S. R. Yu and S. C. Shieh (2009). "Equivalence between the Trefftz method and the method of fundamental solution for the annular Green's function using the addition theorem and image concept." *Engineering Analysis with Boundary Elements* **33**(5): 678–688.

[164] Chen, J. T., Y. T. Lee and Y. J. Lin (2010). "Analysis of multiple-sphere radiation and scattering problems by using a null-field integral equation approach." *Applied Acoustics* **71**(8): 690–700.

[165] Chen, J. T., J. W. Lee, Y. T. Lee and W. C. Lee (2014). "True and spurious eigensolutions of an elliptical membrane by using the nondimensional dynamic influence function method." *Journal of Vibration and Acoustics* **136**(2): 021018.

[166] Chen, J. T., Y. L. Chang, S. K. Kao and J. Jian (2015). "Revisit of the indirect boundary element method: Necessary and sufficient formulation." *Journal of Scientific Computing* **65**(2): 467–485.

[167] Chen, J. T., S. R. Kuo, Y. L. Huang and S. K. Kao (2020). "Revisit of logarithmic capacity of line segments and double-degeneracy of BEM/BIEM." *Engineering Analysis with Boundary Elements* **120**: 238–245.

[168] Chen, K. H., J. H. Kao, J. T. Chen, D. L. Young and M. C. Lu (2006). "Regularized meshless method for multiply-connected-domain Laplace problems." *Engineering Analysis with Boundary Elements* **30**(10): 882–896.

[169] Chen, K. H., J. H. Kao and J. T. Chen (2009). "Regularized meshless method for antiplane piezoelectricity problems with multiple inclusions." *Computers, Materials and Continua* **9**(3): 253–279.

[170] Chen, S., C. F. N. Cowan and P. M. Grant (1991). "Orthogonal least squares learning algorithm for radial basis function networks." *IEEE Transactions on Neural Networks* **2**(2): 302–309.

[171] Chen, S. L., L. Z. Chen and E. Pan (2007). "Three-dimensional time-harmonic Green's functions of saturated soil under buried loading." *Soil Dynamics and Earthquake Engineering* **27**(5): 448–462.

[172] Chen, W. (2002). "Symmetric boundary knot method." *Engineering Analysis with Boundary Elements* **26**(6): 489–494.

[173] Chen, W. and M. Tanaka (2002). "A meshless, integration-free, and boundary-only RBF technique." *Computers and Mathematics with Applications* **43**(3–5): 379–391.

[174] Chen, W. and Y. C. Hon (2003). "Numerical investigation on convergence of boundary knot method in the analysis of homogeneous Helmholtz, modified Helmholtz, and convection-diffusion problems." *Computer Methods in Applied Mechanics and Engineering* **192**(15): 1859–1875.

[175] Chen, W., Z. Fu and X. Wei (2009). "Potential problems by singular boundary method satisfying moment condition." *Computer Modeling in Engineering and Sciences* **54**(1): 65–85.

[176] Chen, W. and F. Z. Wang (2010). "A method of fundamental solutions without fictitious boundary." *Engineering Analysis with Boundary Elements* **34**(5): 530–532.

[177] Chen, W., Z. J. Fu and C. S. Chen (2014). *Recent Advances in Radial Basis Function Collocation Methods*. Springer.

[178] Chen, W., J. Y. Zhang and Z. J. Fu (2014). "Singular boundary method for modified Helmholtz equations." *Engineering Analysis with Boundary Elements* **44**: 112–119.

[179] Chen, W. Q., K. Yong Lee and H. J. Ding (2004). "General solution for transversely isotropic magneto-electro-thermo-elasticity and the potential theory method." *International Journal of Engineering Science* **42**(13): 1361–1379.

[180] Chen, Y. W., W. C. Yeih, C. S. Liu and J. R. Chang (2012). "Numerical simulation of the two-dimensional sloshing problem using a multi-scaling Trefftz method." *Engineering Analysis with Boundary Elements* **36**(1): 9–29.

[181] Chen, Z., F. Wang, S. Cheng and G. Wu (2022). "Localized MFS for three-dimensional acoustic inverse problems on complicated domains." *International Journal of Mechanical System Dynamics* **2**(1): 143–152.

[182] Chen, Z. and F. Wang (2023). "Localized method of fundamental solutions for acoustic analysis inside a car cavity with sound-absorbing material." *Advances in Applied Mathematics and Mechanics* **15**(1): 182–201.

[183] Cheng, A. H. D. (1984). "Darcy flow with variable permeability—a boundary integral solution." *Water Resources Research* **20**(7): 980–984.

[184] Cheng, A. H. D. and J. A. Liggett (1984). "Boundary integral equation method for linear porous-elasticity with applications to soil consolidation." *International Journal for Numerical Methods in Engineering* **20**(2): 255–278.

[185] Cheng, A. H. D. and J. A. Liggett (1984). "Boundary integral equation method for linear porous-elasticity with applications to fracture propagation." *International Journal for Numerical Methods in Engineering* **20**(2): 279–296.

[186] Cheng, A. H. D. (1987). Heterogeneities in flows through porous media by the boundary element method. Chap. 6 in: *Topics in Boundary Element Research, Vol. 4, Applications in Geomechanics*. (Ed.) C. A. Brebbia, Springer-Verlag: 129–144.

[187] Cheng, A. H. D. and E. Detournay (1988). "A direct boundary element method for plane-strain poroelasticity." *International Journal for Numerical and Analytical Methods in Geomechanics* **12**(5): 551–572.

[188] Cheng, A. H. D., T. Badmus and D. E. Beskos (1991). "Integral equation for dynamic poroelasticity in frequency domain with BEM solution." *Journal of Engineering Mechanics* **117**(5): 1136–1157.

[189] Cheng, A. H. D. and O. K. Morohunfola (1993). "Multilayered leaky aquifer systems: I. Pumping well solution." *Water Resources Research* **29**(8): 2787–2800.

[190] Cheng, A. H. D. and O. K. Morohunfola (1993). "Multilayered leaky aquifer systems: II. Boundary element solution." *Water Resources Research* **29**(8): 2801–2811.

[191] Cheng, A. H. D., H. Antes and N. Ortner (1994). "Fundamental solutions of products of Helmholtz and polyharmonic operator." *Engineering Analysis with Boundary Elements* **14**(2): 187–191.

[192] Cheng, A. H. D., O. Lafe and S. Grilli (1994). "Dual reciprocity BEM based on global interpolation functions." *Engineering Analysis with Boundary Elements* **13**(4): 303–311.

[193] Cheng, A. H. D. and E. Detournay (1998). "On singular integral equations and fundamental solutions of poroelasticity." *International Journal of Solids and Structures* **35**(34–35): 4521–4555.

[194] Cheng, A. H. D. (2000). *Multilayered Aquifer Systems: Fundamentals and Applications*. Marcel Dekker.

[195] Cheng, A. H. D. (2000). "Particular solutions of Laplacian, Helmholtz-type, and polyharmonic operators involving higher order radial basis functions." *Engineering Analysis with Boundary Elements* **24**(7–8): 531–538.

[196] Cheng, A. H. D., D. L. Young and C. C. Tsai (2000). "Solution of Poisson's equation by iterative DRBEM using compactly supported, positive definite radial basis function." *Engineering Analysis with Boundary Elements* **24**(7–8): 549–557.

[197] Cheng, A. H. D., C. S. Chen, M. A. Golberg and Y. F. Rashed (2001). "BEM for theomoelasticity and elasticity with body force—a revisit." *Engineering Analysis with Boundary Elements* **25**(4–5): 377–387.

[198] Cheng, A. H. D., M. A. Golberg, E. J. Kansa and G. Zammito (2003). "Exponential convergence and h-c multiquadric collocation method for partial differential equations." *Numerical Methods for Partial Differential Equations* **19**(5): 571–594.

[199] Cheng, A. H. D. and J. J. S. P. Cabral (2005). "Direct solution of ill-posed boundary value problems by radial basis function collocation method." *International Journal for Numerical Methods in Engineering* **64**(1): 45–64.

[200] Cheng, A. H. D. and D. T. Cheng (2005). "Heritage and early history of the boundary element method." *Engineering Analysis with Boundary Elements* **29**(3): 268–302.

[201]   Cheng, A. H. D. (2012). "Multiquadric and its shape parameter—A numerical investigation of error estimate, condition number, and round-off error by arbitrary precision computation." *Engineering Analysis with Boundary Elements* **36**(2): 220–239.

[202]   Cheng, A. H. D. (2016). *Poroelasticity*. Springer.

[203]   Cheng, A. H. D. and Y. Hong (2020). "An overview of the method of fundamental solutions—Solvability, uniqueness, convergence, and stability." *Engineering Analysis with Boundary Elements* **120**: 118–152.

[204]   Cheung, Y. K., W. G. Jin and O. C. Zienkiewicz (1989). "Direct solution procedure for solution of harmonic problems using complete, non-singular, Trefftz functions." *Communications in Applied Numerical Methods* **5**(3): 159–169.

[205]   Chiba, F. and T. Ushijima (2009). "Exponential decay of errors of a fundamental solution method applied to a reduced wave problem in the exterior region of a disc." *Journal of Computational and Applied Mathematics* **231**(2): 869–885.

[206]   Cho, H. A., M. A. Golberg, A. S. Muleshkov and X. Li (2004). "Trefftz methods for time dependent partial differential equations." *Computers Materials and Continua* **1**(1): 1–37.

[207]   Chow, V. T., D. R. Maidment and L. W. Mays (1988). *Applied Hydrology*. McGraw-Hill.

[208]   Christiansen, S. (1981). "Condition number of matrices derived from two classes of integral equations." *Mathematical Methods in the Applied Sciences* **3**(1): 364–392.

[209]   Christiansen, S. and P. C. Hansen (1994). "The effective condition number applied to error analysis of certain boundary collocation methods." *Journal of Computational and Applied Mathematics* **54**(1): 15–36.

[210]   Christiansen, S. and J. Saranen (1996). "The conditioning of some numerical methods for first kind boundary integral equations." *Journal of Computational and Applied Mathematics* **67**(1): 43–58.

[211]   Cisilino, A. P. and B. Sensale (2002). "Application of a simulated annealing algorithm in the optimal placement of the source points in the method of the fundamental solutions." *Computational Mechanics* **28**(2): 129–136.

[212]   Clements, D. L. and F. J. Rizzo (1978). "Method for numerical-solution of boundary-value problems governed by 2nd-order elliptic systems." *Journal of the Institute of Mathematics and its Applications* **22**(2): 197–202.

[213]   Clements, D. L. (1980). "A boundary integral equation method for the numerical solution of a second order elliptic equation with variable coefficients." *Journal of Australian Mathematical Society. Series B. Applied Mathematics* **22**(2): 218–228.

[214] Clements, D. L. and M. K. Haselgrove (1981). "A method for the numerical-solution of a class of boundary-value-problems governed by 2nd order elliptic-systems." *Letters of Applied Engineering Science* **19**(7): 1017–1029.

[215] Clements, D. L. and A. Larsson (1993). "A boundary-element method for the solution of a class of time-dependent problems for inhomogeneous media." *Communications in Numerical Methods in Engineering* **9**(2): 111–119.

[216] Cochelin, B., N. Damil and M. Potier-Ferry (1994). "Asymptotic–numerical methods and Pade approximants for non-linear elastic structures." *International Journal for Numerical Methods in Engineering* **37**(7): 1187–1213.

[217] Colton, D. and W. Watzlawek (1977). "Complete families of solutions to the heat equation and generalized heat equation in R^n." *Journal of Differential Equations* **25**(1): 96–107.

[218] Copley, L. G. (1967). "Integral equation method for radiation from vibrating bodies." *Journal of the Acoustical Society of America* **41**(4): 807–816.

[219] Copley, L. G. (1968). "Fundamental results concerning integral representations in acoustic radiation." *Journal of the Acoustical Society of America* **44**(1): 28–32.

[220] Corfdir, A. and G. Bonnet (2013). "Degenerate scale for the Laplace problem in the half-plane; Approximate logarithmic capacity for two distant boundaries." *Engineering Analysis with Boundary Elements* **37**(5): 836–841.

[221] Cortes, C. and V. Vapnik (1995). "Support-vector networks." *Machine Learning* **20**(3): 273–297.

[222] Cortez, R. (2001). "The method of regularized stokeslets." *SIAM Journal on Scientific Computing* **23**(4): 1204–1225.

[223] Cortez, R., L. Fauci and A. Medovikov (2005). "The method of regularized Stokeslets in three dimensions: Analysis, validation, and application to helical swimming." *Physics of Fluids* **17**(3): 031504-1-14.

[224] Crain, I. K. (1970). "Computer interpolation and contouring of two-dimensional data: A review." *Geoexploration* **8**(2): 71–86.

[225] Cressie, N. (2015). *Statistics for Spatial Data*, 2nd edn. Wiley.

[226] Cristescu, I. A. (2017). "Numerical resolution of coupled two-dimensional Burgers' equation." *Romanian Journal of Physics* **62**(1–2): Article 103.

[227] Cristescu, I. A. (2018). "Finite element approximation for the coupled two-dimensional Burgers' equation." *Romanian Journal of Physics* **63**(1–2): Article 105.

[228] Crouch, S. L. (1976). "Solution of plane elasticity problems by the displacement discontinuity method. I. Infinite body solution." *International Journal for Numerical Methods in Engineering* **10**(2): 301–343.

[229] Crouch, S. L. and A. M. Starfield (1983). *Boundary Element Methods in Solid Mechanics*. Allen and Unwin.

[230] Cruse, T. A. (1974). "An improved boundary-integral equation method for three dimensional elastic stress analysis." *Computers and Structures* **4**: 741–754.

[231] Cruse, T. A. (2001). "Boundary Integral equations—A personal view." *Engineering Analysis with Boundary Elements* **25**(9): 709–712.

[232] Damil, N. and M. Potier-Ferry (1990). "A New method to compute perturbed bifurcations: Application to the buckling of imperfect elastic structures." *International Journal of Engineering Science* **28**(9): 943–957.

[233] Dangal, T., C. S. Chen and J. Lin (2017). "Polynomial particular solutions for solving elliptic partial differential equations." *Computers and Mathematics with Applications* **73**(1): 60–70.

[234] Daros, C. H. (2008). "A fundamental solution for SH-waves in a class of inhomogeneous anisotropic media." *International Journal of Engineering Science* **46**(8): 809–817.

[235] Dasgupta, S. and D. Sengupta (1990). "A higher-order triangular plate bending element revisited." *International Journal for Numerical Methods in Engineering* **30**(3): 419–430.

[236] Davey, B., J. Hill and S. Mayboroda (2018). "Fundamental matrices and Green matrices for non-homogeneous elliptic systems." *Publicacions Matemàtiques* **62**: 537–614.

[237] Davey, K. R. (1994). "Degaussing with BEM and MFS." *IEEE Transactions on Magnetics* **30**(5): 3451–3454.

[238] Davis, P. J. (1994). *Circulant Matrices*, 2nd edn. AMS Chelsea Publishing.

[239] De Boor, C. and H. O. Kreiss (1986). "On the condition of the linear systems associated with discretized BVPs of ODEs." *SIAM Journal on Numerical Analysis* **23**(5): 936–939.

[240] Deckers, E., O. Atak, L. Coox, R. D'Amico, H. Devriendt, S. Jonckheere, K. Koo, B. Pluymers, D. Vandepitte and W. Desmet (2014). "The wave based method: An overview of 15 years of research." *Wave Motion* **51**(4): 550–565.

[241] Dederichs, P. H. and G. Leibfried (1969). "Elastic green's function for anisotropic cubic crystals." *Physical Review* **188**(3): 1175–1183.

[242] Dehghan, M. and R. Salehi (2011). "The solitary wave solution of the two-dimensional regularized long-wave equation in fluids and plasmas." *Computer Physics Communications* **182**(12): 2540–2549.

[243] Dehghan, M. and R. Salehi (2012). "A method based on meshless approach for the numerical solution of the two-space dimensional hyperbolic telegraph equation." *Mathematical Methods in the Applied Sciences* **35**(10): 1220–1233.

[244] Dehghan, M. and R. Salehi (2012). "A meshless based numerical technique for traveling solitary wave solution of Boussinesq equation." *Applied Mathematical Modelling* **36**(5): 1939–1956.

[245] Dehghan, M. and M. Abbaszadeh (2018). "An upwind local radial basis functions-differential quadrature (RBF-DQ) method with proper orthogonal decomposition (POD) approach for solving compressible Euler equation." *Engineering Analysis with Boundary Elements* **92**: 244–256.

[246] de Medeiros, G. C., P. W. Partridge and J. O. Brandão (2004). "The method of fundamental solutions with dual reciprocity for some problems in elasticity." *Engineering Analysis with Boundary Elements* **28**(5): 453–461.

[247] Demkowicz, L., J. T. Oden, W. Rachowicz and O. Hardy (1989). "Toward a universal h-p adaptive finite element strategy, part 1. Constrained approximation and data structure." *Computer Methods in Applied Mechanics and Engineering* **77**(1): 79–112.

[248] Demmel, J. W. (1997). *Applied Numerical Linear Algebra*. SIAM.

[249] Desmet, W., B. van Hal, P. Sas and D. Vandepitte (2002). "A computationally efficient prediction technique for the steady-state dynamic analysis of coupled vibro-acoustic systems." *Advances in Engineering Software* **33**(7): 527–540.

[250] Detournay, E. and A. H. D. Cheng (1987). "Poroelastic solution of a plane-strain point displacement discontinuity." *Journal of Applied Mechanics* **54**(4): 783–787.

[251] Detournay, E. and A. H. D. Cheng (1993). "Fundamentals of poroelasticity." In: *Comprehensive Rock Engineering: Principles, Practice And Projects, Vol. II, Analysis and Design Method*. (ed.) C. Fairhurst. Pergamon Press: 113–171.

[252] Deutsch, C. V. and A. G. Journel (1998). *GSLIB: Geostatistical Software and User's Guide*, 2nd edn. Oxford University Press.

[253] de Vaucorbeil, A., V. P. Nguyen, S. Sinaie and J. Y. Wu (2020). "Material point method after 25 years: Theory, implementation, and applications." Chap. 2 in *Advances in Applied Mechanics*, S. P. A. Bordas and D. S. Balint, eds. Elsevier. **53**: 185–398.

[254] Dijkstra, W. and R. M. M. Mattheij (2007). "A relation between the logarithmic capacity and the condition number of the BEM-matrices." *Communications in Numerical Methods in Engineering* **23**(7): 665–680.

[255] Ding, B. Y., A. H. D. Cheng and Z. L. Chen (2013). "Fundamental solutions of poroelastodynamics in frequency domain based on wave decomposition." *Journal of Applied Mechanics* **80**(6): 061021.

[256] Ding, H. J., G. Q. Wang and W. Q. Chen (1998). "A boundary integral formulation and 2D fundamental solutions for piezoelectric media." *Computer Methods in Applied Mechanics and Engineering* **158**(1): 65–80.

[257] Ding, H. and J. Liang (1999). "The fundamental solutions for transversely isotropic piezoelectricity and boundary element method." *Computers and Structures* **71**(4): 447–455.

[258] Ding, H., A. Jiang, P. Hou and W. Chen (2005). "Green's functions for two-phase transversely isotropic magneto-electro-elastic media." *Engineering Analysis with Boundary Elements* **29**(6): 551–561.

[259] Ding, J., H. Y. Tian and C. S. Chen (2009). "The recursive formulation of particular solutions for some elliptic pdes with polynomial source functions." *Communications in Computational Physics* **5**(5): 942–958.

[260] Dirac, P. A. M. (1930). *The Principles of Quantum Mechanics*. Oxford University Press.

[261] Divo, E. and A. J. Kassab (2007). "An efficient localized radial basis function meshless method for fluid flow and conjugate heat transfer." *Journal of Heat Transfer* **129**(2): 124–136.

[262] Doicu, A. and M. I. Mishchenko (2020). "An overview of the null-field method. I: Formulation and basic results." *Physics Open* **5**: 100020.

[263] Doicu, A. and M. I. Mishchenko (2020). "An overview of the null-field method. II: Convergence and numerical stability." *Physics Open* **3**: 100019.

[264] Dong, C. F., F. Y. Sun and B. Q. Meng (2007). "A method of fundamental solutions for inverse heat conduction problems in an anisotropic medium." *Engineering Analysis with Boundary Elements* **31**(1): 75–82.

[265] Dong, L. and S. N. Atluri (2012). "A simple multi-source-point Trefftz method for solving direct/inverse SHM problems of plane elasticity in arbitrary multiply-connected domains." *Computer Modeling in Engineering and Sciences* **85**(1): 1–43.

[266] Dou, F., Z. C. Li, C. S. Chen and Z. Tian (2018). "Analysis on the method of fundamental solutions for biharmonic equations." *Applied Mathematics and Computation* **339**: 346–366.

[267] Dravinski, M. and T. K. Mossessian (1987). "Scattering of plane harmonic P, SV, and Rayleigh waves by dipping layers of arbitrary shape." *Bulletin of the Seismological Society of America* **77**(1): 212–235.

[268] Dravinski, M. and T. Zheng (2000). "Numerical evaluation of three-dimensional time-harmonic Green's functions for a nonisotropic full-space." *Wave Motion* **32**(2): 141–151.

[269] Dravinski, M. and Y. Niu (2002). "Three-dimensional time-harmonic Green's functions for a triclinic full-space using a symbolic computation system." *International Journal for Numerical Methods in Engineering* **53**(2): 445–472.

[270] Driscoll, T. A. and B. Fornberg (2002). "Interpolation in the limit of increasingly flat radial basis functions." *Computers and Mathematics with Applications* **43**(3–5): 413–422.

[271] Drombosky, T. W., A. L. Meyer and L. Ling (2009). "Applicability of the method of fundamental solutions." *Engineering Analysis with Boundary Elements* **33**(5): 637–643.

[272] Duarte, C. A. and J. T. Oden (1996). "*H-p* clouds—an *h-p* meshless method." *Numerical Methods for Partial Differential Equations* **12**(6): 673–705.

[273] Duarte, C. A. and J. T. Oden (1996). "An *h-p* adaptive method using clouds." *Computer Methods in Applied Mechanics and Engineering* **139**(1): 237–262.

[274] Duarte, C. A., I. Babuška and J. T. Oden (2000). "Generalized finite element methods for three-dimensional structural mechanics problems." *Computers and Structures* **77**(2): 215–232.

[275] Duffin, R. J. (1961). "The maximum principle and biharmonic functions." *Journal of Mathematical Analysis and Applications* **3**(3): 399–405.

[276] Duffy, D. G. (2001). *Green's Functions with Applications*, 2nd edn. Chapman & Hall; 2015, CRC Press.

[277] Dunn, M. L. and H. A. Wienecke (1996). "Green's functions for transversely isotropic piezoelectric solids." *International Journal of Solids and Structures* **33**(30): 4571–4581.

[278] Ehrenpreis, L. (1954). "Solution of some problems of division: Part I. Division by a polynomial of derivation." *American Journal of Mathematics* **76**(4): 883–903.

[279] Ehrenpreis, L. (1955). "Solution of some problems of division: Part II. Division by a punctual distribution." *American Journal of Mathematics* **77**(2): 286–292.

[280] Elliott, H. A. (1948). "Three-dimensional stress distributions in hexagonal aeolotropic crystals." *Mathematical Proceedings of Cambridge Philosophical Society* **44**(4): 522–533.

[281] England, R., F. J. Sabina and I. Herrera (1980). "Scattering of SH waves by surface cavities of arbitrary shape using boundary methods." *Physics of the Earth and Planetary Interiors* **21**(2): 148–157.

[282] Erickson, L. L. (1990). *Panel Methods—An Introduction*. NASA Technical Paper 2995.

[283] Eymard, R., T. Gallouët and R. Herbin (2000). "Finite volume methods." *Handbook of Numerical Analysis, Vol. 7*. (eds.) P. G. Ciarlet and J. L. Lions: 713–1018.

[284] Fairweather, G. and R. L. Johnston (1982). "The method of fundamental solutions for problems in potential theory." In: *Treatment of Integral Equations by Numerical Methods*. (eds.) C. T. H. Baker and G. F. Miller, Academic Press: 349–359.

[285] Fairweather, G. and A. Karageorghis (1998). "The method of fundamental solutions for elliptic boundary value problems." *Advances in Computational Mathematics* **9**(1–2): 69–95.

[286] Fairweather, G., A. Karageorghis and P. A. Martin (2003). "The method of fundamental solutions for scattering and radiation problems." *Engineering Analysis with Boundary Elements* **27**(7): 759–769.

[287] Fairweather, G., A. Karageorghis and Y. S. Smyrlis (2005). "A matrix decomposition MFS algorithm for axisymmetric biharmonic problems." *Advances in Computational Mathematics* **23**(1): 55–71.

[288] Fallah, N. and M. Delzendeh (2018). "Free vibration analysis of laminated composite plates using meshless finite volume method." *Engineering Analysis with Boundary Elements* **88**: 132–144.

[289] Fan, C. M., D. L. Young and C. L. Chiu (2009). "Method of fundamental solutions with external source for the eigenfrequencies of waveguides." *Journal of Marine Science and Technology* **17**(3): 164–172.

[290] Fan, C. M., H. F. Chan, C. L. Kuo and W. Yeih (2012). "Numerical solutions of boundary detection problems using modified collocation Trefftz method and exponentially convergent scalar homotopy algorithm." *Engineering Analysis with Boundary Elements* **36**(1): 2–8.

[291] Fan, C. M., Y. K. Huang, C. S. Chen and S. R. Kuo (2019). "Localized method of fundamental solutions for solving two-dimensional Laplace and biharmonic equations." *Engineering Analysis with Boundary Elements* **101**: 188–197.

[292] Fan, Z., Y. Liu, A. Hong, F. Xu and F. Wang (2022). "The localized method of fundamental solution for two dimensional Signorini problems." *Computer Modeling in Engineering and Sciences* **132**(1): 341–355.

[293] Farin, G. (2002). *Curves and Surfaces for CAGD: A Practical Guide*, 5th edn. Academic Press.

[294] Farin, G., J. Hoschek and M. S. Kim, Eds. (2002). *Handbook of Computer Aided Geometric Design*. North Holland.

[295] Farrell, W. E. (1972). "Deformation of the Earth by surface loads." *Reviews of Geophysics* **10**(3): 761–797.

[296] Fasshauer, G. E. (1998). "On smoothing for multilevel approximation with radial basis functions." In: *Approximation Theory IX*. (eds.) C. K. Chui and L. L. Schumaker, Vanderbilt University Press: 1–8.

[297] Fasshauer, G. E. (1999). "Solving differential equations with radial basis functions: multilevel methods and smoothing." *Advances in Computational Mathematics* **11**(2–3): 139–159.

[298] Fasshauer, G. E. and J. G. Zhang (2007). "On choosing "optimal" shape parameters for RBF approximation." *Numerical Algorithms* **45**(1): 345–368.

[299] Favata, A. (2012). "On the Kelvin problem." *Journal of Elasticity* **109**(2): 189–204.

[300] Ferreira, A. J. M. (2003). "A formulation of the multiquadric radial basis function method for the analysis of laminated composite plates." *Composite Structures* **59**(3): 385–392.

[301] Fichera, G. (1961). "Linear elliptic equations of higher order in two independent variables and singular integral equations, with applications to anisotropic inhomogeneous elasticity." In: *Proc. Symp. Partial Differential Equations and Continuum Mechanics.* (ed.) R. E. Langer. University of Wisconsin Press: 55–80.

[302] Firoozjaee, A. R. and M. H. Afshar (2009). "Discrete least squares meshless method with sampling points for the solution of elliptic partial differential equations." *Engineering Analysis with Boundary Elements* **33**(1): 83–92.

[303] Fleming, M., Y. A. Chu, B. Moran and T. Belytschko (1997). "Enriched element-free galerkin methods for crack tip fields." *International Journal for Numerical Methods in Engineering* **40**(8): 1483–1504.

[304] Fontes, E. F., J. A. F. Santiago and J. C. F. Telles (2015). "An iterative coupling between meshless methods to solve embedded crack problems." *Engineering Analysis with Boundary Elements* **55**: 52–57.

[305] Fornberg, B., G. Wright and E. Larsson (2004). "Some observations regarding interpolants in the limit of flat radial basis functions." *Computers and Mathematics with Applications* **47**(1): 37–55.

[306] Fornberg, B. and C. Piret (2007). "A stable algorithm for flat radial basis functions on a sphere." *SIAM Journal on Scientific Computing* **30**(1): 60–80.

[307] Fornberg, B., E. Larsson and N. Flyer (2011). "Stable computations with gaussian radial basis functions." *SIAM Journal on Scientific Computing* **33**(2): 869–892.

[308] Fornberg, B., E. Lehto and C. Powell (2013). "Stable calculation of Gaussian-based RBF-FD stencils." *Computers and Mathematics with Applications* **65**(4): 627–637.

[309] Fornberg, B. and N. Flyer (2015). *A Primer on Radial Basis Functions with Applications to the Geosciences.* SIAM.

[310] Fosdick, R. L. (1965). "On the vector potential and the representation of a polyharmonic function in n-dimensions." *Journal of Mathematics and Mechanics* **14**(4): 573–587.

[311] Fosdick, R. L. (1970). "On the complete representation of biharmonic functions." *SIAM Journal on Applied Mathematics* **19**(1): 243–250.

[312] Fox, L. and I. B. Parker (1968). *Chebyshev Polynomials in Numerical Analysis*, Oxford University Press.

[313] Franke, R. (1979). "A critical comparison of some methods for interpolation of scattered data." Technical Report NPS 53-79-003, Naval Postgraduate School, Monterey, Calif.

[314] Franke, R. (1982). "Scattered data interpolation–Tests of some methods." *Mathematics of Computation* **38**(157): 181–200.

[315] Fu, Z. J., W. Chen and Q. H. Qin (2012). "Three boundary meshless methods for heat conduction analysis in nonlinear FGMs with Kirchhoff and Laplace transformation." *Advances in Applied Mathematics and Mechanics* **4**(5): 519–542.

[316] Fu, Z. J., Q. Xi, W. Chen and A. H. D. Cheng (2018). "A boundary-type meshless solver for transient heat conduction analysis of slender functionally graded materials with exponential variations." *Computers and Mathematics with Applications* **76**(4): 760–773.

[317] Fu, Z. J., Q. Xi, Y. Gu, J. Li, W. Qu, L. Sun, X. Wei, F. Wang, J. Lin, W. Li, W. Xu and C. Zhang (2023). "Singular boundary method: A review and computer implementation aspects." *Engineering Analysis with Boundary Elements* **147**: 231–266.

[318] Gander, M. J. and G. Wanner (2012). "From Euler, Ritz, and Galerkin to modern computing." *SIAM Review* **54**(4): 627–666.

[319] Garnir, H. G. (1951). "Sur les distributions résolvantes des opérateurs de la physique mathématique, I & II (On resolving distributions of operators of mathematical physics, I & II)." *Bulletin de la Société Royale des Sciences de Liége.* **20**: 174–194, 271–296.

[320] Garnir, H. G. (1958). *Les Problémes aux Limites de la Physique Mathématique. Introduction à leur étude"* (Problems at the Limits of Mathematical Physics. Introduction to Their Study). Birkhäuser.

[321] Garvin, W. W. and R. Stoneley (1956). "Exact transient solution of the buried line source problem." *Proceedings of the Royal Society of London. Ser. A.* **234**(1199): 528–541.

[322] Gáspár, C. (2015). "A regularized multi-level technique for solving potential problems by the method of fundamental solutions." *Engineering Analysis with Boundary Elements* **57**: 66–71.

[323] Gáspár, C. (2019). "A multi-level technique for the Method of Fundamental Solutions without regularization and desingularization." *Engineering Analysis with Boundary Elements* **103**: 145–159.

[324] Gasser, T. C. and G. A. Holzapfel (2005). "Modeling 3D crack propagation in unreinforced concrete using PUFEM." *Computer Methods in Applied Mechanics and Engineering* **194**(25): 2859–2896.

[325] Geertsma, J. (1957). "A remark on the analogy between thermoelasticity and the elasticity of saturated porous media." *Journal of the Mechanics and Physics of Solids* **6**(1): 13–16.

[326] Genechten, B. V., B. Bergen, D. Vandepitte and W. Desmet (2010). "A Trefftz-based numerical modelling framework for Helmholtz problems with complex multiple-scatterer configurations." *Journal of Computational Physics* **229**(18): 6623–6643.

[327] Georgia Institute of Technology (2022). *Large geometric models archive.* https://www.cc.gatech.edu/projects/large_models/.

[328] Ghassemi, A. and X. Zhou (2011). "A three-dimensional thermoporoelastic model for fracture response to injection/extraction in enhanced geothermal systems." *Geothermics* **40**(1): 39–49.

[329] Ghimire, B. K., X. Li, C. S. Chen and A. R. Lamichhane (2020). "Hybrid Chebyshev polynomial scheme for solving elliptic partial differential equations." *Journal of Computational and Applied Mathematics* **364**: 112324.

[330] Gilbarg, D. and N. S. Trudinger (1983). *Elliptic Partial Differential Equations of Second Order*, 3rd edn. Springer.

[331] Gingold, R. A. and J. J. Monaghan (1977). "Smoothed particle hydrodynamics: theory and application to non-spherical stars." *Monthly Notices of the Royal Astronomical Society* **181**(3): 375–389.

[332] Godino, L., A. P. Mendes, A. Tadeu, A. Cadena-Isaza, F. J. Sánchez-Sesma, C. Smerzini, R. Madec and D. Komatitsch (2009). "Numerical simulation of ground rotations along 2D topographical profiles under the incidence of elastic plane waves." *Bulletin of the Seismological Society of America* **99**(2 B): 1147–1161.

[333] Godinho, L., D. Soares and P. G. Santos (2016). "Efficient analysis of sound propagation in sonic crystals using an ACA-MFS approach." *Engineering Analysis with Boundary Elements* **69**: 72–85.

[334] Golberg, M. A. (1995). "The method of fundamental solutions for Poisson's equation." *Engineering Analysis with Boundary Elements* **16**(3): 205–213.

[335] Golberg, M. A., C. S. Chen and S. R. Karur (1996). "Improved multiquadric approximation for partial differential equations." *Engineering Analysis with Boundary Elements* **18**(1): 9–17.

[336] Golberg, M. A., C. S. Chen, H. Bowman and H. Power (1998). "Some comments on the use of radial basis functions in the dual reciprocity method." *Computational Mechanics* **21**(2): 141–148.

[337] Golberg, M. A. and C. S. Chen (1999). "The method of fundamental solutions for potential, Helmholtz and diffusion problems." In: *Boundary Integral Methods: Numerical and Mathematical Aspects.* (ed.) M. A. Golberg, WIT Press: 103–176.

[338] Golberg, M. A., C. S. Chen and H. Bowman (1999). "Some recent results and proposals for the use of radial basis functions in the BEM." *Engineering Analysis with Boundary Elements* **23**(4): 285–296.

[339] Golberg, M. A., C. S. Chen and Y. F. Rashed (1999). "The annihilator method for computing particular solutions to partial differential equations." *Engineering Analysis with Boundary Elements* **23**(3): 275–279.

[340] Golberg, M. A., A. S. Muleshkov, C. S. Chen and A. H. D. Cheng (2003). "Polynomial particular solutions for certain partial differential operators." *Numerical Methods for Partial Differential Equations* **19**(1): 112–133.

[341] Golub, G. H. and C. F. van Loan (2013). *Matrix Computations*, 4th edn. The Johns Hopkins University Press.

[342] Gordon, W. J. and J. A. Wixom (1978). "Shepard's method of 'metric interpolation' to bivariate and multivariate interpolation." *Mathematics of Computation* **32**: 253–264.

[343] Gorzelańczyk, P. (2011). "Method of fundamental solution and genetic algorithms for torsion of bars with multiply connected cross sections." *Journal of Theoretical and Applied Mechanics* **49**(4): 1059–1078.

[344] Gourgeon, H. and I. Herrera (1981). "Boundary methods. C-complete systems for the biharmonic equations." In: *Boundary Element Methods*. (ed.) C. A. Brebbia, Springer-Verlag. **3**: 431–441.

[345] Goursat, E. (1898). "Sur l'equation $\Delta^2 = 0$." *Bulletin de la Société Mathématique de France* **26**: 236–237.

[346] Gray, J. (1994). "On the history of the Riemann mapping theorem." *Rendiconti del Circolo Matematico di Palermo*, Serie II Supplemento N. 34: 47–94.

[347] Green, G. (1828). *An Essay on the Application of Mathematical Analysis to the Theories of Electricity and Magnetism*, printed for the author, by T. Wheelhouse, Nottingham: 72 pp.

[348] Greengard, L. and V. Rokhlin (1987). "A fast algorithm for particle simulations." *Journal of Computational Physics* **73**(2): 325–348.

[349] Greengard, L. F. (1988). *The Rapid Evaluation of Potential Fields in Particle Systems*. MIT Press.

[350] Groetsch, C. W. (1984). *The Theory of Tikhonov Regularization for Fredholm Equations of the First Kind*. Pitman.

[351] Gu, M. H., C. M. Fan and D. L. Young (2011). "The method of fundamental solutions for the multi-dimensional wave equations." *Journal of Marine Science and Technology* **19**(6): 586–595.

[352] Gu, Y., W. Chen and C. Z. Zhang (2011). "Singular boundary method for solving plane strain elastostatic problems." *International Journal of Solids and Structures* **48**(18): 2549–2556.

[353] Gu, Y., W. Chen and X. Q. He (2012). "Singular boundary method for steady-state heat conduction in three dimensional general anisotropic media." *International Journal of Heat and Mass Transfer* **55**(17): 4837–4848.

[354] Gu, Y. and W. Chen (2013). "Infinite domain potential problems by a new formulation of singular boundary method." *Applied Mathematical Modelling* **37**(4): 1638–1651.

[355] Gu, Y., W. Chen, Z. J. Fu and B. Zhang (2014). "The singular boundary method: Mathematical background and application in orthotropic elastic problems." *Engineering Analysis with Boundary Elements* **44**: 152–160.

[356] Gu, Y., C. M. Fan, W. Qu and F. Wang (2019). "Localized method of fundamental solutions for large-scale modelling of three-dimensional anisotropic heat conduction problems—Theory and MATLAB code." *Computers and Structures* **220**: 144–155.

[357] Gu, Y., C. M. Fan, W. Qu, F. Wang and C. Zhang (2019). "Localized method of fundamental solutions for three-dimensional inhomogeneous elliptic problems: Theory and MATLAB code." *Computational Mechanics* **64**(6): 1567–1588.

[358] Gu, Y., C. M. Fan and R. P. Xu (2019). "Localized method of fundamental solutions for large-scale modeling of two-dimensional elasticity problems." *Applied Mathematics Letters* **93**: 8–14.

[359] Gu, Y., C. M. Fan and Z. Fu (2021). "Localized method of fundamental solutions for three-dimensional elasticity problems: Theory." *Advances in Applied Mathematics and Mechanics* **13**(6): 1520–1534.

[360] Gu, Y., M. V. Golub and C. M. Fan (2021). "Analysis of in-plane crack problems using the localized method of fundamental solutions." *Engineering Fracture Mechanics* **256**: 107994.

[361] Gu, Y., J. Lin and C. M. Fan (2023). "Electroelastic analysis of two-dimensional piezoelectric structures by the localized method of fundamental solutions." *Advances in Applied Mathematics and Mechanics* **15**(4): 880–900.

[362] Gu, Y. T. and G. R. Liu (2002). "A boundary point interpolation method for stress analysis of solids." *Computational Mechanics* **28**(1): 47–54.

[363] Gutmann, H. M. (2001). "A radial basis function method for global optimization." *Journal of Global Optimization* **19**(3): 201–227.

[364] Hansen, P. C. (1994). "Regularization tools: A Matlab package for analysis and solution of discrete ill-posed problems." *Numerical Algorithms* **6**(1): 1–35.

[365] Hansen, P. C. (2007). "Regularization tools version 4.0 for Matlab 7.3." *Numerical Algorithms* **46**(2): 189–194, available at www.netlib.org/num eralgo.

[366] Hansen, P. C. (2010). *Discrete Inverse Problems: Insight and Algorithms.* SIAM.

[367] Hao, F., H. Lv and X. Liu (2018). "Numerical comparison of the LOOCV-MFS and the MS-CTM for 2D equations." *Advances in Applied Mathematics and Mechanics* **10**(1): 41–61.

[368] Hardy, R. L. (1971). "Multiquadric equations of topography and other irregular surfaces." *Journal of Geophysical Research* **76**(8): 1905–1915.

[369] Hardy, R. L. (1990). "Theory and applications of the multiquadric biharmonic method: 20 years of discovery 1968–1988." *Computers and Mathematics with Applications* **19**(8–9): 163–208.

[370] Hardy, R. L. (1992). "Comments on the discovery and evolution of the multiquadric method." *Computers and Mathematics with Applications* **24**(12): R11–R12.

[371] He, W. J., H. J. Ding and H. C. Hu (1996). "Degenerate scales and boundary element analysis of two dimensional potential and elasticity problems." *Computers and Structures* **60**(1): 155–158.

[372] Head, A. K. (1979). "The Galois unsolvability of the sextic equation of anisotropic elasticity." *Journal of Elasticity* **9**(1): 9–20.

[373] Head, A. K. (1979). "The monodromic Galois groups of the sextic equation of anisotropic elasticity." *Journal of Elasticity* **9**(3): 321–324.

[374] Head, A. K. (2009). "Explicit Galoisian solutions in anisotropic elasticity." *European Journal of Applied Mathematics* **2**(3): 191–197.

[375] Helmholtz, H. (1860). "Theorie der Luftschwingungen in Röhren mit offenen Enden (Theory of air vibrations in pipes with open ends)." *Journal für die reine und angewandte Mathematik* **57**: 1–72.

[376] Herrera, I. (1980). "Boundary methods: A criterion for completeness." *Proceedings of National Academy of Sciences* **77**(8): 4395–4398.

[377] Herrera, I. and H. Gourgeon (1982). "Boundary methods, C-complete systems for Stokes problems." *Computer Methods in Applied Mechanics and Engineering* **30**(2): 225–241.

[378] Herrera, I. (1984). "Trefftz method." In: *Topics in Boundary Element Research, Vol. 1 Basic Principles and Applications*, C. A. Brebbia, ed. Springer: 225–253.

[379] Herrera, I. (1997). "Trefftz-Herrera method." *Computer Assisted Mechanics and Engineering Sciences* **4**(3–4): 369–382.

[380] Herrera, I. and M. Díaz (1999). "Indirect methods of collocation: Trefftz-Herrera collocation." *Numerical Methods for Partial Differential Equations* **15**(6): 709–738.

[381] Herrera, I. (2000). "Trefftz method: A general theory." *Numerical Methods for Partial Differential Equations* **16**(6): 561–580.

[382] Hess, J. L. and A. M. O. Smith (1964). "Calculation of nonlifting potential flow about arbitrary three-dimensional bodies." *Journal of Ship Research* **8**: 22–44.

[383] Hess, J. L. and A. M. O. Smith (1967). "Calculation of potential flow about arbitrary bodies." *Progress in Aerospace Sciences* **8**: 1–138.

[384] Hess, J. L. (1975). "Review of integral-equation techniques for solving potential-flow problems with emphasis on the surface-source method." *Computer Methods in Applied Mechanics and Engineering* **5**: 145–196.

[385] Higham, N. J. (1987). "Error analysis of the Björck-Pereyra algorithms for solving Vandermonde systems." *Numerische Mathematik* **50**(5): 613–632.

[386] Higham, N. J. (2002). *Accuracy and Stability of Numerical Algorithms*, 2nd edn. SIAM.

[387] Higham, N. J. (2017). "A multiprecision world." *SIAM News* October: 2–3.

[388] Hill, J. A. (2017). "Fundamental solutions and Green functions for nonhomogeneous second order elliptic operators with discontinuous coeffcients." Ph.D. dissertation, University of Minnesota.

[389] Himanshi, A. S. Rana and V. K. Gupta (2024) "A viewpoint on thermally-induced transport in rarefied gases through the method of fundamental solutions." *Journal of Computational and Theoretical Transport*: 1–23.

[390] Hinton, E. and J. S. Campbell (1974). "Local and global smoothing of discontinuous finite element functions using a least squares method." *International Journal for Numerical Methods in Engineering* **8**(3): 461–480.

[391] Hon, Y. C. and T. Wei (2002). "A meshless computational method for solving inverse heat conduction problems." *Boundary Elements XXIV*. (eds.) C. A. Brebbia, A. Tadeu and V. Popov, WIT Press: 10 p.

[392] Hon, Y. C. and W. Chen (2003). "Boundary knot method for 2D and 3D Helmholtz and convection-diffusion problems under complicated geometry." *International Journal for Numerical Methods in Engineering* **56**(13): 1931–1948.

[393] Hon, Y. C. and T. Wei (2003). "A meshless scheme for solving inverse problems of Laplace equation." In: *Recent Development in Theories and Numerics*. (eds.) Y. C. Hon, M. Yamamoto and J. Cheng, World Scientific: 291–300.

[394] Hon, Y. C. and T. Wei (2004). "A fundamental solution method for inverse heat conduction problem." *Engineering Analysis with Boundary Elements* **28**(5): 489–495.

[395] Hon, Y. C. and T. Wei (2005). "The method of fundamental solution for solving multidimensional inverse heat conduction problems." *Computer Modeling in Engineering and Sciences* **7**(2): 119–132.

[396] Hon, Y. C. and R. Schaback (2008). "Solvability of partial differential equations by meshless kernel methods." *Advances in Computational Mathematics* **28**(3): 283–299.

[397] Hörmander, L. (1969). *Linear Partial Differential Operators*. Springer.

[398] Hou, P. F., H. R. Chen and S. He (2009). "Three-dimensional fundamental solution for transversely isotropic electro-magneto-thermo-elastic materials." *Journal of Thermal Stresses* **32**(9): 887–904.

[399] Houbolt, J. C. (1950). "A recurrence matrix solution for the dynamic response of elastic aircraft." *Journal of the Aeronautical Sciences* **17**(9): 540–550.

[400] Hsiao, G. and R. C. Maccamy (1973). "Solution of boundary value problems by integral equations of the first kind." *SIAM Review* **15**(4): 687–705.

[401] Hsiao, G. and W. L. Wendland (2008). *Boundary Integral Equations.* Springer.

[402] Hu, Y., Y. Fang, Z. Ge, Z. Qu, Y. Zhu, A. Pradhana and C. Jiang (2018). "A moving least squares material point method with displacement discontinuity and two-way rigid body coupling." *ACM Transactions on Graphics* **37**(4).

[403] Huang, C. S., C. F. Lee and A. H. D. Cheng (2007). "Error estimate, optimal shape factor, and high precision computation of multiquadric collocation method." *Engineering Analysis with Boundary Elements* **31**(7): 614–623.

[404] Huang, C. S., H. D. Yen and A. H. D. Cheng (2010). "On the increasingly flat radial basis function and optimal shape parameter for the solution of elliptic PDEs." *Engineering Analysis with Boundary Elements* **34**(9): 802–809.

[405] Huang, H. T. and Z. C. Li (2006). "Effective condition number and super-convergence of the Trefftz method coupled with high order FEM for singularity problems." *Engineering Analysis with Boundary Elements* **30**(4): 270–283.

[406] Ingber, M. S., C. S. Chen and J. A. Tanski (2004). "A mesh free approach using radial basis functions and parallel domain decomposition for solving three-dimensional diffusion equations." *International Journal for Numerical Methods in Engineering* **60**(13): 2183–2201.

[407] Itagaki, M. and C. A. Brebbia (1993). "Generation of higher order fundamental solutions to the two-dimensional modified Helmholtz equation." *Engineering Analysis with Boundary Elements* **11**(1): 87–90.

[408] Itagaki, M. (1995). "Higher order three-dimensional fundamental solutions to the Helmholtz and the modified Helmholtz equations." *Engineering Analysis with Boundary Elements* **15**(3): 289–293.

[409] Jackson, J. D. (1999). *Classical Electrodynamics*, 3rd edn. Wiley.

[410] Janssen, H. L. and H. L. Lambert (1992). "Recursive construction of particular solutions to nonhomogeneous linear partial differential equations of elliptic type." *Journal of Computational and Applied Mathematics* **39**(2): 227–242.

[411] Jaswon, M. A. (1963). "Integral equation methods in potential theory. I." *Proceedings of the Royal Society of London, Series A: Mathematical and Physical Sciences* **275**(1360): 23–32.

[412] Jaswon, M. A. and G. T. Symm (1977). *Integral Equation Methods in Potential Theory and Elastostatics.* Academic Press.

[413] Jiang, X., W. Chen and C. S. Chen (2013). "A fast method of fundamental solutions for solving Helmholtz-type equations." *International Journal of Computational Methods* **10**(2): 1341008.

[414] Jin, B. (2004). "A meshless method for the Laplace and Biharmonic equations subjected to noisy boundary data." *Computer Modeling in Engineering and Sciences* **6**(3): 253–261.

[415] Jin, B. and Y. Zheng (2006). "A meshless method for some inverse problems associated with the Helmholtz equation." *Computer Methods in Applied Mechanics and Engineering* **195**(19): 2270–2288.

[416] Jin, B. and L. Marin (2007). "The method of fundamental solutions for inverse source problems associated with the steady-state heat conduction." *International Journal for Numerical Methods in Engineering* **69**(8): 1570–1589.

[417] Jin, B. T. and Y. Zheng (2005). "Boundary knot method for some inverse problems associated with the Helmholtz equation." *International Journal for Numerical Methods in Engineering* **62**(12): 1636–1651.

[418] Jin, B. T. and Y. Zheng (2005). "Boundary knot method for the Cauchy problem associated with the inhomogeneous Helmholtz equation." *Engineering Analysis with Boundary Elements* **29**(10): 925–935.

[419] Jin, H., A. Samuelsson and O. Tullberg (1985). "Boundary element method for the injection-moulding filling process of thin cavities." *Engineering Analysis* **2**(1): 9–15.

[420] Jin, W. G., Y. K. Cheung and O. C. Zienkiewicz (1990). "Application of the Trefftz method in plane elasticity problems." *International Journal for Numerical Methods in Engineering* **30**(6): 1147–1161.

[421] Jirousek, J. (1978). "Basis for development of large finite elements locally satisfying all field equations." *Computer Methods in Applied Mechanics and Engineering* **14**(1): 65–92.

[422] Jirousek, J. and A. Wróblewski (1996). "T-elements: State of the art and future trends." *Archives of Computational Methods in Engineering* **3**(4): 323–434.

[423] Johansson, B. T. and D. Lesnic (2008). "A method of fundamental solutions for transient heat conduction." *Engineering Analysis with Boundary Elements* **32**(9): 697–703.

[424] Johansson, B. T., D. Lesnic and T. Reeve (2011). "A method of fundamental solutions for two-dimensional heat conduction." *International Journal of Computer Mathematics* **88**(8): 1697–1713.

[425] Johansson, B. T., D. Lesnic and T. Reeve (2011). "A comparative study on applying the method of fundamental solutions to the backward heat conduction problem." *Mathematical and Computer Modelling* **54**(1–2): 403–416.

[426] Johansson, B. T., D. Lesnic and T. Reeve (2011). "A method of fundamental solutions for the one-dimensional inverse Stefan problem." *Applied Mathematical Modelling* **35**(9): 4367–4378.

[427] Johansson, B. T. (2017). "Properties of a method of fundamental solutions for the parabolic heat equation." *Applied Mathematics Letters* **65**: 83–89.

[428] Johansson, B. T. (2021). "On non-denseness for a method of fundamental solutions with source points fixed in time for parabolic equations." *Comptes Rendus. Mathématique* **359**(6): 733–738.

[429] Johnston, P. R. (2018). "Accuracy of electrocardiographic imaging using the method of fundamental solutions." *Computers in Biology and Medicine* **102**: 433–448.

[430] Johnston, R. L. and G. Fairweather (1984). "The method of fundamental solutions for problems in potential flow." *Applied Mathematical Modelling* **8**(4): 265–270.

[431] Jopek, H. and J. A. Kolodziej (2008). "Application of genetic algorithms for optimal positions of source points in the method of fundamental solutions." *Computer Assisted Mechanics and Engineering Sciences* **15**(3–4): 215–224.

[432] Kahnert, F. M. (2003). "Numerical methods in electromagnetic scattering theory." *Journal of Quantitative Spectroscopy and Radiative Transfer* **79–80**: 775–824.

[433] Kamal, M. A., Y. F. Rashed, J. T. Katsikadelis and C. S. Chen (2024). "Fundamental solutions and Cayley-Hamilton formula." *Engineering Analysis with Boundary Elements* **165**: 105780.

[434] Kang, S. W., J. M. Lee and Y. J. Kang (1999). "Vibration analysis of arbitrarily shaped membranes using non-dimensional dynamic influence function." *Journal of Sound and Vibration* **221**(1): 117–132.

[435] Kang, S. W. and J. M. Lee (2000). "Application of free vibration analysis of membranes using the non-dimensional dynamic influence function." *Journal of Sound and Vibration* **234**(3): 455–470.

[436] Kang, S. W. and J. M. Lee (2001). "Free vibration analysis of arbitrarily shaped plates with clamped edges using wave-type functions." *Journal of Sound and Vibration* **242**(1): 9–26.

[437] Kang, S. W., I. S. Kim and J. M. Lee (2008). "Free vibration analysis of arbitrarily shaped plates with smoothly varying free edges using NDIF method." *Journal of Vibration and Acoustics* **130**(4): 041010.

[438] Kang, S. W., S. N. Atluri and S. H. Kim (2012). "Application of the nondimensional dynamic influence function method for free vibration analysis of arbitrarily shaped membranes." *Journal of Vibration and Acoustics* **134**(4): 041008.

[439] Kang, Z. and Y. Wang (2011). "Structural topology optimization based on non-local Shepard interpolation of density field." *Computer Methods in Applied Mechanics and Engineering* **200**(49): 3515–3525.

[440] Kansa, E. J. (1990). "Multiquadrics—A scattered data approximation scheme with applications to computational fluid-dynamics. 1. Surface approximations and partial derivative estimates." *Computers and Mathematics with Applications* **19**(8–9): 127–145.

[441] Kansa, E. J. (1990). "Multiquadrics—A scattered data approximation scheme with applications to computational fluid-dynamics. 2. Solutions to parabolic, hyperbolic and elliptic partial differential equations." *Computers and Mathematics with Applications* **19**(8–9): 147–161.

[442] Kanwal, R. P. (1971). *Linear Integral Equations: Theory and Technique.* Academic Press.

[443] Karageorghis, A. and G. Fairweather (1987). "The method of fundamental solutions for the numerical solution of the biharmonic equation." *Journal of Computational Physics* **69**(2): 434–459.

[444] Karageorghis, A. (1992). "The method of fundamental solutions for the solution of steady-state free boundary problems." *Journal of Computational Physics* **98**: 119–128.

[445] Karageorghis, A. and G. Fairweather (1998). "The method of fundamental solutions for axisymmetric acoustic scattering and radiation problems." *Journal of the Acoustical Society of America* **104**(6): 3212–3218.

[446] Karageorghis, A. and Y. S. Smyrlis (2004). "A matrix decomposition MFS algorithm for Helmholtz problems." *Symposium Vibrations in Physical Systems XXI*, Poznań-Kierz, Polska, Poznan University of Technology, Institute of Applied Mechanics, 47–54.

[447] Karageorghis, A., A. Poullikkas and J. R. Berger (2006). "Stress intensity factor computation using the method of fundamental solutions." *Computational Mechanics* **37**(5): 445–454.

[448] Karageorghis, A., Y. S. Smyrlis and T. Tsangaris (2006). "A matrix decomposition MFS algorithm for certain linear elasticity problems." *Numerical Algorithms* **43**(2): 123.

[449] Karageorghis, A. and I. Kyza (2007). "Efficient algorithms for approximating particular solutions of elliptic equations using Chebyshev polynomials." *Communications in Computational Physics* **2**(3): 501–521.

[450] Karageorghis, A. and Y. S. Smyrlis (2007). "Matrix decomposition MFS algorithms for elasticity and thermo-elasticity problems in axisymmetric domains." *Journal of Computational and Applied Mathematics* **206**(2): 774–795.

[451]   Karageorghis, A. and D. Lesnic (2008). "Steady-state nonlinear heat conduction in composite materials using the method of fundamental solutions." *Computer Methods in Applied Mechanics and Engineering* **197**(33): 3122–3137.

[452]   Karageorghis, A. and Y. S. Smyrlis (2008). "Conformal mapping for the efficient MFS solution of Dirichlet boundary value problems." *Computing* **83**(1): 1–24.

[453]   Karageorghis, A. (2009). "Efficient MFS algorithms in regular polygonal domains." *Numerical Algorithms* **50**(2): 215–240.

[454]   Karageorghis, A. and D. Lesnic (2009). "Detection of cavities using the method of fundamental solutions." *Inverse Problems in Science and Engineering* **17**(6): 803–820.

[455]   Karageorghis, A. (2010). "Efficient Kansa-type MFS algorithm for elliptic problems." *Numerical Algorithms* **54**(2): 261–278.

[456]   Karageorghis, A. (2011). "Efficient MFS algorithms for inhomogeneous polyharmonic problems." *Journal of Scientific Computing* **46**(3): 519–541.

[457]   Karageorghis, A. and D. Lesnic (2011). "Application of the MFS to inverse obstacle scattering problems" *Engineering Analysis with Boundary Elements* **35**(4): 631–638.

[458]   Karageorghis, A., D. Lesnic and L. Marin (2011). "A survey of applications of the MFS to inverse problems." *Inverse Problems in Science and Engineering* **19**(3): 309–336.

[459]   Karageorghis, A., D. Lesnic and L. Marin (2011). "The MFS for inverse geometric problems." In: *Inverse Problems and Computational Mechanics.* (eds.) L. Marin, L. Munteanu and V. Chiroiu. Bucharest, Romania, Editura Academiei. **1**: 191–216.

[460]   Karageorghis, A., B. T. Johansson and D. Lesnic (2012). "The method of fundamental solutions for the identification of a sound-soft obstacle in inverse acoustic scattering." *Applied Numerical Mathematics* **62**(12): 1767–1780.

[461]   Karageorghis, A., D. Lesnic and L. Marin (2012). "The method of fundamental solutions for the detection of rigid inclusions and cavities in plane linear elastic bodies." *Computers and Structures* **106-107**: 176–188.

[462]   Karageorghis, A., D. Lesnic and L. Marin (2013). "A moving pseudo-boundary method of fundamental solutions for void detection." *Numerical Methods for Partial Differential Equations* **29**(3): 935–960.

[463]   Karageorghis, A., D. Lesnic and L. Marin (2013). "A moving pseudo-boundary MFS for three-dimensional void detection." *Advances in Applied Mathematics and Mechanics* **5**(4): 510–527.

[464]   Karageorghis, A., D. Lesnic and L. Marin (2014). "The method of fundamental solutions for an inverse boundary value problem in static thermo-elasticity." *Computers and Structures* **135**: 32–39.

[465] Karageorghis, A., D. Lesnic and L. Marin (2014). "A moving pseudo-boundary MFS for void detection in two-dimensional thermoelasticity." *International Journal of Mechanical Sciences* **88**: 276–288.

[466] Karageorghis, A., D. Lesnic and L. Marin (2015). "The method of fundamental solutions for solving direct and inverse Signorini problems." *Computers and Structures* **151**: 11–19.

[467] Karageorghis, A., D. Lesnic and L. Marin (2016). "The method of fundamental solutions for three-dimensional inverse geometric elasticity problems." *Computers and Structures* **166**: 51–59.

[468] Karageorghis, A., D. Lesnic and L. Marin (2018). "The method of fundamental solutions for the identification of a scatterer with impedance boundary condition in interior inverse acoustic scattering." *Engineering Analysis with Boundary Elements* **92**: 218–224.

[469] Karageorghis, A. and D. Lesnic (2019). "The method of fundamental solutions for the Oseen steady-state viscous flow past obstacles of known or unknown shapes." *Numerical Methods for Partial Differential Equations* **35**(6): 2103–2119.

[470] Karageorghis, A. (2021). "Singular matrices arising in the MFS from certain boundary and pseudo-boundary symmetries." *Engineering Analysis with Boundary Elements* **125**: 135–156.

[471] Karageorghis, A., D. Lesnic and L. Marin (2023). "An efficient moving pseudo-boundary MFS for void detection." *Engineering Analysis with Boundary Elements* **147**: 90–111.

[472] Karageorghis, A., D. Lesnic and L. Marin (2024). "Solution of inverse geometric problems using a non-iterative MFS." *Communications in Computational Physics* **35**(3): 553–578.

[473] Karageorghis, A. and C. S. Chen (2025). "Multi-level method of fundamental solutions for solving polyharmonic problems." *Journal of Computational and Applied Mathematics* **456**: 116220.

[474] Katsikadelis, J. T. and M. S. Nerantzaki (1994). "Non-linear analysis of plates by the Analog Equation Method." *Computational Mechanics* **14**(2): 154–164.

[475] Katsikadelis, J. T. and M. S. Nerantzaki (1999). "The boundary element method for nonlinear problems." *Engineering Analysis with Boundary Elements* **23**(5–6): 365–373.

[476] Katsikadelis, J. T. (2011). "The BEM for numerical solution of partial fractional differential equations." *Computers and Mathematics with Applications* **62**(3): 891–901.

[477] Katsurada, M. and H. Okamoto (1988). "A mathematical study of the charge simulation method I." *Journal of Faculty of Science. University of Tokyo. Section IA. Mathematics* **35**(3): 507–518.

[478] Katsurada, M. (1989). "A mathematical study of the charge simulation method II." *Journal of Faculty of Science. University of Tokyo. Section IA. Mathematics* **36**(1): 135–162.

[479] Katsurada, M. (1990). "Asymptotic error analysis of the charge simulation method in a Jordan region with an analytic boundary." *Journal of Faculty of Science. University of Tokyo. Section IA. Mathematics* **37**(3): 635–657.

[480] Katsurada, M. and H. Okamoto (1996). "The collocation points of the fundamental solution method for the potential problem." *Computers and Mathematics with Applications* **31**(1): 123–137.

[481] Kellogg, O. D. (1953). *Foundations of Potential Theory*. Dover.

[482] Kelly, D. W., J. P. De S. R. Gago, O. C. Zienkiewicz and I. Babuška (1983). "A posteriori error analysis and adaptive processes in the finite element method: Part I—error analysis." *International Journal for Numerical Methods in Engineering* **19**(11): 1593–1619.

[483] Khoshroo, M., M. R. Hematiyan and Y. Daneshbod (2020). "Two-dimensional elastodynamic and free vibration analysis by the method of fundamental solutions." *Engineering Analysis with Boundary Elements* **117**: 188–201.

[484] Kikuta, M., H. Togoh and M. Tanaka (1987). "Boundary element analysis of nonlinear transient heat conduction problems." *Computer Methods in Applied Mechanics and Engineering* **62**(3): 321–329.

[485] Kim, S. (2013). "An improved boundary distributed source method for two-dimensional Laplace equations." *Engineering Analysis with Boundary Elements* **37**(7): 997–1003.

[486] Kita, E. and N. Kamiya (1995). "Trefftz method: An overview." *Advances in Engineering Software* **24**(1–3): 3–12.

[487] Kita, E., N. Kamiya and Y. Ikeda (1995). "Boundary-type sensitivity analysis scheme based on indirect Trefftz formulation." *Advances in Engineering Software* **24**(1–3): 89–96.

[488] Kita, E., Y. Kataoka and N. Kamiya (1997). "Application of element-free Trefftz method to second-order design sensitivity analysis of two-dimensional elastic problem." *JSME International Journal Series A* **40**(4): 375–381.

[489] Kitagawa, T. (1988). "On the numerical stability of the method of fundamental solution applied to the Dirichlet problem." *Japan Journal of Applied Mathematics* **5**(1): 123–133.

[490] Kitagawa, T. (1991). "Asymptotic stability of the fundamental solution method." *Journal of Computational and Applied Mathematics* **38**(1): 263–269.

[491] Kline, M. (1972). *Mathematical Thought from Ancient to Modern Times*. Oxford University Press.

[492]   Kołodziej, J. A. and A. P. Zieliński (2009). *Boundary Collocation Techniques and Their Application in Engineering.* WIT Press.

[493]   Kołodziej, J. A. and P. Gorzelańczyk (2012). "Application of method of fundamental solutions for elasto-plastic torsion of prismatic rods." *Engineering Analysis with Boundary Elements* **36**(2): 81–86.

[494]   Kołodziej, J. A. and J. K. Grabski (2015). "Application of the method of fundamental solutions and the radial basis functions for viscous laminar flow in wavy channel." *Engineering Analysis with Boundary Elements* **57**: 58–65.

[495]   Kondapalli, P. S., D. J. Shippy and G. Fairweather (1992). "The method of fundamental solutions for transmission and scattering of elastic waves." *Computer Methods in Applied Mechanics and Engineering* **96**(2): 255–269.

[496]   Kondapalli, P. S., D. J. Shippy and G. Fairweather (1992). "Analysis of acoustic scattering in fluids and solids by the method of fundamental solutions." *Journal of Acoustical Society of America* **91**(4): 1844–1854.

[497]   Kresse, O., X. Weng, H. Gu and R. Wu (2013). "Numerical modeling of hydraulic fractures interaction in complex naturally fractured formations." *Rock Mechanics and Rock Engineering* **46**(3): 555–568.

[498]   Krige, D. G. (1951). "A statistical approach to some basic mine valuation problems on the Witwatersrand." *Journal of the Southern African Institute of Mining and Metallurgy* **52**(6): 119–139.

[499]   Krongauz, Y. and T. Belytschko (1997). "A Petrov-Galerkin diffuse element method (PG DEM) and its comparison to EFG." *Computational Mechanics* **19**(4): 327–333.

[500]   Ku, C. Y., C. L. Kuo, C. M. Fan, C. S. Liu and P. C. Guan (2015). "Numerical solution of three-dimensional Laplacian problems using the multiple scale Trefftz method." *Engineering Analysis with Boundary Elements* **50**: 157–168.

[501]   Kuo, C. L., W. Yeih, C. S. Liu and J. R. Chang (2015). "Solving Helmholtz equation with high wave number and ill-posed inverse problem using the multiple scales Trefftz collocation method." *Engineering Analysis with Boundary Elements* **61**: 145–152.

[502]   Kuo, L. H., J. Uvah and C. S. Chen (2022). "Residual-Error Cross-Validation method for selecting a suitable shape parameter for RBF interpolation." *Engineering Analysis with Boundary Elements* **143**: 331–339.

[503]   Kuo, S. R., J. T. Chen and S. K. Kao (2013). "Linkage between the unit logarithmic capacity in the theory of complex variables and the degenerate scale in the BEM/BIEMs." *Applied Mathematics Letters* **26**(9): 929–938.

[504]   Kuo, S. R., S. K. Kao, Y. L. Huang and J. T. Chen (2019). "Revisit of the degenerate scale for an infinite plane problem containing two circular holes using conformal mapping." *Applied Mathematics Letters* **92**: 99–107.

[505] Kupradze, V. D. (1964). "A method for the approximate solution of limiting problems in mathematical physics." *USSR Computational Mathematics and Mathematical Physics* **4**(6): 199–205.

[506] Kupradze, V. D. and M. A. Aleksidze (1964). "The method of functional equations for the approximate solution of certain boundary value problems." *USSR Computational Mathematics and Mathematical Physics* **4**(4): 86–126.

[507] Kupradze, V. D. (1965). *Potential Methods in the Theory of Elasticity.* Israel Program for Scientific Translations, Jerusalem.

[508] Kupradze, V. D. (1967). "On the approximate solution of problems in mathematical physics." *Russian Mathematical Survey* **22**(2): 58–108.

[509] Kupradze, V. D., T. G. Gezelia, M. O. Basheleishvili and T. V. Burchuladze (1979). *Three-Dimensional Problems of the Mathematical Theory of Elasticity and Thermoelasticity.* North-Holland.

[510] Kythe, P. K. (1996). *Fundamental Solutions for Differential Operators and Applications.* Birkhäuser.

[511] Ladyzhenskaya, O. A. (1963). *The Mathematical Theory of Viscous Incompressible Flow.* Gordon and Breach.

[512] Lagrange, J. L. (1773). *Mémoires de l'Académie Royale des Sciences de Paris*, VII, Savants étrangèrs.

[513] Lai, S. J., B. Z. Wang and Y. Duan (2008). "Meshless radial basis function method for transient electromagnetic computations." *IEEE Transactions on Magnetics* **44**(10): 2288–2295.

[514] Lamb, H. (1904). "On the propagation of tremors over the surface of an elastic solid." *Philosophical Transactions of the Royal Society of London. Ser. A,* **203**(359–371): 1–42.

[515] Lamichhane, A. R. and C. S. Chen (2015). "The closed-form particular solutions for Laplace and biharmonic operators using a Gaussian function." *Applied Mathematics Letters* **46**: 50–56.

[516] Lamichhane, A. R., D. L. Young and C. S. Chen (2016). "Fast method of approximate particular solutions using Chebyshev interpolation." *Engineering Analysis with Boundary Elements* **64**: 290–294.

[517] Lancaster, P. and K. Salkauskas (1981). "Surfaces generated by moving least squares methods." *Mathematics of Computation* **37**(155): 141–158.

[518] Laplace, P. S. (1782). *Histoire de l'Académie des Sciences de Paris,* **85**: 135; (1787) **89**: 252.

[519] Larsson, E. and B. Fornberg (2005). "Theoretical and computational aspects of multivariate interpolation with increasingly flat radial basis functions." *Computers and Mathematics with Applications* **49**(1): 103–130.

[520] Lax, P. D. (2007). *Linear Algebra and Its Applications,* 2nd edn. Wiley.

[521] Leblanc, A., A. Lavie and R. Ing (2011). "The method of fundamental solutions for the impulse responses reconstruction in arbitrarily shaped plates." *Acta Acustica united with Acustica* **97**: 919–925.

[522] Lee, J. K. and J. A. Kong (1983). "Dyadic green's functions for layered anisotropic medium." *Electromagnetics* **3**(2): 111–130.

[523] Lee, J. S. and L. Z. Jiang (1994). "A boundary integral formulation and 2D fundamental solution for piezoelastic media." *Mechanics Research Communications* **21**(1): 47–54.

[524] Lee, J. W., J. T. Chen and C. F. Nien (2019). "Indirect boundary element method combining extra fundamental solutions for solving exterior acoustic problems with fictitious frequencies." *Journal of the Acoustical Society of America* **145**(5): 3116–3132.

[525] Lei, B., C. M. Fan and M. Li (2018). "The method of fundamental solutions for solving non-linear Berger equation of thin elastic plate." *Engineering Analysis with Boundary Elements* **90**: 100–106.

[526] Leitão, V. M. A. (1997). "On the implementation of a multi-region Trefftz-collocation formulation for 2-D potential problems." *Engineering Analysis with Boundary Elements* **20**(1): 51–61.

[527] Lejček, L. (1969). "The Green function of the theory of elasticity in an anisotropic hexagonal medium." *Czechoslovak Journal of Physics B* **19**(6): 799–803.

[528] Lesnic, D. (2021). *Inverse Problems with Applications in Science and Engineering*. Chapman and Hall/CRC Press.

[529] LeVeque, R. J. (2002). *Finite Volume Methods for Hyperbolic Problems*. Cambridge University Press.

[530] Leviatan, Y. and A. Boag (1987). "Analysis of electromagnetic scattering from dielectric cylinders using a multifilament current model." *IEEE Transactions on Antennas and Propagation* **35**(10): 1119–1127.

[531] Lewy, H. (1957). "An example of a smooth linear partial differential equation without solution." *Annals of Mathematics* **66**(1): 155–158.

[532] Li, G. and N. R. Aluru (2002). "Boundary cloud method: a combined scattered point/boundary integral approach for boundary-only analysis." *Computer Methods in Applied Mechanics and Engineering* **191**(21–22): 2337–2370.

[533] Li, G. and N. R. Aluru (2003). "A boundary cloud method with a cloud-by-cloud polynomial basis." *Engineering Analysis with Boundary Elements* **27**(1): 57–71.

[534] Li, J., W. Chen, Z. J. Fu and L. Sun (2016). "Explicit empirical formula evaluating original intensity factors of singular boundary method for potential and Helmholtz problems." *Engineering Analysis with Boundary Elements* **73**: 161–169.

[535] Li, J., Z. J. Fu and W. Chen (2016). "Numerical investigation on the obliquely incident water wave passing through the submerged breakwater by singular boundary method." *Computers and Mathematics with Applications* **71**(1): 381–390.

[536] Li, J. and W. Chen (2018). "A modified singular boundary method for three-dimensional high frequency acoustic wave problems." *Applied Mathematical Modelling* **54**: 189–201.

[537] Li, J., Z. J. Fu, W. Chen and X. Liu (2019). "A dual-level method of fundamental solutions in conjunction with kernel-independent fast multipole method for large-scale isotropic heat conduction problems." *Advances in Applied Mathematics and Mechanics* **11**(2): 501–517.

[538] Li, M., C. S. Chen and A. Karageorghis (2013). "The MFS for the solution of harmonic boundary value problems with non-harmonic boundary conditions." *Computers and Mathematics with Applications* **66**(11): 2400–2424.

[539] Li, M., C. S. Chen, C. C. Chu and D. L. Young (2014). "Transient 3D heat conduction in functionally graded materials by the method of fundamental solutions." *Engineering Analysis with Boundary Elements* **45**: 62–67.

[540] Li, S. F. and W. K. Liu (1999). "Reproducing kernel hierarchical partition of unity, Part I—Formulation and theory." *International Journal for Numerical Methods in Engineering* **45**(3): 251–288.

[541] Li, W. (2021). "Localized method of fundamental solutions for 2D harmonic elastic wave problems." *Applied Mathematics Letters* **112**: 106759.

[542] Li, X. (2005). "On convergence of the method of fundamental solutions for solving the Dirichlet problem of Poisson's equation." *Advances in Computational Mathematics* **23**(3): 265–277.

[543] Li, X. and J. Zhu (2009). "A Galerkin boundary node method and its convergence analysis." *Journal of Computational and Applied Mathematics* **230**(1): 314–328.

[544] Li, X. and J. Zhu (2009). "The method of fundamental solutions for non-linear elliptic problems." *Engineering Analysis with Boundary Elements* **33**(3): 322–329.

[545] Li, X. and J. Zhu (2009). "A Galerkin boundary node method for biharmonic problems." *Engineering Analysis with Boundary Elements* **33**(6): 858–865.

[546] Li, Y., H. Dang, G. Xu, C. Fan and M. Zhao (2016). "Extended displacement discontinuity boundary integral equation and boundary element method for cracks in thermo-magneto-electro-elastic media." *Smart Materials and Structures* **25**(8): 085048.

[547] Li, Y. L., C. H. Liu and S. J. Franke (1990). "Three-dimensional Green's function for wave propagation in a linearly inhomogeneous medium—the exact analytic solution." *Journal of the Acoustical Society of America* **87**(6): 2285–2291.

[548] Li, Z. C., T. T. Lu, H. Y. Hu and A. H. D. Cheng (2005). "Particular solutions of Laplace's equations on polygons and new models involving mild singularities." *Engineering Analysis with Boundary Elements* **29**(1): 59–75.

[549] Li, Z. C., C. S. Chien and H. T. Huang (2007). "Effective condition number for finite difference method." *Journal of Computational and Applied Mathematics* **198**(1): 208–235.

[550] Li, Z. C., T. T. Lu, H. T. Huang and A. H. D. Cheng (2007). "Trefftz, collocation, and other boundary methods—A comparison." *Numerical Methods for Partial Differential Equations* **23**(1): 93–144.

[551] Li, Z. C. (2008). "The Trefftz method for the Helmholtz equation with degeneracy." *Applied Numerical Mathematics* **58**(2): 131–159.

[552] Li, Z. C. and H. T. Huang (2008). "Effective condition number for numerical partial differential equations." *Numerical Linear Algebra with Applications* **15**(7): 575–594.

[553] Li, Z. C. and H. T. Huang (2008). "Effective condition number for simplified hybrid Trefftz methods." *Engineering Analysis with Boundary Elements* **32**(9): 757–769.

[554] Li, Z. C. and H. T. Huang (2008). "Study on effective condition number for collocation methods." *Engineering Analysis with Boundary Elements* **32**(10): 839–848.

[555] Li, Z. C., H. T. Huang and J. Huang (2008). "Effective condition number of the Hermite finite element methods for biharmonic equations." *Applied Numerical Mathematics* **58**(9): 1291–1308.

[556] Li, Z. C., Z. Z. Lu, H. Y. Hu and A. H. D. Cheng (2008). *Trefftz and Collocation Methods*. WIT Press.

[557] Li, Z. C., H. T. Huang and Y. Wei (2011). "Ill-conditioning of the truncated singular value decomposition, Tikhonov regularization and their applications to numerical partial differential equations." *Numerical Linear Algebra with Applications* **18**(2): 205–221.

[558] Li, Z. C., M. G. Lee, J. Y. Chiang and Y. P. Liu (2011). "The Trefftz method using fundamental solutions for biharmonic equations." *Journal of Computational and Applied Mathematics* **235**(15): 4350–4367.

[559] Li, Z. C., H. T. Huang, Y. M. Wei and A. H. D. Cheng (2015). *Effective Condition Number for Numerical Partial Differential Equations*, 2nd edn. Science Press, Beijing.

[560] Liao, S. J. and A. T. Chwang (1999). "General boundary-element method for unsteady nonlinear heat transfer problems." *Numerical Heat Transfer, Part B: Fundamentals* **35**(2): 225–242.

[561] Liao, S. J. (2002). "A direct boundary element approach for unsteady nonlinear heat transfer problems." *Engineering Analysis with Boundary Elements* **26**(1): 55–59.

[562] Liao, S. J. (2004). "On the homotopy analysis method for nonlinear problems." *Applied Mathematics and Computation* **147**(2): 499–513.

[563] Liao, S. J. and Y. Tan (2007). "A general approach to obtain series solutions of nonlinear differential equations." *Studies in Applied Mathematics* **119**(4): 297–354.

[564] Liao, S. J. (2011). *Homotopy Analysis Method in Nonlinear Differential Equations*, Springer.

[565] Lie, K. H. C. and J. S. Koehler (1968). "The elastic stress field produced by a point force in a cubic crystal." *Advances in Physics* **17**(67): 421–478.

[566] Liew, K. M., Y. M. Cheng and S. Kitipornchai (2006). "Boundary element-free method (BEFM) and its application to two-dimensional elasticity problems." *International Journal for Numerical Methods in Engineering* **65**(8): 1310–1332.

[567] Lifshitz, I. M. and L. N. Rosenzweig (1947). "Construction of the Green tensor for the fundamental equation of elasticity theory in the case of unbounded elastically anisotropic medium." *Journal of Experimental and Theoretical Physics* **17**: 783–791.

[568] Lin, J., W. Chen and F. Wang (2011). "A new investigation into regularization techniques for the method of fundamental solutions." *Mathematics and Computers in Simulation* **81**(6): 1144–1152.

[569] Lin, J., C. S. Chen, C. S. Liu and J. Lu (2016). "Fast simulation of multidimensional wave problems by the sparse scheme of the method of fundamental solutions." *Computers and Mathematics with Applications* **72**(3): 555–567.

[570] Lin, J., A. R. Lamichhane, C. S. Chen and J. Lu (2018). "The adaptive algorithm for the selection of sources of the method of fundamental solutions." *Engineering Analysis with Boundary Elements* **95**: 154–159.

[571] Lin, J., C. Zhang, L. Sun and J. Lu (2018). "Simulation of seismic wave scattering by embedded cavities in an elastic half-plane using the novel singular boundary method." *Advances in Applied Mathematics and Mechanics* **10**(2): 322–342.

[572] Lin, Z. and S. J. Liao (2011). "The scaled boundary FEM for nonlinear problems." *Communications in Nonlinear Science and Numerical Simulation* **16**(1): 63–75.

[573] Ling, L. and R. Schaback (2009). "An improved subspace selection algorithm for meshless collocation methods." *International Journal for Numerical Methods in Engineering* **80**(13): 1623–1639.

[574] Liu, C. S. (2007). "An effectively modified direct Trefftz method for 2D potential problems considering the domain's characteristic length." *Engineering Analysis with Boundary Elements* **31**(12): 983–993.

[575] Liu, C. S. (2008). "A modified collocation Trefftz method for the inverse Cauchy problem of Laplace equation." *Engineering Analysis with Boundary Elements* **32**(9): 778–785.

[576] Liu, C. S. (2012). "An equilibrated method of fundamental solutions to choose the best source points for the Laplace equation." *Engineering Analysis with Boundary Elements* **36**(8): 1235–1245.

[577] Liu, C. S. (2014). "The pre/post equilibrated conditioning methods to solve Cauchy problems." *Engineering Analysis with Boundary Elements* **40**: 62–70.

[578] Liu, G. R. and Y. T. Gu (2001). "A point interpolation method for two-dimensional solids." *International Journal for Numerical Methods in Engineering* **50**(4): 937–951.

[579] Liu, G. R. and Y. T. Gu (2001). "A local radial point interpolation method (LRPIM) for free vibration analyses of 2-D solids." *Journal of Sound and Vibration* **246**(1): 29–46.

[580] Liu, G. R., L. Yan, J. G. Wang and Y. T. Gu (2002). "Point interpolation method based on local residual formulation using radial basis functions." *Structural Engineering and Mechanics* **14**(6): 713–732.

[581] Liu, G. R. and Y. T. Gu (2004). "Boundary meshfree methods based on the boundary point interpolation methods." *Engineering Analysis with Boundary Elements* **28**(5): 475–487.

[582] Liu, G. R., K. Y. Dai and T. T. Nguyen (2007). "A smoothed finite element method for mechanics problems." *Computational Mechanics* **39**(6): 859–877.

[583] Liu, G. R., T. Nguyen-Thoi and K. Y. Lam (2009). "An edge-based smoothed finite element method (ES-FEM) for static, free and forced vibration analyses of solids." *Journal of Sound and Vibration* **320**(4): 1100–1130.

[584] Liu, Q. G. and B. Šarler (2013). "Non-singular method of fundamental solutions for two-dimensional isotropic elasticity problems." *Computer Modeling in Engineering and Sciences* **91**(4): 235–266.

[585] Liu, Q. G. and B. Šarler (2014). "Non-singular method of fundamental solutions for anisotropic elasticity." *Engineering Analysis with Boundary Elements* **45**: 68–78.

[586] Liu, Q. G. and B. Šarler (2017). "A non-singular method of fundamental solutions for two-dimensional steady-state isotropic thermoelasticity problems." *Engineering Analysis with Boundary Elements* **75**: 89–102.

[587] Liu, Q. G. and B. Šarler (2018). "Non-singular method of fundamental solutions for elasticity problems in three-dimensions." *Engineering Analysis with Boundary Elements* **96**: 23–35.

[588] Liu, Q. G., C. M. Fan and B. Šarler (2021). "Localized method of fundamental solutions for two-dimensional anisotropic elasticity problems." *Engineering Analysis with Boundary Elements* **125**: 59–65.

[589] Liu, S., P. W. Li, C. M. Fan and Y. Gu (2021). "Localized method of fundamental solutions for two- and three-dimensional transient convection-diffusion-reaction equations." *Engineering Analysis with Boundary Elements* **124**: 237–244.

[590] Liu, W. K., S. Jun, S. F. Li, J. Adee and T. Belytschko (1995). "Reproducing kernel particle methods for structural dynamics." *International Journal for Numerical Methods in Engineering* **38**(10): 1655–1679.

[591] Liu, W. K., S. Jun and Y. F. Zhang (1995). "Reproducing kernel particle methods." *International Journal for Numerical Methods in Fluids* **20**(8–9): 1081–1106.

[592] Liu, Y. J. (2000). "On the simple-solution method and non-singular nature of the BIE/BEM—a review and some new results." *Engineering Analysis with Boundary Elements* **24**(10): 789–795.

[593] Liu, Y. J., N. Nishimura and Z. H. Yao (2005). "A fast multipole accelerated method of fundamental solutions for potential problems." *Engineering Analysis with Boundary Elements* **29**(11): 1016–1024.

[594] Liu, Y. J. (2010). "A new boundary meshfree method with distributed sources." *Engineering Analysis with Boundary Elements* **34**(11): 914–919.

[595] Liu, Y. X. and A. H. Barnett (2016). "Efficient numerical solution of acoustic scattering from doubly-periodic arrays of axisymmetric objects." *Journal of Computational Physics* **324**: 226–245.

[596] Liu, Z., J. Liang and C. Wu (2016). "The diffraction of Rayleigh waves by a fluid-saturated alluvial valley in a poroelastic half-space modeled by MFS." *Computers and Geosciences* **91**: 33–48.

[597] Liu, Z., J. Liang, C. Wu, R. Zhao and Y. Li (2017). "The method of fundamental solution for elastic wave scattering and dynamic stress concentration in a fluid-saturated poroelastic layered half-plane." *Engineering Analysis with Boundary Elements* **84**: 154–167.

[598] Liu, Z., Z. Wang, A. H. D. Cheng, J. Liang and C. Wang (2018). "The method of fundamental solution for 3-D wave scattering in a fluid-saturated poroelastic infinite domain." *International Journal for Numerical and Analytical Methods in Geomechanics* **42**(15): 1866–1889.

[599] Liu, Z., Z. Wang, A. H. D. Cheng and X. Zhang (2021). "The method of fundamental solutions for the elastic wave scattering in a double-porosity dual-permeability medium." *Applied Mathematical Modelling* **97**: 721–740.

[600] Loeb, L. B., J. H. Parker, E. E. Dodd and W. N. English (1950). "The choice of suitable gap forms for the study of corona breakdown and the field along the axis of a hemispherically capped cylindrical point-to-plane gap." *Review of Scientific Instruments* **21**(1): 42–47.

[601] Lorentz, H. A. (1907). "Ein allgemeiner Satz, die Bewegung einer reibenden Flüssigkeit betreffend, nebst einigen Anwendungen desselben (A general theorem relating to the movement of viscous liquid, with applications)." *Abhandlungen über theoretische Physik*. B. G. Teubner, Lipzig: 23–42.

[602] Lu, Y. Y., T. Belytschko and L. Gu (1994). "A new implementation of the element free Galerkin method." *Computer Methods in Applied Mechanics and Engineering* **113**(3–4): 397–414.

[603] Lucy, L. B. (1977). "A numerical approach to the testing of the fission hypothesis." *Astronomical Journal* **82**(12): 1013–1024.

[604] Lv, H., F. Hao, Y. Wang and C. S. Chen (2017). "The MFS versus the Trefftz method for the Laplace equation in 3D." *Engineering Analysis with Boundary Elements* **83**: 133–140.

[605] Madych, W. R. (1992). "Miscellaneous error-bounds for multiquadric and related interpolators." *Computers and Mathematics with Applications* **24**(12): 121–138.

[606] Madych, W. R. and S. A. Nelson (1992). "Bounds on multivariate polynomials and exponential error-estimates for multiquadric interpolation." *Journal of Approximation Theory* **70**(1): 94–114.

[607] Mahoney, P. and A. Povitsky (2024). "Modeling of chemical vapor infiltration for fiber-reinforced silicon carbide composites using meshless method of fundamental solutions." *Mathematical and Computational Applications* **29**(2): 27.

[608] Maier, G. and C. Polizzotto (1987). "A Galerkin approach to boundary element elastoplastic analysis." *Computer Methods in Applied Mechanics and Engineering* **60**(2): 175–194.

[609] Malgrange, B. (1956). "Existence et approximation des solutions des équations aux dérivées partielles et des équations de convolution." *Annales de l'Institut Fourier* **6**: 271–355.

[610] Malik, N. H. (1989). "A review of the charge simulation method and its applications." *IEEE Transactions on Electrical Insulation* **24**(1): 3–20.

[611] Manolis, G. D., P. S. Dineva, T. V. Rangelov and F. Wuttke (2017). *Seismic Wave Propagation in Non-Homogeneous Elastic Media by Boundary Elements*. Springer.

[612] Maplesoft (2022). *Maple*. https://www.maplesoft.com/.

[613] Marin, L. and D. Lesnic (2004). "The method of fundamental solutions for the Cauchy problem in two-dimensional linear elasticity." *International Journal of Solids and Structures* **41**(13): 3425–3438.

[614] Marin, L. (2005). "Numerical solution of the Cauchy problem for steady-state heat transfer in two-dimensional functionally graded materials." *International Journal of Solids and Structures* **42**(15): 4338–4351.

[615] Marin, L. (2005). "A meshless method for solving the Cauchy problem in three-dimensional elastostatics." *Computers and Mathematics with Applications* **50**(1–2): 73–92.

[616] Marin, L. (2005). "A meshless method for the numerical solution of the Cauchy problem associated with three-dimensional Helmholtz-type equations." *Applied Mathematics and Computation* **165**(2): 355–374.

[617] Marin, L. and D. Lesnic (2005). "The method of fundamental solutions for the Cauchy problem associated with two-dimensional Helmholtz-type equations." *Computers and Structures* **83**(4–5): 267–278.

[618] Marin, L. and D. Lesnic (2005). "The method of fundamental solutions for inverse boundary value problems associated with the two-dimensional biharmonic equation." *Mathematical and Computer Modelling* **42**(3–4): 261–278.

[619] Marin, L. and D. Lesnic (2007). "The method of fundamental solutions for nonlinear functionally graded materials." *International Journal of Solids and Structures* **44**(21): 6878–6890.

[620] Marin, L. and B. T. Johansson (2010). "Relaxation procedures for an iterative MFS algorithm for the stable reconstruction of elastic fields from Cauchy data in two-dimensional isotropic linear elasticity." *International Journal of Solids and Structures* **47**(25): 3462–3479.

[621] Marin, L. and A. Karageorghis (2013). "The MFS-MPS for two-dimensional steady-state thermoelasticity problems." *Engineering Analysis with Boundary Elements* **37**(7): 1004–1020.

[622] Marin, L., A. Karageorghis and D. Lesnic (2015). "A numerical study of the SVD–MFS solution of inverse boundary value problems in two-dimensional steady-state linear thermoelasticity." *Numerical Methods for Partial Differential Equations* **31**(1): 168–201.

[623] Marin, L. (2016). "An invariant method of fundamental solutions for two-dimensional steady-state anisotropic heat conduction problems." *International Journal of Heat and Mass Transfer* **94**: 449–464.

[624] Marin, L., A. Karageorghis and D. Lesnic (2016). "Regularized MFS solution of inverse boundary value problems in three-dimensional steady-state linear thermoelasticity." *International Journal of Solids and Structures* **91**: 127–142.

[625] Martin, P. A. (1982). "Acoustic scattering and radiation problems, and the null-field method." *Wave Motion* **4**(4): 391–408.

[626] Martin, P. A. and F. J. Rizzo (1995). "Partitioning, boundary integral equations, and exact Green's functions." *International Journal for Numerical Methods in Engineering* **38**(20): 3483–3495.

[627] Martin, P. A., J. D. Richardson, L. J. Gray and J. R. Berger (2002). "On Green's function for a three-dimensional exponentially graded

elastic solid." *Proceedings of the Royal Society of London. Series A: Mathematical, Physical and Engineering Sciences* **458**(2024): 1931–1947.

[628] Mason, J. C. and D. Handscomb (2003). *Chebyshev Polynomials*. Chapman & Hall.

[629] Matheron, G. (1963). "Principles of geostatistics." *Economic Geology* **58**(8): 1246–1266.

[630] Mathon, R. and R. L. Johnston (1977). "Approximate solution of elliptic boundary-value problems by fundamental solutions." *SIAM Journal on Numerical Analysis* **14**(4): 638–650.

[631] Mathworks Inc. (2022). *MATLAB*, https://www.mathworks.com/.

[632] Matthys, L., H. Lambert and G. DeMey (1996). "A recursive construction of particular solutions to a system of coupled linear partial differential equations with polynomial source term." *Journal of Computational and Applied Mathematics* **69**(2): 319–329.

[633] Mayer, C. D. (2000). *Matrix Analysis and Applied Linear Algebra*. SIAM.

[634] McLain, D. H. (1974). "Drawing contours from arbitrary data points." *Computer Journal* **17**(4): 318–324.

[635] Melan, E. (1932). "Der Spannungszustand der durch eine Einzelkraft im Innern beanspruchten Halbscheibe." *Zeitschrift für Angewandte Mathematik und Mechanik* **12**: 343–346.

[636] Melenk, J. M. and I. Babuška (1996). "The partition of unity finite element method: Basic theory and applications." *Computer Methods in Applied Mechanics and Engineering* **139**(1–4): 289–314.

[637] Melnikov, Y. A. and M. Y. Melnikov (2012). *Green's Functions Construction and Applications*. De Gruyter.

[638] Mera, N. S. (2005). "The method of fundamental solutions for the backward heat conduction problem." *Inverse Problems in Science and Engineering* **13**(1): 65–78.

[639] Mera, N. S. and D. Lesnic (2005). "A three-dimensional boundary determination problem in potential corrosion damage." *Computational Mechanics* **36**(2): 129–138.

[640] Meyer, W. L., W. A. Bell, B. T. Zinn and M. P. Stallybrass (1978). "Boundary integral solutions of three dimensional acoustic radiation problems." *Journal of Sound and Vibration* **59**(2): 245–262.

[641] Micchelli, C. A. (1986). "Interpolation of scattered data—distance matrices and conditionally positive definite functions." *Constructive Approximation* **2**(1): 11–22.

[642] Mierzwiczak, M. and J. A. Kołodziej (2011). "The determination temperature-dependent thermal conductivity as inverse steady heat conduction problem." *International Journal of Heat and Mass Transfer* **54**(4): 790–796.

[643] Mikhlin, S. G. (1964). *Integral Equations and Their Applications to Certain Problems in Mechanics, Mathematical Physics and Technology*. Pergamon.

[644] Miller, R. D., E. T. M. Jr., H. Huang and H. Überall (1991). "A comparison between the boundary element method and the wave superposition approach for the analysis of the scattered fields from rigid bodies and elastic shells." *Journal of the Acoustical Society of America* **89**(5): 2185–2196.

[645] Mindlin, R. D. (1936). "Note on the Galerkin and Papkovitch stress functions." *Bulletin of American Mathematical Society* **42**: 373–376.

[646] Mindlin, R. D. (1936). "Force at a point in the interior of a semi-infinite solid." *Physics* **7**: 195–202.

[647] Mindlin, R. D. (1953). "Force at a point in the interior of a semi-infinite solid." *Proceedings of the First Midwestern Conference on Solid Mechanics*. Urbana: 56–59.

[648] Mirzaei, D., R. Schaback and M. Dehghan (2011). "On generalized moving least squares and diffuse derivatives." *IMA Journal of Numerical Analysis* **32**(3): 983–1000.

[649] Mirzaei, D. and R. Schaback (2013). "Direct Meshless Local Petrov-Galerkin (DMLPG) method: A generalized MLS approximation." *Applied Numerical Mathematics* **68**: 73–82.

[650] Mishchenko, M. I., L. D. Travis and D. W. Mackowski (1996). "T-matrix computations of light scattering by nonspherical particles: A review." *Journal of Quantitative Spectroscopy and Radiative Transfer* **55**(5): 535–575.

[651] Moës, N., J. Dolbow and T. Belytschko (1999). "A finite element method for crack growth without remeshing." *International Journal for Numerical Methods in Engineering* **46**(1): 131–150.

[652] Moës, N. and T. Belytschko (2002). "Extended finite element method for cohesive crack growth." *Engineering Fracture Mechanics* **69**(7): 813–833.

[653] Mohebbi, A., M. Abbaszadeh and M. Dehghan (2013). "The use of a meshless technique based on collocation and radial basis functions for solving the time fractional nonlinear Schrödinger equation arising in quantum mechanics." *Engineering Analysis with Boundary Elements* **37**(2): 475–485.

[654] Moosavi, M. R., F. Delfanian and A. Khelil (2011). "Orthogonal meshless finite volume method in elasticity." *Thin-Walled Structures* **49**(6): 708–712.

[655] Morse, P. M. and H. Feshbach (1953). *Methods of Theoretical Physics, Part I and II*. McGraw-Hill.

[656] Movahedian, B., B. Boroomand and S. Soghrati (2013). "A Trefftz method in space and time using exponential basis functions: Application to direct and inverse heat conduction problems." *Engineering Analysis with Boundary Elements* **37**(5): 868–883.

[657] Mukherjee, Y. X. and S. Mukherjee (1997). "The boundary node method for potential problems." *International Journal for Numerical Methods in Engineering* **40**(5): 797–815.

[658] Muleshkov, A. S., M. A. Golberg and C. S. Chen (1999). "Particular solutions of Helmholtz-type operators using higher order polyharmonic splines." *Computational Mechanics* **23**(5–6): 411–419.

[659] Muleshkov, A. S., C. S. Chen, M. A. Golberg and A. H. D. Cheng (2000). "Analytic particular solutions for inhomogeneous Helmholtz-type equations." In: *Advances in Computational Engineering and Sciences, Vol. 1*, (eds.) S. N. Atluri and F. W. Brust. Tech Science Press: 27–32.

[660] Muleshkov, A. S. and M. A. Golberg (2007). "Particular solutions of the multi-Helmholtz-type equation." *Engineering Analysis with Boundary Elements* **31**(7): 624–630.

[661] Mura, T. (1987). *Micromechanics of Defects in Solids*. Springer.

[662] Muskhelishvili, N. I. (1954). *Some Basic Problems of the Mathematical Theory of Elasticity*. Noordhoff.

[663] Nakamura, G. and K. Tanuma (1997). "A formula for the fundamental solution of anisotropic elasticity." *Quarterly Journal of Mechanics and Applied Mathematics* **50**(2): 179–194.

[664] Nakayama, T. and K. Washizu (1981). "The boundary element method applied to the analysis of two-dimensional nonlinear sloshing problems." *International Journal for Numerical Methods in Engineering* **17**(11): 1631–1646.

[665] Narcowich, F. J. and J. D. Ward (1991). "Norms of inverses and condition numbers for matrices associated with scattered data." *Journal of Approximation Theory* **64**(1): 69–94.

[666] Narcowich, F. J. and J. D. Ward (1992). "Norm estimates for the inverses of a general-class of scattered-data radial-function interpolation matrices." *Journal of Approximation Theory* **69**(1): 84–109.

[667] Nayroles, B., G. Touzot and P. Villon (1992). "Generalizing the finite element method: Diffuse approximation and diffuse elements." *Computational Mechanics* **10**(5): 307–318.

[668] Nečas, J. and M. Štípl (1976). "A paradox in the theory of linear elasticity." *Aplikace Matematiky* **21**(6): 431–433.

[669] Nedelec, J. C. (1986). "The double layer potential for periodic elastic waves in R^3." In: *Boundary Elements*. (ed.) Q. H. Du, Pergamon: 439–448.

[670] Nennig, B., E. Perrey-Debain and J. D. Chazot (2011). "The method of fundamental solutions for acoustic wave scattering by a single and a periodic array of poroelastic scatterers." *Engineering Analysis with Boundary Elements* **35**(8): 1019–1028.

[671] Nerantzaki, M. S. and J. T. Katsikadelis (1996). "Buckling of plates with variable thickness—an analog equation solution." *Engineering Analysis with Boundary Elements* **18**(2): 149–154.

[672] Netuzhylov, H. (2008). "Enforcement of boundary conditions in meshfree methods using interpolating moving least squares." *Engineering Analysis with Boundary Elements* **32**(6): 512–516.

[673] Neumann, C. (1877). *Untersuchungen über das logarithmische und Newton'sche Potential* (Investigations into the logarithmic and Newtonian potential). Teubner, Leipzig.

[674] Neves, A. C. and C. A. Brebbia (1991). "The multiple reciprocity boundary element method in elasticity—a new approach for transforming domain integrals to the boundary." *International Journal for Numerical Methods in Engineering* **31**(4): 709–727.

[675] Neves, A. M. A., A. J. M. Ferreira, E. Carrera, M. Cinefra, C. M. C. Roque, R. M. N. Jorge and C. M. M. Soares (2013). "Static, free vibration and buckling analysis of isotropic and sandwich functionally graded plates using a quasi-3D higher-order shear deformation theory and a meshless technique." *Composites Part B: Engineering* **44**(1): 657–674.

[676] Newton, I. (1726). *Philosophiæ Naturalis Principia Mathematica (Mathematical Principles of Natural Philosophy)*, 3rd edn. James Maclehouse, Glasgow.

[677] Niu, Y. and M. Dravinski (2003). "Three-dimensional BEM for scattering of elastic waves in general anisotropic media." *International Journal for Numerical Methods in Engineering* **58**(7): 979–998.

[678] Niwa, Y., S. Kobayashi and T. Fukui (1974). "An application of the integral equation method to plate–bending problems." *Memoirs of the Faculty of Engineering, Kyoto University* **36**: 140–158.

[679] Noma, A. A. and M. G. Misulia (1959). "Programming topographic maps for automatic terrain model construction." *Surveying and Mapping* **19**: 355–366.

[680] Noorizadegan, A., C.-S. Chen, D. L. Young and C. S. Chen (2022). "Effective condition number on the selection of the shape parameter of RBFs with fictitious point method." *Applied Numerical Mathematics* **178**: 280–295.

[681] Nowacki, W. (1986). *Thermoelasticity*, 2nd edn. Pergamon.

[682] Nowak, A. J. and C. A. Brebbia (1989). "The multiple-reciprocity method. A new approach for transforming BEM domain integrals to the boundary." *Engineering Analysis with Boundary Elements* **6**(3): 164–167.

[683] Oberhettinger, F. (1973). *Fourier Transforms of Distributions and Their Inverses: A Collectionn of Tables*. Academic Press.

[684] Oden, J. T., L. Demkowicz, W. Rachowicz and T. A. Westermann (1989). "Toward a universal h-p adaptive finite element strategy, part 2. A posteriori error estimation." *Computer Methods in Applied Mechanics and Engineering* **77**(1): 113–180.

[685] Ogata, H., F. Chiba and T. Ushijima (2011). "A new theoretical error estimate of the method of fundamental solutions applied to reduced wave problems in the exterior region of a disk." *Journal of Computational and Applied Mathematics* **235**(12): 3395–3412.

[686] Oh, J., H. Zhu and Z. Fu (2019). "An adaptive method of fundamental solutions for solving the Laplace equation." *Computers and Mathematics with Applications* **77**(7): 1828–1840.

[687] Oliveira, E. R. A. (1968). "Plane stress analysis by a general integral method." *Journal of Engineering Mechanics Division, ASCE* **94**(1): 79–101.

[688] Olver, F. W. J., D. W. Lozier, R. F. Boisvert and C. W. Clark (2010). *NIST Handbook of Mathematical Functions*. NIST & Cambridge University Press.

[689] Oñate, E., S. Idelsohn, O. C. Zienkiewicz and R. L. Taylor (1996). "A finite point method in computational mechanics. Applications to convective transport and fluid flow." *International Journal for Numerical Methods in Engineering* **39**(22): 3839–3866.

[690] Oñate, E., S. Idelsohn, O. C. Zienkiewicz, R. L. Taylor and C. Sacco (1996). "A stabilized finite point method for analysis of fluid mechanics problems." *Computer Methods in Applied Mechanics and Engineering* **139**(1): 315–346.

[691] Oñate, E., F. Perazzo and J. Miquel (2001). "A finite point method for elasticity problems." *Computers and Structures* **79**(22): 2151–2163.

[692] Onsager, L. (1931). "Reciprocal relations in irreversible process I." *Physical Review* **37**(4): 405–426.

[693] Onsager, L. (1931). "Reciprocal relations in irreversible process II." *Physical Review* **38**(12): 2265–2279.

[694] Ortner, N. and P. Wagner (1997). "A survey on explicit representation formulæ for fundamental solutions of linear partial differential operators." *Acta Applicandae Mathematicae* **47**(1): 101–124.

[695] Ortner, N. and P. Wagner (2015). *Fundamental Solutions of Linear Partial Differential Operators: Theory and Practice*. Springer.

[696] Osgood, W. F. (1900). "On the existence of the Green's function for the most general simply connected plane region." *Transactions of the American Mathematical Society* **1**(3): 310–314.

[697] Osman, I. H. and G. Laporte (1996). "Metaheuristics: A bibliography." *Annals of Operations Research* **63**: 513–623.

[698] Paget, D. F. (1981). "The numerical evaluation of Hadamard finite-part integrals." *Numerische Mathematik* **36**: 447–453.

[699] Pan, E. (1997). "Static Green's functions in multilayered half spaces." *Applied Mathematical Modelling* **21**(8): 509–521.

[700] Pan, E. (1999). "A BEM analysis of fracture mechanics in 2D anisotropic piezoelectric solids." *Engineering Analysis with Boundary Elements* **23**(1): 67–76.

[701] Pan, E. (1999). "Green's functions in layered poroelastic half-spaces." *International Journal for Numerical and Analytical Methods in Geomechanics* **23**(13): 1631–1653.

[702] Pan, E. and F. Tonon (2000). "Three-dimensional Green's functions in anisotropic piezoelectric solids." *International Journal of Solids and Structures* **37**(6): 943–958.

[703] Pan, E. (2002). "Three-dimensional Green's functions in anisotropic magneto-electro-elastic bimaterials." *Zeitschrift für angewandte Mathematik und Physik* **53**(5): 815–838.

[704] Pan, E. and F. Han (2004). "Green's functions for transversely isotropic piezoelectric multilayered half-spaces." *Journal of Engineering Mathematics* **49**(3): 271–288.

[705] Pan, E. and W. Q. Chen (2015). *Static Green's Functions in Anisotropic Media*. Cambridge University Press.

[706] Pan, E. (2019). "Green's functions for geophysics: a review." *Reports on Progress in Physics* **82**(10): 106801.

[707] Pan, E., C. P. Lin and J. Zhou (2021). "Fundamental solution of general time-harmonic loading over a transversely isotropic, elastic and layered half-space: An efficient and accurate approach." *Engineering Analysis with Boundary Elements* **132**: 309–320.

[708] Pan, Y. C. and T. W. Chou (1976). "Point force solution for an infinite transversely isotropic solid." *Journal of Applied Mechanics* **43**(4): 608–612.

[709] Paoluzzi, A. (2003). *Geometric Programming For Computer Aided Design*. Wiley.

[710] Papakanellos, P. J., N. L. Tsitsas and H. T. Anastassiu (2024). "The method of auxiliary sources (MAS) in computational electromagnetics: A comprehensive review of advancements over the past two decades." *Electronics* **13**(17): 3520.

[711] Papanicolopulos, S. A., A. Zervos and I. Vardoulakis (2009). "A three-dimensional C^1 finite element for gradient elasticity." *International Journal for Numerical Methods in Engineering* **77**(10): 1396–1415.

[712] Partridge, P. W. and C. A. Brebbia (1989). "Computer implementation of the BEM dual reciprocity method for the solution of Poisson type equations." *Software for Engineering Workstations* **5**: 199.

[713] Partridge, P. W., C. A. Brebbia and L. C. Wrobel (1992). *The Dual Reciprocity Boundary Element Method*. Computational Mechanics Publications.

[714] Partridge, P. W. and B. Sensale (2000). "The method of fundamental solutions with dual reciprocity for diffusion and diffusion-convection using subdomains." *Engineering Analysis with Boundary Elements* **24**(9): 633–641.

[715] Pei, X., C. S. Chen and F. Dou (2017). "The MFS and MAFS for solving Laplace and biharmonic equations." *Engineering Analysis with Boundary Elements* **80**: 87–93.

[716] Peltier, W. R. (1974). "The impulse response of a Maxwell Earth." *Reviews of Geophysics* **12**(4): 649–669.

[717] Peng, M. and Y. Cheng (2009). "A boundary element-free method (BEFM) for two-dimensional potential problems." *Engineering Analysis with Boundary Elements* **33**(1): 77–82.

[718] Perne, M., B. Šarler and F. Gabrovšek (2012). "Calculating transport of water from a conduit to the porous matrix by boundary distributed source method." *Engineering Analysis with Boundary Elements* **36**(11): 1649–1659.

[719] Perrone, N. and R. Kao (1975). "A general finite difference method for arbitrary meshes." *Computers and Structures* **5**(1): 45–57.

[720] Petrovsky, I. G. (1954). *Lectures on Partial Differential Equations*. Interscience (Dover, 1992).

[721] Pipher, J. and G. Verchota (1993). "A maximum principle for biharmonic functions in Lipschitz and C^1 domains." *Commentarii Mathematici Helvetici* **68**(1): 385–414.

[722] Pipher, J. and G. C. Verchota (1995). "Maximum principles for the polyharmonic equation on Lipschitz domains." *Potential Analysis* **4**(6): 615–636.

[723] Pluymers, B., B. Van Hal, D. Vandepitte and W. Desmet (2007). "Trefftz-based methods for time-harmonic acoustics." *Archives of Computational Methods in Engineering* **14**(4): 343–381.

[724] Poisson, S. D. (1813). "Remarques sur une équation qui se présente dans la théorie des attractions des sphéroïdes (Remarks on an equation which occurs in the theory of the attractions of spheroids)." *Nouveau Bulletin de la Soclété Philomathique de Paris* **3**: 388–392.

[725] Poullikkas, A., A. Karageorghis and G. Georgiou (1998). "Methods of fundamental solutions for harmonic and biharmonic boundary value problems." *Computational Mechanics* **21**(4–5): 416–423.

[726] Poullikkas, A., A. Karageorghis and G. Georgiou (1998). "The method of fundamental solutions for inhomogeneous elliptic problems." *Computational Mechanics* **22**(1): 100–107.

[727] Poullikkas, A., A. Karageorghis and G. Georgiou (1998). "The method of fundamental solutions for Signorini problems." *IMA Journal of Numerical Analysis* **18**(2): 273–285.

[728] Poullikkas, A., A. Karageorghis, G. Georgiou and J. Ascough (1998). "The Method of Fundamental Solutions for Stokes flows with a free surface." *Numerical Methods for Partial Differential Equations* **14**(5): 667–678.

[729] Poullikkas, A., A. Karageorghis and G. Georgiou (2002). "The method of fundamental solutions for three-dimensional elastostatics problems." *Computers and Structures* **80**(3–4): 365–370.

[730] Powell, M. J. D. (1981). *Approximation Theory and Methods*. Cambridge University Press.

[731] Pozrikidis, C. (2000). "On the method of functional equations and the performance of desingularized boundary element methods." *Engineering Analysis with Boundary Elements* **24**(1): 3–16.

[732] Press, W. H., S. A. Teukolsky, W. T. Vetterling and B. P. Flannery (2007). *Numerical Recipes: The Art of Scientific Computing*, 3rd edn. Cambridge University Press.

[733] Protter, M. H. and H. F. Weinberger (1967). *Maximum Principles in Differential Equations*. Prentice-Hall.

[734] Providakis, C. P. and D. E. Beskos (1999). "Dynamic analysis of plates by boundary elements." *Applied Mechanics Reviews* **52**(7): 213–236.

[735] Puswewala, U. G. A. and R. K. N. D. Rajapakse (1988). "Axisymmetric fundamental solutions for a completely saturated porous elastic solid." *International Journal of Engineering Science* **26**(5): 419–436.

[736] Qin, Q. H. (2000). *The Trefftz Finite and Boundary Element Method*. WIT Press.

[737] Qin, X., J. Zhang, G. Li, X. Sheng, Q. Song and D. Mu (2010). "An element implementation of the boundary face method for 3D potential problems." *Engineering Analysis with Boundary Elements* **34**(11): 934–943.

[738] Qu, W. and W. Chen (2015). "Solution of two-dimensional Stokes flow problems using improved singular boundary method." *Advances in Applied Mathematics and Mechanics* **7**(1): 13–30.

[739] Qu, W., C. M. Fan and Y. Gu (2019). "Localized method of fundamental solutions for interior Helmholtz problems with high wave number." *Engineering Analysis with Boundary Elements* **107**: 25–32.

[740] Qu, W., C. M. Fan, Y. Gu and F. Wang (2019). "Analysis of three-dimensional interior acoustic fields by using the localized method of fundamental solutions." *Applied Mathematical Modelling* **76**: 122–132.

[741] Qu, W., C. M. Fan and X. Li (2020). "Analysis of an augmented moving least squares approximation and the associated localized method of fundamental solutions." *Computers and Mathematics with Applications* **80**(1): 13–30.

[742] Qu, W., L. Sun and P. W. Li (2021). "Bending analysis of simply supported and clamped thin elastic plates by using a modified version of the LMFS." *Mathematics and Computers in Simulation* **185**: 347–357.

[743] Rajapakse, R. K. N. D. and T. Senjuntichai (1993). "Fundamental solutions for a poroelastic half-space with compressible constituents." *Journal of Applied Mechanics* **60**(4): 847–856.

[744] Ramachandran, P. A. (2002). "Method of fundamental solutions: singular value decomposition analysis." *Communications in Numerical Methods in Engineering* **18**(11): 789–801.

[745] Rangelov, T. V., G. D. Manolis and P. S. Dineva (2005). "Elastodynamic fundamental solutions for certain families of 2d inhomogeneous anisotropic domains: Basic derivations." *European Journal of Mechanics—A/Solids* **24**(5): 820–836.

[746] Rankine, W. J. M. (1864). "On plane water-lines in two dimensions." *Philosophical Transactions of Royal Society of London* **154**: 369–391.

[747] Rankine, W. J. M. (1871). "On the mathematical theory of stream-lines, especially those with four foci and upwards." *Philosophical Transactions of Royal Society of London* **161**: 267–306.

[748] Rashed, Y. F., M. H. Aliabadi and C. A. Brebbia (1998). "The boundary element method for thick plates on a Winkler foundation." *International Journal for Numerical Methods in Engineering* **41**(8): 1435–1462.

[749] Redekop, D. (1982). "Fundamental solutions for the collation method in planar elastostatics." *Applied Mathematical Modelling* **6**(5): 390–393.

[750] Reutskiy, S. Y. and C. S. Chen (2006). "Approximation of multivariate functions and evaluation of particular solutions using Chebyshev polynomial and trigonometric basis functions." *International Journal for Numerical Methods in Engineering* **67**(13): 1811–1829.

[751] Reutskiy, S. Y. (2007). "The method of fundamental solutions for problems of free vibrations of plates." *Engineering Analysis with Boundary Elements* **31**(1): 10–21.

[752] Reutskiy, S. Y., C. S. Chen and H. Y. Tian (2008). "A boundary meshless method using Chebyshev interpolation and trigonometric basis function for solving heat conductions problems." *International Journal for Numerical Methods in Engineering* **74**(10): 1621–1644.

[753] Rice, J. R. (1983). *Matrix Computation and Mathematical Software.* McGraw Hill.

[754] Riemann, B. (1867). *Grundlagen für eine allgemeine Theorie der Functionen einer veränderlichen complexen Grösse, Inauguráldissertation* (Basis for a general theory of functions of a variable complex quantity, Inaugural dissertation). Rente, Göttingen.

[755] Rippa, S. (1999). "An algorithm for selecting a good value for the parameter c in radial basis function interpolation." *Advances in Computational Mathematics* **11**(2–3): 193–210.

[756] Ritz, W. (1908). "Über eine neue methode zur Lösung gewissen variations—Problems der mathematischen physik (Using a new method for solving certain variations of mathematical physics problems)." *Journal für die reine und angewandte Mathematik* **135**: 1–61.

[757] Rizzo, F. J. and D. J. Shippy (1970). "A method for stress determination in plane anisotropic elastic bodies." *Journal of Composite Materials* **4**: 36–61.

[758] Rizzo, F. J. (1989). "The boundary element method, some early history—a personal view." In: *Boundary Element Methods in Structural Analysis.* (ed.) D. E. Beskos, ASCE: 1–16.

[759] Rokhlin, V. (1985). "Rapid solution of integral equations of classical potential theory." *Journal of Computational Physics* **60**(2): 187–207.

[760] Roque, C. M. C., A. J. M. Ferreira and J. N. Reddy (2011). "Analysis of Timoshenko nanobeams with a nonlocal formulation and meshless method." *International Journal of Engineering Science* **49**(9): 976–984.

[761] Saavedra, I. and H. Power (2003). "Multipole fast algorithm for the least-squares approach of the method of fundamental solutions for three-dimensional harmonic problems." *Numerical Methods for Partial Differential Equations* **19**(6): 828–845.

[762] Sadd, M. H. (2014). *Elasticity: Theory, Applications, and Numerics.* Academic Press.

[763] Sakakibara, K. (2020). "Bidirectional numerical conformal mapping based on the dipole simulation method." *Engineering Analysis with Boundary Elements* **114**: 45–57.

[764] Salam, N., A. Haddade, D. L. Clements and M. I. Azis (2017). "A boundary element method for a class of elliptic boundary value problems of functionally graded media." *Engineering Analysis with Boundary Elements* **84**: 186–190.

[765] Salazar, F. M. V. and I. D. P. Arcila (2024). "In silico study about the influence of electroporation parameters on the cellular internalization, spatial uniformity, and cytotoxic effects of chemotherapeutic drugs using the Method of Fundamental Solutions." *Medical & Biological Engineering & Computing* **62**(3): 713–749.

[766] Sánchez-Sesma, F. J. and E. Rosenblueth (1979). "Ground motion at canyons of arbitrary shape under incident SH waves." *Earthquake Engineering and Structural Dynamics* **7**(5): 441–450.

[767] Sánchez-Sesma, F. J. (1981). "A boundary method applied to elastic scattering problems." *Archives of Mechanics* **33**(2): 167–179.

[768] Sánchez-Sesma, F. J., I. Herrera and J. Avilés (1982). "A boundary method for elastic wave diffraction: Application to scattering of SH waves by surface irregularities." *Bulletin of the Seismological Society of America* **72**(2): 473–490.

[769] Sánchez-Sesma, F. J., U. Iturrarán-Viveros and E. Kausel (2013). "Garvin's generalized problem revisited." *Soil Dynamics and Earthquake Engineering* **47**: 4–15.

[770] Santos, W. J., J. A. F. Santiago and J. C. F. Telles (2012). "An application of genetic algorithms and the method of fundamental solutions to simulate cathodic protection systems." *Computer Modeling in Engineering and Sciences* **87**(1): 23–40.

[771] Šarler, B. (2002). "Towards a mesh-free computation of transport phenomena." *Engineering Analysis with Boundary Elements* **26**(9): 731–738.

[772] Šarler, B. (2006). "Solution of a two-dimensional bubble shape in potential flow by the method of fundamental solutions." *Engineering Analysis with Boundary Elements* **30**(3): 227–235.

[773] Šarler, B. and R. Vertnik (2006). "Meshfree explicit local radial basis function collocation method for diffusion problems." *Computers and Mathematics with Applications* **51**(8): 1269–1282.

[774] Šarler, B. (2009). "Solution of potential flow problems by the modified method of fundamental solutions: Formulations with the single layer and the double layer fundamental solutions." *Engineering Analysis with Boundary Elements* **33**(12): 1374–1382.

[775] Schaback, R. (1995). "Error estimates and condition numbers for radial basis function interpolation." *Advances in Computational Methematics* **3**: 251–264.

[776] Schaback, R. (1995). "Multivariate interpolation and approximation by translates of a basis function." *Approximation Theory VIII, Vol. 1: Approximation and Interpolation*. (eds) C. Chui and L. Schumaker, World Scientific Publishing: 491–514.

[777] Schaback, R. (2005). "Multivariate interpolation by polynomials and radial basis functions." *Constructive Approximation* **21**(3): 293–317.

[778] Schaback, R. (2007). "On the numerical solution of MFS systems." Presentation at First International Workshop on the Method of Fundamental Solutions, Ayia Napa, Cyprus, June 11–13, 2007.

[779] Schaback, R. (2008). "Adaptive numerical solution of MFS systems." In: *The Method of Fundamental Solutions—A Meshless Method*, C. S. Chen, A. Karageorghis and Y. S. Smyrlis, eds. Dynamic Publishers: 1–27.

[780] Schaback, R. (2008). "Limit problems for interpolation by analytic radial basis functions." *Journal of Computational and Applied Mathematics* **212**(2): 127–149.

[781] Schaback, R. (2021). "An Approximation Theorist's view on solving operator equations—With special attention to Trefftz, MFS, MPS, and DRM methods." *Computers and Mathematics with Applications* **88**: 70–77.

[782] Schenck, H. A. (1968). "Improved integral formulation for acoustic radiation problems." *Journal of the Acoustical Society of America* **44**(1): 41–58.

[783] Schoenberg, I. J. (1937). "On certain metric spaces arising from Euclidean space by a change of metric and their imbedding in Hilbert space." *Annals of Mathematics* **38**: 787–793.

[784] Schoenberg, I. J. (1938). "Metric spaces and completely monotone functions." *Annals of Mathematics* **39**: 811–841.

[785] Schreck, C., C. Hafner and C. Wojtan (2019). "Fundamental solutions for water wave animation." *ACM Transactions on Graphics* **38**(4): Article 130.

[786] Schumaker, L. L. (1976). Fitting surfaces to scattered data. In: *Approximation Theory II*. (eds.) G. G. Lorentz, C. K. Chui and L. L. Schumaker, Academic Press: 203–268.

[787] Schwartz, L. (1950–51). *Théorie des distributions*, Vols. 1 & 2. Hermann, Paris.

[788] Seremet, V. (2002). *Handbook of Green's Functions and Matrices*. WIT Press.

[789] Seybert, A. F. and T. K. Rengarajan (1987). "The use of CHIEF to obtain unique solutions for acoustic radiation using boundary integral equations." *Journal of the Acoustical Society of America* **81**(5): 1299–1306.

[790] Shao, K. R. and J. D. Lavers (1994). "The extended method of fundamental solution based on the second order vector potential formulation for 3D transient eddy current problems." *IEEE Transactions on Magnetics* **30**(5): 3036–3039.

[791] Shao, Y. L. and O. M. Faltinsen (2014). "A harmonic polynomial cell (HPC) method for 3D Laplace equation with application in marine hydrodynamics." *Journal of Computational Physics* **274**: 312–332.

[792] Shaw, R. P. and N. Makris (1992). "Green functions for Helmholtz and Laplace equations in heterogeneous media." *Engineering Analysis with Boundary Elements* **10**(2): 179–183.

[793] Shepard, D. (1968). "A two-dimensional interpolation function for irregularly-spaced data." In: *Proceedings of the 1968 23rd ACM National Conference*, Association for Computing Machinery: 517–524.

[794] Shiah, Y. C., C. L. Tan and C. Y. Wang (2012). "An improved numerical evaluation scheme of the fundamental solution and its derivatives for 3D anisotropic elasticity based on Fourier series." *Computer Modeling in Engineering and Sciences* **87**(1): 1–22.

[795] Shiah, Y. C., C. L. Tan and C. Y. Wang (2012). "Efficient computation of the Green's function and its derivatives for three-dimensional anisotropic elasticity in BEM analysis." *Engineering Analysis with Boundary Elements* **36**(12): 1746–1755.

[796] Shigeta, T. and D. L. Young (2009). "Method of fundamental solutions with optimal regularization techniques for the Cauchy problem of the Laplace equation with singular points." *Journal of Computational Physics* **228**(6): 1903–1915.

[797] Shigeta, T. and D. L. Young (2011). "Regularized solutions with a singular point for the inverse biharmonic boundary value problem by the method of fundamental solutions." *Engineering Analysis with Boundary Elements* **35**(7): 883–894.

[798] Shigeta, T., D. L. Young and C. S. Liu (2012). "Adaptive multilayer method of fundamental solutions using a weighted greedy QR decomposition for the Laplace equation." *Journal of Computational Physics* **231**(21): 7118–7132.

[799] Shou, K. J. and S. L. Crouch (1995). "A higher order displacement discontinuity method for analysis of crack problems." *International Journal of Rock Mechanics and Mining Sciences and Geomechanics Abstracts* **32**(1): 49–55.

[800] Shu, C. and B. E. Richards (1992). "Application of generalized differential quadrature to solve two-dimensional incompressible Navier-Stokes equations." *International Journal for Numerical Methods in Fluids* **15**(7): 791–798.

[801] Shu, C., H. Ding and K. S. Yeo (2003). "Local radial basis function-based differential quadrature method and its application to solve two-dimensional incompressible Navier-Stokes equations." *Computer Methods in Applied Mechanics and Engineering* **192**(7–8): 941–954.

[802] Shu, C., H. Ding and K. S. Yeo (2004). "Solution of partial differential equations by a global radial basis function-based differential quadrature method." *Engineering Analysis with Boundary Elements* **28**(10): 1217–1226.

[803] Shu, C., H. Ding, H. Q. Chen and T. G. Wang (2005). "An upwind local RBF-DQ method for simulation of inviscid compressible flows." *Computer Methods in Applied Mechanics and Engineering* **194**(18–20): 2001–2017.

[804] Simpson, R. N., M. A. Scott, M. Taus, D. C. Thomas and H. Lian (2014). "Acoustic isogeometric boundary element analysis." *Computer Methods in Applied Mechanics and Engineering* **269**: 265–290.

[805] Singer, H., H. Steinbigler and P. Weiss (1974). "A charge simulation method for the calculation of high voltage fields." *IEEE Transactions on Power Apparatus and Systems* **93**(5): 1660–1668.

[806] Sládek, J., P. Stanak, Z. D. Han, V. Sládek and S. N. Atluri (2013). "Applications of the MLPG method in engineering and sciences: A review." *Computer Modeling in Engineering and Sciences* **92**(5): 423–475.

[807] Sládek, V. and J. Sládek (1991). "Why use double nodes in BEM?" *Engineering Analysis with Boundary Elements* **8**(2): 109–112.

[808] Smyrlis, Y. S. and A. Karageorghis (2001). "Some aspects of the method of fundamental solutions for certain harmonic problems." *Journal of Scientific Computing* **16**(3): 341–371.

[809] Smyrlis, Y. S. and A. Karageorghis (2003). "Some aspects of the method of fundamental solutions for certain biharmonic problems." *Computer Modeling in Engineering and Sciences* **4**(5): 535–550.

[810] Smyrlis, Y. S. and A. Karageorghis (2004). "A matrix decomposition MFS algorithm for axisymmetric potential problems." *Engineering Analysis with Boundary Elements* **28**(5): 463–474.

[811] Sokolnikoff, I. S. (1956). *Mathematical Theory of Elasticity*. McGraw-Hill.

[812] Stanford Computer Graphics Laboratory (2022). *The Stanford 3D scanning repository*. http://graphics.stanford.edu/data/3Dscanrep/.

[813] Steinbigler, H. (1969). "Anfangsfeldstärken und Ausnutzungsfaktoren rotationssymmetrischer Elektrodenanordnungen in Luft (Initial field strengths and utilization factors of rotationally symmetric electrode configurations in air)." Doctoral Dissertation, Technical University of Munich.

[814] Sternberg, W. J. and T. L. Smith (1944). *The Theory of Potential and Spherical Harmonics*. University of Toronto Press.

[815] Štípl, M. (1978). "On the maximum principle in the linear-elasticity theory." *Acta Universitatis Carolinae. Mathematica et Physica* **19**(2): 65–68.

[816] Strang, G. (2006). *Linear Algebra and Its Applications*, 4th edn. Cengage Learning.

[817] Stroh, A. N. (1958). "Dislocations and cracks in anisotropic elasticity." *Philosophical Magazine: A Journal of Theoretical Experimental and Applied Physics* **3**(30): 625–646.

[818] Strouboulis, T., I. Babuška and K. Copps (2000). "The design and analysis of the Generalized Finite Element Method." *Computer Methods in Applied Mechanics and Engineering* **181**(1): 43–69.

[819] Sulsky, D., S. J. Zhou and H. L. Schreyer (1995). "Application of a particle-in-cell method to solid mechanics." *Computer Physics Communications* **87**(1): 236–252.

[820] Sulsky, D. and H. L. Schreyer (1996). "Axisymmetric form of the material point method with applications to upsetting and Taylor impact problems." *Computer Methods in Applied Mechanics and Engineering* **139**(1): 409–429.

[821] Sun, L., W. Chen and A. H. D. Cheng (2016). "Method of fundamental solutions without fictitious boundary for plane time harmonic linear elastic and viscoelastic wave problems." *Computers and Structures* **162**: 80–90.

[822] Sun, L., Z. J. Fu and Z. Chen (2023). "A localized collocation solver based on fundamental solutions for 3D time harmonic elastic wave propagation analysis." *Applied Mathematics and Computation* **439**: 127600.

[823] Symm, G. T. (1963). "Integral equation methods in potential theory II." *Proceedings of the Royal Society of London, Series A: Mathematical and Physical Sciences* **275**(1360): 33–46.

[824] Symm, G. T. (1967). "Numerical mapping of exterior domains." *Numerische Mathematik* **10**(5): 437–445.

[825] Symm, G. T. (1969). "Conformal mapping of doubly-connected domains." *Numerische Mathematik* **13**(5): 448–457.

[826] Tai, G. R. C. and R. P. Shaw (1974). "Helmholtz-equation eigenvalues and eigenmodes for arbitrary domains." *Journal of the Acoustical Society of America* **56**(3): 796–804.

[827] Tan, J. Y., L. H. Liu and B. X. Li (2006). "Least-squares collocation meshless approach for coupled radiative and conductive heat transfer." *Numerical Heat Transfer, Part B: Fundamentals* **49**(2): 179–195.

[828] Tang, Z., Z. Fu, D. Zheng and J. Huang (2018). "Singular boundary method to simulate scattering of shwave by the canyon topography." *Advances in Applied Mathematics and Mechanics* **10**(4): 912–924.

[829] Tankelevich, R., G. Fairweather, A. Karageorghis and Y. S. Smyrlis (2006). "Potential field based geometric modelling using the method of fundamental solutions." *International Journal for Numerical Methods in Engineering* **68**(12): 1257–1280.

[830] Tankelevich, R., G. Fairweather and A. Karageorghis (2009). "Three-dimensional image reconstruction using the PF/MFS technique." *Engineering Analysis with Boundary Elements* **33**(12): 1403–1410.

[831] Tanuma, K. (1996). "Surface-impedance tensors of transversely isotropic elastic materials." *Quarterly Journal of Mechanics and Applied Mathematics* **49**(1): 29–48.

[832] Taylor, Z. (2022). "Find 3D normals and curvature." *MATLAB Central File Exchange*.

[833] Telles, J. C. F. and C. A. Brebbia (1981). "Boundary element solution for half-plane problems." *International Journal of Solids and Structures* **17**(12): 1149–1158.

[834] Thomson, W. (Lord Kelvin) (1848). "Note on the integrations of the equations of equilibrium of an elastic solid." *Cambridge Dublin Mathematical Journal* **3**: 76–99.

[835] Tian, Z., Z. C. Li, H. T. Huang and C. S. Chen (2017). "Analysis of the method of fundamental solutions for the modified Helmholtz equation." *Applied Mathematics and Computation* **305**: 262–281.

[836] Tikhonov, A. N., A. V. Goncharsky, V. V. Stepanov and A. G. Yagola (1995). *Numerical Methods for the Solution of Ill-posed Problems*. Kluwer.

[837] Timoshenko, S. P. and J. N. Goodier (1970). *Theory of Elasticity*. McGraw-Hill.

[838] Ting, T. C. T. (1996). *Anisotropic Elasticity: Theory and Applications*. Oxford University Press.

[839] Ting, T. C. T. and V. G. Lee (1997). "The three-dimensional elastostatic green's function for general anisotropic linear elastic solids." *Quarterly Journal of Mechanics and Applied Mathematics* **50**(3): 407–426.

[840] Tonon, F., E. Pan and B. Amadei (2001). "Green's functions and boundary element method formulation for 3D anisotropic media." *Computers and Structures* **79**(5): 469–482.

[841] Tran-Cong, T. and J. R. Blake (1982). "General solutions of the Stokes' flow equations." *Journal of Mathematical Analysis and Applications* **90**(1): 72–84.

582   *An Introduction to the Method of Fundamental Solutions*

[842] Trefftz, E. (1926). "Ein Gegenstück zum Ritz'schen verfahren (A counterpart to Ritz method)." Proc. 2nd Int. Congress Applied Mechanics. Zurich: 131–137.

[843] Trefftz, E. (1928). "Konvergenz und fehlerabschatzung beim Ritz'schen verfahre (Convergence and error evaluation in the Ritz procedure)." *Mathematische Annalen* **100**: 503–521.

[844] Treves, F. (1962). "Fundamental solutions of linear partial differential equations with constant coefficients depending on parameters." *American Journal of Mathematics* **84**(4): 561–577.

[845] Tri, A., H. Zahrouni and M. Potier-Ferry (2011). "Perturbation technique and method of fundamental solution to solve nonlinear Poisson problems." *Engineering Analysis with Boundary Elements* **35**(3): 273–278.

[846] Tri, A., H. Zahrouni and M. Potier-Ferry (2012). "High order continuation algorithm and meshless procedures to solve nonlinear Poisson problems." *Engineering Analysis with Boundary Elements* **36**(11): 1705–1714.

[847] Tri, A., O. Askour, B. Braikat, H. Zahrouni and M. Potier-Ferry (2019). "Fundamental solutions and asymptotic numerical methods for bifurcation analysis of nonlinear bi-harmonic problems." *Numerical Methods for Partial Differential Equations* **35**(6): 2091–2102.

[848] Tsai, C. C., D. L. Young and A. H. D. Cheng (2002). "Meshless BEM for three-dimensional Stokes flows." *Computer Modeling in Engineering and Sciences* **3**(1): 117–128.

[849] Tsai, C. C., Y. C. Lin, D. L. Young and S. N. Atluri (2006). "Investigations on the accuracy and condition number for the method of fundamental solutions." *Computer Modeling in Engineering and Sciences* **16**(2): 103–114.

[850] Tsai, C. C., D. L. Young, C. M. Fan and C. W. Chen (2006). "MFS with time-dependent fundamental solutions for unsteady Stokes equations." *Engineering Analysis with Boundary Elements* **30**(10): 897–908.

[851] Tsai, C. C. (2008). "Particular solutions of Chebyshev polynomials for polyharmonic and poly-Helmholtz equations." *Computer Modeling in Engineering and Sciences* **27**(3): 151–162.

[852] Tsai, C. C. (2009). "The method of fundamental solutions with dual reciprocity for three-dimensional thermoelasticity under arbitrary body forces." *Engineering Computations* **26**(3): 229–244.

[853] Tsai, C. C., C. S. Chen and T. W. Hsu (2009). "The method of particular solutions for solving axisymmetric polyharmonic and poly-Helmholtz equations." *Engineering Analysis with Boundary Elements* **33**(12): 1396–1402.

[854] Tsai, C. C., A. H. D. Cheng and C. S. Chen (2009). "Particular solutions of splines and monomials for polyharmonic and products of Helmholtz operators." *Engineering Analysis with Boundary Elements* **33**(4): 514–521.

[855] Tsai, C. C. (2012). "Homotopy method of fundamental solutions for solving certain nonlinear partial differential equations." *Engineering Analysis with Boundary Elements* **36**(8): 1226–1234.

[856] Tsai, C. C. (2013). "Analytical particular solutions of multiquadrics associated with polyharmonic operators." *Mathematical Problems in Engineering* **2013**: 613082.

[857] Tsai, C. C. and T. W. Hsu (2013). "Multiquadric and Chebyshev approximation to three-dimensional thermoelasticity with arbitrary body forces." *Engineering Analysis with Boundary Elements* **37**(10): 1259–1266.

[858] Tsai, C. C. (2015). "Polyharmonic multiquadric particular solutions for Reissner/Mindlin plate." *Mathematical Problems in Engineering* **2015**: 246159.

[859] Tsai, C. C. (2015). "Generalized polyharmonic multiquadrics." *Engineering Analysis with Boundary Elements* **50**: 239–248.

[860] Tsangaris, T., Y. S. Smyrlis and A. Karageorghis (2004). "A matrix decomposition MFS algorithm for biharmonic problems in annular domains." *Computers, Materials and Continua* **1**(3): 245–258.

[861] Tsangaris, T., Y. S. Smyrlis and A. Karageorghis (2006). "Numerical analysis of the method of fundamental solution for harmonic problems in annular domains." *Numerical Methods for Partial Differential Equations* **22**(3): 507–539.

[862] Tsangaris, T., Y. S. Smyrlis and A. Karageorghis (2006). "A matrix decomposition MFS algorithm for problems in hollow axisymmetric domains." *Journal of Scientific Computing* **28**(1): 31–50.

[863] Tschumperlé, D. and R. Deriche (2005). "Vector-valued image regularization with PDEs: a common framework for different applications." *IEEE Transactions on Pattern Analysis and Machine Intelligence* **27**(4): 506–517.

[864] Tsiatas, G. C. (2009). "A new Kirchhoff plate model based on a modified couple stress theory." *International Journal of Solids and Structures* **46**(13): 2757–2764.

[865] Turk, G. and J. F. O'Brien (2002). "Modelling with implicit surfaces that interpolate." *ACM Transactions on Graphics* **21**(4): 855–873.

[866] Twomey, S. (1963). "On numerical solution of Fredholm integral equations of first kind by inversion of linear system produced by quadrature." *Journal of ACM* **10**(1): 97–101.

[867] Uścilowska, A. (2015). "The MFS as a basis for the PIM or the HAM—comparison of numerical methods." *Engineering Analysis with Boundary Elements* **57**: 72–87.

[868] Ushijima, T. and F. Chiba (2003)."A fundamental solution method for the reduced wave problem in a domain exterior to a disc." *Journal of Computational and Applied Mathematics* **152**(1): 545–557.

[869] Ushijima, T. and F. Chiba (2003). "Error estimates for a fundamental solution method applied to reduced wave problems in a domain exterior to a disc." *Journal of Computational and Applied Mathematics* **159**(1): 137–148.

[870] Valle, M. F., M. J. Colaço and F. S. Neto (2008). "Estimation of the heat transfer coefficient by means of the method of fundamental solutions." *Inverse Problems in Science and Engineering* **16**(6): 777–795.

[871] Valtchev, S. S. (2018). "A meshfree method with plane waves for elastic wave propagation problems." *Engineering Analysis with Boundary Elements* **92**: 64–72.

[872] Vandamme, L., E. Detournay and A. H. D. Cheng (1989). "A two-dimensional poroelastic displacement discontinuity method for hydraulic fracture simulation." *International Journal for Numerical and Analytical Methods in Geomechanics* **13**(2): 215–224.

[873] Vanmaele, C., D. Vandepitte and W. Desmet (2007). "An efficient wave based prediction technique for plate bending vibrations." *Computer Methods in Applied Mechanics and Engineering* **196**(33): 3178–3189.

[874] Vekua, I. N. (2013). *On Metaharmonic Functions.* Tbilisi University Press.

[875] Velho, L., J. Gomes and L. H. de Figueiredo (2002). *Implicit Objects in Computer Graphics.* Springer.

[876] Vertnik, R. and B. Šarler (2006). "Meshless local radial basis function collocation method for convective-diffusive solid-liquid phase change problems." *International Journal of Numerical Methods for Heat and Fluid Flow* **16**(5): 617–640.

[877] Vodička, R. and M. Petrík (2015). "Degenerate scales for boundary value problems in anisotropic elasticity." *International Journal of Solids and Structures* **52**: 209–219.

[878] von Kármán, T. (1927). "Berechnung der druckverteilung an luftschiffkörpern (Calculation of pressure distribution on airship hulls)." *Abhandlungen aus dem Aerodynamischen Institut an der Technischen Hochschule Aachen* **6**: 3–13.

[879] von Kármán, T. (1930). *Calculation of Pressure Distribution on Airship Hulls.* Technical Memorandum No. 574, National Advisory Committee on Aeronautics.

[880] von Kármán, T. (1963). *Aerodynamics.* McGraw-Hill.

[881] Walsh, J. L. (1973). "History of the Riemann mapping theorem." *American Mathematical Monthly* **80**(3): 270–276.

[882] Wang, C. Y. (1994). "Two-dimensional elastostatic Green's functions for general anisotropic solids and generalization of Stroh's formalism." *International Journal of Solids and Structures* **31**(19): 2591–2597.

[883] Wang, C. Y. and J. D. Achenbach (1994). "Elastodynamic fundamental solutions for anisotropic solids." *Geophysical Journal International* **118**(2): 384–392.

[884] Wang, C. Y. and J. D. Achenbach (1995). "Three-dimensional time-harmonic elastodynamic Green's functions for anisotropic solids." *Proceedings of the Royal Society of London. A. Mathematical and Physical Sciences* **449**: 441–458.

[885] Wang, C. Y. (1997). "Elastic fields produced by a point source in solids of general anisotropy." *Journal of Engineering Mathematics* **32**(1): 41–52.

[886] Wang, F., W. Chen and X. Jiang (2010). "Investigation of regularized techniques for boundary knot method." *International Journal for Numerical Methods in Biomedical Engineering* **26**(12): 1868–1877.

[887] Wang, F., W. Chen, A. Tadeu and C. G. Correia (2017). "Singular boundary method for transient convection-diffusion problems with time-dependent fundamental solution." *International Journal of Heat and Mass Transfer* **114**: 1126–1134.

[888] Wang, F., C. S. Liu and W. Qu (2018). "Optimal sources in the MFS by minimizing a new merit function: Energy gap functional." *Applied Mathematics Letters* **86**: 229–235.

[889] Wang, F., W. Cai, B. Zheng and C. Wang (2020). "Derivation and numerical validation of the fundamental solutions for constant and variable-order structural derivative advection-dispersion models." *Zeitschrift für angewandte Mathematik und Physik* **71**(4): 135.

[890] Wang, F., C. M. Fan, Q. Hua and Y. Gu (2020). "Localized MFS for the inverse Cauchy problems of two-dimensional Laplace and biharmonic equations." *Applied Mathematics and Computation* **364**: 124658.

[891] Wang, F., Y. Gu, W. Qu and C. Zhang (2020). "Localized boundary knot method and its application to large-scale acoustic problems." *Computer Methods in Applied Mechanics and Engineering* **361**: 112729.

[892] Wang, F., Y. C. Liu and H. Zheng (2022). "A localized method of fundamental solution for numerical simulation of nonlinear heat conduction." *Mathematics* **10** DOI: 10.3390/math10050773.

[893] Wang, G., L. Dong and S. N. Atluri (2018). "A Trefftz collocation method (TCM) for three-dimensional linear elasticity by using the Papkovich-Neuber solutions with cylindrical harmonics." *Engineering Analysis with Boundary Elements* **88**: 93–103.

[894] Wang, H. and Q. H. Qin (2006). "A meshless method for generalized linear or nonlinear Poisson-type problems." *Engineering Analysis with Boundary Elements* **30**(6): 515–521.

[895] Wang, H. and Q. H. Qin (2008). "Meshless approach for thermo-mechanical analysis of functionally graded materials." *Engineering Analysis with Boundary Elements* **32**(9): 704–712.

[896] Wang, H. and Q. Qin (2019). *Method of Fundamental Solutions in Solid Mechanics*. Elsevier.

[897] Wang, J. G. and G. R. Liu (2002). "A point interpolation meshless method based on radial basis functions." *International Journal for Numerical Methods in Engineering* **54**(11): 1623–1648.

[898] Wang, J. G. and G. R. Liu (2002). "On the optimal shape parameters of radial basis functions used for 2-D meshless methods." *Computer Methods in Applied Mechanics and Engineering* **191**(23–24): 2611–2630.

[899] Wang, Q. X., H. Li and K. Y. Lam (2005). "Development of a new meshless—point weighted least-squares (PWLS) method for computational mechanics." *Computational Mechanics* **35**(3): 170–181.

[900] Wang, Y. and Y. Rudy (2006). "Application of the method of fundamental solutions to potential-based inverse electrocardiography." *Annals of Biomedical Engineering* **34**(8): 1272–1288.

[901] Waterman, P. C. (1965). "Matrix formulation for electromagnetic scattering." *Proceedings of Institute of Electrical and Electronics Engineers* **53**(8): 805–812.

[902] Waterman, P. C. (1969). "New formulation of acoustic scattering." *Journal of Acoustical Society of America* **45**(6): 1417–1429.

[903] Waterman, P. C. (1976). "Matrix theory of elastic wave scattering." *Journal of Acoustical Society of America* **60**(3): 567–580.

[904] Watson, J. O. (2003). "Boundary elements from 1960 to the present day." *Electronic Journal of Boundary Elements* **1**(1): 34–46.

[905] Wazwaz, A. M. (2011). "The regularization method for Fredholm integral equations of the first kind." *Computers and Mathematics with Applications* **61**(10): 2981–2986.

[906] Wei, T., Y. C. Hon and L. Ling (2007). "Method of fundamental solutions with regularization techniques for Cauchy problems of elliptic operators." *Engineering Analysis with Boundary Elements* **31**(4): 373–385.

[907] Wei, X., W. Chen and B. Chen (2014). "An ACA accelerated MFS for potential problems." *Engineering Analysis with Boundary Elements* **41**: 90–97.

[908] Wen, P. H. (1987). "Point intensity method of solving circular plate resting on elastical subgrade." *Engineering Mechanics*, **4**: 18–26.

[909] Wen, P. H., M. Adetoro and Y. Xu (2008). "The fundamental solution of Mindlin plates with damping in the Laplace domain and its applications." *Engineering Analysis with Boundary Elements* **32**(10): 870–882.

[910] Wen, P. H. and C. S. Chen (2010). "The method of particular solutions for solving scalar wave equations." *International Journal for Numerical Methods in Biomedical Engineering* **26**(12): 1878–1889.

[911] Wendland, H. (1995). "Piecewise polynomial, positive definite and compactly supported radial functions of minimal degree." *Advances in Computational Mathematics* **4**: 389–396.

[912] Wendland, H. (1998). "Error estimates for interpolation by compactly supported radial basis functions of minimal degree." *Journal of Approximation Theory* **93**(2): 258–272.

[913] Wendland, H. (2005). *Scattered Data Approximation*. Cambridge University Press.

[914] Wikipedia (2022). *List of arbitrary-precision arithmetic software*. https://en.wikipedia.org/wiki/Listofarbitrary-precisionarithmeticsoftware.

[915] Wilkinson, J. H. (1965). *Algebraic Eigenvalue Problem*. Oxford University Press.

[916] Willis, J. R. (1965). "The elastic interaction energy of dislocation loops in anisotropic media." *Quarterly Journal of Mechanics and Applied Mathematics* **18**(4): 419–433.

[917] Wilson, R. B. and T. A. Cruse (1978). "Efficient implementation of anisotropic three dimensional boundary-integral equation stress analysis." *International Journal for Numerical Methods in Engineering* **12**(9): 1383–1397.

[918] Wolfram Research Inc. (2022). *Mathematica*. https://www.wolfram.com/mathematica.

[919] Wong, H. L. (1982). "Effect of surface topography on the diffraction of P, SV, and Rayleigh waves." *Bulletin of the Seismological Society of America* **72**(4): 1167–1183.

[920] Wright, G. B. and B. Fornberg (2006). "Scattered node compact finite difference-type formulas generated from radial basis functions." *Journal of Computational Physics* **212**(1): 99–123.

[921] Wright, G. B. and B. Fornberg (2017). "Stable computations with flat radial basis functions using vector-valued rational approximations." *Journal of Computational Physics* **331**: 137–156.

[922] Wrobel, L. C. and C. A. Brebbia (1987). "The dual reciprocity boundary element formulation for nonlinear diffusion problems." *Computer Methods in Applied Mechanics and Engineering* **65**(2): 147–164.

[923] Wrobel, L. C. (2002). *The Boundary Element Method, Volume 1: Applications in Thermo-Fluids and Acoustics*. Wiley.

[924] Wu, K. and J. E. Olson (2014). "Simultaneous multifracture treatments: Fully coupled fluid flow and fracture mechanics for horizontal wells." *SPE Journal* **20**(2): 337–346.

[925] Wu, K. C. (1998). "Generalization of the Stroh formalism to 3-dimensional anisotropic elasticity." *Journal of Elasticity* **51**(3): 213–225.

[926] Wu, N. J., T. K. Tsay and D. L. Young (2006). "Meshless numerical simulation for fully nonlinear water waves." *International Journal for Numerical Methods in Fluids* **50**(2): 219–234.

[927] Wu, N. J. and T. K. Tsay (2008). "Applicability of the method of fundamental solutions to 3-D wave-body interaction with fully nonlinear free surface." *Journal of Engineering Mathematics* **63**(1): 61–78.

[928] Wu, W. C. and N. J. Altiero. (1979). "A boundary integral method applied to plates of arbitrary plan form and arbitrary boundary conditions." *Computers and Structures*, **10**: 703–707.

[929] Wu, Y., J. Zhang, S. Ding and Y. C. Liu (2022). "Localized boundary knot method for solving two-dimensional inverse Cauchy problems." *Mathematics* **10**(8): 1324.

[930] Wu, Y. Y., S. J. Liao and X. Y. Zhao (2005). "Some notes on the general boundary element method for highly nonlinear problems." *Communications in Nonlinear Science and Numerical Simulation* **10**(7): 725–735.

[931] Xi, Q., Z. Fu, T. Rabczuk and D. Yin (2021). "A localized collocation scheme with fundamental solutions for long-time anomalous heat conduction analysis in functionally graded materials." *International Journal of Heat and Mass Transfer* **180**: 121778.

[932] Xiao, J. E., C. Y. Ku, C. Y. Liu, C. M. Fan and W. Yeih (2017). "On solving free surface problems in layered soil using the method of fundamental solutions." *Engineering Analysis with Boundary Elements* **83**: 96–106.

[933] Xie, L., C. Hwu and C. Zhang (2016). "Advanced methods for calculating Green's function and its derivatives for three-dimensional anisotropic elastic solids." *International Journal of Solids and Structures* **80**: 261–273.

[934] Yan, L., C. L. Fu and F. L. Yang (2008). "The method of fundamental solutions for the inverse heat source problem." *Engineering Analysis with Boundary Elements* **32**(3): 216–222.

[935] Yan, L., F. L. Yang and C. L. Fu (2009). "A meshless method for solving an inverse spacewise-dependent heat source problem." *Journal of Computational Physics* **228**(1): 123–136.

[936] Yan, L. and F. Yang (2013). "A Kansa-type MFS scheme for two-dimensional time fractional diffusion equations." *Engineering Analysis with Boundary Elements* **37**(11): 1426–1435.

[937] Yan, L. and F. Yang (2014). "Efficient Kansa-type MFS algorithm for time-fractional inverse diffusion problems." *Computers and Mathematics with Applications* **67**(8): 1507–1520.

[938] Yan, Y. and I. H. Sloan (1988). "On integral equations of the first kind with logarithmic kernels." *Journal of Integral Equations and Applications* **1**(4): 549–580.

[939] Yang, F., L. Yan and T. Wei (2009). "Reconstruction of the corrosion boundary for the Laplace equation by using a boundary collocation method." *Mathematics and Computers in Simulation* **79**(7): 2148–2156.

[940] Yao, G., C. S. Chen and C. C. Tsai (2009). "A revisit on the derivation of the particular solution for the differential operator $\Delta^2 \pm \lambda^2$." *Advances in Applied Mathematics and Mechanics* **1**(6): 750–768.

[941] Yao, G. (2010). Local radial basis function methods for solving partial differential equations. Ph.D. Dissertation, University of Southern Mississippi.

[942] Ye, L., S. Hu, G. Xu and T. Yan (2023). "A meshless regularized method of fundamental solution for electromagnetic scattering problems of three-dimensional perfect electric conductor targets." *Engineering Analysis with Boundary Elements* **155**: 401–406.

[943] Ymeri, H., B. Nauwelaers and K. Maex (2001). "On the modelling of multiconductor multilayer systems for interconnect applications." *Microelectronics Journal* **32**(4): 351–355.

[944] Young, D. L., C. C. Tsai and C. M. Fan (2004). "Direct approach to solve nonhomogeneous diffusion problems using fundamental solutions and dual reciprocity methods." *Journal of the Chinese Institute of Engineers* **27**(4): 597–609.

[945] Young, D. L., C. C. Tsai, K. Murugesan, C. M. Fan and C. W. Chen (2004). "Time-dependent fundamental solutions for homogeneous diffusion problems." *Engineering Analysis with Boundary Elements* **28**(12): 1463–1473.

[946] Young, D. L., K. H. Chen and C. W. Lee (2005). "Novel meshless method for solving the potential problems with arbitrary domain." *Journal of Computational Physics* **209**(1): 290–321.

[947] Young, D. L. and J. W. Ruan (2005). "Method of fundamental solutions for scattering problems of electromagnetic waves." *Computer Modeling in Engineering and Sciences* **7**(2): 223–232.

[948] Young, D. L., S. J. Jane, C. M. Fan, K. Murugesan and C. C. Tsai (2006). "The method of fundamental solutions for 2D and 3D Stokes problems." *Journal of Computational Physics* **211**(1): 1–8.

[949] Young, D. L., C. C. Tsai, Y. C. Lin and C. S. Chen (2006). "The method of fundamental solutions for eigenfrequencies of plate vibrations." *Computers, Materials and Continua* **4**(1): 1–10.

[950] Young, D. L., K. H. Chen, J. T. Chen and J. H. Kao (2007). "A modified method of fundamental solutions with source on the boundary for solving Laplace equations with circular and arbitrary domains." *Computer Modeling in Engineering and Sciences* **19**(3): 197–221.

[951] Young, D. L., C. M. Fan, S. P. Hu and S. N. Atluri (2008). "The Eulerian-Lagrangian method of fundamental solutions for two-dimensional unsteady Burgers' equations." *Engineering Analysis with Boundary Elements* **32**(5): 395–412.

[952] Young, D. L., C. C. Tsai, C. W. Chen and C. M. Fan (2008). "The method of fundamental solutions and condition number analysis for inverse problems of Laplace equation." *Computers and Mathematics with Applications* **55**(6): 1189–1200.

[953] Young, D. L., M. H. Gu and C. M. Fan (2009). "The time-marching method of fundamental solutions for wave equations." *Engineering Analysis with Boundary Elements* **33**(12): 1411–1425.

[954] Young, D. L., Y. C. Lin, C. M. Fan and C. L. Chiu (2009). "The method of fundamental solutions for solving incompressible Navier-Stokes problems." *Engineering Analysis with Boundary Elements* **33**(8): 1031–1044.

[955] Young, D. L., S. R. Lin, C. S. Chen and C. S. Chen (2021). "Two-step MPS-MFS ghost point method for solving partial differential equations." *Computers and Mathematics with Applications* **94**: 38–46.

[956] Zaridze, R. S., R. G. Jobava, G. G. Bit-Banik, D. D. Karkasbadze, D. P. Economou and N. K. Uzunoglu (1998). "The method of auxiliary sources and scattered field singularities (Caustics)." *Journal of Electromagnetic Waves and Applications* **12**(11): 1491–1507.

[957] Zaridze, R. S., V. A. Tabatadze, I. M. Petoev-Darsavelidze and G. V. Popov (2019). "Determination of the location of field singularities using the method of auxiliary sources." *Journal of Communications Technology and Electronics* **64**(11): 1170–1178.

[958] Zeb, A., D. B. Ingham and D. Lesnic (2008). "The method of fundamental solutions for a biharmonic inverse boundary determination problems." *Computational Mechanics* **42**: 371–379.

[959] Zhang, A., Y. Gu, Q. Hua, W. Chen and C. Zhang (2018). "A regularized singular boundary method for inverse Cauchy problem in three-dimensional elastostatics." *Advances in Applied Mathematics and Mechanics* **10**(6): 1459–1477.

[960] Zhang, J., C. Yang, H. Zheng, C. M. Fan and M. F. Fu (2022)."The localized method of fundamental solutions for 2-D and 3D inhomogeneous problems." *Mathematics and Computers in Simulation* **200**: 504–524.

[961] Zhang, J., H. Zheng, C. M. Fan and M. F. Fu (2022). "The localized method of fundamental solutions for two–dimensional inhomogeneous inverse Cauchy problems." *Mathematics* **10**: 1464 (22 pages).

[962] Zhang, J. M., Z. H. Yao and H. Li (2002). "A hybrid boundary node method." *International Journal for Numerical Methods in Engineering* **53**(4): 751–763.

[963] Zhang, J. M., X. Qin, X. Han and G. Li (2009). "A boundary face method for potential problems in three dimensions." *International Journal for Numerical Methods in Engineering* **80**(3): 320–337.

[964] Zhang, L. P., Z. C. Li, H. T. Huang and Y. Wei (2019). "The modified method of fundamental solutions for exterior problems of the Helmholtz equation; spurious eigenvalues and their removals." *Applied Numerical Mathematics* **145**: 236–260.

[965] Zhang, X., X. H. Liu, K. Z. Song and M. W. Lu (2001). "Least-squares collocation meshless method." *International Journal for Numerical Methods in Engineering* **51**(9): 1089–1100.

[966] Zhao, S., Y. Gu, C. M. Fan and X. Wang (2022). "The localized method of fundamental solutions for 2D and 3D second-order nonlinear boundary value problems." *Engineering Analysis with Boundary Elements* **139**: 208–220.

[967] Zheng, C. J., H. B. Chen, H. F. Gao and L. Du (2015). "Is the Burton-Miller formulation really free of fictitious eigenfrequencies?" *Engineering Analysis with Boundary Elements* **59**: 43–51.

[968] Zheng, H., F. Wang, C. S. Chen, M. Lei and Y. Wang (2020). "Improved 3D surface reconstruction via the method of fundamental solutions." *Numerical Mathematics: Theory, Methods and Applications* **13**(4): 973–985.

[969] Zhou, H. and C. Pozrikidis (1995). "Adaptive singularity method for Stokes flow past particles." *Journal of Computational Physics* **117**(1): 79–89.

[970] Zhu, H., H. Shu and M. Ding (2010). "Numerical solutions of partial differential equations by discrete homotopy analysis method." *Applied Mathematics and Computation* **216**(12): 3592–3605.

[971] Zhu, T., J. Zhang and S. N. Atluri (1998). "A meshless local boundary integral equation (LBIE) method for solving nonlinear problems." *Computational Mechanics* **22**(2): 174–186.

[972] Zhu, X., F. F. Dou, A. Karageorghis and C. S. Chen (2020). "A fictitious points one-step MPS-MFS technique." *Applied Mathematics and Computation* **382**: 125332.

[973] Zieliński,, A. P. and O. C. Zienkiewicz (1985). "Generalized finite-element analysis with T-complete boundary solution functions." *International Journal for Numerical Methods in Engineering* **21**(3): 509–528.

[974] Zieliński, A. P. and I. Herrera (1987). "Trefftz method: Fitting boundary conditions." *International Journal for Numerical Methods in Engineering* **24**(5): 871–891.

[975] Zienkiewicz, O. C. and J. Z. Zhu (1992). "The superconvergent patch recovery and a posteriori error estimates. Part 1: The recovery technique." *International Journal for Numerical Methods in Engineering* **33**(7): 1331–1364.

[976] Zienkiewicz, O. C. and J. Z. Zhu (1992). "The superconvergent patch recovery and a posteriori error estimates. Part 2: Error estimates and adaptivity." *International Journal for Numerical Methods in Engineering* **33**(7): 1365–1382.

[977] Zienkiewicz, O. C. (1997). "Trefftz type approximation and the generalized finite element method: History and development." *Computer Assisted Mechanics and Engineering Sciences* **4**(3–4): 305–316.

[978] Zienkiewicz, O. C., R. L. Taylor and J. Z. Zhu (2013). *The Finite Element Method: Its Basis and Fundamentals*, 7th edn. Elsevier.

Index